C000311771

NATURAL SELECTION AND BEYOND

Woodcut of Wallace. From Edward Drinker Cope's work *Alfred Russel Wallace* (1891). Out of copyright.

Natural Selection and Beyond

The Intellectual Legacy of Alfred Russel Wallace

Edited by Charles H. Smith
and George Beccaloni

UNIVERSITY PRESS

OXFORD
UNIVERSITY PRESS

Great Clarendon Street, Oxford OX2 6DP

Oxford University Press is a department of the University of Oxford.
It furthers the University's objective of excellence in research, scholarship,
and education by publishing worldwide in

Oxford New York

Auckland Cape Town Dar es Salaam Hong Kong Karachi
Kuala Lumpur Madrid Melbourne Mexico City Nairobi
New Delhi Shanghai Taipei Toronto

With offices in

Argentina Austria Brazil Chile Czech Republic France Greece
Guatemala Hungary Italy Japan Poland Portugal Singapore
South Korea Switzerland Thailand Turkey Ukraine Vietnam

Published in the United States
by Oxford University Press Inc., New York

First published 2008

British Library Cataloguing in Publication Data
Data available

Library of Congress Cataloging in Publication Data
Data available

Typeset by SPI Publisher Services, Pondicherry, India
Printed in Great Britain
on acid-free paper by
CPI Antony Rowe, Chippenham, Wiltshire

ISBN 978–0–19–923916–0

1 3 5 7 9 10 8 6 4 2

Contents

Foreword

Peter Bowler

Alfred Russel Wallace is often depicted as a "forgotten" figure—although he has been rediscovered many times by authors who have contrived to ignore their predecessors' efforts. He is known to the general public, if at all, as the co-discoverer of natural selection, and several biographers have used this point as a means of trying to undermine Darwin's reputation, suggesting that Wallace was deliberately marginalized by Darwin's supporters and by the scientific elite. Despite his achievements in science Wallace is often portrayed as an outsider, someone devoted to eccentric ideas and beliefs including spiritualism, an opposition to vaccination, and a then deeply unfashionable support of socialism. Yet these views should not be dismissed as mere eccentricities. They link Wallace to a broader current of opposition to the materialism of mainstream Victorian culture which is now increasingly acknowledged by historians. His accounts of his expeditions to South America and the Far East were widely praised at the time and still appeal to modern travel writers. Like his political and philosophical works, they reveal his remarkable humanity as well as his love of nature.

Providing alternatives to the triumphalist story of Darwinism is laudable enough in this age of obsessive celebrity-worship, but Wallace deserves better than the routine use of his name by iconoclasts seeking to undermine Darwin's position in the pantheon of science. He deserves our attention for two very different reasons. First, he made major contributions to evolutionary biology, probing the logic of the Darwinian theory and extending the range of its applications in areas such as the explanation of the geographical distribution of species and the evolution of animal coloration. Second, he can be seen, not as an outsider, but as an original thinker who questioned many of the assumptions on which the ideologies of Victorian progressionism and scientific naturalism were based. His apparent eccentricities followed from his efforts to create an alternative philosophy of life, and we owe it to him to rediscover the core of his vision. This book will go a long way to providing the sympathetic but sophisticated reassessment of his life we so badly need.

In science, Wallace became one of Darwin's staunchest supporters—yet also one of the few who could be trusted to understand and evaluate the theory's structure and implications. He differed from Darwin on several key issues, never accepting the analogy between artificial and natural selection, and remaining suspicious of the extension of the model into the area of sexual selection. Even here, their debates helped both of them to refine their thinking, while in other areas Wallace's knowledge of distribution and the complexities of adaptation allowed him to explore the theory in ways which considerably extended its range. His work on animal coloration and on geographical speciation was of major significance to the debate. Equally significant was the project which led to the publication of his two-volume survey *The Geographical Distribution of Animals* in 1876. Here Wallace did, for once, interact with the scientific community in a coordinated way as he gathered information on the distribution of species. And this book, along with his *Island Life*, seems to have triggered an explosion of research into historical biogeography in the later nineteenth century.

Outside of science—yet in many ways there is no separation—we are beginning to see how Wallace's so-called eccentricities fitted together to make a coherent philosophy of life. His much commented-on rejection of the material origins of the human mind, linked to his growing enthusiasm for spiritualism, identify him with a major strand of Victorian thought which resisted the rise of scientific naturalism. His socialism—along with his various other campaigns on social issues—marks him out also as a member of a small but ultimately influential body of opinion critical of the rush into free-enterprise capitalism. Here was a man who saw the world as a whole with a spiritual purpose, and for all his endorsement of natural selection we should remember that his last book, *The World of Life* of 1911, depicted the whole sweep of evolution as the unfolding of a divine plan. This book explores many aspects of Wallace's work, including his views on the possibility of extraterrestrial life, and his adventures into a wide range of cultural and social debates ranging from spiritualism to land nationalization. It reveals that behind the apparent diversity lies a unified world view, a view with which many will want to sympathize.

PREFACE

Alfred Wallace, Field Collector

Earl of Cranbrook

Alfred Russel Wallace became my guide and mentor when, in July 1956, a little over a century after he began his epic travels in the Malay Archipelago, a small vessel of the Blue Funnel line negotiated the docks of Liverpool, embarking me on a passage of twenty-eight days to Singapore, gateway to the lands of orang-utans and birds of paradise. Like Wallace, I was duly befuddled by the donkey drivers and gully-gully men in Egypt, spent a day at "desolate, volcanic Aden" and, in the monsoon rollers of the Indian Ocean, found myself a bad sailor. I, too, made landfall at "Pulo Penang," with its picturesque mountains, its spice trees, and its famous waterfall (S729 1905a, 1:333–35). I have subsequently followed his trail to Malacca, Java, and Bali. Ultimately, in Papua New Guinea, like the "natives of Aru" (S715 1883, plate 34), I climbed a tree and watched entranced as birds of paradise (*Paradisaea raggiana* in my case) performed in its branches. For me in many aspects, as for Wallace, this eastern venture "constituted the central and controlling incident of my life" (S729 1905a, 1:336).

In Singapore, Wallace met Rajah James Brooke (who was then enduring an official Commission of Inquiry into his activities and status) and, with the Rajah's offer of every assistance, rearranged his plans. Thus Borneo became the first island of the Malay Archipelago that he visited, and his base Kuching (then still generally called "Sarawak"). Kuching was already my destination, but in Singapore I met my future employer at a curry tiffin. Arrangements settled, I shipped onwards and, after three more days at sea, reached Muara Tebas, the eastern mouth of the Sarawak delta. From there, as in Wallace's day, the river meandered a dozen muddy miles to the dock at Kuching.

Wallace eventually spent a longer continuous period in what was then the small territory of Sarawak than at any other location in the Archipelago (1 November 1854 to 25 January 1856). Sarawak provided his most exciting catches of insects: "during my whole twelve years' collecting in the western and eastern tropics, I never enjoyed such advantages in this respect as at the Simunjon coal-works" (S715 1883, 36), and (of night-flying moths) "during the six succeeding years I was never once able to make any collections at all approaching those at Sarawak" (S715

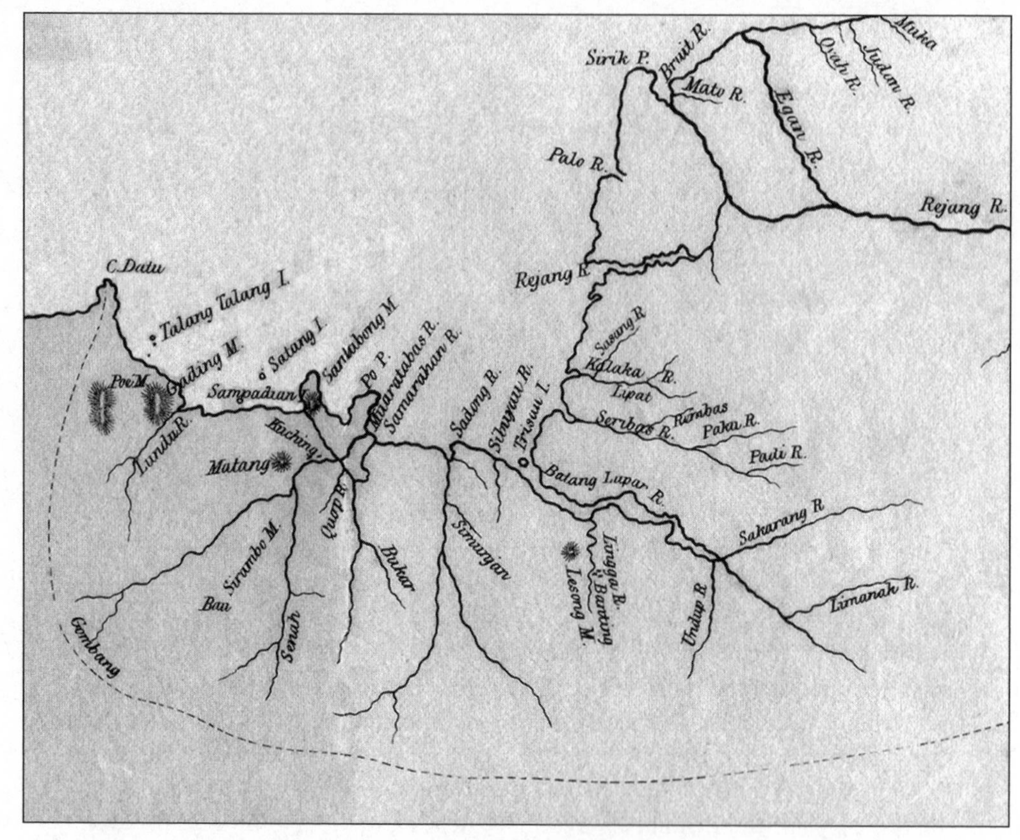

Figure 1 Map of the western part of Sarawak from the time when Rajah Brooke was the ruler. Portion of a map from Volume 1 of Spenser St John's *Life in the Forests of the Far East* (1862). Out of copyright.

1883, 86–87). More significantly in the history of biological theory, it was from Sarawak in 1855 that he despatched the first formulation of his ideas on the origin of species (S20 1855). A century later, the premise of natural selection underlay all biological teaching and research. During my fourteen years of residence in Sarawak, Indonesia, and Peninsular Malaysia, like Wallace I was continually amazed and inspired by the natural biodiversity of the region. For me, my students, professional colleagues, and naturalist friends, the challenge was to find an explanation for this equatorial richness in the framework of "Darwinism"—the term that Wallace (S724 1889) himself generously used.

Latterly, with colleagues at the Natural History Museum (NHM), London,[1] I have spent time tracing his Sarawak specimens (Cranbrook *et al.* 2005; Cranbrook *et al.* in prep.). Thanks to Wallace himself, his family and friends, much manuscript material survives, including the journals of his travels in the Malay Archipelago, contemporary notebooks and correspondence, and, notably, the so-called *Species Registry* held in the NHM Zoology Library (Z MSS 89, O WAL). But it is the excitement of handling Wallace's specimens themselves—material objects to augment the written word—that has given new perceptions of the man and his aptitudes (and weaknesses), and illuminated his achievements as a naturalist and collector in the humid equatorial environment.

By his own estimate, Wallace sent to London more than twenty-five thousand insect specimens from Sarawak. Some examples that he retained in his own private collection are illustrated on the NHM website (see **http://www.nhm.ac.uk/nature-online/collections-at-the-museum/wallace-collection/collecting.jsp**) but, through his agent, Samuel Stevens, most of this huge assemblage was rapidly distributed among specialists. Although more than one thousand new species were ultimately described, the role of types was poorly defined at the time and it is unrealistic to hope to track down, enumerate, and reassemble even these items of the Sarawak collection (Polaszek and Cranbrook 2006). His mollusc shells, similarly, were dispersed into the hands of private and institutional purchasers. Some were later acquired by the museum, but it would be unrewarding and excessively time-consuming to hunt for all among the vast, only partly registered collections.

Wallace's Sarawak vertebrates, however, amounted to a moderate number. Although there have been difficulties and surprises in the search, almost all vertebrate specimens from Sarawak attributed to him in the British Museum (BM) register have been found; a few have also been located in other museums. For the later investigator, it has proved a great boon that Wallace obtained a stock of small parchment labels for use during his travels in the Malay Archipelago. The printed blank reads:

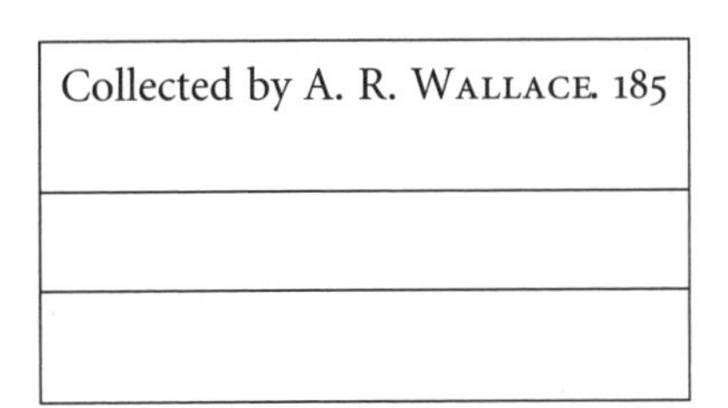

On these labels, the final digit of the year date is entered by hand. On the second line, a systematic name may be written, including author's name (abbreviated) in the case of birds. The locality, "Sarawak," is written on the bottom line, sometimes the sex (♂ or ♀) and, in the lower right corner, a serial number. Separate numbered series were created for mammals, birds, and reptiles. Labels of this type (AW) are found on many of the mammals and bird skins registered in Wallace's name and, if present, can be taken as unassailable authentication of any specimen to which they are attached.

Other types include printed or blank British Museum (BM) parchment labels on which there is written attribution to Wallace that, along with registration details, provides satisfactory provenance. Attached to some specimens are small, square pieces of card which appear (from remnant wording on the reverse) to have been cut from invitations to meetings of the Royal Geographical Society. These hand-made tags bear numbers that can be matched with serial numbers in the *Species Registry* notebook.

Inscriptions on the AW labels often include a name that can only have been added later. This is puzzling, but Wallace himself provided the answer. In October 1854, after two energetic months collecting in Malacca, he wrote home to say: "I am glad to be safe in Singapore with my collections, as from here they can be insured. I have now a fortnight's work to arrange, examine, and pack them, and four months hence there will be work for Mr. Stevens" (S729 1905a, 1:341); *i.e.*, Samuel Stevens, the London agent who had served Wallace so well during and after his Amazon adventure. It seems that, while in the field, Wallace attached hand-made numbered tags to his specimens and later, as part of general curation and preparation for despatch to London, he substituted his pre-printed labels onto which he transcribed essential details, including an identification when possible.

The cramped conditions under which he worked are wholly familiar to me from field experience in the region, including the all-important, versatile, ant-proof box. Describing conditions in Lombok, he wrote: "My principal piece of furniture was a box, which served me as a dining table, a seat while skinning birds, and as the receptacle of the birds when skinned and dried. To keep them free from ants we borrowed, with some difficulty, an old bench, the four legs of which being placed in cocoa-nut shells filled with water kept us tolerably free from these pests. The box and the bench... were generally well occupied by two insect boxes and about a hundred birds' skins in process of drying" (S715 1883, 155).

The gun was Wallace's main collecting tool for mammals and birds. His skill and persistence are shown by his story of hunting "the beautiful ground thrushes" (*Pitta concinna*) of Lombok:

> They were so shy that it was difficult to get a shot at them, and it was only after a good deal of practice that I discovered how to do it. The habit of these birds is to hop about on the ground picking up insects, and on the least alarm to run into the densest thicket or take flight close along the ground. At intervals they utter a peculiar cry of two notes, which when once heard is easily recognised, and they can also be heard hopping along among the dried leaves. My practice was, therefore, to walk cautiously along the narrow pathways with which the country abounded, and, on detecting any sign of a Pitta's vicinity to stand motionless and, give a gentle whistle occasionally, imitating the notes as near as possible. After half an hour's waiting, I was often rewarded by seeing the pretty bird hopping along in the thicket. Then I would perhaps lose sight of it again, till, having my gun raised and ready for a shot, a second glimpse would enable me to secure my prize, and admire its soft puffy plumage and lovely colours (S715 1883, 158).

Mimicking their calls is still a useful method to lure into view many of the understorey birds of Malaysian rainforest. Thanks to Wallace and other collectors of the past, however, the field ornithologist no longer needs dead bodies but is satisfied by sight records!

Wallace also relied on assistants, both to obtain and to prepare specimens. Thus, reflecting on his time in Sarawak, he regretted that he "had no hunter to shoot for me regularly, and, being myself fully occupied with insects, I did not succeed in obtaining a very good collection of the birds or Mammalia" (S715 1883, 39). Referring to Charles Allen, "a boy of sixteen whom I had brought with me from London as he wished to become a collector," Wallace observed that he "learned to shoot and catch insects pretty well, but not to prepare them properly" (S729 1905a, 1:340). When Wallace left Sarawak, Charles preferred to stay in Kuching (though some years later Allen decided to rejoin the expedition). His successor was "the Malay lad named Ali," originally recruited in December 1855. Ali "soon learnt to shoot birds, to skin them properly, and latterly to put up the skins very neatly" (S729, 1:384–85), and accompanied Wallace as a valued assistant throughout the rest of his time in the Archipelago.

Wallace brought two guns with him. At the start of his travels, in Malacca, he found them "both very good" (S729, 1:339) and, in 1862, among his parting gifts to Ali he described them as "my two double-barrelled guns" (S729, 1:383). When hunting orang-utan in Sarawak, he carried the larger of these guns, both barrels of which were loaded with a single ball: "I got a shot at it, and the second barrel caused it to fall down almost dead, the two balls having entered the body." Shortly afterwards, on his third encounter with an orang-utan, he was armed only with "a small 80-bore gun."[2] This he fired but succeeded only in enraging the ape which escaped, wounded (S715 1883, 40).

Wallace was not the first naturalist to collect vertebrates in Sarawak, having been preceded by Hugh Low (1848) and, indeed, by Rajah James Brooke who had presented the British Museum with eighteen mammals in 1845 (Thomas 1906) and 104 birds in 1845 and 1850 (Sharpe 1906). With supplementary information from the Dutch expeditions (Müller 1839–40), the mammal fauna of Borneo was reasonably well known. As already noted, apart from his collection of orang-utans which "succeeded beyond my expectations" (S715 1883, 39) and was financially rewarding, Wallace was dissatisfied with his collection of mammals in Sarawak. In a paper submitted soon after his return to Singapore, over the dateline March 10, 1856, he claimed to have collected about thirty-five species, "among which are two species of orang-utan, five other Quadrumania, the rare and curious otter *Potamophilus barbatus*, the no less interesting *Gymnurus Rafflesii*, and several curious Rodentia and Insectivora" (S25 1856, 5113). The list in the *Species Registry* numbers thirty-two species (excluding orang-utans), including the rarities named above, but only one monkey (= "Quadrumania"). No monkey specimens can be found in NHM, or elsewhere, and this discrepancy is unresolved.

The first mammal collected in Sarawak was, indeed, a monkey, "*Macacus cynomolgus*" = *Macaca fascicularis*, a female long-tailed macaque, obtained at "Serambo mountain Nov. 1854." This note in the *Species Registry* is the only indication that, in addition to the famous December 1855 and January 1856 visits to Rajah Brooke's hilltop bungalow "Peninjau" on Serambu hill (S715 1883, 84–86), he also made an unaccompanied visit at the start of his stay in Sarawak. Initial enthusiasm led Wallace to describe this monkey and the next seven specimens in some detail in his notebook (*Species Registry*), recording standard measurements, coloration, and field observations of behaviour. For much of the time in Sarawak, however, he was preoccupied with his insect collections (S715 1883, 39). His lack of attention to mammals is reflected by the increasingly casual nature of entries in the *Species Registry*. Only the first twenty-two specimens were given a systematic or vernacular name, and (with the exception of the colugo or flying lemur, the digestive system of which he dissected and sketched) descriptions falter as the list lengthens until the final entry casually aggregates "about 10 [un-named] species from Peter," with no further explanation.

Wallace's Sarawak mammals in the NHM comprise thirty-one specimens representing nineteen species (including the orang-utans, now considered to be a single species). Twelve skins (plus one unregistered) bear pre-printed AW labels, with his field collection numbers. There are, however, notable absentees from the *Species Registry* list, including all those highlighted later by Wallace for their particular interest or curiosity. The BM did not receive his first mammal, the female long-tailed macaque, nor his second, "*Sciurus ephippium*" = giant squirrel *Ratufa affinis*, nor the two "*Gymnurus Rafflesii*" = moonrat *Echinosorex gymnurus*, the otter-civet, a mouse deer *Tragulus* sp. (*Species Registry* note: "Admirable eating"), nor "*Galeopithecus volans*" = colugo *Galeopithecus variegates*, and acquired only one of four felids listed. The fate of these specimens is unknown, but they are likely to have been sold by Stevens to private collectors. For instance, nine of the authenticated Wallace mammals from Sarawak in NHM were finally acquired with one such collection, that of Robert F. Tomes.

Combining those named in the *Species Registry* with the NHM collection, Wallace's mammals comprised a reasonable representation of forms that can best be obtained by shooting, notably the monkey, colugo, squirrels, and treeshrew. Descriptive field notes in the *Species Registry* confirm that Wallace saw the squirrels alive, and was probably the collector. Other techniques were also used, perhaps by local assistants: a "tiger cat" was caught in a snare, and the otter-civet was "caught in a trap at foot of a mountain one mile from a river."

The condition of the specimens is variable. The orang-utan skeleton is wired and suspended in a standing posture, fully articulated, and the two stuffed orangutans are mounted in elaborate poses. Since all orang-utan skins were preserved in casks of alcohol ("arrack"), and shipped to London in this medium, the mounting may have been part of the "work" of Mr Stevens. In his preface to *The Malay Archipelago*, Wallace acknowledged his indebtedness to Stevens "for the care he took of my collections," a phrase which evidently included the final preparation of these key specimens.

The flying squirrel *Iomys horsfieldii* (not listed in *Species Registry*) is also mounted, in a crudely lifelike posture, with skull inside the head and glass eyes; this was probably the work of a later preparator, neither Wallace nor Stevens. A bear *Ursus (Helarctos) malayanus* is represented only by a skull. The skin of the Borneo bay cat *Catopuma badia* was received at the museum in a tattered condition, and was never stuffed. The skins of other small mammals are in the main neatly prepared, but some are poorly filled. The slender squirrel *Callosciurus tenuis* and one (unregistered) of two pigmy squirrels *Nannosciurus melanotis* retain the skulls in the heads. From other skins, the skulls have been removed. All skulls are damaged in the occipital region, indicating that (as was general practice at the time) Wallace or his assistant left the skull in the head, opening it at the back to clean out the brain. The extraction of the skulls was a later event, probably undertaken at the museum.

Wallace showed greater enthusiasm for birds, which he was better equipped to identify. While preparing for the expedition, still in London, he invested in Prince Lucien Bonaparte's *Conspectus generum avium*, "a large octavo volume of 800 pages, containing a well-arranged catalogue of all the known species of birds up to 1850, with references to descriptions and figures, and the native country and distribution of each species" (S729 1905a, 1:327). In the margins of this book, he added notes on the distinguishing characters of the species expected in the Malay Archipelago. As a result, "during my whole eight years' collecting in the East, I could almost always identify every bird already described, and if I could not do so, was pretty sure that it was a new or undescribed species" (S729, 1:327). This book was later acquired by Thomas Henry Riches who presented it to the Linnean Society, London (shelf mark 598.c BON).

He also made arrangements with Samuel Stevens to ensure that a good representative sample of birds was kept for his own use. Although from 1857 onwards Stevens regularly supplied birds sent by Wallace from other destinations in the Archipelago, it was not until 1873, when the BM purchased Wallace's retained personal holding of 2,474 skins (Sharpe 1906), that representatives of his Sarawak collection reached the museum. Among this large acquisition, 109 skins were registered as Wallace specimens from "Sarawak" or "Borneo." We have since found one unregistered specimen: a skinned head (only) of the velvet-fronted nuthatch *Sitta frontalis*, authenticated by a Wallace label. There is also, at the Cambridge University Museum of Zoology, a detached head of a male bushy-crested hornbill *Anorrhinus galeritus* from Sarawak (reg. no. 25/Buc/2/a/7), not listed in the *Species Registry* but authenticated by a Wallace label.

Among the skins bearing AW labels, there is good correlation with his provisional identifications and field numbers. Of the ninety-eight numbered entries in the *Species Registry*, eighty-nine corresponding species are in NHM. Fifteen bird species are represented by more than one specimen, but only four by more than two. This assemblage is a fair sample of the Borneo avifauna of 622 species, representing around one-fourth of the lowland forest, riparian, and coastal birds that could have been accessible to him. The diversity is impressive, including many arboreal birds, from huge hornbills to minute flowerpeckers, along with waterfowl, shorebirds, large diurnal raptors, and owls.

Wallace managed to attach a name to most bird specimens, evidently being well served by his favoured reference book. With the few exceptions noted below, replicates bear the same collection number. As with mammals, birds that he judged to be one species were tagged with a common number, matching the sequential numbering of the *Species Registry.* He was confused by the similarity of two pale-bellied, olivaceous bulbuls: hairy-backed bulbul *Tricholestes criniger* and buff-vented bulbul *Iole olivacea.* Both were given two numbers: the former as #45 (original name deleted) and # 58 (no identification), and the latter as #59 and #61, in both cases named "*Trichastoma.*" He could only provide a generic name (*Edolius*) for all drongos, and he gave different field numbers to skins of greater racket-tailed drongos *Dicrurus paradiseus* with at least one intact racket, as opposed to a bird that had lost both rackets. Creditably, he distinguished (although he could not name) the three lookalike brown bulbuls, giving them different collection numbers: #10 red-eyed bulbul *Pycnonotus brunneus*, #44 cream-vented bulbul *Pycnonotus simplex* (two skins), and #72 spectacled bulbul *Pycnonotus erythrophthalmus.*

Was the commendable diversity of his Sarawak birds achieved by planned collecting, or by selection from a wider variety of specimens? Wallace himself, as we have seen, was a discriminating collector, but it is unlikely that his assistants or local hunters would have distinguished between common birds and those that were new or unusual. A century later, at the Sarawak Museum, when sorting and identifying incoming bird skins from rural out-station collectors, I found that each consignment contained duplicates of common species and rarely provided surprises. I suspect that Wallace chose only a proportion of the birds brought in by his assistants, guided by his growing familiarity with the local ornithology. That he made use of local expertise is sure. He may have paid for birds, as he did for insects at Simunjan. For example, his collection included scarce, ground-living forest birds that are particularly susceptible to traditional snares and traps used in Sarawak, such as pheasants, partridges, and the rare endemic Bornean ground cuckoo *Carpococcyx radiatus*—a bird I have never seen alive!

A few AW labels attached to bird skins carry information on soft-parts coloration, showing at least that Wallace saw the fresh carcass, although he did not necessarily prepare the skin himself. Some skins were not filled at all, mainly of larger species including the type of Wallace's hawk eagle *Spizaetus nanus.* A fine thread is looped through the nostrils of some, suggesting that they were hung up to dry. Four of these skins are dated 1856, and were therefore obtained in the hurried days of January, shortly before Wallace's return to Singapore, when he himself was preoccupied with curating his huge moth collection from Peninjau.

Perhaps the variation in the quality of the skins of mammals and birds reflects the different hands involved, including the under-performing Charles and the neophyte Ali. If so, the poorly prepared 1856 bird skins were Ali's first efforts.

Wallace paid scant attention to the collection of reptiles and amphibians. The *Species Registry* (p. 24) contains only a one-page list of six numbered collections of "tortoises," apparently amounting to seventeen stuffed or dried specimens. Representatives of these were bought by the BM from Stevens in 1856. One is an Asian leaf turtle *Cyclemys dentata*, a small shell to which a dried head is attached by thread, bearing an AW label and a number (#5) corresponding with the *Species Registry* entry "#5 *Emys* sp. 2 young shells and a head." Two other members of this series are marked on the carapace and/or plastron in black ink with provenance "Sar." or "Sarawak" and numbers that correspond with entries in the *Species Registry*, and are thereby authenticated.

Other reptiles and a frog can be matched with Wallace's story how in August 1855, after a bout of illness at Simunjan, he and Charles Allen went up a branch of the Sadong river, "to a place called Menyille, where there were several small Dyak houses and one large one. Here the landing place was a bridge of rickety poles, over a considerable distance of water; and I thought it safer to leave my cask of arrack securely placed in the fork of a tree. To prevent the natives from drinking it, I let several of them see me put in a number of snakes and lizards" (S715 1883, 54). This incident accounts for a group of specimens in spirit bought from Stevens by the BM in 1856, comprising a gecko *Aeluroscalabotes felinus*, two lizards *Gonocephalus liogaster* and *Dasia olivacea*, small snakes *Cylindrophis ruffus*, *Amphiesma petersii*, and *Calamaria lumbricoidea*, a Malayan box turtle *Cuora amboinensis*, and one harlequin tree frog *Rhacophorus pardalis*, all presumably collected on the journey, if not on the spot!

This was the only frog collected by Wallace in Sarawak. He wrote a vivid description of the species commonly called after him, Wallace's flying frog, including an illustration (see Plate 1) and the assessment that it was a new *Rhacophorus* (S715 1883, 38–39), but this remained unnamed until described as *Rhacophorus nigropalmatus* (Boulenger 1895), type specimen collected by Charles Hose.

* * *

Superficially, conditions in Sarawak of 1956 were not strikingly different from those when Wallace left, a century earlier. The walls of the government bungalow that I shared for a while were sheets of pandan leaf (*kajang*) and its roof was palm thatch. Roads were few, and travel mainly by boat. Lumbering was confined to the swamps; hill forest was still inviolate, except where cleared for swidden farms. Streams ran clear in the bamboo-rich countryside inland of Kuching.

Wallace's collections, however, show that there had been some environmental change. With few exceptions, his specimens are dated only by year. Knowing from his own account that the first four months of his sojourn "were spent in various parts of the Sarawak River, from Santubong at its mouth up to the picturesque limestone Mountains and Chinese gold-fields of Bow and Bedé" (S715 1883, 35), any record dated 1854 must derive from this riparian tract. Only one mammal received by BM, a plantain squirrel, is so dated, but the *Species Registry* adds #1, ♀ long-tailed macaque, Nov. 1854 at Serambo mountain. Among birds there is a larger sample dated 1854: crested partridge *Rollulus rouroul*, band-bellied crake *Porzana paykulli*, pink-necked green pigeon *Treron vernans*, brown hawk owl *Ninox scutulata*, brown barbet *Caloramphus fuliginosus*, red-throated barbet *Megalaima mystacophanos*, blue-eared barbet *Megalaima australis*, checker-throated woodpecker *Picus mentalis*, common iora *Aegithina tiphia*, lesser green leafbird *Chloropsis cyanopogon*, red-eyed bulbul *Pycnonotus brunneus*, oriental magpie robin *Copsychus saularis*, fluffy-backed tit babbler *Macronus ptilosus*, white-bellied yuhina *Yuhina zantholeuca*, pied fantail *Rhipidura javanica*, long-billed spiderhunter *Arachnothera robusta*, Asian glossy starling *Aplonis panayensis*, dusky munia *Lonchura fuscans*, and black-headed munia *Lonchura malacca*.

Many of these birds are characteristic of tall forest, with disturbance and edge habitat. Additionally, the crake is migratory, wintering in wet paddy fields and grasslands; the partridge favours bamboo groves and orchards; the green pigeon flocks in old cleared land with orchards, second growth, and riparian forest, and the tit babbler is found in disturbed forest and mature tree plantations. Although normally frequenting submontane elevations, the yuhina has been recorded at Semongok, in the Sarawak valley. The iora is a familiar bird of trees in gardens, the magpie robin is a characteristic ground living bird of gardens, tree plantations, and riparian forest, and only munias are dependent on grassy clearings. This assemblage therefore accords with a varied landscape of original forest patches, old cleared land with tall trees, mature second growth, orchards, bamboo groves, and a few open, grassy patches or rice paddies.

Absentees from this list include birds now ubiquitous in open country and scrub in lowland Sarawak: spotted-neck dove *Streptopelia chinensis*, yellow-vented bulbul *Pycnonotus goiavier*, yellow-bellied prinia *Prinia flaviventris*, and little spiderhunter *Arachnothera longirostra*, for example. Unless Wallace and his assistants collected selectively, we must judge that these open country birds were rare or absent in Sarawak in the mid-nineteenth century. Since 1956, the avifauna of open country has been further altered by well-documented twentieth-century invasions of tree sparrow *Passer montanus* and mynas *Acridotheres* spp.; the latter have been followed, in turn, by their nest parasite, the koel *Eudynamis scolopacea* (Davison 1999). Doves are notoriously invasive, and the spotted-neck dove may have been an unrecorded arrival in the intervening century. Other species of this short list could originally have been confined to restricted natural areas of limited extent, perhaps semi-open strand habitat along sandy shores.

Among Wallace's small collection of reptiles and amphibians, the Malayan box turtle frequents wet rice-fields, ditches, and drains but, with this exception, species characteristic of open country, cultivated land, or human habitation are again lacking. A quick modern collection from such habitats in Sarawak would include, for instance, the village brown skink *Mabuya multifasciata* and green tree lizard *Bronchocoela cristatella*, with the house gecko *Hemidactylus frenatus*, round-tailed gecko *Gehyra mutilata*, and a variety of common amphibians. Once again, the absence of these open country or semi-commensal species points to the wider prevalence of a richer, more pristine habitat in 1854–56. The Sarawak landscape enjoyed by Wallace, and by myself a century later, is now irrecoverable, and its vertebrate fauna has been both diluted by species loss, and polluted by the addition of invasive and commensal species from elsewhere.

But Sarawak was only a beginning for Wallace. His further travels in the Malay Archipelago refined his ideas on the origin of species through natural selection, culminating in the 1858 letter from Ternate that brought Darwin into the open. Neither man was present when their papers were read at the meeting of the Linnean Society but, on his return to England in 1862, Wallace found his future set on a new trajectory. The interactions of a man of his background and experiences with the scientific and social community of his time had profound effects. The bigger story unfolds through the twenty-one chapters of this commemorative book, presented by authors whose studies cover the wide scope of Wallace's own interests and pursuits during his long life. This volume will be valued not just as a memorial to Wallace, the man, but also the vehicle for new messages from a worldwide group of scholars who have themselves felt the impact of his ideas, and been intrigued by the course of his intellectual development and its legacy. Along with the authors of these essays, I congratulate the editors, Charles H. Smith and George Beccaloni, for their drive and enthusiasm, to which we owe the rich content and timely appearance of this book.

Notes

1. Originally the British Museum (BM), then the British Museum (Natural History) BM(NH), and now the Natural History Museum (NHM).
2. The gauge is denominated by the number of solid lead shot of the diameter of the bore (*i.e.*, 80, in this case) that would together weigh 1 lb (=454 g). This is therefore an extremely small bore, about 1/3 inch (~ 8 mm) diameter, and it is likely that the barrel was rifled (Keith Honeycombe, pers. comm.).

List of Figures

List of Plates

List of Tables

List of Contributors

Editors

Charles H. Smith, Science Librarian and Professor of Library Public Services, Western Kentucky University, Bowling Green, KY USA.

George Beccaloni, Curator of Orthopteroid Insects, Department of Entomology, The Natural History Museum, London UK.

Contributors

George Beccaloni, Department of Entomology, The Natural History Museum, London UK.

Ted Benton, Department of Sociology, University of Essex, Wivenhoe Park, Colchester UK.

Andrew J. Berry, Museum of Comparative Zoology, Harvard University, Cambridge, MA USA.

Peter Bowler, School of History and Anthropology, Queen's University Belfast, Belfast, UK.

Tim Caro, Department of Wildlife, Fish, and Conservation Biology and Center for Population Biology, University of California, Davis, CA USA.

Gregory Claeys, Department of History, Royal Holloway, University of London, London UK.

Gathorne Cranbrook (Earl of Cranbrook), Glemham House, Great Glemham, Saxmundham UK.

Steven J. Dick, NASA History Division, NASA, Washington, DC USA.

Melinda B. Fagan, Department of Philosophy, Rice University, Houston, TX USA.

Martin Fichman, Division of Humanities, York University, Toronto, Ontario, Canada.

Geoffrey Hill, Department of Biological Sciences, Auburn University, Auburn, AL USA.

Norman A. Johnson, Department of Plant, Soil and Insect Sciences, University of Massachusetts, Amherst, MA USA.

Sandra Knapp, Department of Botany, The Natural History Museum, London UK.

Leena Lindström, Department of Biological and Environmental Science, University of Jyväskylä, Jyväskylä, Finland.

James Mallet, Galton Laboratory, Department of Biology, University College London, London UK.

Sami Merilaita, Department of Zoology, Stockholm University, Stockholm, Sweden.

Bernard Michaux, Private Bag, Kaukapakapa 1250, New Zealand.

James R. Moore, Department of the History of Science, Technology and Medicine, The Open University, Milton Keynes UK.

Diane B. Paul, Professor Emerita, University of Massachusetts Boston, Boston, MA USA.

Peter Raby, Homerton College, University of Cambridge, Cambridge UK.

Charles H. Smith, Department of Library Public Services, Western Kentucky University, Bowling Green, KY USA.

Michael P. Speed, School of Biological Sciences, Liverpool University, Liverpool UK.

David A. Stack, Department of History, School of Humanities, University of Reading, Reading UK.

Martin Stevens, Behavioural Ecology Group, Department of Zoology, University of Cambridge, Cambridge UK.

Keith Tinkler, Department of Earth Sciences, Brock University, St Catharines, Ontario, Canada.

Introduction

• • • • • •

I am a socialist because I believe that the highest law for mankind is justice. I therefore take for my motto, "Fiat Justitia, Ruat Coelum"; and my definition of socialism is, "The use by every one of his faculties for the common good, and the voluntary organization of labour for the equal benefit of all." That is absolute social justice; that is ideal socialism. It is, therefore, the guiding star for all true social reform.

(S729 1905a, 2:274)

We are here, your two editors and twenty-three other contributors, to pay our respects to the work and legacy of British polymath Alfred Russel Wallace (8 January 1823 to 7 November 1913). The short quotation presented above, excerpted from Wallace's two-volume autobiography *My Life*, introduces his personal motto, *Fiat justitia, ruat coelum*—roughly, "let justice prevail, though the heavens fall." From these words one can recognize a certain quiet yet solid nobility in Wallace's makeup, a nobility which is more than hinted at in words penned by another noted polymath, Charles Peirce, in a review of Wallace's *Studies Scientific and Social* (S727) Peirce contributed to *The Nation* in 1901 (Peirce 1901, 36):

> Not quite a typical man of science is Wallace; not a man who observes and studies only because he is eager to learn, because he is conscious that his actual conceptions and theories are inadequate, and feels a need of being set right; nor yet one of those men who are so dominated by a sense of the tremendous importance of a truth in their possession that they are borne on to propagate it by all means that God and nature have put into their hands—no matter what, so long as it be effective. He is rather a man conscious of superior powers of sound and solid reasoning, which enable him to find paths to great truths that other men could not, and also to put the truth before his fellows with a demonstrative evidence that another man could not bring out; and along with this there is a moral sense, childlike in its candor, manly in its vigor, which will not allow him to approve anything illogical or wrong, though it be upon his own side of a question which stirs the depths of his moral nature.

Truly, there was something in Wallace's makeup that attracted him to the excavation of truths—whether these lay hidden in the wilds of some tropical

Figure 2 Wallace in 1848, aged twenty-five.
Reproduced from S729. Out of copyright.

cul-de-sac where no Westerner had ever trod, cryptic as abstractions of the natural processes responsible for evolving existence as we know it, or latent within the swirling political and social miasmas that alternately favored or eroded the basic rights of individual persons. Not surprisingly, his efforts often resulted in conflict: confrontations with the differing agendas or expectations of others, especially those of a more conservative or controlling bent. Wallace often showed considerable courage at these times, and across several fronts: countering dangers to his physical person (natural history field work in the nineteenth century was often a risky enterprise), challenges to his creative process as a practicing intellectual (who else, as G. K. Chesterton once referred to him for his concurrent acceptance of evolutionary theory and spiritualism, has been both "the leader of a revolution and the leader of a counter revolution"?), and the scorn heaped on by those powerful figures whose priorities lay more with maintaining the status quo than with improving the lot of the common man.

Perhaps, one might claim, this was just a more naive age, one in which windmill-tilting was more in vogue, and so too dilettantism. But Wallace was no mere dilettante, as this collection clearly illustrates. The insight Wallace poured into a variety of subjects is both amazing in itself, and remarkable for its lasting influence. In the following pages an attempt is made to survey the better part of what we feel to be Wallace's intellectual legacy, and it is a journey that traverses both the past and the present. Some of the essays dwell more on Wallace in historical context, whereas others use that context as a starting point only, seeing his influence through to scientific or social developments of the present day. There will even be some hinting at possible future directions for thought and research: that is, by identifying some arguably unfulfilled potentials suggested by Wallace's many-directioned originality.

Accordingly, we the editors have not asked our contributors to adhere rigidly to any one analytical or literary style in carrying out their basic charge. How could we? Wallace's influence has been felt in many ways: as an important innovator within the world of basic science, as an inspiration to the traveler and field biologist, as a vivid writer who successfully engaged a wide readership, as a provider of countless thousands of specimens of new and other species for the practicing specialist, as a model for the humane understanding of underprivileged and otherwise marginalized human populations, and, not least, as a man who was not afraid to state publicly his deepest convictions, even when these ran counter to established opinions. It is only fitting that the essayists here should have been allowed to approach their task in a way allowing these many different sides to show through.

We have organized the essays into two general parts: "In the World of Nature," and "In the World of Man, and Worlds Beyond." This ends up being a somewhat artificial arrangement, but is to a degree chronological, and also helps highlight his main emphases. After the Earl of Cranbrook's Preface, written from the perspective of a veteran field worker and zoologist, co-editor George Beccaloni introduces the subject of our attention from a unique biographical perspective: an account of his many places of residence. Part I begins with historical studies related to Wallace's work in the field and thoughts on species (Berry, Fagan, Beccaloni, and Mallet), then studies in some detail his work in particular areas of evolutionary biology (Johnson, Caro *et al.* and Caro *et al.*). The part concludes with essays on Wallace's biogeography (Michaux), interests (perhaps surprising to some) in glaciology (Tinkler), and conservation (Knapp). In Part II, Raby leads off with a reminder of Wallace's stature as a literary figure. This is followed by four analyses of his engagements with the societal milieu of his time (Claeys, Paul, Stack, and Fichman). Then we're off to the more esoteric realms: Wallace as an astrobiologist (Dick), systems thinker (Smith), spiritualist (Moore), and philosopher in matters related to the evolution of the human spirit (Benton and Smith). These last two essays dwell on two particular interpretations of Wallace's personal intellectual development, the "change of mind" and "no change of mind" models.

Conventions on form . . . After some discussion we decided to combine the use of Harvard-style citations (*e.g.*, "Smith 2000") within the essays with an endnote convention; both within-text citations and within-endnote citations refer to items listed in a single References Cited compilation at the end of the book. One other convention should be duly noted. To eliminate the need for constant mention of

Figure 3 Wallace in old age.
Original in private collection. Copyright George Beccaloni.

Wallace's name throughout (or awkward referrals such as "Wallace, 1871h"), we are making use of the numbering system for Wallace publications introduced in Charles Smith's *Alfred Russel Wallace: An Anthology of His Shorter Works* in 1991, and continued and extended on his Wallace website (Smith, 2000–). Thus, if one sees "(S43 1858, 54)" in the text, the referral is to a passage on page 54 of the 1858 essay "On the Tendency of Varieties to Depart Indefinitely From the Original Type," bibliographic details for which will be found in the section of Wallace's works in the References Cited compilation, between listings for S42 and S44 (if they also are cited here). The employment of this convention makes possible one other potential convenience: just about all of the works by Wallace referred to here are available in html full-text at Smith's "Alfred Russel Wallace Page" (http://www.wku.edu/~smithch/index1.htm), where they may be accessed by "S" number through a single menu found under its "Wallace Writings" feature. Thus, as one is reading along it is possible to simultaneously sample further from Wallace source material as desired.

The Editors would like to thank a number of individuals for their generosity in helping us in various ways, starting with Peter Bowler and Lord Cranbrook for contributing our opening statements, and then Richard Burkhardt, Colin Thorn, Meagan Miles, Christie Henry, Fern Elsdon-Baker, Nancy Stepan, Michael Cremo, Gregory Radick, Jane Camerini, and Peter Raby. Also, Western Kentucky University Libraries (especially their interlibrary loan group), The Natural History

Museum (London), several individuals who suggested names of potential contributors, and several others who provided anonymous reviews of publication proposals. Special thanks to Peter Raby and Wallace's grandsons, Richard and John, for allowing us to reproduce several images they own the copyright of.

Charles H. Smith and
George Beccaloni, April 2008

1

Homes Sweet Homes: A Biographical Tour of Wallace's Many Places of Residence*

George Beccaloni

Introduction

Alfred Russel Wallace lived in a great many houses during his long life, and even in his later years he moved house frequently, always choosing properties with pleasant views over the surrounding countryside (Richard Russel Wallace, pers. commun. 2006). In this chapter I discuss the houses he lived in for more than a few weeks, plus all the buildings he is known to have designed, even if he never occupied them. I will concentrate on his houses in Britain because of the difficulty of documenting the dozens of usually ephemeral dwellings he occupied during his travels in South America, South East Asia, and elsewhere. I have, however, made an exception for the houses in which he wrote his famous "Sarawak Law" and "Ternate" papers whilst in South East Asia. The information I present is largely restricted to descriptions of the properties plus brief summaries of any particularly interesting events which took place in them. The descriptions of the houses have been taken from the two-volume American edition of Wallace's autobiography *My Life: A Record of Events and Opinions* (S729 1905b), except where noted. I have also tried to include information on the current condition of the properties.

Without doubt the most interesting of all of Wallace's houses are the three he built for himself and his family: The Dell, Nutwood Cottage, and Old Orchard. Sadly only one of these, The Dell in Grays, Essex, has survived destruction by modern development. Fortunately, this house now has some legal protection, as it was designated a Grade II listed building in April 2000. Ironically, it was listed

* I would like to thank the following people for providing me with help and useful information: Mike Brooke, Lord Cranbrook, Tom Gladwin, Judith Magee, Jim Moore, the late Ken Parker, Peter Raby, Sister Rita, John Wallace, Richard Wallace, and John Webb. I am also grateful to Charles Smith and Judith Magee for their comments on an earlier draft of this chapter, and to my wife Jan for her help with my Wallace work.

Figure 4 Wallace's tiny house at Bessir, Waigiou (now Besir, Waigeo Island, Indonesia) where he lived for six weeks in 1860. The figure under the hut is Wallace sitting in a wicker chair beside the table at which he worked.

Reproduced from S715. Out of copyright.

on the basis of its architectural merit, rather than any link with Wallace. This contrasts with Charles Darwin's house, Down House in Kent, which was made a Grade I listed building largely because of the association with its former illustrious owner.

Wallace also designed (and in one case built) a number of other buildings, some of which still survive. He and his brother John designed the Mechanics' Institute in the town of Neath, Wales, which was completed in 1847, and still exists. The brothers also designed and built a cottage for a client in the late 1840s in or near Neath, but it has not been possible to establish whether this has survived. Intriguingly, there is good evidence that Wallace designed two other houses: one for his daughter and wife, and one for his wife's family. These buildings still exist, but in order to conclusively prove that Wallace designed them it would be necessary to examine their title deeds, which to date has not been possible.

A list of the houses discussed in this chapter is given in Table 1.1.

Table 1.1. A chronological list of the houses and/or places where Wallace lived

Dates	House and/or place
8 January 1823–1828	Lives in Kensington Cottage, Usk, Monmouthshire (Wallace was born here).
1828–early 1837	Five childhood homes in Hertford and Hertford Grammar School.
Early 1837–summer 1837	Lives with his brother John in Robert Street, London.
Summer 1837–December 1843	Works as a land surveyor and travels widely in England and Wales.
Early 1844–Easter 1845	Boards in the Headmaster's House at the Collegiate School, Leicester.
Easter 1845–early 1848	Lives in Neath, Wales (he initially lodges with Mr Thomas Sims; then rents Llantwit Cottage).
26 April 1848[a]–1 October 1852	Spends four years travelling in South America.
Christmas 1852–March 1854	Rents 44 Upper Albany Street, London.
March 1854–April 1862	Spends eight years travelling in South East Asia and whilst there writes his famous "Sarawak Law" and "Ternate" papers.
April 1862–April 1865	Stays with his sister and her husband at 5 Westbourne Grove Terrace, London.
April 1865–mid 1867; summer 1868[b]–March 1870	Rents 9 St Mark's Crescent, Regent's Park, London.
Mid 1867–summer 1868[b]	Stays at Treeps in Hurstpierpoint, Sussex (Wallace's father-in-law's house).
25 March 1870[a]–25 March 1872	Rents Holly House, Tanner Street, Barking.
25 March 1872[a]–July 1876	Builds and lives in The Dell, Grays, Essex.
July 1876–March 1878	Rents Rose Hill, Dorking, Surrey.
March 1878–1880	Rents Waldron Edge, Duppas Hill, Croydon.
1880–May 1881	Rents Pen-y-Bryn, Croydon.
May 1881–June 1889	Designs, builds and lives in Nutwood Cottage, Godalming, Surrey.
June 1889–December 1902	Rents, then buys Corfe View, Parkstone, Dorset.
December 1902–7 November 1913	Designs, builds and lives in Old Orchard, Broadstone, Dorset (Wallace dies here).

[a] Raby (2001).
[b] Peter Raby pers. commun. 2002.

Kensington Cottage, Usk, Monmouthshire, England (8 January 1823–1828)

Alfred Russel Wallace was born in Kensington Cottage (Plate 2), Monmouthshire, England (originally Gwent, Wales—later Gwent again, and most recently Monmouthshire again, but as one of twenty-two "principal areas" of Wales) on 8 January 1823 to Thomas Vere Wallace and Mary Ann Wallace (née Greenell), a middle-class English couple of modest means. He was the eighth of nine children, three of whom did not survive to adulthood. Wallace's father was of Scottish descent (reputedly, of a lineage leading back to the famous William Wallace), whilst the Greenells were a respectable Hertford family.

Kensington Cottage is situated beside the river Usk, half a mile or so from the town of Usk on a road leading to the village of Llanbadoc. Wallace lived here until he was about six and when he was in his eighties he could still remember "the little house and room we chiefly occupied, with a French window opening to the garden, a steep wooded bank on the right, the road, river, and distant low hills to the left." He continues:

> The house itself was built close under this bank, which was quite rocky in places, and a little back yard between the kitchen and a steep bit of rock has always been clearly pictured before me ... In the house, I recollect the arrangement of the rooms, the French window to the garden, and the blue-papered room in which I slept ... so far as I remember, only one servant was kept [the cook], and my father did most of the garden work himself, and provided the family with all the vegetables and most of the fruit which was consumed. Poultry, meat, fish, and all kinds of dairy produce were especially cheap; my father taught the children himself; the country around was picturesque and the situation healthy ... (S729).

Wallace recalls fishing for small lampreys from large slabs of rock which jutted into the river Usk not far from the house. These had been flung into the river from a nearby stone quarry many years before. He also remembers seeing "men fishing in coracles, the ancient form of boat made of strong wicker-work, somewhat the shape of the deeper half of a cockle-shell, and covered with bullock's hide" (S729). Wallace was "half-baptized" on 19 January 1823 (in case he died suddenly) and fully baptized in the nearby Llanbadoc church on 16 February 1823 (Raby 2001).

Kensington Cottage (now named Kensington House) still survives, although there have been some structural alterations and the houses which used to be to either side of it have been demolished. The bank of the Usk in front of the house has been built up to protect against winter floods and on the part of the bank nearest the house is a metal bench with a stainless steel plaque dedicated to Wallace's memory. No plaque has been put on the house itself as it is set back too far from the road for one to be seen. On 20 May 2006 a monument sponsored by the Alfred Russel Wallace Memorial Fund[1] was unveiled by Wallace's grandson

Richard, outside the yard of Llanbadoc church not far from the cottage. The monument is made from Carboniferous limestone with fossils on its surface (best seen when the rock is wet) and it has a black granite plaque on it commemorating Wallace.[2]

Childhood Houses in Hertford (1828–early 1837)

Wallace writes in his autobiography that "In the year 1828[3] my mother's mother-in-law, Mrs. Rebecca Greenell, died at Hertford, and I presume it was in consequence of this event that the family left Usk in that year . . ." Wallace's father had gone on ahead to "make arrangements for the family at Hertford" and in the meantime Wallace was taken to "a children's school at Ongar, in Essex, kept by two ladies—the Misses Marsh." Wallace then travelled to Hertford, where he lived "for eight or nine years, and where I had the whole of my school education" (S729).

Wallace lived in five houses in Hertford. He describes the first as

> . . . a small house, the first of a row of four at the beginning of St. Andrew's Street, and I must have been a little more than six years old when I first remember myself in this house, which had a very narrow yard at the back, and a dwarf wall, perhaps five feet high, between us and the adjoining house. The very first incident which I remember, which happened, I think, on the morning after my arrival, was of a boy about my own age looking over this wall, who at once inquired, "Hullo! who are you?" I told him that I had just come, and what my name was, and we at once made friends. The stand of a water-butt enabled me to get up and sit upon the wall, and by means of some similar convenience he could do the same, and we were thus able to sit side by side and talk, or get over the wall and play together when we liked. Thus began the friendship of George Silk and Alfred Wallace, which, with long intervals of absence at various periods, has continued to this day (S729).

The Wallace family lived in this house for a year or two (S729).

The second house which the Wallace family occupied was situated "beyond the Old Cross, nearly opposite to the lane leading to Hartham" and Wallace recalled that it

> . . . was a more commodious one, and besides a yard at one side, it had a small garden at the back with a flower border at each side, where I first became acquainted with some of our common garden flowers. The gable end of the house in the yard, facing nearly south, had few windows, and was covered over with an old vine which not only produced abundance of grapes, but enabled my father to make some gallons of wine from the thinnings. But the most interesting feature of the premises to us two boys [Wallace and his brother John] was a small stable with a loft over it, which, not being used except to store garden-tools and odd lumber, we had practically to ourselves. The loft especially was most delightful to us (S729).

Due to financial difficulties the family then moved (probably in 1835: Gander 1998), into half a "rambling old house near All Saints' Church" (S729), which had a large garden. Part of the house was used as a post office when Wallace lived in it and it had formerly been the Vicarage for All Saints' church (S729). At about this time, "when the family was temporarily broken up, for some reason I do not remember," Wallace went to board in Mr Cruttwell's house for about six months (Clement Henry Cruttwell was headmaster of Hertford Grammar School) (S729).

Early in 1836 Wallace's family moved again, into another house in St Andrew's Street and finally in midsummer 1836 they moved into "part of a house next to St Andrew's Church" (S729) which they shared with the Silk family. The Silks lived in "the larger half of the house. They also had most of the garden, on the lawn of which was a fine old mulberry tree, which in the late summer was so laden with fruit that the

Figure 5 11 St Andrew's Street, the first house in Hertford that Wallace lived in, as it is today. Copyright George Beccaloni.

ground was covered beneath it, and I and my friend George used to climb up into the tree, where we could gather the largest and ripest fruit and feast luxuriously" (S729).

Wallace left Hertford in early 1837 (when his parents moved to Rawdon Cottage, Hoddesdon, Hertford) and moved to London to live with his brother John (Raby 2001).

The first house in which the Wallace family lived in Hertford is the house which still exists at 11 St Andrew's Street (Fig. 5). It is now a doctors' surgery called "Wallace House" and has a plaque on the outside wall which reads "In this house lived Alfred Russel Wallace OM. LLD DCL. FRS. FLS.; Born 1823–Died 1913; Naturalist, Author, Scientist; Educated at Hertford Grammar School." Green (1995) states that the position and location of the second house (*i.e.*, the one located "beyond the Old Cross") "agrees with the present 23 Old Cross." The third house near All Saints' church was demolished many years ago and the site is now occupied by St John's Hall, built in 1939 (Tom Gladwin, pers. commun. 2008), whilst the fourth house on St Andrew's Street can not be identified as Wallace does not provide enough information about it. The fifth house adjacent to St Andrew's church is thought to be the building on St Andrew's Street to the west of the church. It once had an old mulberry tree in the garden behind it (Tom Gladwin, pers. commun. 2008), but this garden no longer exists as a number of other houses have been built on it.

Hertford Grammar School (late 1830–18 March 1837)

In his autobiography Wallace wrote:

> My recollections of life at our first house in St. Andrew's Street are very scanty. My father had about half a dozen small boys to teach, and we used to play together; but I think that when we had been there about a year or two, I went to the Grammar School with my brother John, and was at once set upon that most wearisome of tasks, the Latin grammar (S729).

Wallace probably started school late in 1830, when he was nearly eight (the usual age of young entrants at the Grammar School) and left on 18 March 1837 when he was fourteen (Gander 1998; Raby 2001). He describes the school as follows:

> The school itself was built in the year 1617, when the school was founded. It consisted of one large room, with a square window at each end and two on each side. In the centre of one side was a roomy porch, and opposite to it a projecting portion, with a staircase leading to two rooms above the schoolroom and partly in the roof. The schoolroom was fairly lofty. Along the sides were what were termed porches—desks and seats against the wall with very solid, roughly carved ends of black oak, much cut with the initials of names of many generations of schoolboys. In the central space were two rows of desks with forms on each side. There was a master's desk at each end, and two others on the sides, and two open fireplaces equidistant from the ends.

Every boy had a desk, the sloping lid of which opened, to keep his school-books and anything else he liked, and between each pair of desks at the top was a leaden ink-pot, sunk in a hole in the middle rail of the desks. As we went to school even in winter at seven in the morning, and three days a week remained till five in the afternoon, some artificial light was necessary, and this was effected by the primitive method of every boy bringing his own candles or candle-ends with any kind of candlestick he liked. An empty ink-bottle was often used, or the candle was even stuck on to the desk with a little of its own grease. So that it enabled us to learn our lessons or do our sums, no one seemed to trouble about how we provided the light.

The school was reached by a path along the bottom of All Saints' Church-yard, and entered by a door in the wall which entirely surrounded the school playground and master's garden. Over this door was a Latin motto—'Inter umbras Academi studere delectat' [*i.e.*, 'It is a pleasure to study beneath the shadow of Academus' (the Greek hero in whose grove was situated the Academy of Plato) (Gander 1998)].

This was appropriate, as the grounds were surrounded by trees, and at the north end of the main playground there were two very fine old elms . . . (S729).

Figure 6 Hertford Grammar School in the early nineteenth century (a watercolour by Eliza Dobinson *c.* 1815).

Hertford Grammar School is now called Richard Hale School and it has moved to new premises off Peggs Lane. The original building in Hale Street still exists but it has been extended and is now used to house Longmores Education Support Centre (Wilson 2000). Wilson (2000, 13) describes a visit he and his wife made to the old school building in 1998:

> The staircase leading to the rooms above is still used and the original oak is well worn. The present school has a new front door, as the old one was removed and placed in the new school, The Richard Hale School, just inside the front door in 1930. We could recognise features of the old school from photographs, such as the doorway and porch, the windows from the outside and even the cupola on top. Perhaps the only change immediately obvious was the presence of a skylight on the roof, visible when the building was viewed from above, as we were able to do later from the top of a nearby car park. The trees shown in the old photographs have gone and it was not long before we could see why.
>
> Just outside the front porch there is now a brick wall and the noise of traffic is very evident. The school that Wallace described has been cut in two by the construction of a main highway, Gascoyne Way, which keeps traffic away from the centre of the town. All Saints Church has also been cut off from the centre of the town. The master's house, where Wallace was once a boarder for six months, and which once was connected to the school by a long garden, is now on the other side of the highway, and borders on to Fore Street. The building is occupied by a firm of estate agents [it is soon to be converted into a hotel]. A bird's-eye view of all this can be better appreciated from the top of a multi-storey car park opposite the school, on the other side of the "relief road" How sad it is to see the school and its once lovely grounds sacrificed for a roadway.

Robert Street, London (early 1837–summer 1837)

In early 1837 Wallace left Hertford and moved to London to stay with his brother John in Robert Street, off Hampstead Road near Regent's Park in London. John lived in the house of his employer, a builder called Mr Webster, and Wallace shared John's room (Raby 2001). I do not know whether this house survives.

Surveying Work in England and Wales (summer 1837–December 1843)

In the summer of 1837, Wallace began training as a land surveyor with his older brother William—a job he held for the next six and a half years. In consequence of this work the brothers moved from job to job, and lived in many inns and lodgings in Bedford (Beds.), Hereford and Worcester (Heref./Worcs.), Shropshire, and Wales.

The first place the Wallaces found work was in the parish of Higham Gobion (Beds.), where they stayed in the Coach and Horses public-house in the village of Barton-in-the-Clay (Barton-le-Clay). Wallace recalls that when he was alone here he frequently used to sit in "the tap-room with the tradesmen and labourers for a little conversation or to hear their songs or ballads." In January 1838 the brothers moved to an inn called The Tinker of Turvey in the village of Turvey (Beds.); then in May/June to an inn in the village of Silsoe (Beds.); then a school house in the village of Soulbury (Buckinghamshire); and then to Leighton Buzzard (Beds.) where they "lodged in the house of a tin-and-copper smith in the middle of the town," before going home to Hoddesdon for Christmas. Early in 1839 Wallace was hired by a watchmaker, Mr William Matthews, in Leighton Buzzard, and lodged in his house. He worked for Matthews for nine months, then in the autumn of 1839 rejoined his brother for more surveying work. The brothers travelled to Kington (Heref./Worcs.), where they boarded in the house of a gunmaker, Mr Samuel Wright. During the winter, Wallace went alone to New Radnor (Powys, Wales), where he stayed in a small inn. In February 1840 the Wallace brothers moved to Rhayader (Powys), but Alfred developed a serious chest infection after falling into a bog and had to go back to his parents' home in Hoddesdon to recuperate for two months. He then returned to work with his brother in Kington. Shortly afterwards, they moved to the village of Llanbister (Powys), where they stayed in the local public-house. Next they moved to Llandrindod Wells (Powys), where they lodged in a small hotel. The brothers must then have returned to Kington, since Wallace next recounts in his autobiography how in the summer/early autumn of 1841 they left this town and travelled to the village of Trallong (Powys), where they boarded in the house of a shoemaker for several months. They then moved on to Senni Bridge (Sennybridge, Powys), but soon after Wallace alone went to stay in a small public-house in the chapelry of Senni, before meeting up with his brother again and moving back to Kington for a short time (S729 1908).

In late autumn 1841 the brothers moved to a farmhouse called Bryn-coch about two miles north of the town of Neath (West Glamorgan, Wales), and lodged with the farmer David Rees and his family for over a year. It was here that his interest in botany developed. Desiring to identify the plants he found in the surrounding countryside he bought two books, a "shilling paper-covered book published by the Society for the Diffusion of Useful Knowledge" and then a costly copy of "Lindley's 'Elements of Botany'" (S729). However, these did not prove satisfactory so he asked a local bookseller, Mr Hayward:

> ... if he knew of any book that would help me. To my great delight he said he had Loudon's "Encyclopaedia of Plants," which contained all the British plants, and he would lend it to me, and I could copy the characters of the British species. I therefore took it home to Bryn-coch, and for some weeks

> spent all my leisure time in first examining it carefully, finding that I could make out both the genus and the species of many plants by the very condensed but clear descriptions, and I therefore copied out the characters of every British species there given ... But I soon found that by merely identifying the plants I found in my walks I lost much time in gathering the same species several times, and even then not being always quite sure that I had found the same plant before. I therefore began to form a herbarium, collecting good specimens and drying them carefully between drying papers and a couple of boards weighted with books or stones ... Now, I have some reason to believe that this was the turning point of my life, the tide that carried me on, not to fortune, but to whatever reputation I have acquired, and which has certainly been to me a never-failing source of much health of body and supreme mental enjoyment (S729).

Probably in late 1842 Wallace and his brother moved into an old, roomy cottage nearer the town of Neath, where they boarded with a colliery surveyor, Samuel Osgood, and his wife. During the year that they lived here they had, in addition to their usual surveying jobs, "a little architectural and engineering work, in designing and superintending the erection of warehouses with powerful cranes." Paid work was, however, scarce, and in December 1843, William told his brother he should leave and find a job elsewhere. Wallace therefore left Wales in mid-December, returning to spend Christmas in Hoddesdon with his sister Frances (Fanny) and mother, before getting a job in the New Year as a teacher at the Collegiate School in Leicester.

I do not know whether any of the properties mentioned above still exist, with the exception of Bryn-coch farmhouse. This survives, although it has been much modified (Steve Griffiths, pers. commun. 2007). The farm which it was originally part of is now a separate property (called Bryncoch), which has recently been the focus of a campaign to protect it from a proposed housing development.[4]

Collegiate School, Leicester (early 1844–Easter 1845)

Early in 1844, Wallace obtained a job teaching junior classes at the Collegiate School in Leicester. Wallace recounts how he was only required to

> ... take the junior classes in English reading, writing, and arithmetic, teach a very few boys surveying, and beginners in drawing ... I was to live in the house, preside over the evening preparation of the boarders (about twenty in number), and to have, I think, thirty or forty pounds a year, with which I was quite satisfied. My employer was the Rev. Abraham Hill, Headmaster of the Collegiate School ... After a few weeks, finding I knew a little Latin, Mr. Hill asked me to take the lowest class, and even that required some preparation in the evening (S729 1908).

Wallace lodged with Revd Hill and his wife, and he recalls that he "had a very comfortable bedroom, where a fire was lit every afternoon in winter, so that with the exception of one hour with the boys and half an hour at supper with Mr. and Mrs. Hill, my time after four or five in the afternoon was my own" (S729). Leicester had a good library and Wallace was able to study several important works on natural history and travel. It was probably in this library that he first met amateur naturalist Henry Walter Bates, who got Wallace very interested in the study of entomology (especially beetles). It was Bates who in 1848 would travel with Wallace to the Amazon in Brazil.

In March 1845[5] Wallace's brother William died suddenly and at Easter Wallace left Leicester and moved to Neath with his brother John in order to wind up William's affairs.

The Collegiate School on College Street was built in 1835 and opened in 1836. It closed in 1866 and the building was then partly used by the Wycliffe Congregational Church.[6] It is now owned by Leicester City Council and is a Grade II listed building. Its " 'Gothic' windows in Perpendicular style, its statuary in niches and its moulded stone copings and pinnacles is a high point of South Highfields' townscape." The Headmaster's House (where Wallace lived) still exists on College Street and is also Grade II listed. It is described as being of "Domestic Gothic style in white brick with stone dressings."[7]

Neath, Wales (Easter 1845–early 1848)

In Neath, Wallace initially boarded with Mr Thomas Sims (who married his sister Fanny in 1849) on "the main street," and it was whilst living here in 1845, that Wallace first read Robert Chambers's controversial book *Vestiges of the Natural History of Creation*, which had been published anonymously the year before. This book convinced Wallace of the reality of evolution (then known as transmutation) (Slotten 2004) and in a letter to Bates dated 28 December 1845 he remarked:

> I have rather a more favourable opinion of the "Vestiges" than you appear to have. I do not consider it as a hasty generalisation, but rather as an ingenious hypothesis strongly supported by some striking facts and analogies but which remains to be proved by more facts & the additional light which future researches may throw upon the subject—it at all events furnishes a subject for every observer of nature to turn his attention to; every fact he observes must make either for or against it, and it thus furnishes both an incitement to the collection of facts & an object to which to apply them when collected. (Original in the Natural History Museum (NHM), London, catalogue number WP1/3/17.)

In the autumn of 1846, Wallace, his mother, and his brother John moved into "a small cottage close to Llantwit Church, and less than a mile from the middle of the town [of Neath]. It had a nice little garden and yard, with fowl-house, shed, etc., going down to the Neath Canal, immediately beyond which was the river Neath, with a pretty view across the valley to Cadoxton and the fine Drumau Mountain" (S729). His brother Herbert also moved in (Raby 2001), as did later his sister Fanny, who returned to England from Georgia, USA in September 1847. This was to be the last time that the whole of his remaining family was to live together (Wallace's father had died in April 1843: Raby [2001], says May). Wallace lived in Llantwit Cottage until early 1848, when he and his friend Bates departed on their collecting trip to the Amazon.

Llantwit Cottage (which now has a built-up road in front of it) was badly damaged in a fire in the late twentieth century, but as of 1996 it was being restored (Wilson 2000).

Whilst living in Neath, Wallace recalls how he and his brother John "had a little building and architectural work. A lady wanted us to design a cottage for her, with six or seven rooms, I think, for £200. Building with the native stone was cheap in the country, but still, what she wanted was impossible, and at last she agreed to go £250, and with some difficulty we managed to get one built for her for this amount" (S729). Unfortunately, Wallace does not say where this cottage was located and I do not know whether it still survives.

Wallace and his brother also designed the Mechanics' Institute in the town of Neath. It was built by J. Townsend of Swansea and the Wallace brothers supervised

Figure 7 The former Mechanics' Institute in Neath, Wales, as it is today.
Copyright George Beccaloni.

the construction work. The building was completed in 1847 and it was officially opened in 1848 (Raby 2001). In his autobiography Wallace says that

> ... a building was required at Neath for a Mechanics' Institute, for which £600 was available. It was to be in a narrow side street, and to consist of two rooms only, a reading room and library below, and a room above for classes and lectures. We were asked to draw the plans and supervise the execution, which we did, and I think the total cost did not exceed the sum named by more than £50. It was, of course, very plain, but the whole was of local stone, with door and window-quoins, cornice, etc., hammer-dressed; and the pediments over the door and windows, arched doorway, and base of squared blocks gave the whole a decidedly architectural appearance. It is now used as a free library ...

Despite experiencing a severe fire in 1903 the building still survives: situated in Church Place, it overlooks the graveyard. The Neath Museum used to be housed in it and currently it provides office space for museum staff. A plaque on it reads "Neath Borough Council; Alfred Russell [*sic*] Wallace; 1823–1913; Designed this building. He lived in Neath from 1841 to 1848 during which period he worked as a surveyor and studied natural history. In his lifetime he collaborated [*sic*!] with Charles Darwin in the study of the laws of natural selection. And with him presented the first paper on the subject in 1858."

Travels in South America (26 April 1848–1 October 1852)

Whilst living in Neath Wallace was inspired by a recently published book by William Henry Edwards entitled *A Voyage Up the River Amazon*, and in late 1847/early 1848 (S729) he suggested to Bates that they travel to Brazil to collect specimens of insects, birds, and other animals, both for their private collections and to sell to collectors and museums in Europe. The primary aim of the expedition, as far as Wallace was concerned, was to seek evidence for evolution and attempt to discover its mechanism (Slotten 2004). Bates liked the idea and the two young men (at the time Wallace was twenty-five and Bates twenty-three) set off by ship from Liverpool to Pará (Belém) on 26 April 1848. At first they worked as a team, but after a few months they split up in order to collect in different areas (Raby 2001). Wallace centred his activities in the middle Amazon and Rio Negro, journeying up the latter river further than anyone else had up to that point. He drafted a map of the Rio Negro using the skills he had learnt when he was a land surveyor. This was published by the Royal Geographical Society, London (S11 1853), and it proved accurate enough to become the standard for many years (Wilson 2000).

By early 1852 Wallace was in poor health and in no condition to continue travelling (S729). He decided to return to Britain, and began the long trip back down the Rio Negro and Amazon to Pará. Passing through Barra (Manaus),

Wallace found to his dismay that most of his specimens from the preceding two years which he had been gradually sending down the Amazon, had been delayed by the officials there because they were worried that the boxes might contain contraband. After declaring their contents, he collected the six large cases and finally reached Pará on 2 July. There he visited the grave of his younger brother Herbert, who had died in the town on 8 June 1851 (S729; Raby 2001). Herbert had travelled to Brazil in June 1849 to work with Wallace, but in May 1851 he wanted to return to England and had gone to Pará to find a ship back home, when he caught yellow fever and died.

Wallace sailed for England on 12 July (Raby 2001) on the brig *Helen*. Tragically, on 6 August, twenty-six days into the voyage, the ship caught fire and sank, taking with it his irreplaceable specimens, field notes, and collection of live animals. All he managed to rescue was a tin box containing a few shirts, into which he put his watch, some money, his drawings of fish and palms, the diary he kept on the Rio Negro, plus some notes and observations of the Rio Negro and Uaupés (Slotten 2004). Luckily, Samuel Stevens, Wallace's agent in London, had the foresight to insure his collections for £200 (Wallace estimated that they had been worth £500: Raby 2001).[8] Wallace and the crew struggled to survive in a pair of barely seaworthy lifeboats, and fortunately after ten days drifting in the open sea they were picked up by a passing cargo ship making its way back to England. They landed at Deal, England, on 1 October 1852.

44 Upper Albany Street, London (Christmas 1852–March 1854)

In late 1852, soon after his return from his trip to South America, Wallace rented a house near Regent's Park in London close to the Zoological Gardens (now London Zoo). In his autobiography he writes:

> As I wished to be with my sister and mother during my stay in England, I took a house then vacant in Upper Albany Street (No. 44), where there was then no photographer, so that we might all live together. While it was getting ready I took lodgings next door, as the situation was convenient, being close to the Regent's Park and Zoological Gardens, and also near the Society's offices in Hanover Square, and with easy access to Mr. Stevens's office close to the old British Museum. At Christmas we were all comfortably settled, and I was able to begin the work which I had determined to do before again leaving England.

Wallace lived in this house with his mother, his sister Fanny, and her husband, Thomas Sims, for a little over a year. He wrote his books *Palm Trees of the Amazon and Their Uses* (S713, published October 1853) and *A Narrative of Travels on the Amazon and Rio Negro* (S714, published December 1853) there. Upper Albany Street was the northern half of what is now Albany Street. I do not know whether the house that Wallace rented survives.

The Malay Archipelago (March 1854–April 1862)

Wallace left Britain again in 1854 on a collecting expedition to the Malay Archipelago (Malaysia and Indonesia). He spent nearly eight years in the region, and undertook sixty or seventy separate journeys resulting in a combined total of around 14,000 miles of travel. He visited every important island in the archipelago at least once, and several on multiple occasions, and collected almost 110,000 insects, 7,500 shells, 8,050 bird skins, and 410 mammal and reptile specimens, including over a thousand species new to science (S715 1989). The book he later wrote describing his work and experiences there, *The Malay Archipelago* (S715), is the most celebrated of all travel writings on this region, and ranks with a few other works as one of the best scientific travel books of the nineteenth century. It was in the Malay Archipelago, halfway through his trip, that Wallace made his greatest discovery: the primary mechanism of evolutionary change, natural selection.

Santubong, Sarawak, Borneo (early 1855)

In February 1855 whilst staying in a small house in Sarawak, Borneo, Wallace wrote what was to become one of the most important papers on evolution prior to the discovery of natural selection (Beddall 1988a). In *My Life* Wallace relates how this paper

> ... was written during the wet season, while I was staying in a little house at the mouth of the Sarawak river, at the foot of the Santubong mountain. I was quite alone, with one Malay boy as cook, and during the evenings and wet days I had nothing to do but to look over my books and ponder over the problem which was rarely absent from my thoughts. Having always been interested in the geographical distribution of animals and plants, having studied Swainson and Humboldt, and having now myself a vivid impression of the fundamental differences between the Eastern and Western tropics; and having also read through such books as Bonaparte's "Conspectus," already referred to, and several catalogues of insects and reptiles in the British Museum (which I almost knew by heart), giving a mass of facts as to the distribution of animals over the whole world, it occurred to me that these facts had never been properly utilized as indications of the way in which species had come into existence. The great work of Lyell [*Principles of Geology*] had furnished me with the main feature of the succession of species in time, and by combining the two I thought that some valuable conclusions might be reached. I accordingly put my facts and ideas on paper, and the result seeming to me to be of some importance, I sent it to *The Annals and Magazine of Natural History*, in which it appeared in the following September (1855). Its title was "On the Law which has regulated the Introduction of New Species," which law was briefly stated (at the end) as follows: "*Every species has come into existence coincident both in space and time* [this should read 'time and space'] *with a pre-existing closely-allied species.*" This clearly pointed to some kind of evolution. It suggested the *when* and the

> *where* of its occurrence, and that it could only be through natural generation, as was also suggested in the "Vestiges"; but the *how* was still a secret only penetrated some years later.
>
> Soon after this article appeared, Mr. Stevens [Wallace's agent in London] wrote me that he had heard several naturalists express regret that I was "theorizing," when what we had to do was to collect more facts. After this, I had in a letter to Darwin expressed surprise that no notice appeared to have been taken of my paper, to which he replied that both Sir Charles Lyell and Mr. Edward Blyth, two very good men, specially called his attention to it.

In fact, Wallace's "Sarawak Law" paper (S20) made such an impression on Charles Lyell that in November 1855, soon after reading it, he opened his "species notebook" in which he began to contemplate the implications of evolutionary change (Beddall 1988a). Then, shortly after Darwin had explained his theory of natural selection to him for the first time (during a visit he made to Down House in April 1856) Lyell sent a letter to Darwin urging him to publish the theory lest someone beat him to it. Darwin heeded his advice and in May 1856 he began to write a "sketch" of his ideas for publication. This "sketch" was abandoned in about October 1856 and Darwin instead began to write an extensive book about natural selection—an abstract of which would be published as the *Origin of Species* in November 1859 (Beddall 1988a).

Many authors, including Williams-Ellis (1966) and Wilson (2000), have assumed that the Santubong house belonged to Wallace's friend Rajah James Brooke, the ruler of Sarawak. Wallace was Brooke's guest on several occasions, and given the fact that Brooke had a small house at the foot of Santubong mountain, this assumption seems justified. In her book *Sketches of Our Life at Sarawak* first published in 1882, McDougall (1992) describes this house as follows:

> As early as the year 1848, the Rajah had a little Dyak house built on high poles, under the mountain of Santubong. It was an inconvenient little place, into which you climbed up a steep ladder—only one room, in fact, with a verandah; but we spent some happy days there, for the beauty of that shore made the house a secondary consideration. A small Malay village nestled in cocoa-nut palms at the foot of Santubong; in front lay a smooth stretch of sand, and a belt of casuarina-trees always whispering, without any apparent wind to move their slender spines. The deer in those days stole out of the jungle at night to eat the sea-foam which lay in flakes along the sand, and wild pigs could often be shot in a moonlight stroll under the trees.

Rajah Brooke's Santubong house did not face east across Buntal Bay to Bako National Park as Raby (2001) suggested, but westwards out to sea (Gathorne Cranbrook, pers. commun. 2008). The original building was replaced by a substantial government rest house (now derelict) which still exists on a promontory informally known as "Wallace Point." There are plans to refurbish this building and possibly make it into a field studies centre (Gathorne Cranbrook, pers. commun. 2008).

Ternate (several stays between early January 1858 and June/July 1861)[9]

In his *Malay Archipelago* (S715 1989) Wallace writes:

> On the morning of the 8th of January, 1858, I arrived at Ternate, the fourth of a row of fine conical volcanic islands which skirt the west coast of the large and almost unknown island of Gilolo. The largest and most perfectly conical mountain is Tidore, which is over four thousand feet high—Ternate being very nearly the same height, but with a more rounded and irregular summit. The town of Ternate is concealed from view till we enter between the two islands, when it is discovered stretching along the shore at the very base of the mountain. Its situation is fine, and there are grand views on every side. Close opposite is the rugged promontory and beautiful volcanic cone of Tidore; to the east is the long mountainous coast of Gilolo, terminated towards the north by a group of three lofty volcanic peaks, while immediately behind the town rises the huge mountain, sloping easily at first, and covered with thick groves of fruit trees, but soon becoming steeper, and furrowed with deep gullies. Almost to the summit, whence issue perpetually faint wreaths of smoke, it is clothed with vegetation, and looks calm and beautiful, although beneath are hidden fires which occasionally burst forth in lava-streams, but more frequently make their existence known by the earthquakes which have many times devastated the town.
>
> I brought letters of introduction to Mr. Duivenboden, a native of Ternate, of an ancient Dutch family, but who was educated in England, and speaks our language perfectly. He was a very rich man, owned half the town, possessed many ships, and above a hundred slaves. He was, moreover, well educated, and fond of literature and science—a phenomenon in these regions. He was generally known as the king of Ternate, from his large property and great influence with the native Rajahs and their subjects. Through his assistance I obtained a house, rather ruinous, but well adapted to my purpose, being close to the town, yet with a free outlet to the country and the mountain. A few needful repairs were soon made, some bamboo furniture and other necessaries obtained, and, after a visit to the Resident and police magistrate, I found myself an inhabitant of the earthquake-tortured island of Ternate, and able to look about me and lay down the plan of my campaign for the ensuing year. I retained this house for three years, as I found it very convenient to have a place to return to after my voyages to the various islands of the Moluccas and New Guinea, where I could pack my collections, recruit my health, and make preparations for future journeys ...
>
> A description of my house (the plan of which is here shown) will enable the reader to understand a very common mode of building in these islands. There is of course only one floor. The walls are of stone up to three feet high; on this are strong squared posts supporting the roof, everywhere except in the verandah filled in with the leaf-stems of the sago palm, fitted neatly in wooden framing. The floor is of stucco, and the ceilings are like the walls. The house is forty feet square, consists of four rooms, a hall, and two

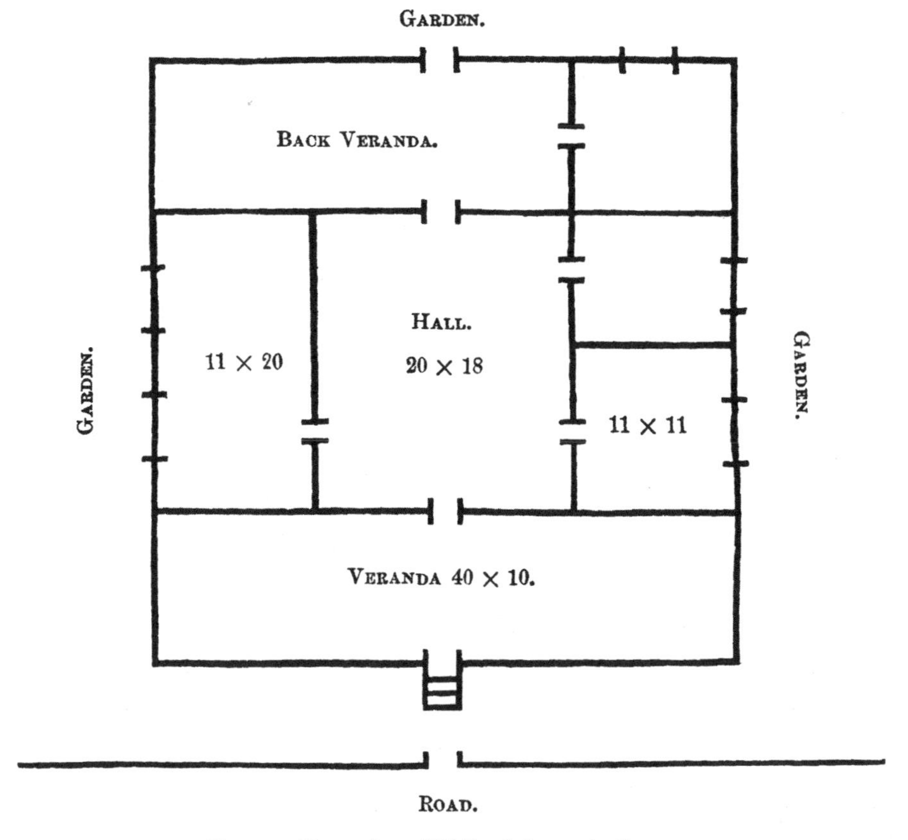

Figure 8 Floor plan of Wallace's house in Ternate.

Reproduced from S715. Out of copyright.

> verandahs, and is surrounded by a wilderness of fruit-trees. A deep well supplied me with pure cold water—a great luxury in this climate. Five minutes' walk down the road brought me to the market and the beach, while in the opposite direction there were no more European houses between me and the mountain. In this house I spent many happy days. Returning to it after a three or four months' absence in some uncivilized region, I enjoyed the unwonted luxuries of milk and fresh bread, and regular supplies of fish and eggs, meat and vegetables, which were often sorely needed to restore my health and energy. I had ample space and convenience for unpacking, sorting, and arranging my treasures, and I had delightful walks in the suburbs of the town, or up the lower slopes of the mountain, when I desired a little exercise, or had time for collecting.

It was in this house that Wallace said he wrote his famous "Ternate Essay" (S43 1858), in which he proposed his independently conceived theory of evolution by natural selection. Later he recalled:

> After writing the preceding paper [the "Sarawak Law"] the question of *how* changes of species could have been brought about was rarely out of my mind, but no satisfactory conclusion was reached till February 1858. At that time I was suffering from a rather severe attack of intermittent fever ... and one day while lying on my bed during the cold fit, wrapped in blankets, though the thermometer was at 88° F., the problem again presented itself to me, and something led me to think of the "positive checks" described by Malthus in his "Essay on Population," a work I had read several years before, and which had made a deep and permanent impression on my mind. These checks—war, disease, famine and the like—must, it occurred to me, act on animals as well as on man. Then I thought of the enormously rapid multiplication of animals, causing these checks to be much more effective in them than in the case of man; and while pondering vaguely on this fact there suddenly flashed upon me the *idea* of the survival of the fittest—that the individuals removed by these checks must be on the whole inferior to those that survived. In the two hours that elapsed before my ague fit was over I had thought out almost the whole of the theory, and the same evening I sketched the draft of my paper, and in the two succeeding evenings wrote it out in full, and sent it by the next post to Mr. Darwin (S725 1891, 20).

Wallace's letter to Darwin containing his essay about natural selection was posted from Ternate (now in Indonesia) on 9 March 1858 and it probably arrived at Darwin's house in Downe, Kent on 18 June 1858 (suggestions by some authors of an earlier arrival date are not supported by convincing evidence: Beddall 1988a). Unbeknownst to Wallace, Darwin had in fact discovered natural selection about twenty years earlier, and was partway through writing his "big book" on the subject, a project which, ironically, had been prompted by Wallace's "Sarawak Law" paper almost three years earlier (Beddall 1988a). Darwin was therefore horrified when he received Wallace's letter, and appealed to his influential friends Charles Lyell and Joseph Hooker for advice on what to do. As the famous story goes, Lyell and Hooker decided to present Wallace's essay, along with two unpublished excerpts from Darwin's writings on the subject, to a meeting of the Linnean Society of London on 1 July 1858. These documents were published together on 20 August of the same year as the paper "On the Tendency of Species to Form Varieties; And On the Perpetuation of Varieties and Species by Natural Means of Selection" in the Society's journal (Darwin and Wallace 1858). Darwin's contributions were placed before Wallace's essay, thus emphasizing Darwin's priority to the idea. Wallace later remarked that the paper "was printed without my knowledge, and of course without any correction of proofs" (Meyer 1895), contradicting Lyell and Hooker's statement in their introduction to the paper that "both authors ... [have] ... unreservedly placed their papers in our hands" (Lyell and Hooker 1858). It was, of course, this event which prompted Darwin to produce an "abstract" of his "big species book," which was published fifteen months later in November 1859 as *On the Origin of Species.*

Although Wallace's essay was marked as having being written on Ternate in February 1858, this cannot have been the case since Wallace's unpublished *Malay Field Journal* in the Linnean Society shows that he was on Gilolo during the whole of February, only returning to the neighbouring island of Ternate on 1 March. It is probable that he wrote "Ternate" on the essay simply because this was the island where he had his base, and because it was his postal address. Alternatively, he got the month wrong and should have written "March" instead of "February." However, this would have been a curious error to make as he wrote a letter to Frederick Bates which he dated 2 March 1858 (original in NHM, catalogue number WP1/3/42), and which was posted to Britain on the same mail ship as his Ternate Essay. Why he never corrected either the date or the place of his discovery in his published accounts of this event is curious. For more analysis of this situation see McKinney (1972a).

Wilson (2000) searched for Wallace's Ternate house in 1997. He was first shown a partly ruined dwelling, the floor plan of which did not exactly match the one that Wallace drew (see above) and which was not situated in the correct area of the town. He then went to what he thought was the correct area of the town but was not successful in finding the house. The first house Wilson was shown may be the one located at 16 Jalan Sultan Hairun that the Sultan of Ternate claims to be the one in which Wallace stayed.[10] However, it is far from certain that this is correct.

Dodinga, Gilolo (February 1858)

Wallace spent most of February 1858 living in the village of Dodinga on the island of Gilolo (now Halmahera Island, Indonesia). He noted that village is "situated at the head of a deep bay exactly opposite Ternate, and a short distance up a little stream which penetrates a few miles inland. The village is a small one, and is completely shut in by low hills." He continues:

> As soon as I arrived, I applied to the head-man of the village for a house to live in, but all were occupied, and there was much difficulty in finding one. In the mean time I unloaded my baggage on the beach and made some tea, and afterwards discovered a small hut which the owner was willing to vacate if I would pay him five guilders for a month's rent. As this was something less than the fee-simple value of the dwelling, I agreed to give it him for the privilege of immediate occupation, only stipulating that he was to make the roof water-tight. This he agreed to do, and came every day to talk and look at me; and when I each time insisted upon his immediately mending the roof according to contract, all the answer I could get was, "Ea nanti" (Yes, wait a little). However, when I threatened to deduct a quarter-guilder from the rent every day it was not done, and a guilder extra if any of my things were wetted, he condescended to work for half an hour, which did all that was absolutely necessary (S715 1989).

In the same source Wallace mentions that he was ill for most of the time he was at Dodinga, and it was here, in the hut he rented, where he may well have had his famous "ague fit" and discovered natural selection. Dodinga village still exists, but the hut in which he lived must have rotted away a very long time ago.

5 Westbourne Grove Terrace, London (April 1862–April 1865)

After returning to London from his collecting trip to the Malay Archipelago in the spring of 1862 Wallace says that he

> … went to live with my brother-in-law, Mr. Thomas Sims, and my sister Mrs. Sims, who had a photographic business in Westbourne Grove. Here, in a large empty room at the top of the house, I brought together all the collections which I had reserved for myself and which my agent, Mr. Stevens, had taken care of for me. I found myself surrounded by a quantity of packing-cases and storeboxes, the contents of many of which I had not seen for five or six years, and to the examination and study of which I looked forward with intense interest.
>
> From my first arrival in the East I had determined to keep a complete set of certain groups from every island or distinct locality which I visited for my own study on my return home, as I felt sure they would afford me very valuable materials for working out the geographical distribution of animals in the archipelago, and also throw light on various other problems. These various sets of specimens were sent home regularly with the duplicates for sale, but either packed separately or so distinctly marked "Private" that they could be easily put aside till my return home. The groups thus reserved were the birds, butterflies, beetles, and land-shells, and they amounted roughly to about three thousand bird skins of about a thousand species, and, perhaps, twenty thousand beetles and butterflies of about seven thousand species.
>
> As I reached home in a very weak state of health, and could not work long at a time without rest, my first step was to purchase the largest and most comfortable easy-chair I could find in the neighbourhood, and then engage a carpenter to fit up one side of the room with movable deal shelves, and to make a long deal table, supported on trestles, on which I could unpack and assort my specimens (S729).

This large property still exists (Judith Magee, pers. commun. 2008).

9 St Mark's Crescent, Regent's Park, London (April 1865–mid-1867 and summer 1868–March 1870)

In *My Life* Wallace writes that

> … in the spring of 1865 I took a small house for myself and my mother, in St. Mark's Crescent, Regent Part [*sic*], quite near the Zoological Gardens, and within a pleasant walk across the park of the society's library in Hanover Square, where I had to go very often to consult books of reference. Here

> I lived five years, having Dr. W. B. Carpenter for a near neighbour, and it was while living in this house that I saw most of my few scientific friends.

Friends who visited Wallace in this house included Charles Lyell, who if he "had any special subject on which he wished for information ... would sometimes walk across the park to St. Mark's Crescent for an hour's conversation," and Charles Darwin, who "also sometimes called on me in St. Mark's Crescent for a quiet talk or to see some of my collections" (S729).

Wallace married Annie Mitten in the parish church at Hurstpierpoint, Sussex, in April 1866, and in mid-1867, a few weeks after she had given birth to her first child, Herbert Spencer (Bertie) (born 22 June), they let the house for a year and moved to Treeps, Hurstpierpoint. Wallace and Annie returned to St Mark's Crescent in the summer of 1868, and on 15 November of this year Wallace's mother died. On 25 January 1869[11] Annie gave birth to her second child, Violet Isabel, and in March 1870 Wallace and family moved to Holly House in Barking, which he regarded as "a kind of half-way house" whilst trying to find a place to live in the countryside.

The house in St Mark's Crescent still exists (James Moore, pers. commun. 2006).

Treeps, Hurstpierpoint, Sussex (mid-1867–summer 1868)

Treeps (Plate 3) was Annie Wallace's family home. Wallace and Annie moved to Treeps in mid-1867 for a year so that Annie's mother and sisters could help look after baby Bertie (Raby 2001), and Wallace could work on his book *The Malay Archipelago* (S715, published 9 March 1869). Annie's father, William Mitten, was a chemist and a respected amateur botanist. On 24 April 1874[12] Wallace's son Bertie (who had been ill for some time) died at Treeps, aged six.

Treeps still exists and on 24 September 2005 a commemorative plaque was installed on the garden wall by the front gate. It reads "Alfred Russel Wallace O.M.; Naturalist; 1823–1913; wrote *The Malay Archipelago* whilst staying in Treeps 1867–1868."

Holly House, Tanner Street, Barking, London (25 March 1870–25 March 1872)

Wallace moved to Barking as a temporary measure until he could find a property to purchase in the countryside to the east of London. The reasons for moving were first, that he "had a great longing for life in the country" where he could devote much of his time "to gardening and rural walks," and second, because he hoped to secure a job as the director of a proposed museum of art and natural history in Bethnal Green in East London (S729).

Wallace recounts that he

> ... took an old cottage at Barking—Holly Lodge [it was actually called Holly House]—to which we moved in March, 1870, and where I was still

> almost in London. Though Barking was a miserable kind of village, surrounded by marshes and ugly factories, there were yet some pleasant walks along the Thames and among the meadows, while within a quarter of a mile of us was a well-preserved tumulus close to an old farmhouse (S729).

The Census taken on 3 April 1871 shows that he lived in the house with his wife, two children, a nurse and domestic servant, Rachel Nichol, and a cook, Maria Hocking (George 2001). Their last child, William Greenell, was born there on 30 December 1871.

Holly House has been demolished (George 2001).

The Dell, Grays, Essex (25 March 1872–July 1876)

In 1870 Wallace found,

> ... near the village of Grays, on the Thames, twenty miles from London, a picturesque old chalk-pit which had been disused so long that a number of large elms and a few other trees had grown up in its less precipitous portions. The chalk here was capped by about twenty feet of Thanet sand and pleistocene gravel, and from the fields at the top there was a beautiful view over Erith to the Kent hills and down a reach of the Thames to Gravesend, forming a most attractive site for a house. After some difficulty I obtained a lease for ninety-nine years of four acres, comprising the pit itself, an acre of the field on the plateau above, and about an equal amount of undulating cultivable ground between the pit and the lane which gave access to it. I had to pay seven pounds an acre rent, as the owner could not sell it, and though I thought it very dear, as so much of it was unproductive, the site was so picturesque, and had such capabilities of improvement, that I thought it would be a fair investment ...
>
> As there was a deep bed of rough gravel on my ground and there were large cement works at Grays, I thought it would be economical to build of concrete, and I found an architect of experience, Mr. Wonnacott, of Farnham, who made the plans and specifications, while I myself saw that the gravel was properly washed. In order to obtain water in ample quantity for building and also for the garden and other purposes, I had a well sunk about a hundred feet into a water-bearing stratum of the chalk, and purchased a small iron windmill with a two-inch force pump to obtain the water. I made two small concrete ponds in the garden—one close to the windmill—and had a large tank at the top of a low tower to supply house water ...
>
> With the help of another labourer I also myself laid down 1¼-inch galvanized water-pipes to the house, with branches and taps where required in the garden. I also built concrete walls round the acre of ground at top, the part facing south about nine feet high for fruit trees, the rest about five feet; and also laid out the garden, planted mounds for shelter, made a winding road from below, which, when the shrubs had grown up, became exceedingly picturesque; and helped to shift out hundreds of cubic yards of gravel to

> improve the land for the kitchen garden. All this work was immensely interesting, and I have seldom enjoyed myself more thoroughly, especially as my friend Geach was a continual visitor, was always ready with his help and advice, and took as much interest in the work as I did myself. We got into the house in March, 1872, and I began to take that pleasure in gardening, and especially in growing uncommon and interesting as well as beautiful plants, which in various places, under many difficulties and with mingled failures and successes, has been a delight and solace to me ever since (S729).

Raby (2001, 210) gives the following description of The Dell:

> ... the house was built on a grand scale, and even had a four-roomed entrance lodge for the gardener. There was a hall, drawing room, dining room, library and conservatory on the ground floor, four principal bedrooms, dressing room and bathroom on the first floor, and four more bedrooms, or nurseries above—plenty of room for children, and servants. The rooms were spacious, high ceilinged, full of light; the style plain, but with well-chosen decorations, such as the tiles and coloured glass in the hall. Outside, the grounds were beautifully laid out, both in terms of economy—walled gardens, greenhouses, a fowl house—and pleasure: walks and terraces, ponds, a fountain, a croquet lawn, and, eventually, a rich variety of flowers and shrubs and trees.

In a letter to Charles Darwin dated 24 November 1870 (Marchant 1916a, vol. 1), Wallace indicated that he hoped to settle in The Dell for the rest of his life, but in the event he only lived in the house for about four years. He sold the property by auction in June 1876 for a variety of reasons, including, perhaps, the fact that his

Figure 9 The Dell, Grays, Essex, at the time Wallace lived there.
Reproduced from S729. Out of copyright.

young son Bertie had died (in Treeps) whilst the Wallace family were living in The Dell, and also because of financial difficulties (Chase 1979). In a letter to Darwin dated 7 June 1876 (Marchant 1916a, vol. 1) Wallace gave the following explanation:

> For two years I have made up my mind to leave this place—mainly for two reasons: drought and wind prevent the satisfactory growth of all delicate plants; and I cannot stand being unable to attend evening meetings and being obliged to refuse every invitation in London. But I was obliged to stay till I had got it into decent order to attract a customer. At last it is so, and I am offering it for sale, and as soon as it is disposed of I intend to try the neighbourhood of Dorking, whence there are late trains from Cannon Street and Charing Cross.

Whilst Wallace was living in The Dell, he released *On Miracles and Modern Spiritualism* (S717, published March 1875) and wrote *The Geographical Distribution of Animals* (S718, published May 1876). The architect's drawings for The Dell are in the NHM library (catalogue number WP4/1) (Plate 4).

Figure 10 The Dell, Grays, Essex, as it is today.
Copyright Janet Beccaloni.

The Dell is notable in being one of the earliest surviving shuttered concrete houses in Britain (*i.e.*, as contrasted with prefabricated concrete). Its architect Thomas Wonnacott[13] was an early exponent of the use of this material, and had delivered a brief paper on Cement Concrete to the Royal Institute of British Architects on his election as an Associate of this organization in 1870 (John Webb, pers. commun. 2001). The Dell is also significant in being the only one of the three houses Wallace built for himself and his family to have survived until the present time.

The property has had several owners[14] since 1876 when Wallace sold it, including William Winch Hughes, founder of the Victoria Wine Company (which is still a successful UK business). It is currently a convent owned by the La Sainte Union Order of nuns and is not open to the public.

Since Wallace's time, the original estate of four acres has been subdivided and the entrance is now off College Avenue (house number 25), rather than Dell Road. Although the main fabric of the house has remained largely unchanged, the building was reroofed in the mid 1990s and the water tower and chimneys have been removed. The original conservatory on the west side of the main building was replaced in 1998 (Sister Rita, pers. commun. 1999). The entrance lodge on Dell Road has been demolished (John Webb, pers. commun. 2002).

In 1999 Thurrock Council made a tree preservation order on a number of trees in the grounds of The Dell (some of which may have been planted by Wallace), and in April 2000 Thurrock Local History Society, Thurrock Council, Thurrock Museum, and Thurrock Heritage Forum succeeded in getting The Dell designated a Grade II listed building.

On 14 September 2002 a cast aluminium plaque (which reads "Alfred Russel Wallace; Naturalist; 1823–1913; Built this house and lived here from 1872 to 1876") was unveiled by Wallace's grandson Richard, on the wall of The Dell to the left of the front door. The Wallace Memorial Fund designed and paid for the plaque, which forms part of a commemorative plaques scheme being run by the Thurrock Local History Society, the Heritage Forum, and Thurrock Council.

Should The Dell be sold in the future, it is to be hoped that an organization might purchase it and turn it into a museum which commemorates Wallace's life and work.

Rose Hill, Dorking, Surrey (July 1876–March 1878)

After moving from Grays, Wallace rented a large house on the desirable middle-class Rose Hill residential estate in Dorking. This property has not been identified, although it is likely that it still survives as most of the original houses on the estate still exist (Brigham 1997).

In his autobiography Wallace gives his reason for leaving Dorking as follows:

> ... when we were living at Dorking, my little boy [William], then five years old, became very delicate, and seemed pining away without any perceptible ailment. At that time I was being treated myself for a chronic complaint by

> an American medium, in whom I had much confidence; and one day, when in his usual trance, he told me, without any inquiry on my part, that the boy was in danger, and that if we wished to save him we must leave Dorking, go to a more bracing place, and let him be out-of-doors as much as possible and "have the smell of the earth." I then noticed that we were all rather languid without knowing why, and therefore removed in the spring to Croydon, where we all felt stronger, and the boy at once began to get better and has had fair health ever since (S729).

Waldron Edge, Duppas Hill Lane, Croydon (March 1878–1880)

In *My Life* Wallace recounts that the reason he chose to move to Croydon was "chiefly in order to send our children first to kindergarten, and then to a high school . . ."

Waldron Edge was one of two houses he lived in whilst in Croydon. He probably wrote his book *Island Life* (S721, published October 1880) whilst living here. This house was apparently demolished many years ago, and the site it was on now lies under the western end of the Croydon flyover (Sowan and Byatt 1974).

Pen-y-Bryn, St Peter's Road, Croydon (1880–May 1881)

Wallace moved to Pen-y-Bryn (subsequently numbered 44) in St Peter's Road, Croydon, London sometime between 9 January and 11 October 1880 (Sowan and Byatt 1974). The house still exists and has a Blue Plaque on it (erected in 1979) which reads "ALFRED RUSSEL WALLACE, 1823–1913, Naturalist, lived here."

Nutwood Cottage, Godalming, Surrey (May 1881–June 1889)

In his autobiography Wallace wrote:

> In the year 1881 I removed to Godalming, where I had built a small cottage near the water-tower and at about the same level as the Charterhouse School . . . We had here about half an acre of ground with oak trees and hazel bushes (from which I named our place "Nutwood Cottage"), and during the eight years we lived there I thoroughly enjoyed making a new garden, in which, and a small greenhouse, I cultivated at one time or another more than a thousand species of plants. The soil was a deep bed of the Lower Greensand formation, with a thin surface layer of leaf-mould, and it was very favourable to many kinds of bulbous plants as well as half-hardy shrubs, several of which grew there more freely and flowered better than in any of my other gardens (S729).

Wallace designed this house himself and his architectural drawings are in the NHM library (catalogue number WP4/2). Whilst living in Nutwood Cottage he wrote the books *Land Nationalisation* (S722, published May 1882), *Bad Times* (S723, published November 1885), and *Darwinism* (S724, published May 1889).

Figure 11 Nutwood Cottage, Godalming, Surrey.
Reproduced from S729. Out of copyright.

On 9 October 1886 he travelled to America on a successful lecture tour, returning to England on 19 August 1887 (Raby 2001).

He sold the property in 1901 (S729), just before he started to build Old Orchard in Broadstone, Dorset.

Nutwood Cottage was knocked down in 1970 and in 1971 a terrace of houses was built on the site. The development is called "Nutwood" and it is situated on Frith Hill Road.

Corfe View, Parkstone, Dorset (June 1889–December 1902)

In *My Life* Wallace recounts:

> Finding my house at Godalming in an unsatisfactory situation, with a view almost confined to the small garden, the south sun shut off by a house and by several oak trees, while exposed to north and east winds, and wishing for a generally milder climate, I spent some weeks in exploring the country between Godalming and Portsmouth, and then westward to Bournemouth and Poole. I had let my house from Lady Day, and had moved temporarily into another, and therefore wished to decide quickly. We were directed by some friends to Parkstone as a very pretty and sheltered place, and here we found a small house to be let, which suited us tolerably well, with the option of purchase at a moderate price. The place attracted us because we saw abundance of great bushes of the evergreen purple veronicas, which must have been a dozen or twenty years old, and also large specimens of eucalyp-

> tus; while we were told that there had been no skating there for twenty years. We accordingly took the house, and purchased it in the following year; and by adding later a new kitchen and bedroom, and enlarging the drawing-room, converted it from a cramped, though very pretty cottage, into a convenient, though still small house. The garden on the south side was in a hollow on the level of the basement, while on the north it was from ten to thirty feet higher, there being on the east a high bank, with oak trees and pines, producing a very pretty effect. This bank, as well as the lower part of the garden, was peat or peaty sand, and as I knew this was good for rhododendrons and heaths, I was much pleased to be able to grow these plants.

Wallace constructed a small pond in the garden in which to grow water-lilies and other aquatic plants and a small orchid house "in three divisions so as to get different temperatures." However, owing to

> ... the entrance to the orchid house being on a different floor from my study, the constant attention orchids require in shading, ventilating, and keeping up a moist atmosphere, involved such an amount of running up and down stairs, or up and down steps or slopes in the garden, that I found it seriously affected my health, as I was at that time subject to palpitations and to attacks of asthma, which were brought on by any sudden exertion. I was therefore obliged to give up growing them, as I found it impossible to keep them in a satisfactory condition. This was partly owing to the position of my houses, which were exposed to an almost constant wind or draught of air, which rendered it quite impossible to keep up the continuously moist atmosphere and uniform temperature which are essential conditions for successful orchid-growing (S729).

During his residence in Corfe View (Plate 5), Wallace was interviewed by several journalists. The following is taken from one of the resulting articles (S738 1898):

> Although quite close to the railway station, the spot where Dr. and Mrs. Wallace have had their home for the last eight years is secluded and picturesque. Standing on sloping ground, and surrounded by a garden, the pretty creeper-covered cottage commands a fine view across Poole Harbour to the Purbeck Hills; Corfe Castle is discernible in clear weather, hence the name of the house—"Corfe View."

The following extract from an anonymous interview with Wallace in 1893, gives more details about the house:

> Three miles of lonesome road, cut through a pine forest, separates the home of Dr. Wallace, at Parkstone, from fashionable Bournemouth. The house itself, standing on a slight elevation, commands a fine view across the sea to Swanage and the Purbeck Hills. "Look at our lovely view," were almost the first words which Dr. Wallace said to me upon entering his house. And so for

> a few minutes I stood in the spacious drawing room window, beside the tall, erect figure with silver hair and beard, drinking in the beauty of the scene, until the rosy tints of sunset had faded away, leaving the hills and the water dull and grey. "Let us go down to the study now," said the doctor, and following him I entered a cosy retreat, in the lower part of the house, ranged around with books and pictures, the chairs suggestive of comfort and the well-littered tables of much study and research. "Your study lies in a wood, Dr. Wallace," I involuntarily exclaimed, as I looked through the stretch of windows flanking the outer side of the room to the garden beyond, rising gradually upwards until it joined the distant wood. Then the lamp was lighted, the blinds drawn down, and the great scientist seated himself in his special armchair, drawn up close to the blazing fire, and proceeded to discourse upon the subject of natural selection, in which, as an original thinker, he stands unequalled save by Darwin (S736 1893).

Wallace wrote his book *The Wonderful Century* (S726, published 10 June 1898) whilst living here. Unfortunately, this interesting house was demolished twenty or more years ago (Alfred John Russel Wallace, pers. commun. 2008). The site it occupied is situated on Sandringham Road.

Old Orchard, Broadstone, Dorset (December 1902–7 November 1913)

In 1905 Wallace wrote of this residence:

> About the year 1899 our house at Parkstone became no longer suitable owing to the fact that building had been going on all around us and what had been pretty open country when we came there had become streets of villas, and in every direction we had to walk a mile or more to get into any open country. I therefore began to search about various parts of the southern counties for a suitable house ... after almost giving up the attempt in despair, we accidentally found a spot within four miles of our Parkstone home and about half a mile from a station, with such a charming distant view and pleasant surroundings that we determined, if we could get two or three acres at a moderate price, to build a small house upon it.
>
> After a rather long negotiation I obtained three acres of land, partly wood, at the end of the year 1901; sold my cottage at Godalming at a fair price, began at once making a new garden and shrubbery, decided on plans, and began building early in the new year. The main charm of the site was a small neglected orchard with old much-gnarled apple, pear, and plum trees, in a little grassy hollow sloping to the southeast, with a view over moors and fields towards Poole harbour, beyond which were the Purbeck hills to the right, and a glimpse of the open sea to the left. In the foreground were clumps of gorse and broom, with some old picturesque trees, while the orchard was sheltered on both sides by patches of woodland. The house was nearly finished in about a year, and we got into it at Christmas, 1902, when we decided to call it Old Orchard.

> Being so near to our former house, I was able to bring all our choicer plants to the new ground, and there was, fortunately, a sale of the whole stock of a small nursery near Poole in the winter, at which I bought about a thousand shrubs and trees at very low prices, which enabled us at once to plant some shrubberies and flower borders, and thus to secure something like a well-stocked garden by the time we got into the house (S729).

An interview of Wallace in 1909 (S745) includes more details about the property:

> From the summit of the hill, under the brow of which the house stands, you dip down, with the waters of the Channel always before you, and turn aside along a winding fir-lined pathway that leads to "Old Orchard." The house stands in four or five acres of land—half cultivated and half wild, in all its Dorset beauty. It is built after the design of the Doctor himself, who, past his eightieth birthday, felt equal to such a task, and also to that of reducing the wilderness of the little estate to a semblance of cultivation. The garden has the order of disorder; nothing of sharply defined paths and trimly kept lawns, but wild firs, bunches of bracken, hosts of evergreens and subtropical plants, and here a pool with broken irregular edge to add a mirror of nature to the rustic scene.
>
> Immediately as you enter the house you are conscious of the intellectual atmosphere pervading the place. There are suggestive hints of the owner and his tastes—a large-scale map of the district on the wall, and a staircase lined with orchid pictures that recall the floral splendours amid which so much of his time was spent fifty-odd years ago ...

Figure 12 Old Orchard, Broadstone, Dorset.

"I am always at work," said Dr. Wallace, in reply to an early question as to how he was spending the evening of his days. "As a rule I manage two steady hours every morning. In the afternoon I take a quiet doze, or content myself with watching the harbour, which you can see from my window there; and in the evening I am ready for another spell of writing or study."

The room, lighted from the south and west, bore every trace that it was meant for use rather than for ornament. Most of the walls were covered by the shelves of what Sir Walter Besant delighted to call a "working library," every book being intended for use, and showing that it was fulfilling its purpose. There was a strong array of scientific works, many of them presentation copies, a "file" of a well-known periodical devoted to gardening, an assertive row in blazing red of a certain much-advertised history in twenty-odd volumes, many novels, poetry down to the latest editions of Barnes, the Dorset poet, and solemn *Fortnightlies* mingling with the latest penny productions of the Socialistic press. Here and there I detected, nevertheless, the touch of a woman's hand, and I found Mrs. Wallace, when I met her and her daughter at lunch, a bright, intellectual woman keenly interested in her husband's work. There was a system of arrangement about the library, too, which suggested that its owner knew every corner of it, and could, if need be, find anything he wanted in the dark. To call it a library would suggest uniformity, with machine-made rows of books;

Figure 13 Wallace's study in Old Orchard.
From Marchant (1916). Out of copyright.

> and who in such a place ever saw a tea canister—not a tobacco-jar, mind you—peeping up from behind a pile of papers?
> ... Adjoining the study is a small conservatory, fitting into a right angle of the house. Here Dr. Wallace loves to experiment, for he still hears the east a calling, though no longer able to obey the summons, and is as keenly interested as ever in his fine collection of tropical plants.

Wallace's architectural drawings of Old Orchard are in the NHM library (catalogue number WP4/4). He wrote the following books whilst living there: *Man's Place in the Universe* (S728, published October 1903; possibly written at Corfe View), *My Life* (S729, published October 1905), *Is Mars Habitable?* (S730, published December 1907), *The World of Life* (S732, published December 1910), *Social Environment and Moral Progress* (S733, published March 1913), and *The Revolt of Democracy* (S734, published October 1913). He also edited Richard Spruce's *Notes of a Botanist on the Amazon and Andes* (S731, published December 1908).

Wallace died in his sleep at Old Orchard at 9.25 a.m. on 7 November 1913, and three days later he was buried in a public cemetery nearby (Broadstone Cemetery off Dunyeats Road). His grave is marked by an unusual and striking monument—a seven-foot-tall fossil tree trunk from the Portland beds mounted on a cubic base of Purbeck limestone. This monument was restored in 2000 and a bronze plaque commemorating his achievements was installed on the grey granite surround.[15]

Unfortunately Old Orchard has been destroyed. Brackman (1980) recorded its demolition:

> With Wallace's death, his Civil List Pension ended abruptly. An ill Annie could not afford to retain Old Orchard and the family was forced to sell it in early 1914 [this date is probably incorrect: see entry for Tulgey Wood below].
>
> In 1964 Old Orchard was bulldozed to make way for a housing development. An effort was made to save the site for posterity—the developer had paid £16,000 for the house and three acres—but as an official of the Linnean Society said in a letter to the Wallace family, "I cannot see that much can be done about this unfortunate matter."
>
> At this writing, rubble still litters the spot, bricks and bits of glass from Wallace's plant rooms, an odd beam from his study. The road running through the development is named Wallace Court [it is actually named Wallace Road]. More likely than not, none of the residents have ever heard of him.

There used to be a rectangular commemorative plaque on the wall of the house which read "ALFRED RUSSEL WALLACE O.M. F.R.S., Naturalist and Explorer, LIVED HERE 1902–1913." This plaque was made by Poole Pottery, and when the house was demolished it was given by the Linnean Society to the Bournemouth Natural Science Society who probably still have it (the late Barbara Waterman, pers. commun. 1999). The only items saved from the house itself were some wooden panels with plants carved on them, which were apparently from Wallace's study, and the terracotta monogram of Wallace's initials which had been set into the chimney by the front door (Plate 6) (the position of this on the house is given in a letter from Wallace to his daughter Violet in 1902 [original in NHM, catalogue number WP1/2/129]). The items mentioned are currently in a private collection. The site of Old Orchard is located at the end of Benridge Close off Wallace Road, somewhere under the houses on the west side of the Close. The former garden of the house, to the south of the property, still appears to be largely intact and is reputed to still contain trees planted by Wallace (Douglas Rodenhurst, pers. commun. 2008).

Tulgey Wood, Broadstone, Dorset

In September 1999, the late Mrs Barbara Waterman of Broadstone, Dorset sent the author a fascinating letter about a house near the site of Old Orchard in Broadstone, which she believed had been built by Wallace and which still exists. Interestingly, Mrs, Waterman used to live in Old Orchard when she was young, and she remembered the "remarkable, and sometimes decidedly 'quirky,' interior of the house which was unique." An extract from her letter follows:

> ... there is a small house in Broadstone called Tulgey Wood, 149 Lower Blandford Rd., very near to where Old Orchard once stood and believed to have been built by Wallace as the Dower House; he thought he would die before his much younger wife.
>
> I am now in my 87th year and have retained a clear memory of my early childhood in W.W.I. which includes arriving to live in Old Orchard at the age of three years old! None of my generation in the village now remains.
>
> My parents, Brig. General & Mrs J. D. T. Tyndale-Biscoe, rented Old Orchard from Dr Wallace's son in 1916 and bought it from him in 1920 on my father's retirement from the Army. It was our much-loved family home for over 40 years, until my father, approaching ninety, was forced to sell it. During those years my parents took great care of the house & garden, which had been neglected during the three years after Wallace's death in 1913 ... When we arrived in 1916 it [Tulgey Wood] was no longer part of Old Orchard but our front drive was used by the occupants as it passed their front door & ended in our gate on the main road to Poole. At that time Tulgey Wood was either rented or bought from young Mr. Wallace [Wallace's son William] by a Captain Grant-Dalton & his wife; they were followed by the Morgan-Singers & eventually by Captain and Mrs Desborough who got permission from my father to close the drive when they bought an area of heathland from Lord Wimborne between their front door and what is now Wallace Rd. in which they made an outstanding garden with many cuttings and seeds from our valuable collection at Old Orchard. My father turned the back gate into our front entrance onto the lane [which] eventually [became] Wallace Rd., to the north of us ...

Tulgey Wood was built in 1908 (Douglas Rodenhurst, pers. commun. 2008) and it must have been named after "the tulgey wood" in Lewis Carroll's poem "The Jabberwocky" from *Alice Through the Looking-Glass.* Carroll was a favourite of Wallace's, and his daughter Violet remembered her father's jokes about Boojums and Jabberwocky (Raby 2001). Wallace's grandson John recalls his father William mentioning that Violet lived in Tulgey Wood (Alfred John Russel Wallace, pers. commun. 2008) and Raby (2001) says that she started a small school there and lived there for a time with her pupils. Wallace's wife Annie, who was already very ill at the time of Wallace's death in November 1913, is known to have died in this house on 10 December 1914, aged 68.[16]

Unfortunately, it has not yet been possible to prove that Tulgey Wood was designed and/or built by Wallace, and this can probably only be resolved by examining the title deeds of the property. It is very likely to be the case, however, especially considering the following statement Wallace made in a letter to his son William dated 22 June 1905:

> We have two or three places where we can build a bungalow on our *own ground.* I think therefore it will be better to give up the idea of buying more land, but to spend what money we can spare on a little house for Ma & Violet after I am gone to another country, as you suggested (original in NHM, catalogue number WP1/1/107).

The interior of Tulgey Wood was renovated recently (Mike Brooke, pers. commun. 2008) and there is a rumour that the area it is in may soon be redeveloped.

Culver Croft, Hurstpierpoint, Sussex

Wallace's grandson Richard informed the author that Wallace may have designed a house called Culver Croft for Wallace's sisters-in-law, Rose, Flora, and Bessie, to live in after their father William Mitten's death in July 1906. This house still exists and it is situated behind William Mitten's former home, Treeps, in Hurstpierpoint. Again, whether or not Wallace designed this house could be proved or disproved by examining the title deeds of the property.

Notes

1. See **http://www.wallacefund.info/**
2. See **http://www.wallacefund.info/2006-llanbadoc-monument**
3. Raby (2001) notes that Rebecca Greenell died on 18 October 1826, but the family may not have moved to Hertford immediately.
4. See **http://www.bryncochfarm.org.uk/index.html**
5. In *My Life* (S729) Wallace says that William died in February 1846, but his Death Certificate indicates that he died in March 1845.
6. **http://british-history.ac.uk/report.aspx?compid=66570#s7**
7. **http://www.leicester.gov.uk/EasySite/lib/serveDocument.asp?doc=1330&pgid=5703**

8. Curiously, in his autobiography (S729 1908, 357) Wallace states that the collections that were lost with the ship "... would probably have sold for £200. My agent, Mr. Stevens, had fortunately insured them for £150, which enabled me to live a year in London, and get a good outfit and a sufficient cash balance for my Malayan journey."
9. Baker (2001).
10. See **http://www.iht.com/articles/2004/06/19/edsoch_ed3_.php**
11. Handwritten entry in front of the Wallace family prayerbook owned by Richard Wallace.
12. Death Certificate of Herbert Spencer Wallace.
13. For more information about Thomas Wonnacott (1834–1918) see **http://wallacefund.info/thomas-wonnacott**
14. For a chronological list of the owners of The Dell see **http://wallacefund.info/owners-dell**
15. For more information see **http://wallacefund.info/1999-wallace-s-grave**
16. Death Certificate of Annie Wallace.

I

······

In the World of Nature

2

"Ardent Beetle-Hunters": Natural History, Collecting, and the Theory of Evolution*

Andrew Berry

In 1908, to mark the fiftieth anniversary of the reading of the original Darwin-Wallace paper on evolution by natural selection in its meeting rooms on Piccadilly, the Linnean Society issued its Darwin-Wallace medal. With a profile of Darwin on one side and a full-face image of Wallace on the other, it was awarded in 1908 and in 1958. It will be awarded again in 2008.[1] In 1958 all the recipients, and in 1908 all but one, received silver strikings of the medal. The 1958 contingent included the likes of J. B. S. Haldane, R. A. Fisher, and Ernst Mayr (from a total of twenty awardees); the 1908 silvers (six awardees) included Francis Galton, Ernst Haeckel, and Joseph Dalton Hooker. The only gold medal ever awarded went to Wallace. In 1908, he was still going strong at eighty-five, with five years yet to live and three books yet to publish.

Not surprisingly, 1908 was a busy year for Wallace as the scientific world scrambled to commemorate the 1858 publication. He received the Royal Society's Copley Medal (forty-four years after Darwin, twenty-one years after Hooker, and twenty years after T. H. Huxley) and was made Order of Merit (OM) by King Edward VII. Wallace's reaction to being feted in this way was, typically, one of mild bemusement: "Is it not awful—two more now! I should think very few men have had three such honours within six months! I have never felt myself worthy of the Copley medal—and as to the Order of Merit—to be given to a red-hot Radical, Land Nationaliser, Socialist, Anti-Militarist, etc., etc., etc., is quite astounding and unintelligible!" (letter to Arabella Fisher, Marchant 1916b, 447). Wallace was never keen on ceremony, turning down an honorary degree from the University of Wales in 1902 because of "The bother, the ceremony, the having perhaps to get a blue or yellow or scarlet gown! and at all events new black clothes and a new topper!" (letter to Dora Best, Marchant 1916b, 446). Wallace's aversion to pomp and circumstance even extended to his pleading successfully that he was, as Slotten

* I thank Janet Browne and Naomi Pierce for helpful discussion and comments.

(2004, 479; my italics) puts it, "*conveniently* suffering from some sort of illness" when summoned to Buckingham Palace for the OM investiture. The hardware was eventually hand-delivered by an equerry. However, Wallace apparently felt differently about the Linnean Society occasion, a grand reception held at the Institute of Civil Engineers in London on the very day, 1 July, of the anniversary of the Darwin-Wallace paper, and he travelled up to town from Dorset to make a rare public appearance at the event. Perhaps the Linnean Society's role in elevating Wallace in one giant stride from obscure collector to member of the scientific elite made Wallace feel especially loyal to the institution.

The great and good in the scientific community were assembled for the occasion along with various European ambassadors (Marchant 1916b). After an introduction by the President of the Linnean Society, Wallace took the floor. The speech (S656 1909) is vintage Wallace, a mix of self-deprecation and insight. He starts by insisting that the theory of natural selection should be regarded primarily as Darwin's: "... it was only Darwin's extreme desire to perfect his work that allowed me to come in, as very bad second ..." From here, however, Wallace moves into less familiar territory.

> And this brings me to the very interesting question: Why did so many of the greatest intellects fail, while Darwin and myself hit upon the solution of this problem—a solution which this Celebration proves to have been (and still to be) a satisfying one to a large number of those best able to form a judgment on its merits? As I have found what seems to me a good and precise answer to this question, and one which is of some psychological interest, I will, with your permission, briefly state what it is.
>
> On a careful consideration, we find a curious series of correspondences, both in mind and in environment, which led Darwin and myself, alone among our contemporaries, to reach identically the same theory.
>
> First (and most important, as I believe), in early life both Darwin and myself became ardent beetle-hunters. Now there is certainly no group of organisms that so impresses the collector by the almost infinite number of its specific forms, the endless modifications of structure, shape, colour, and surface-markings that distinguish them from each other, and their innumerable adaptations to diverse environments. These interesting features are exhibited almost as strikingly in temperate as in tropical regions, our own comparatively limited island-fauna possessing more than 3000 species of this one order of insects.
>
> Again, both Darwin and myself had, what he terms "the mere passion of collecting,"—not that of studying the minutiæ of structure, either internal or external. I should describe it rather as an intense interest in the mere variety of living things—the variety that catches the eye of the observer even among those which are very much alike, but which are soon found to differ in several distinct characters.
>
> Now it is this superficial and almost child-like interest in the outward forms of living things, which, though often despised as unscientific, happened to be

> the only one which would lead us towards a solution of the problem of species. For nature herself distinguishes her species by just such characters—often exclusively so, always in some degree—very small changes in outline, or in the proportions of appendages, as give a quite distinct and recognisable facies to each, often aided by slight peculiarities in motions or habits; while in a large number of cases differences of surface-texture, of colour, or in the details of the same general scheme of colour-pattern or of shading, give an unmistakable individuality to closely allied species.
>
> It is the constant search for and detection of these often unexpected differences between very similar creatures, that gives such an intellectual charm and fascination to the mere collection of these insects; and when, as in the case of Darwin and myself, the collectors were of a speculative turn of mind, they were constantly led to think upon the "why" and the "how" of all this wonderful variety in nature—this overwhelming, and, at first sight, purposeless wealth of specific forms among the very humblest forms of life.
>
> Then, a little later (and with both of us almost accidentally) we became travellers, collectors, and observers, in some of the richest and most interesting portions of the earth; and we thus had forced upon our attention all the strange phenomena of local and geographical distribution, with the numerous problems to which they give rise. Thenceforward our interest in the great mystery of how species came into existence was intensified, and—again to use Darwin's expression—"haunted" us (S656, 7–9).

Wallace's reduction here of his and Darwin's multi-dimensional talents to those of "beetle-hunters" has typically been received by historians of science as another instance of Wallace's almost pathological modesty. However, though the statement does indeed bear the imprint of Wallace's signature self-deprecation, it should not be dismissed merely as a mildly flippant solution to the problem of having to come up with something to say on a grand retrospective occasion. My goal, in this paper, is to show that, contrary to appearances, Wallace's "beetle-hunting" comment was in fact a remarkably deft summary of a rich tradition, part intellectual, part artisan, whose significance in the development of biological theory has too often been overlooked by historians of science in their examination of social, political, and institutional factors influencing the development of scientific thought.

Beetles: Biodiversity Encapsulated

Wallace was a late bloomer by the standards of naturalists, many of whom are passionately engaged with all or some part of the natural world from an early age. In his late teens, working with his brother William as a surveyor, Wallace spent most days outdoors, typically in rural areas. It was during this period that he "first began to feel the influence of nature and to wish to know more of the various flowers, shrubs, and trees I daily met with, but of which for the most part I did not even know the English names" (S729 1905a, 1:109). To start with, Wallace was very much the neophyte: with no training, formal or informal, he "hardly realized that there was such a science as systematic botany, that every flower and every meanest and most insignificant weed had been accurately described and classified" (S729, 1:110). The story he tells about how he first became interested is curious:

> This wish to know the names of wild plants, to be able even to speak of them, and to learn anything that was known about them, had arisen from a chance remark I had overheard about a year before. A lady, who was a governess in a Quaker family we knew at Hertford, was talking to some friends in the street when I and my father met them, and stayed for a few minutes to greet them. I then heard the lady say, "We found quite a rarity the other day—the Monotropa—it had not been found here before." This I pondered over, and wondered what the Monotropa[2] was. All my father could tell me was that it was a rare plant; and I thought how nice it must be to know the names of rare plants when you found them. However, as I did not even know there were books that described every British plant, and as my brother appeared to take no interest in native plants or animals, except as fossils, nothing came of this desire for knowledge till a few years later (S729, 1:110).

Wallace's early career as a field botanist was a stumbling affair. He went to some trouble and expense to procure a reference book, John Lindley's *Elements of Botany,* that he assumed would allow him to identify the plants he was finding and was disappointed to find that he had acquired what was essentially a systematic textbook with "hardly any reference to British plants" (S729, 1:192). Fortunately, Wallace was able to copy much of the content of a borrowed copy of J. C. Loudon's *Encyclopaedia of Plants* to the margins of his Lindley, and his botanical career was launched.

Until he met Henry Walter Bates in Leicester in 1844 (Wallace was twenty-one, Bates nineteen), Wallace's natural history had been a solitary preoccupation. Bates was also largely an autodidact but had benefited from a rather more settled upbringing than Wallace. He opened Wallace's eyes to a whole new scientific universe.

> I found that his specialty was beetle collecting, though he also had a good set of British butterflies. Of the former I had scarcely heard, but as I already knew the fascinations of plant life I was quite prepared to take an interest in any other department of nature. He asked me to see his collection, and I was amazed to find the great number and variety of beetles, their many strange forms and often beautiful markings or colouring, and was even more surprised when I found that almost all I saw had been collected around Leicester, and that there were still many more to be discovered. If I had been asked before how many different kinds of beetles were to be found in any small district near a town, I should probably have guessed fifty or at the outside a hundred, and thought that a very liberal allowance. But I now learnt that many hundreds could easily be collected, and there were probably a thousand different kinds within ten miles of the town (S729, 1:236–37).

From then on, Wallace was a coleopterist.

Beetles comprise approximately one-quarter of all named species. Globally, some 350,000 species have been named; for comparison, birds weigh in at about 10,000 species and mammals a paltry 5,400. It is likely too that the figures for birds and mammals are close to final—occasional species will be added over the years as ever more obscure and remote regions are biologically explored and as ever more nuanced, usually genetic, taxonomic methods are used to split single species into two or more—but inconceivable that the figure for beetles is anywhere near final. Many beetle species inhabit the canopies of tropical forest trees, an environment that is largely unexplored because of the difficulty of access. Whether, as some suggest, concerted inventories of the rain forest canopy will swell beetle species numbers by literally millions remains to be seen (Erwin 2004); it is clear, though, that the 350,000 figure is destined to increase substantially. J. B. S. Haldane's famous, if perhaps apocryphal, quip in response to a question about what a lifetime of studying the natural world had taught him about the Creator—that He has an inordinate fondness for beetles (see Gould 1993)—is becoming truer by the day as tropical forest canopies are explored ever more intensively.

Small and northerly, Britain is generally depauperate in species. Wallace's eye may have been caught by Bates's butterfly collection, but he would not have been impressed by the number of species on display: there are only about sixty species of butterfly (the vagueness stems from whether or not one counts occasional migrants) native to the UK. In contrast, Wallace gives a figure of 3,000 British beetle species in the 1908 Linnean Society address; today's estimate is 4,000 species (National Biodiversity Network 2008). The discrepancy between these two figures for British beetle diversity highlights one reason why beetles were an excellent group for a trainee naturalist to specialize on: beetle discovery, even in a thoroughly explored biota like Britain's, was ongoing. Again the comparison to butterflies is instructive. Bates notes, when exalting in the richness of the Amazon fauna, that the number of British butterflies "does not exceed 66" (Bates 1892, 52). Butterflies in Britain had been *done*; beetles, however, were an unfinished project.

As Wallace was serving his biological apprenticeship in Britain, beetles were thus the ideal group to focus on. Here, even in Britain, was a chance to encounter genuine biodiversity, possibly even to discover new species. Also, with a view to transferring natural history skills learnt in a temperate environment to the tropics, beetles were again a fortuitous choice. The seat of beetle biodiversity is the tropical rain forest, and Wallace's familiarity with the group—its basic systematics and ecology—would have helped him through that daunting first encounter with the species richness of the Neotropics.

The breadth of the beetle scientific niche (as opposed to, say, the narrow, already-filled butterfly one) was, I suggest, key to Wallace's first tentative steps as a professional scientist. Wallace would surely have been impressed that, by the time they met, Bates had already contributed to the scientific literature. In 1843, Bates published in *The Zoologist*, a journal that acted as a clearing house for entomological information: "Note on Coleopterous Insects Frequenting Damp Places." The very first formal scientific foray of the Wallace-Bates pairing was a celebration of the adaptability of beetles—of their having successfully colonized an environment that has been exploited by only a limited range of insect groups. "Many a long day, in sunshine and in shower, has seen me wading in those miry paradises, in the praiseworthy endeavour to effect my little towards the advancement of our favourite science" (Bates 1843, 114). Beetle diversity, then, was the foundation of Bates's precocious publishing career. The world of beetles was expansive enough for Edward Newman, *The Zoologist*'s editor, to value as new the hard-won observations of a seventeen-year-old. Had Bates specialized exclusively on butterflies, I suspect he would have had a harder time publishing: the field was smaller and the competition more intense.

For all his modesty, Wallace was a driven competitor. His decision, later on, to head to South East Asia within about eighteen months of his disastrous return from the Amazon bespeaks a powerful desire to make his mark in the world of science, and I suspect that one of the reasons for the Wallace-Bates split early in their Amazon travels was the realization that continuing as a duo would impair each man's ability to excel individually. It is, therefore, not surprising that Wallace's name also subsequently appears in the columns of *The Zoologist.* One gets a sense of how far behind Bates he was, however, from Newman's editorial comment. "Capture of *Trichius fasciatus*[3] near Neath. I took a single specimen of this beautiful insect on a blossom of *Carduus heterophyllus* near the falls at the top of Neath Vale. Alfred R. Wallace, Neath. [The other insects in my correspondent's list are scarcely worth publishing. E. Newman]" (S2 1847, 1676). Not the most auspicious of scientific debuts, but it was surely a source of pride to Wallace: he was now a legitimate, published member of the entomological community.

Darwin too began his formal scientific career as a coleopterist. Admittedly, Darwin presented papers on bryozoans and seaweeds at the Plinian Society in Edinburgh during his brief stint as medical student, and he is also mentioned in a published paper on marine leeches by his Edinburgh mentor Robert Grant (1827), but it is clear from his autobiography that he considered a series of new records of beetles to be his first formal contribution to science. "No poet ever felt more delight at seeing his first poem published than I did at seeing, in Stephens' *Illustrations of British Insects* the magic words, 'captured by C. Darwin, Esq.'"[4] (Darwin 1887a, 51).

Precisely why beetles are so hyper-diverse remains a controversial focus of ongoing research. Many models entail coevolution between beetles and the plants they feed on whereby the plants evolve mechanisms to inhibit or prohibit beetle herbivory and the beetles, in turn, evolve to negate the plants' defences. Farrell (1998) has argued that this coevolutionary process has been particularly effective in driving beetle diversification in the interaction between herbivorous beetles and flowering plants, angiosperms. Recently, however, Hunt *et al.* (2007) have questioned the centrality of herbivory's role. Lepidoptera, after all, are herbivorous insects as well, and yet at 165,000 species globally they do not rival beetles in overall species numbers. Moreover, whole major groups of beetles, such as the carabids (an estimated 40,000 species globally; Lovei and Sunderland 1996), are primarily carnivorous. Hunt *et al.* suggest that the key to beetle diversity is the group's facility for making ecological shifts: their phylogenetic analysis reveals, for example, that the shift from a terrestrial habit to an aquatic one has happened at least ten times in the history of the group. For Hunt *et al.*, then, beetles are simply remarkable ecological opportunists.

Over evolutionary time, the propensity of beetles for exploiting different aspects of the environment has possibly been facilitated by their limited dispersal abilities. A characteristic of virtually all beetles is that their forewings are hardened into elytra that cover the two folded flying wings beneath and yield a rigid carapace. Because the elytra must be raised to permit the hind wings to act, beetles are typically weak flyers, meaning that populations are often highly localized. Population genetics dictates that isolated, local populations will diverge genetically from each other, with speciation the ultimate outcome. Speciation, the generation of new species, underpins all biological diversification.

That beetle populations tend to be local can be a challenge to the naturalist. If I have two similar-but-not-identical-looking individuals from separate populations,

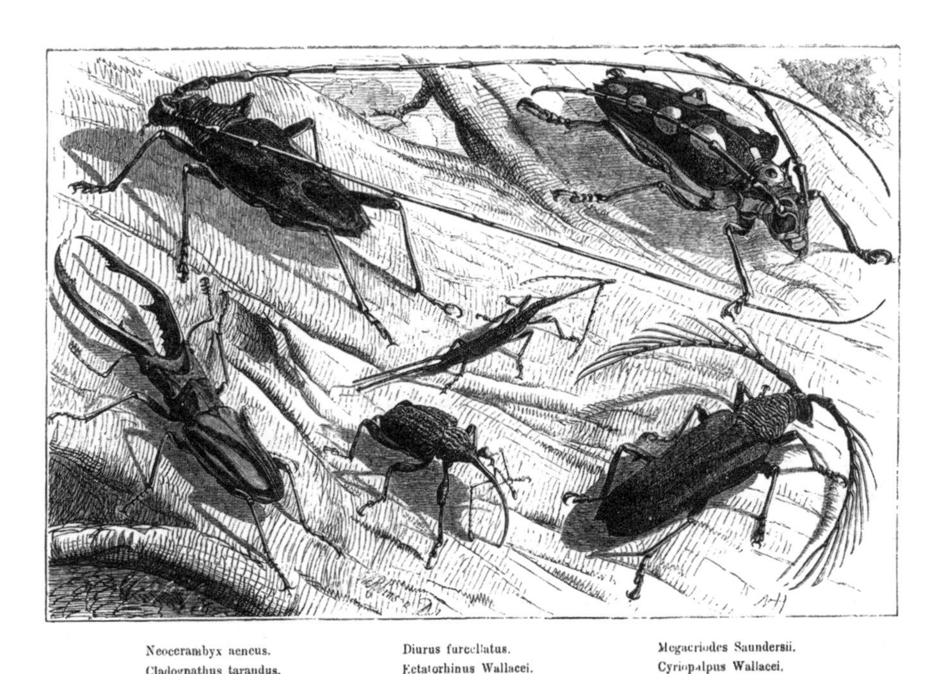

Figure 14 A selection of beetle species collected by Wallace at the Simunjan coalworks, Borneo in 1865.

From S715. Out of copyright

should I regard them as members of the same species or merely geographic races (or subspecies) of a single species? Wallace, as an apprentice coleopterist, had to confront this issue, which lies at the core of the evolutionary process. He would have been forced to think about what delineates a species; when is a species merely a variety and *vice versa*? In addition, he would have been forced to appreciate the significance of geography, whether aspects of topography or simple distance, in determining the distribution of species in nature. I suggest that Wallace acquired the perspective that would eventually yield his masterwork on biogeography, *The Geographical Distribution of Animals* (S718 1876), while discovering the Leicester area's beetles under the tutelage of Bates.

T. H. Huxley makes this point eloquently, expressing the blurriness of species versus variant designations in his review in *The Times* of *The Origin*, using beetles as the capstone of his argument.

> Let the botanist or the zoologist examine and describe the productions of a country, and one will pretty certainly disagree with the other as to the number, limits, and definitions of the species into which he groups the very same things. In these islands, we are in the habit of regarding mankind as of one species, but a fortnight's steam will land us in a country where divines and savants, for once in agreement, vie with one another in loudness of assertion, if not in cogency of proof, that men are of different species; and, more particularly, that the species negro is so distinct from our own that the Ten Commandments have actually no reference to him. Even in the calm region of entomology, where, if anywhere in this sinful world, passion and prejudice should fail to stir the mind, one learned coleopterist will fill ten attractive volumes with descriptions of species of beetles, nine-tenths of which are immediately declared by his brother beetle-mongers to be no species at all (Huxley 1859).

Beetles are a wonderful case study in ecological diversification in the face of anatomical constraint. Despite the huge number of taxa, most beetles are instantly recognizable as beetles, even to the inexperienced eye: the basic beetle architecture is highly conserved over evolution. Many of the modifications are subtle yet significant. For example, members of the aquatic family Dytiscidae trap air between their abdomen and the elytra when diving. Much of evolution is effectively about variation upon a theme in just this way. Think of the canonical example of the evolutionary process, Darwin's finches in the Galapagos: here we see an array of feeding variations (*i.e.*, adaptations) based on a single finch theme. Because of his work on mimicry, Bates was to become best known for his studies of butterflies. He wrote about butterfly wings (1892, 353) that "It may be said, therefore, that on these expanded membranes nature writes, as on a tablet, the story of the modifications of species, so truly do all changes of the organisation register themselves thereon." Bates's statement applies equally to the morphology of beetles.

Collecting: Mental Databasing

Wallace's beetle-hunting argument does not stop at an interest in, or passion for, beetles. Critically, he and Darwin were *collectors*. This was no casual pursuit: it was their obsession. Their mentality was I suspect rather similar to that of today's "twitchers," bird watchers for whom attempts to add to their "life list" of birds seen are all-consuming. Wallace's early letters to Bates are all about the excitement of finding beetles (and the frustrations of not finding them):

> I have really not had time, or even when I could get a little leisure the low state of my Entomological finances prevented me—all the finest weather of the latter part of the summer has been spent by me in a manner which you might consider very favourable to Entomolical [*sic*] pursuits—viz. in surveying & levelling among the most romantic parts of this beautiful & highly interesting district—But it is really astonishing the few insects which come unsought upon ones view: and though I keep a constant look out at all times when any attention is not otherwise engaged yet but scanty are my gleanings (3 October 1845 letter to Bates, WP1/3/13).[5]

The real measure, though, of Wallace's collecting drive lies in his willingness to add specimens to his collection that he himself had not collected. Collecting was thus not for him a mere pretext for spending time engaging with the natural world; rather, the collection was an end in itself. Writing again to Bates (13 October 1845):

> I shall be much obliged to you for any of the following of which you can send me good specimens.
> *Chlaenius vestitus*
> *Ophonus azureus*
> *Chrysomela hyperici*
> *Brachinus crepitans*
> *Donacia dentata* ...

These letters reveal the speed with which Wallace had become a beetle obsessive. He had, after all, met Bates and been introduced to the joys of beetles just the year before. Beetles had already become things "to have and to hold" to borrow the borrowed title of Phlipp Blom's study of collectors and collecting (2003). Wallace had joined that peculiar subset of humanity, collectors, who indulge with varying degrees of obsession in "the selecting, gathering, and keeping of objects of subjective value" (Muensterberger 1994, 4). One of the features of the collector's mentality is that it is transferable. In his psychoanalytic approach to collecting, Muensterberger (1994) notes that some collectors will start with one obsession, then lose it altogether, transferring their collecting focus to an entirely new domain. Think of the children who collect stamps but graduate, as adults, to collecting porcelain figurines. The stamps are long forgotten but the acquisitive urge remains.

Did Wallace, Bates, and Darwin remain faithful to their early beetle fixations or were beetles their equivalent of an early stamp collection? It's certainly true in Wallace's and Darwin's cases that the beetle fixation peaked early in life and was never fully rekindled. There are plenty of passages in *The Malay Archipelago* in which Wallace exalts in the diversity of beetles he is collecting, but he writes with equal enthusiasm about other taxa, especially butterflies. For all three naturalists, the contrast between the depauperate British fauna they knew and the species-rich tropics was an ever-present theme. The sheer numbers of species they encountered in the tropics were a constant reminder of the contrast. Wallace in Singapore:

> In about two months I obtained no less than 700 species of beetles, a large proportion of which were quite new, and among them were 130 distinct kinds of the elegant Longicorns (Cerambycidae), so much esteemed by collectors. Almost all these were collected in one patch of jungle, not more than a square mile in extent, and in all my subsequent travels in the East I rarely if ever met with so productive a spot (S715 1869a, 1:37–38).

And at the Simunjan coalworks in Borneo:

> When I arrived at the mines, on the 14th of March, I had collected in the four preceding months, 320 different kinds of beetles. In less than a fortnight I had doubled this number, an average of about 24 new species every day. On one day I collected 76 different kinds, of which 34 were new to me. By the end of April I had more than a thousand species, and they then went on increasing at a slower rate, so that I obtained altogether in Borneo about two thousand distinct kinds, of which all but about a hundred were collected at this place, and on scarcely more than a square mile of ground. The most numerous and most interesting groups of beetles were the Longicorns and Rhynchophora, both pre-eminently wood-feeders. The former, characterised by their graceful forms and long antenna, were especially numerous, amounting to nearly three hundred species, nine-tenths of which were entirely new, and many of them remarkable for their large size, strange forms, and beautiful colouring. The latter correspond to our weevils and allied groups, and in the tropics are exceedingly numerous and varied, often swarming upon dead timber, so that I sometimes obtained fifty or sixty different kinds in a day. My Bornean collections of this group exceeded five hundred species (S715 1869a, 1:57–58).

Darwin's beetle fixation seems to have been exorcized by his experiences on the voyage of the Beagle. Possibly, however, we can regard his barnacle period, all eight years of it, as a transfer, or resurrection, of the collecting impulse. Darwin recalls that, though beetles had long since lost the obsessive allure they once held for him, he still had the ability both to be excited by a new find and to identify even the most obscure taxa:

> I am surprised what an indelible impression many of the beetles which I caught at Cambridge have left on my mind. I can remember the exact

> appearance of certain posts, old trees and banks where I made a good capture. The pretty *Panagaus crux-major* was a treasure in those days, and here at Down I saw a beetle running across a walk, and on picking it up instantly perceived that it differed slightly from *P. crux-major*, and it turned out to be *P. quadripunctatus*, which is only a variety or closely allied species, differing from it very slightly in outline. I had never seen in those old days *Licinus* alive, which to an uneducated eye hardly differs from many of the black Carabidous beetles; but my sons found here a specimen, and I instantly recognised that it was new to me; yet I had not looked at a British beetle for the last twenty years (Darwin 1902, 20–21).

His son Francis recalls that beetles could be guaranteed to excite his father: "I have a vivid recollection of the pleasure of turning out my bottle of dead beetles for my father to name, and the excitement, in which he fully shared, when any of them proved to be uncommon ones" (Darwin 1902, 194). But perhaps the most telling of Darwin Senior's late life encounters with beetles can be found in a paper of which he is not a named author. In June 1859, while hard at work on the *Origin of Species*, Darwin submitted a short note to the *Entomologists' Intelligencer Weekly*:

> Coleoptera at Down. We three very young collectors have lately taken, in the parish of Down, six miles from Bromley, Kent, the following beetles, which we believe to be rare, namely, *Licinus silphoides, Panagus 4-pustulatus* and *Clytus mysticus*. As this parish is only fifteen miles from London, we have thought you might think it worth while to insert this short notice in the "Intelligencer." Francis, Leonard, and Horace Darwin (Darwin *et al.* 1859).

Clearly this note was penned by a proud father. At the time, Francis was eleven, Leonard nine, and Horace eight.

Bates's most famous contribution, so-called Batesian mimicry, stemmed from his studies of butterflies, but he remained loyal to beetles throughout his professional life. Wallace records in his obituary of Bates for *Nature* (S446 1892, 399) that, "finding that his circumstances and the time at his disposal did not allow him to keep up and study two such extensive groups as the Coleoptera and Lepidoptera, he parted with his fine collection of South American butterflies." Bates published several major monographic works on Neotropical beetles (see Clodd 1892).

Perhaps the importance of beetle-hunting for these Victorian naturalists was that it stimulated the construction of a mental database. Note Darwin's ability, twenty years on, to identify accurately an obscure beetle species. This database served them in two ways. First, it gave them the ability to make the geographic and biological distinctions among species and populations that are the key to evolutionary thinking. The ability to make these distinctions is a remarkable skill, requiring extraordinary visual memory and total command of the related details. There is no point being able to remember that a new specimen is similar to a species originally encountered many years previously if you cannot remember the latter's name. As Jared Diamond (2007, 659) recalls, Ernst Mayr, a giant of twentieth-century evolutionary biology, had an on-board ornithological database.

> Returning from an expedition to New Guinea in 1965, John Terborgh and I laid out our hundreds of bird specimens in the Harvard Museum of Comparative Zoology for Ernst Mayr to identify. Ernst had made only one collecting trip to New Guinea 36 years previously, and his last publication on New Guinea birds had appeared in 1954. Nevertheless, as he walked along the shelf and glanced at one specimen after another, he quickly identified each by its Latin species name and then by its subspecies name; he told us which zoologist had described it, in what year and in which journal; gave the alternative names under which other zoologists had discussed it; and explained its broader biological significance (for example, "Check that one for altitudinal hybridization"). He hesitated only at one obscurely mottled specimen: "See if that's a female *Rhagologus*." We found later that it was indeed a female *Rhagologus*, a whistler whose relatives are usually banded black and gold.

Second, the naturalist's mental database provides an empirical foundation for biological inference. Both Darwin and Wallace had wonderful synthetic minds; they had the ability to draw together many independent strands of evidence into a single coherent whole. But so did Herbert Spencer and T. H. Huxley, both of whom were interested in the same problems as Wallace and Darwin. Spencer was famously theoretical in his approach (it was with reference to him that Huxley once said that nothing could be more tragic than "a beautiful theory, killed by a nasty, ugly little fact" [quoted by Gould 1994, 440]), and Huxley, despite his travels, was more an engineer than naturalist, happiest when confronted with functional puzzles presented by anatomy. Huxley's response to reading the *Origin* says it all: "How extremely stupid not to have thought of that!" (Huxley 1901, 183). There are of course also many naturalists who lack the synthetic ability of a Darwin, Wallace, or Spencer. Their appreciation of the natural world never moves beyond their life list of birds or butterflies or beetles. Wallace's "beetle-hunting" claim, then, comes down a combination: the on-board database of the collector and the synthetic ability of the scientist.

Naturalists: Born Not Made?

How does the collecting impulse arise? Blom (2002, 165) points out that "the collected objects have a value for the individual collector that only other collectors can understand." There is an extensive literature on collectors and collecting but very little of it can reasonably be applied to the coleopterophilia of Wallace, Bates, and Darwin. Blom argues that the collector is moved in some way to bring order to the chaos of the world around them. Muensterberger, in his explicitly Freudian analysis of the problem, notes that "collecting is much more than the simple experience of pleasure. If that was the case, one butterfly, or one painting, would be enough. Instead, repetition is mandatory. Repeated acquisitions serve as a vehicle to cope with inner uncertainty, a way of dealing with the dread of renewed anxiety, with confusing problems of need and longing" (1994, 11). Baudrillard has a slightly more nuanced spin on the psychology of collecting: "This is why one invests in objects all that one finds impossible to invest in human relationships" (1994, 11). But it is difficult, in this post-Freudian era, to see Victorian entomologists as being driven by the shortcomings of their upbringings or by their social difficulties.

In his study of what he has called "multiple intelligences," Howard Gardner insists that the classical idea of IQ, the notion that it is possible to condense into a single value an individual's mental capability, is woefully incomplete. Gardner includes a "naturalist intelligence": "Naturalist intelligence enables human beings to recognize, categorize and draw upon certain features of the environment. It combines a description of the core ability with a characterization of the role that many cultures value" (Gardner 1999, 48).

Perhaps Wallace's 1908 claim about beetle-hunting was just his way of proclaiming that both he and Darwin possessed, in Gardner's language, high Naturalist IQs. Is "naturalist intelligence" a real phenomenon? I can only illustrate by personal example. As a professional population biologist raised in a pack-the-binoculars-for-the-summer-holidays kind of family, I have had every opportunity to develop my naturalist intelligence. I am however no naturalist. A walk through a tropical forest leaves me confused: is this the same species of tree as that one? Is this a *Cecropia*? Or is *that* a *Cecropia*? In contrast, a genuine naturalist has an amazing facility for knowing what everything is as he or she walks down that same trail. That they have seen distantly related plants a world away (in a European deciduous forest) is sufficient for them to determine their general taxonomic affinities. E. O. Wilson's 1994 autobiography, *Naturalist*, is a book-length paean to naturalist intelligence. Wallace too had ample gifts in this area. His feats as a naturalist were extraordinary, but difficult for us today to appreciate fully. Most of the natural world, or at least most of its more prominent components, is now catalogued, much of it neatly collapsed into field guides. We rely on field guides, keys, monographs, or, sometimes, recourse to museum specimens. Few of these were available to Wallace and Bates as they ventured for the first time, young and incredibly green, into the forests around Belem.

Their lack of preparation is remarkable. Wallace had only converted to beetles in 1844, and here he is, just four years later, embarking on an expedition into one of the most unexplored and biologically diverse parts of the planet. In those four years between Leicester and Brazil, Wallace's opportunities for scientific self-improvement were limited in the extreme. Even once Wallace had been bitten by the beetle-collecting bug, he found himself isolated, complaining to Bates in a letter (11 April 1846, WP1/3/11) that "I quite envy you who have friends near you attached to the same pursuits. I know not a single person in this little town

[Neath] who studies any one branch of natural history so that I am all alone in my glory in this respect."

Wallace and Bates made every effort they could to prepare themselves for their expedition, but their opportunities were decidedly limited. They spent time looking over the collections at the British Museum, and it was this cursory review of the museum's South American holdings that served as the taxonomic foundation for their years of exploration. The East India Company Museum's Thomas Horsfield, who had himself collected for Sir Stamford Raffles in South East Asia, showed them the boxes he had used for shipping specimens. Their agent Samuel Stevens presumably had to find ways to help them fill the vast gaps in their collecting expertise. They were evolving, more or less overnight, from amateur entomologists into full spectrum expedition naturalists: knowledge of how to kill and prepare vertebrate specimens—skills in trapping, shooting, preserving, and taxidermy that many take years to acquire—had to be picked up over just a few days at the end of March 1848. Such was their preparation: beetles, butterflies, and some plants around Leicester and the Welsh borders, a few hours in the British Museum, and a few rushed lessons with London's experts. And yet, within months, Bates and Wallace were collecting in Brazil and making sense of the collections they made.

This is not to suggest that Bates and Wallace became instant authorities on the Amazon biota. Wallace tells a charming story of how their education proceeded at the hands of none other than their cook:

> Our old conductor, though now following the domestic occupation of cook and servant of all work to two foreign gentlemen, had worked much in the forest, and was well acquainted with the various trees, could tell their names, and was learned in their uses and properties. He was of rather a taciturn disposition, except when excited by our exceeding dullness in understanding what he wanted, when he would gesticulate with a vehemence and perform dumb-show with a minuteness worthy of a more extensive audience; yet he was rather fond of displaying his knowledge on a subject of which we were in a state of the most benighted ignorance, and at the same time quite willing to learn. His method of instruction was by a series of parenthetical remarks on the trees as he passed them, appearing to speak rather to them than to us, unless we elicited by questions further information (S714 1853, 31–32).

Regardless, the rapidity with which Bates and Wallace transformed from neophytes to seasoned Amazon naturalists is testimony to the power of their naturalist intelligences.

When we look at the geneses of their careers as naturalists the contrast here between Wallace and Darwin is interesting. Darwin, according to his own recollection, was born a collector. Long before he succumbed to the siren call of beetles while at Cambridge, he had collected whatever came to hand. Darwin's father,

Robert, collected plants for his garden. Presumably the magpie in young Charles was actively encouraged.

> By the time I went to this day-school my taste for natural history, and more especially for collecting, was well developed. I tried to make out the names of plants, and collected all sorts of things, shells, seals, franks, coins, and minerals. The passion for collecting which leads a man to be a systematic naturalist, a virtuoso, or a miser, was very strong in me, and was clearly innate, as none of my sisters or brother ever had this taste (Darwin 1902, 2–3).

Wallace, as we have seen, came relatively late to natural history. He, of course, was at a major disadvantage: money was always tight (and it would have been difficult for the pre-Bates Wallace to conceive of how natural history could ever be profitable) and his family unsupportive. He recalls that his brother William, with whom he roamed the countryside as a surveyor, actively disapproved of his enthusiasm for natural history (S729 1905a, 1:194). Wallace's lonely sojourn in Neath, isolated from other entomologists, contrasts strongly with Darwin's experience. Introduced to the surprisingly fashionable pursuit of beetle-hunting at Cambridge by his cousin William Fox, Darwin seems to have spent his Cambridge years in a kind of miasma of entomological companionship. And, through his Cambridge connections, he was introduced to several of Britain's leading entomologists.

Despite all they had in common, beetles included, Wallace and Darwin were remarkably different men. Darwin was an agreeable companion, a popular figure in Cambridge undergraduate society (Browne 1995). Even in his reclusive years at Down House, he entertained regularly and was at the centre of a large and happy household. Wallace, on the other hand, seems to have got on rather better with the local people he encountered on his travels than with his fellow Victorians. He was socially awkward. His inability to land a job, despite obvious qualifications, surely reflects this (Berry 2002). Raby (2001) tells the story of Wallace's encounter with Alfred Lord Tennyson in 1884. The Poet Laureate had been made a peer earlier that year but Wallace nevertheless chose this occasion to lambast the House of Lords and the hereditary principle. Wallace, at that stage close to declaring himself a socialist (he did so in 1889), had strong opinions on the matter, but surely even a hint of social empathy would have prevented him from unburdening himself on the subject in front of the new peer?

Similarly, I suspect that Wallace's openness to spiritualism—his gullibility on the subject—may have been exacerbated by the same lack of empathy. His tendency to take phenomena he observed at face value indicates an inability (or unwillingness) to dig beneath the surface, to try to gauge another's motivation. A medium, for Wallace, was as advertised: a conduit to a spirit world. Wallace would never dissemble; why, then, would anyone else? One gets the sense here that Wallace has some of the logic-only characteristics of Star Trek's Mr Spock. This same psychological trait is a feature of a number of the more unfortunate episodes in Wallace's life. His disastrous run-in (see Berry 2002) with Mr Hampden, a flat-earther, over whether it was possible to demonstrate that the earth was not in fact flat, stemmed from a failure to comprehend that there was a lot more to Mr Hampden than met the eye. Wallace assumed (with a singular lack of perception) that others, Mr Hampden included, would apply logic to an engineering problem in the same rigorous fashion as he. It did not occur to him that anyone clinging to flat earth notions in mid-Victorian times might not be entirely rational. This same failure to probe into people lies behind Wallace's unfortunate relationship with money. Wallace's career as an investor was catastrophically short-lived because he again failed to do due diligence on his sources of financial advice (Berry 2002).

Applying retrospective psychiatric diagnoses to historical figures is typically unilluminating: there is not enough information either to prove or disprove the diagnosis and the range of behaviour exhibited by people with many of the more widely diagnosed psychiatric conditions is anyway so broad that having a diagnosis provides little predictive power (to the physician) or historical insight (to the historian). Despite these caveats, I still think it worth suggesting that Wallace may have had a correlated group of personality traits that today are given a name, Asperger's Syndrome. Interestingly, Blom (2002, 170–71) tentatively suggests that Asperger's might predispose people to be collectors, especially in light of males being over-represented among collectors (and among people with Asperger's):

> The whole phenomenon of retreat into a world of predictable patterns and away from an environment of social complexity and competing claims for attention and for love brings to mind autism, and, indeed, the majority of those suffering from this condition are boys and men. While the autistic spectrum reaches from mild eccentricity to severe disability, one clinical condition in particular, Asperger's Syndrome, the least severe of the autistic disorders, serves to illustrate the point. This syndrome is characterized by a whole range of symptoms: a resistance to change, relying on repeating patterns, stilted speech, immersion into arcane topics, such as transport timetables, which assume great importance, and collecting series of objects worthless to others ... This, of course, is not to say that collecting is inherently autistic any more than it is inherently male, or that collectors cannot be rounded human beings with thriving personal relationships, but the similarity is arresting.

Collecting is, much to feminist Naomi Schor's irritation (1994, 262), "generally theorised as a masculine activity" but, Schor insists, is perhaps not as gender-biased as is suggested by a stereotypical vision of collectors: train-spotters (male) lined up in the rain along British railway platforms or American baseball card aficionados (male) poring over their collections. Nevertheless, Martin (1999, 70) offers an analysis of gender differences in collecting:

> Men's collecting is indeed often consciously competitive, functional and driven by a need for control. It therefore assumes a rigid focus or tunnel

> vision. Subconsciously, however, there is often greater meaning and depth to their collecting, whilst women's collecting is more often consciously self-referential. This is perhaps because women are more open to, and capable of, exploring emotions and feelings than men.

A current psychological theorist of autism, Simon Baron-Cohen, emphasizes just this distinction in his analysis of the so-called Autism Spectrum Disorders (ASDs, which include Aspergers). For Baron-Cohen, "the female brain is predominantly wired for empathy. The male brain is predominantly wired for understanding and building systems" (Baron-Cohen 2003, 1), and people with ASDs tend to be at or near what Baron-Cohen calls the "systemizing" extreme. Given the congruence between male traits and ASD ones, Baron-Cohen is not surprised that ASDs are vastly more common in males than in females. Newschaffer *et al.* (2007) report that the male:female ratio for ASDs is about 4:1.

Though some aspects of Wallace's personality seem consistent with the vision of the empathy-bereft, mildly-ASD male collector put forward here, I hasten to emphasize that such uni-dimensional analyses of personality should not be given undue weight in the analysis of a historical subject. Insofar, however, as the history of science entails advancing explanations for how and why a particular scientific event occurred at a particular time, such psychological considerations should be added to the mix. Indeed, scientists are the products of their social environments but they are also individuals, each one a mass of psychological quirks. Those quirks are part of the story (*cf.* France and St Clair 2002; Shermer 2002).

Conclusion: Beetle-Collecting, a Forgotten Art

My emphasis on the psychological peculiarities of both Darwin and Wallace and on how these may have predisposed both men to view natural phenomena in a particular way represents an attempt to answer a very limited question. There were many historical strands—theological, social, imperial, institutional, political ... to name a few—that came together in nineteenth-century Britain to create an intellectual climate conducive to thinking about the evolutionary process. Clearly, it is these considerations that are the ultimate answer to the "Why was it that Darwin and I made the discovery and not someone else?" question that motivates Wallace's 1908 paper. However, there is also a more local question. There were, after all, many individuals in Britain and its colonies at that time who had the mental tools and intellectual desire to draw the same conclusions as Wallace and Darwin. Wallace himself alludes to Huxley and Spencer among others in this context. As I hope I have shown, Wallace's own solution, beetle-hunting, to the local question is richer and more insightful than may at first sight appear.

Many influential scientists have followed the same path: from naturalist-collector to synthesizer. E. O. Wilson, already mentioned, is an especially distinguished example. His passion for and knowledge of ants has underpinned a career that has reached out far beyond ants (Farber 2000). He will be remembered for his work on biogeography and on social behaviour, but the foundation of it all has been a life of ant-hunting. He too has one of those extraordinary on-board databases, one that allows him to identify at a glance any species of *Pheidole*, a vast, taxonomically diverse group containing over 625 described species. W. D. Hamilton, feted after his death in 2000 as the most important evolutionary biologist since Darwin, having extended Darwin's notion of individual fitness to include relatives, was also at heart a naturalist and collector, more at home in the forest than in the classroom or the confines of his university office (Berry 2003).

The history of natural history is itself a rich area (Allen 1976, 1996; Farber 2000). Why was it so fashionable among Victorians to assemble insect-filled cabinets and to learn the identities of the denizens of their hedgerows? Darwin and Wallace both clearly benefited from being a part of this, natural history's golden age. Now, however, it is clear that natural history has moved from biology's centre stage to its margins. One is hard pressed to imagine an undergraduate at Cambridge today gaining social kudos from his or her beetle collection. The reasons are many and complex, but three major factors are an increasingly urban population, the wide availability of mass entertainment such as the cinema, and, surely most potently, the extraordinary successes of molecular biology that have refocused biological inquiry predominantly inwards, inside the cell. Today, beetle-hunters are as a result a rare species indeed. E. O. Wilson has written passionately on the need to reawaken the naturalist impulse in the face of the crisis confronting the natural world.

Perhaps it is not surprising that collector-naturalists like Wilson are at the forefront today of the conservation movement. Collectors *value* what they collect. To them, the natural world signifies more than a mere constellation of species. Wallace, ever the naturalist-collector, was one of the first to articulate both the value of biodiversity and the threat it faces:

> ... future ages will certainly look back upon us as a people so immersed in the pursuit of wealth as to be blind to higher considerations. They will charge us with having culpably allowed the destruction of some of those records of Creation which we had it in our power to preserve; and while professing to regard every living thing as the direct handiwork and best evidence of a Creator, yet, with a strange inconsistency, seeing many of them perish irrecoverably from the face of the earth, uncared for and unknown (S78 1863, 234).

The last word contains the collector's plea. We must *know*—the collector's mandate is *to have and to hold*—the natural world if we are to understand it.

Notes

1. Curiously, the medal was not presented on 1 July 2008 as expected. Instead it will be awarded on 12 February 2009—Darwin's 200th birthday.
2. Now classified in the plant family Ericaceae (Heath family), these are indeed curious plants, being parasitic on fungi and lacking chlorophyll, the green photosynthetic pigment.
3. An unusual bee-mimicking scarab.
4. In fact, Darwin misquotes the reference, but he was nevertheless indeed cited by Stephens (Smith 1987).
5. Catalogue number, on-line Wallace archive, Natural History Museum, London: **http://www.nhm.ac.uk/nature-online/collections-at-the-museum/wallace-collection/themeslist.jsp**

3

Theory and Practice in the Field: Wallace's Work in Natural History (1844–1858)*

Melinda Bonnie Fagan

• • • • • •

> **London, June 30th, 1858.**
>
> **My dear sir,—The accompanying papers, which we have the honour of communicating to the Linnean Society, and which all relate to the same subject, viz. the Laws which affect the Production of Varieties, Races, and Species, contain the results of the investigations of two indefatigable naturalists, Mr. Charles Darwin and Mr. Alfred Wallace.**
>
> **These gentlemen having, independently and unknown to one another, conceived the very same ingenious theory to account for the appearance and perpetuation of varieties and specific forms on our planet, may both fairly claim the merit of being original thinkers in this important line of inquiry; but neither of them having published his views, though Mr. Darwin has for many years past been repeatedly urged by us to do so, and both authors having now unreservedly placed their papers in our hands, we think it would best promote the interests of science that a selection from them should be laid before the Linnean Society.**[1]

Introduction

Since their joint publication of 1858, Wallace's contributions to biology have been overshadowed by Darwin's. This is in no small part due to Wallace himself, who

* I thank Charles Smith for the opportunity to contribute to this volume and his archival expertise; Sander Gliboff, Paul Farber, two anonymous reviewers of the *Journal of the History of Biology*, and audiences at Indiana and Rice Universities for valuable comments and criticism; Gina Douglas and her colleagues at the Linnean Society, as well as archivists at the British Museum of Natural History, for permission to examine Wallace's unpublished MSS; and Jane Camerini for advice on Wallace's published and unpublished writings, and valuable comments on an earlier draft of this paper.

amiably conceded priority to Darwin in 1858 and thereafter cast himself in a secondary role as regards the achievement of the *Origin*. But the intensity of scholarly research on Darwin, concentrated particularly on his theory of natural selection, has tended to magnify the asymmetry. Wallace's independent conception of this mechanism of evolutionary change is treated in deflationary fashion by most contemporary historians and philosophers of biology. For some, Wallace is an anomaly to be "explained away," either by saddling him with theoretical confusions and errors, or by reducing his role to credit-seeking within a priority dispute.[2] Most often, Wallace's 1858 essay is simply relegated to a footnote. In what follows, I attempt to redress this asymmetry, by examining the development of Wallace's evolutionary theory on its own terms and in its natural context. Though in no way as comprehensive, rich, or detailed as Ospovat's (1981) study of the development of Darwin's theory, the following makes a start on such an account. And despite their shared culture and similar source materials, the development of Wallace's evolutionary theory contrasts significantly with that of Darwin's. In particular, its context was quite different. Wallace's evolutionary theory emerged from the arduous day-to-day practice of a professional specimen collector, a working man in the tropics.

Placing Wallace's natural history practice at the center of analysis illuminates aspects of his life and thought that narrower contrasts with Darwin leave obscure.[3] A naturalist's routine practice yields diverse products: material collections of specimens, representations of organisms and their environments, and, more abstractly, concepts and theories. The routines that shape these diverse products are in turn shaped by the naturalist's circumstances and motivations, which operate within a particular social and theoretical context. Taking natural history practice as central thus reveals productive connections between biological theories and concepts, the day-to-day work of the naturalist, and features of his or her wider social and historical context. Applying this analytic framework to Wallace reveals that the fevered inspiration articulated in his 1858 essay was no "sudden flash," but the culmination of eight years of intense collecting work in South America and the Malay Archipelago. Wallace's theory of natural selection, no less than his material collections and written reports, was a product of his labors in the field: his working routine of intense collecting and painstaking arrangement of specimens. His fieldwork, in turn, was directed and constrained by two powerful motivations: one economic, and one scientific. Both were rooted in Wallace's social context and circumstances in 1840s England. These motivations reveal the development of Wallace's theory to be a vital part of his natural history practice. Wallace's work in the field thus reveals connections between his socio-economic circumstances, natural history practice, and evolutionary theory, integrating all three into a more comprehensive account of his life and work.[4]

Origins[5]

Wallace made his first forays into natural history in the Welsh border country and the west of England, while working as a land surveyor with his older brother William. The initial choice of trade was based on convenience and opportunity. The surveying and mapping business boomed after Parliament's Act of 1836, requiring farmers to pay rent charges on land based on estimates of its potential productivity as well as actual yield (Kain and Prince 1985). At the same time, Wallace's parents could no longer afford to pay for his education or keep; he was removed from school and sent out to make a living. William, already self-employed as a land surveyor, took his younger brother on. Wallace found the outdoor work agreeable, and was not (yet) troubled by its implications for farmers accustomed to depend on common land. The itinerant lifestyle of surveying, moving several times a year to a new parish, presaged his habits as a collector.

However, surveying itself was not conducive to collecting animal or plant specimens, and William discouraged natural history as frivolous.[6] Nonetheless, Wallace became interested in botany. Curious about the variety of plants seen on surveying projects and on solitary rambles, in 1841 he bought a shilling book on botanical classification published by the Society for the Diffusion of Useful Knowledge. The Society published cheap texts on natural history and the physical sciences, aimed at working men lacking formal education (S729 1906, 1:191). Wallace, clearly within their target audience, had previously purchased Society texts on mechanics and optics. But though the "little book was a revelation to me, and for a year my constant companion . . . ," it did not satisfy his burgeoning enthusiasm for botany. The reason was simple: the introductory text could not identify all the botanical specimens Wallace collected. It described only the orders and major organs of plants; Wallace had no way to identify distinctively British species. So Wallace sought out more advanced (and more expensive) texts with which to identify his specimens: John Lindley's *Elements of Botany* and J. C. Loudon's *An Encyclopaedia of Plants.* The price of the former (ten shillings) was daunting for Wallace, and the latter quite beyond his limited means. A bookseller in Neath, where the Wallace brothers were based from late 1841 to late 1843, allowed him to borrow the Loudon and copy its species descriptions into the margins of the Lindley he had purchased (S729 1906, 1:193). He later applied this annotation strategy when preparing for the field. As land-surveying work grew scarcer, Wallace had more and more leisure to "botanize," and to pursue other intellectual interests at the local Mechanics' Institute. Such institutes provided intellectual forums and libraries for working men, with an emphasis on practical scientific knowledge.

Despite his connections with the local institute, Wallace had no companions who shared his interest in botany. With no formal training or connections, his enthusiasm for natural history might have remained a private hobby, a subject for occasional evening lectures at the institute.[7] Though he retained a lifelong interest in botany, Wallace came to his vocation through entomology. His shift from Welsh wildflowers to beetles and butterflies was as fraught with contingency as his first encounters with natural history. When the languishing surveying business could not support two Wallace brothers (late 1843), Wallace left Neath to find another line of work. He found a post at a boy's school in Leicester, teaching surveying and drawing. The well-stocked Leicester library surpassed any intellectual resource Wallace had encountered outside of London, and he spent several hours each day there, reading avidly.[8] There he met Henry Walter Bates, an apprentice in hosiery manufacture who was, like himself, intellectually ambitious but lacking formal education. A Leicester native, Bates was a dedicated amateur entomologist with a circle of similarly-inclined friends. The pastime was well suited to working men of limited means, requiring only pins, bottles, and boxes for storing specimens, many species of which could be found within a small area. Wallace took it up with characteristic enthusiasm, obtained James Francis Stephens' *A Manual of British Coleoptera*, and formed ambitions for a comprehensive insect collection.[9] Beetles and butterflies, however, were his specialty—a focus that persisted throughout his years in the field.

Wallace's sojourn in Leicester was brief; he returned to land surveying in Wales after William's unexpected death in early 1845. But his entomological collaboration with Bates endured; they regularly exchanged specimens and ideas by mail. Increasingly weary of surveying and business in general, Wallace gradually expanded his knowledge of natural history and his collection of insects and plants. Storage space became a concern—the two amateurs then exchanged plans about economical arrangement of their collections: "I cannot afford to multiply Cabinets indefinitely or I should be glad to adopt your suggestions for arrangement ...— on my plan I can get the Coleoptera very well in a dozen drawers & shall have the rest for Lepidoptera ..."[10] The need to economize was counterbalanced by Wallace's ambition to obtain "a perfect series of every insect ...," a sufficient number of specimens to capture the variation within the species of the order.[11] Though an ideal collection would (Wallace estimated) include about two dozen specimens per species, a "beginner ... one who does not work Entomology the sole study of [*illeg.*] his life ..." could be satisfied with fewer. In practice, Wallace contented himself with between two and six specimens per species.[12] Additional specimens he treated as "duplicates," and traded these with Bates to obtain more new species. Their collections thus grew collaboratively.[13]

Alongside his material collection, Wallace accumulated a "collection of facts" recorded in his "Natural History Journal." This was a "sort of day Book in which I insert all my captures in every branch of Natural Hist. with the day of the Month, locality &c. and [*illeg.*] add any remarks I have to make on specific characters, habits, &c. &c ..."[14] Much of Wallace's correspondence with Bates consisted of such facts: lists of species recently "taken," with remarks on their preferred habitat and abundance at different seasons, and from year to year. Wallace's estimates of abundance or rarity of particular species were generally qualitative:

> The following is a list of any duplicates—
>
> | Coleoptera | Remarks |
> |---|---|
> | Prystorychus tericola— | common |
> | Hopila argentea— | common |
> | Phyllopertha horticola— | very abundant |
> | Cetonia aurata— | nr. London. not taken here |
> | Cryptorhynchus lapathi— | rather common[15] |

These estimates of species abundance did not reflect Wallace's actual collection, but were based on the number of specimens that he judged he *could* have captured at the locality in question (Neath and its environs). They were, that is, estimates based on repeated first-hand observations of insects of various species in the field. This conception of species abundance also figured prominently in Wallace's writings from the field (including his theoretical essays).

Even in his amateur period, Wallace's interest in his "favourite subject—The variations, arrangement, distribution & [*sic*] of species ..."—was not limited to the specimens and facts he eagerly exchanged with Bates, but extended to more abstract questions.[16] His reading in Leicester and Neath included Lyell's *Principles of Geology*, Malthus' *An Essay on the Principle of Population*, and Lawrence's *Lectures on Comparative Anatomy, Physiology, Zoology, and the Natural History of Man*. As the last indicates, Wallace had an abiding interest in the natural history of humans. The theory that influenced him most did not emanate from the British scientific elite, however, but from the anonymous author (Robert Chambers) of the popular *Vestiges of the Natural History of Creation* (1994 [1844]). The work proposed "the developmental hypothesis" that species have their origin in previously existing species, from which they develop by a process of transmutation. Wallace had no sympathy for special creation or the doctrines of natural theology, and saw in *Vestiges* a theory that could guide empirical work:

> I have rather a more favourable opinion of the "Vestiges" than you appear to have—I do not consider it as a hasty generalisation, but rather as an ingenious hypothesis strongly supported by some striking facts and analogies but which remains to be proved by more facts & the additional light which future researchers may throw upon the subject—it at all events furnishes a subject for every [excised: "thing"] observer of nature to turn his attention to; every fact he observes must make either for or against it, and it thus furnishes both an incitement to the collection of facts, and an object to which they can be applied when collected.[17]

By late 1847, Wallace had decided that "the collection of facts" relevant to testing this hypothesis could not be restricted to "a mere local collection," but required broader experience (S729 1906, 1:256). This methodological commitment was probably influenced by Wallace's sense of isolation as a lone amateur collector at Neath and dissatisfaction with the land-surveying business. His reading inclined him toward the tropics, and South America in particular: von Humboldt's *Personal Narrative of Travels in South America*, Darwin's *Journal of Researches*, and W. H. Edwards' *A Voyage Up the River Amazon*—which last imparted the crucial information that it was possible for men of limited means to make a living collecting specimens in the Amazon basin. Wallace and Bates determined to follow in Edwards' footsteps, with additional scientific motivation:

> Wallace ... proposed to me a joint expedition to the river Amazons, for the purpose of exploring the Natural History of its banks; the plan being to make for ourselves a collection of objects, dispose of the duplicates in London to pay expenses, and gather facts, as Mr. Wallace expressed it in one of his letters, "towards solving the problem of the origin of species," a subject on which we had conversed and corresponded much together (Bates 1863, 1:iii).

The two novices put their plan into practice quickly, encouraged by Edwards himself. After stopping in London to acquire an agent (Samuel Stevens) and some hasty training in collection and preservation techniques, Wallace and Bates sailed for Brazil, arriving in Pará (present-day Belém) on 26 May 1848 (S714 1969, 1; Slotten 2004, 42–45).

Key components of Wallace's natural history practice had been established during his amateur period. He combined the itinerant, mapping lifestyle of a land surveyor with the drive to identify species of plants, beetles, and butterflies, and became attentive to individual variation and abundance within species. Traffic in "duplicates," "series" of specimens representing within-species variation, and records of relative species abundance were all hallmarks of Wallace's mature collecting practice. Over the next fourteen years, this practice developed under the impetus and constraint of two powerful motivations: economic and scientific. On the one hand, Wallace was interested in the theory of transmutation of species, and hoped to test it by inventorying species and varieties of particular groups at multiple localities. On the other, he needed to support himself, and worked under constant economic constraint. These two aims jointly influenced Wallace's fieldwork in several ways, structuring the overall intensity and extent of his collecting practice, choice of what to collect, and standards for success.

Wallace in the Field[18]

Wallace's fieldwork was divided into two expeditions: one in the Amazon (May 1848 to July 1852) and the other in the Malay Archipelago (April 1854 to February 1862). The first was not an unqualified success, though the four years' experience in the Amazon were crucial to his development as a naturalist. At first all went well: Wallace and Bates collected as a team around Pará, honing their technique and concentrating on the insect groups with which they had most experience. Their first joint shipment of specimens to England consisted of about 7,000 specimens of over 1,300 species of insect, with butterflies and beetles predominating (S714 1969, 34; Brooks 1984, 20). At three cents per insect sold, their joint collecting venture looked to be a profitable one. But they soon separated, and Wallace thereafter worked alone or with hired assistants.[19] Traveling by river, he collected along the banks of the Amazon, then followed the Rio Negro and its branches into northern Brazil, nearly to the Venezuelan border. This brought him into territory unfrequented by English collectors, who were more densely concentrated near centers like Pará, nodes in the diffuse European "colonial network" that permeated the tropics.[20] Despite the facilitating linkages of the colonial network, travel in the tropics was arduous. Permits to travel in certain areas had to be secured, hired assistants and porters often abandoned expeditions without warning, and illness and injury were constant threats. In the Amazon, Wallace became accustomed to the vicissitudes of collecting.

In territory unfrequented by other Britons, Wallace's goals were more easily combined: a detailed inventory of species at a new locality, in all their variety, was potentially of interest to "museums and amateurs" as well. There was in fact more novelty than Wallace expected. Collecting on major rivers, he noted that species on opposite banks often differed, though habitat appeared identical (S714 1969, 326–28). He also expanded his collecting repertoire to include freshwater fish, river tortoises, and birds. The last required more effort, but specimens with showy plumage fetched high prices; Wallace traveled to several sites specifically to collect such species (S5 1850; S729 1906, 1:284–85). In July 1852, he sailed for England, along with most of his collection (private and "duplicates" for sale). His career was thus seriously set back when the ship caught fire and sank ten days into the voyage (S7 1852, 3641–43; S729 1906, 1:303–15). The loss included all his specimens collected after March 1850, as well as Wallace's journals, drawings, and notes (S7; S729 1906, 1:303–09). All he saved was one "small tin box" containing his notes and drawings of palm trees and fish, and a few Rio Negro notes. More than "any pecuniary loss," Wallace mourned the loss of "all my private collection of insects and birds since I left Para . . . hundreds of new and beautiful species," along with his notes, sketches, and "the three most interesting years of my journal" (S729 1906, 1:306–07). This was not all he had to mourn for. While Wallace was in northern Brazil, his brother Herbert, preparing to return to England, had died of yellow fever. Plausibly this played some role in Wallace's own return the next year, though it seems he also considered his Amazon collection sufficient for his scientific and financial purposes.[21] In any case, though Wallace survived his unlucky voyage back to England, the greater part of his collection and field notes did not. He would have to start again.

That he could do so at all was thanks to Samuel Stevens, his London agent, who had insured Wallace's lost collection for £200. This foresight provided a financial cushion for Wallace, while he extracted what results he could from the remnants of his collection and notes. Stevens' guidance and management was vital throughout Wallace's collecting career (George 1979; Camerini 1996). He was Wallace's principal source of information while in the field, and sole source of money and equipment (contingent, of course, upon satisfactory sale of specimens). He was also Wallace's conduit to the London natural history community during his years abroad. As treasurer of the Entomological Society of London, Stevens exhibited Wallace's specimens at the Society's monthly meetings, bringing his name and activities to the attention of its members, albeit as a "mere collector" (Brooks 1984, 53–54). He also used excerpts from Wallace's letters to advertise his specimens in periodicals such as the *Annals and Magazine of Natural History* and the *Zoologist* (Camerini 1996). During Wallace's second, more successful expedition, Stevens received over twenty shipments from the Malay Archipelago, arranged buyers for "duplicates," and stored specimens Wallace designated as "private" until his return in 1862.

These Malay shipments (sent approximately every few months) added up to a staggering number of specimens: 125,660 in total, over 1,000 of which Wallace judged to be species. The distribution of these specimens into groups is revealing: 310 mammals, 100 reptiles, 8,050 birds, 7,500 shells, 13,100 butterflies, 83,200 beetles, and 13,400 other insects (S715 1869b, viii). Wallace's private collection consisted of approximately 3,000 bird skins of about 1,000 species, and approximately 20,000 beetles and butterflies of about 7,000 species. The groups collected reflect both of Wallace's motivations. He evidently concentrated on groups that fetched good prices: tropical birds, butterflies, and beetles. In addition, their well-established taxonomies focused Wallace's practice, allowing him to target gaps in a stable framework (Desmond and Moore 1991; McOuat 1996, 2001). The preponderance of beetles and butterflies reflects Wallace's abiding interest in these groups. However, botanical specimens, another long-standing enthusiasm, are largely absent. Financial constraints forced him to choose among the branches of natural history: "I cannot afford to collect plants. I have to work for a living, and plants would not pay unless I collect nothing else, which I cannot do, being too much interested in zoology."[22] Shells of land-snails, in contrast, sold reliably and required little effort in pursuit or preparation; this accounts for the large number Wallace collected, despite having little scientific interest in the group.

Regarding birds, Wallace's dual motivations were more easily reconciled. Specimens of tropical species (with plumage intact) commanded high prices. For example, Wallace estimated that one shipment of ~400 birds from the Aru Islands should net £500 in London (£1,000 including his private collection), while a shipment of 4,000 insects would fetch £200.[23] But avian species, genera, and families also exhibited complex distribution patterns, which figured prominently in Wallace's evolutionary and biogeographical theorizing (see below).[24] In birds of paradise, the "most beautiful of all the beautiful winged forms which adorn the earth . . . ," Wallace's dual motivations fused into a prolonged quest to acquire specimens of all existing species from their natural habitat. This ambitious project dominated Wallace's last four years in the Malay Archipelago. He made five different voyages to the Aru Islands and New Guinea, each of which took the better part of a year. Though chronically dissatisfied with the number and quality of his specimens, his enthusiasm occasionally burst out in reports to Stevens; *e.g.*: "I have a new Bird of Paradise! of a new genus!! quite unlike anything yet known, very curious and handsome!!!"[25] Practicality was immediately reasserted, however; "When I can get a couple of pairs, I will send them overland, to see what a new Bird of Paradise will really fetch."

More generally, Wallace's struggle to balance his scientific interests with economic constraints pervades his correspondence and notes from the field. Letters to Stevens often refer to anticipated profits or shortfalls from specimen sales, as does his field notebook recording details of consignments. Yet financial considerations were clearly not the whole story. Wallace chafed at his economic limitations not because he aspired to greater profit, but because they constrained his work in natural history: "a travelling collector of limited means like myself does so much less than might be expected, or than he would himself wish to do."[26] Beset with problems of limited time, storage, and workspace, Wallace wished for "unlimited pecuniary resources ..." but made compromises in practice. Even the numbers of specimens he collected were an impediment to other investigations "highly important to science":

> ... to make any thing like extensive collections of birds & insects, keeping brief notes of the most interesting facts connected with them will fill up the whole time of one person, with two or three native assistants. He absolutely cannot do much else, and is often even obliged to abridge his notes in order to secure the safe preservation of his specimens.[27]

Wallace strove to overcome his economic limitations by focusing on particular (saleable) groups, and collecting intensively within these. The resulting body of evidence would, he hoped, suffice to test the transmutation hypothesis.

As noted above, Wallace's theoretical aims were established before he ever went to the field. His letters to Bates indicate his strategy for testing *Vestiges*' transmutation hypothesis: "I begin to feel rather dissatisfied with a mere local collection; little is to be learnt by it. I should like to take some one family to study thoroughly, principally with a view to the theory of the origin of species. By that means I am strongly of opinion that some definite results might be arrived at" (S729 1906, 1:256–57). These aspirations to thoroughness drove Wallace to collect very intensely. He aimed at a complete inventory of species within a chosen group, not focused on one site but spanning multiple localities. His collecting experience on three continents made him well aware of variation in species composition at different sites, even those with similar habitats. He was also attentive to variation within species, and sought "a good series" of specimens to represent each, as well as to distinguish "true species" from "mere varieties."[28] Accordingly, Wallace collected as many species of butterflies, beetles, and other select groups as he could, at numerous localities. In the Malay Archipelago these included sites in Singapore, Sarawak (in western Borneo), Lombok, Macassar, the Aru Islands, Amboyna, Dorey (on New Guinea), Batchian, Timor, Ceram, Mysol, Waigiou, and, of course, Ternate and Gilolo. His work on New Guinea was a particular triumph: "one of the least known and most promising regions that remain, now that the most remote parts of the earth are ransacked by enterprising collectors" (S40 1858, 5889). But it was not only novelty he sought; Wallace made a point of revisiting localities at different seasons, visiting most of the sites listed above two or three times. Moving back and forth among the islands of the Archipelago, he traveled "about fourteen thousand miles ... [and making] sixty or seventy separate journeys."[29] Between 1854 and 1862, Wallace made more than "eighty movements [of house] averaging one a month ..." and "sixty or seventy separate journeys, each involving some preparation and loss of time, [so] I do not think that more than six years were really occupied in collecting" (Marchant 1916b, 68; S715 1869b, vii–viii). In effect, he adapted from the itinerant life of a land surveyor, to itinerant mapping of species.

At each locality, Wallace attempted to collect specimens of all existing species in his chosen groups. His typical workday was about twelve hours long, "with hardly half an hour's intermission, from 6 am till 6 pm, four or five of the hottest hours being spent entirely out of doors ..." in pursuit of specimens (S45 1859, 111-13). Wallace's typical quarries were insects (beetles or butterflies), hunted with nets. Birds, primates, and other quadrupeds required a gun. Indigenous hunters trapped paradise birds by their feet, to prevent damaging their plumage (Wallace never applied this technique). A typical workday was as follows:

> Get up at half-past five, bath, and coffee. Sit down to arrange and put away my insects of the day before, and set them in a safe place to dry. Charles [Wallace's sometime assistant] mends our insect-nets, fills our pin-cushions, and gets ready for the day. Breakfast at eight; out to the jungle at nine. We have to walk about a quarter mile up a steep hill to reach it, and arrive dripping with perspiration. Then we wander about in the delightful shade along paths made by the Chinese wood-cutters till two or three in the afternoon, generally returning with fifty or sixty beetles, some very rare or beautiful, and perhaps a few butterflies. Change clothes and sit down to kill and pin insects, Charles doing the flies, wasps, and bugs; I do not trust him yet with beetles. Dinner at four, then at work again until six: coffee. Then read or talk, or if insects very numerous, work again till eight or nine. Then to bed.[30]

Morning and early afternoon ("four or five of the hottest hours" [S45 1859, 111–13]) were occupied with capture of specimens: hunting for beetles, butterflies, snails, and (less often) birds, mammals, and reptiles: multiple specimens for each species if possible, for a "good series," and as many different species as could be found. Once Wallace's daily hunt with gun and net was finished, the second half of his workday began. Dozens of specimens had to be prepared and arranged: insects killed and pinned; birds, reptiles, and mammals skinned and dissected, fur or feathers treated with alum or arsenic soap.[31] All had to be identified as known or new species, with the help of a few annotated classification texts.[32] Then came labeling, taking notes, and protecting drying specimens from the ravages of insects. Wallace did not work entirely alone. In the Archipelago he employed Charles Allen, the son of an English carpenter, and Ali, a young Malay hired in 1855, who remained with Wallace for the remainder of his travels.[33] Both young men helped Wallace shoot and prepare specimens, though Ali was apparently much more competent. On occasion he paid local hunters for specimens, though his concern over undamaged specimens and precisely identifying species' habitats and localities (as well as financial constraints) rendered this a secondary strategy.

He was well aware that few collectors shared his ideal of obtaining "a true idea of the Entomology of this country . . . ," and complained about the tendency of amateurs to collect only "large and handsome" species from a given region, neglecting the "smaller, more active, and much more common species." Wallace had no bias toward common species; he found "small and obscure . . . groups being neglected in favour of others" equally lamentable (S25 1856; S44 1858). It was a complete species inventory that he sought, not a collection reflecting relative frequencies (though he did register this information; see below). His ambitions did not blind Wallace to the fallibility of all his estimates, nor how unlikely it was that he achieved a complete species inventory at any site, even with repeated visits. His Natural History notebook includes an entry on "making a systematic calculation of the number of species of Coleop[tera] in countries only partially known . . . [or] from partial collections made at two or more localities in it, or a group of islands only partially explored."[34]

Wallace's ceaseless pursuit of a complete species inventory is reflected in a pronounced emphasis on species in his writings of the period (Fagan 2007). One way this emphasis is revealed is in "species counts": tallies of the number of species in particular families or orders (usually insects or birds) collected at a given locality. Such counts pervade not only Wallace's field notes and reports to Stevens, but also his correspondence with family and friends, and (to a lesser extent) his journal. Lists of consignments to Stevens include numbers of both species and specimens, subdivided into families or orders (with overall totals of higher taxa or common groups). So for the purpose of shipping (and insurance), specimens are counted alongside species. But elsewhere in Wallace's writings from the field, species predominate. His zoology notebook numbers species consecutively; these numerical lists are subdivided into families, and varieties labeled with letters (a, b, c . . .). It is, quite literally, the species that are counted: interspersed with the orderly numbered descriptions are tallies and totals of species collected, with various written emphases: "Ornithoptera <u>20 species</u>"; "Cesothia 24 species!" These counts find their way into Wallace's prose as well. He assesses his work in terms of species "taken" over time. A rich site yielded "thirteen bird species, 194 insect species ..." in six days, while at another "I could only muster 90 species of butterflies and 235 Coleoptera at the end of one month, which had increased to 108 and 340 in two months, with 150 Hymenoptera, 120 Diptera, and other orders scanty, making a total of 850 species of insects" (S38 1857, 484; S40 1858, 5890).

More colorfully, Wallace frequently wrote of "meeting" species or higher taxa exhibiting characteristic behaviors or habits. For example, "Here [in Santarem, Brazil] the Epicalias and Callitheas are to be met with, while numbers of Morphos flap lazily along, and Hesperidæ, sometimes as large as Sphinxes, dart by with the velocity and sound of humming birds." Of the butterflies in Singapore, "[t]he most remarkable is a magnificent Idaea [*Idea*], which is abundant in the forest, sailing or rather floating along ..." In the Aru Islands "The great Fruit Pigeon of the Moluccas ... was abundant, its loud, hoarse cooings constantly resounding through the forest." Writing in Borneo, Wallace attributed orang-utan traits and behaviors to the group "Mias" rather than to individual animals, despite extensive contact with the latter. He affectionately characterized the insects of Borneo:

> Imagine my delight in again meeting with many of my Singapore friends,—beautiful longicorns of the genera Astathes, Glenea and Clytus, the elegant Anthribidae, the pretty little Pericallus and Colliuris, and many other interesting insects. But my pleasure was increased as I daily got numbers of species, and many genera which I had not met with before.

Elusive species were a source of exasperation: "At Macassar I once saw a *Tricondyla* but the villain escaped up a tree & I vainly searched for him for a month afterwards. I shall probably however meet with him when I visit the N. of Celebes ..." "For six weeks I have almost daily seen Papilio Ulysses? or a new closely-allied species, but never a chance of him; he flies high and strong ..." In more informal correspondence, Wallace continually referred to "fresh species," "new species," "handsome species," or "common species."[35]

Many of the above references to species include some mention of their abundance or scarcity. Wallace's writings also include frequent remarks on the abundance of genera, families, and orders. Descriptions of groups as "common" or "rare" figure prominently in Wallace's communiqués from the field, much as in his first notes as an amateur naturalist in Britain. As he gained professional experience, such estimates became central to Wallace's natural history practice, and eventually to his theorizing. Wallace's judgments as to species abundance or rarity were usually expressed qualitatively, in terms of the relative frequency of individual members of a given species present at a collecting site. Such estimates were based on Wallace's own observational powers, and, early in his career, were not particularly nuanced (*e.g.*, "that remarkable bird, the *Opisthocomus cristatus* ... This bird is very abundant on the banks of the Amazon, where we have often observed and shot it" [S28 1856, 213]). But Wallace hunted specimens in the field for twelve years, six hours a day, spending weeks at a time at scores of different sites. Over time and with intensive experience, Wallace's estimates grew more sophisticated and theoretically useful. In his mature practice, he used them to compare relative frequencies of individual members of species both within and among localities: in Batchian "you may see hundreds of the common species [of paradise birds] to perhaps *one* of either of the rarer sorts ..."; while at Santarem "many common insects, such as Heliconia Melpomone and Agraulis Dido, [are] abundant, which we hardly ever saw at Pará."[36] So although Wallace's writings from the field tend to de-emphasize individual organisms, they are obliquely referred to in most of his communiqués, via the concept of species abundance. Frequent references to abundance or rarity indicate that individual organisms were not irrelevant or absent in Wallace's thought, but formed the background for his writings, as well as the material basis for his collection.

The concept of abundance also linked individual organisms with taxonomic levels, allowing Wallace to unify the phenomena of natural history without conflating individuals and groups. Wallace consistently distinguished between the abundance of individuals within a species, and the abundance of species within higher taxonomic groups. At his first collecting site near Pará, he reported:

> The Lepidoptera are numerous in species, but not in individuals; the Coleoptera are exceedingly scarce, and other orders are generally, like the Lepidoptera, sparing in individuals; we attribute it to the uninterrupted extent of monotonous forest over which animal life is sparingly but widely scattered (S3 1849, 74).

Eight years later, writing again to Bates, he drew the same distinction: "The individual abundance of beetles is not, however, so large as the number of species would indicate. I hardly collect on an average more than 50 beetles a day, in which number there will be from 30 to 40 species."[37] At no point did Wallace confuse individuals and groups; indeed, it would have been peculiar had he done so. He was surrounded by individual animals and plants, which provided the medium for his work, suffused his daily routine, and occasionally impeded it (ants were a constant threat to his drying specimens).

Nor did Wallace conflate higher taxa. His field notebooks and reports sent to England include separate discussions of each taxonomic level, beginning with the most inclusive groups (birds, insects, shells), and going on to describe orders, genera, species, and (last of all) individual specimens. An 1856 letter to Stevens contains an abbreviated summary of this sort:

> My collections here consist of birds, shells, and insects. ... The birds are pretty good as containing a good many rare and some new species; but I have been astonished at the want of variety compared with those of the Malayan Island and Peninsula. Whole families and genera are altogether absent, and there is nothing to supply their place ... the result of which is that in about equal time and with greater exertions I have not been able to obtain more than half the number of species I got in Malacca ... You hint

> that in Borneo I neglected Raptores. They are too good to neglect, but there were none. Here in two months I have got fifteen species ... Of these six are represented by single specimens only, but of the rest I send you thirty fine specimens, and they will, I doubt not, contain something new (S33 1857, 5652–53).

Summaries of Wallace's land shell and insect collections from Macassar follow, and are similarly organized, proceeding from higher taxa to species, and lastly to specimens. Such orderly discussion leaves no ambiguity as to which sort of abundance is at issue. Yet the concept facilitated comparisons across taxonomic levels and localities, helping to organize and unify Wallace's collection. At any given locality, families contain more or fewer species, species contain more or fewer individual organisms. These estimates could then be compared and aggregated across localities.

Wallace's writings from the field also highlight his concern to obtain a "good series" to represent each species in his collections. As a novice collector he had found two to six specimens adequate for beetles (see above). But by 1859 his standards had changed; when the *Origin of Species* was published, he remarked to Bates "I now keep more duplicates than I [*sic*] first."[38] From 1856, and perhaps earlier, Wallace aimed to collect both male and female specimens of different stages of maturity (S6 1850, 494–95; S24 1856; S37 1857, 415; S62 1861). For some species of *Paradisaea* (birds of paradise), a "good series" required twenty-five specimens; for certain beetles three were sufficient. (S715 1869b, 537). Of course, Wallace also had financial incentive to collect multiple specimens of exotic, saleable species. But in such cases his dual motivations coincided. His theoretical interests, specifically his distinction of "true species" from "mere varieties" on the basis of character constancy rather than degree of difference, required multiple specimens of each species (S56 1860, 107). More prosaically, his buyers wanted to know what they were getting. To determine what species he had, either for sale or theoretical speculation, Wallace needed to examine multiple individuals. His aim was, quite literally, to capture the variation within species, ontogenetic and varietal. The greater bird of paradise (*Paradisaea apoda*) was ontogenetically variable: "In examining my series of specimens, I find four such well-marked states of the male bird, as to lead me to suppose that three moults are required before it arrives at perfection" (S37 1857, 415). The butterfly *Ornithoptera poseidon* (= *Ornithoptera priamus poseidon*) from the same region, showed extensive variation among individuals:

> The numerous specimens of *Ornithoptera* which I obtained in various parts of New Guinea and the adjacent islands show so much variability of form, colouring & even of neuration, no two individuals being exactly alike, that I am obliged to include them all in one variable species [subdivided into three varieties] ... From these facts I am led to conclude that we have here a variable form spread over an extensive area, and kept so by the continued

> intercrossing of individuals which would otherwise segregate into distinct and sharply defined species.[39]

Capturing intra-specific variation of either sort required an extensive series of specimens.

Key features of Wallace's collecting practice, established during his amateur period, were refined into an intense and rigorous field routine. Throughout his years in the field, the dual motivations that led Wallace to his vocation continued to guide and shape his natural history practice. More than a decade after his departure for Brazil, his reasons had changed very little:

> [Love of] Entomology—the forms & structures & affinities of insects ... but I am engaged in a wider and more general study—that of the relations of animals to time & space or in other words their geographical and geological distribution and its causes ... besides these weighty reasons, there are others quite as powerful—pecuniary ones. I have not yet made enough to live upon & I am likely to make it quicker here than I could in England ... Now though I always liked surveying, I like collecting better.[40]

Theorizing[41]

Collecting dominated Wallace's life in South America and the Malay Archipelago. His theorizing about natural history was entangled with his collecting practice; "the man who works twelve hours every day at his collection ..." has little time for pure reflection (S45 1859, 111). His resources for theorizing were limited in other ways as well. During his years in the field, Wallace was for the most part isolated from English society, and had only sporadic contact with other naturalists. Land around towns was usually clear-cut, so he found it necessary to live outside settlements while collecting, returning every few months to colonial enclaves (such as Ternate and Malacca) to prepare shipments to Stevens. His intellectual as well as economic life was mediated by his agent: Wallace sent specimens and collecting reports back to England, and in return Stevens sent equipment, funds, information about species and localities of interest to potential buyers, and recent issues of natural history periodicals. From 1856, Stevens also conveyed letters to and from Bates, reviving their old collecting correspondence. Though Wallace also discussed natural history with amateur naturalists in the East Indies colonial network (an international and socially fluid mixture of travelers, miners, missionaries, doctors, and businessmen), this seems to have played little role in his theorizing about the origin of species. Intellectual isolation was a frequent complaint in his correspondence:

> The physical privations which must be endured during such journeys are of little importance, except as injuring health and incapacitating from active exertion. Intellectual wants are much more trying: the absence of intimate friends, the craving for intellectual and congenial society, make themselves

> severely felt, and would be unbearable were it not for the constant employment and ever-varying interest of a collector's life ... (S25 1856, 5117).

While in the field, Wallace necessarily participated in the natural history community from a distance, via lengthy and contingent lines of communication. Moreover, the social distance between himself and the "eminent men" of English natural history, such as Owen, Lyell, Hooker, and Darwin, was as important as physical separation. Though well known to the Entomological and Geographical Societies and the British Museum, prior to 1858 Wallace was considered a "mere collector," contributing specimens and observations for the "species-men" (Marchant 1916b, 57; Newman 1854, 144–47; Brooks 1984, 50–55). His theoretical essays of the 1850s were attempts to overcome that barrier and engage with the theoretical side of the natural history community. Evolutionary speculations in his field notebooks are usually responses to texts: articles by Owen, Forbes, Blyth, *etc.*; and standard works such as Lyell's *Principles of Geology* and Darwin's *Journal of Researches.* But his access to such works was limited. Wallace carried few books into the field, and shipments of texts from Stevens reached him irregularly. Much of his theorizing thus occurred "in a place far removed from all means of reference and exact information ..." (S20 1855, 185).[42] Time was also a limiting factor, since collecting and preparing specimens took precedence. Wallace's reading typically was confined to an hour or two at night, after his day's haul of specimens had been processed—and not even that, if his fieldwork had been especially successful.

However, his industry notwithstanding, there were intervals that afforded Wallace more time for reading and speculation. Bouts of bad weather, illness, and injury periodically confined him indoors for days or weeks at a time. Interruption of his collecting distressed Wallace as much as the physical discomfort; for example, in the Aru Islands: "Mosquitoes and minute ticks here attacked me so perseveringly, that my feet and ankles refused to submit, and, breaking out into inflamed ulcers, confined me to the house during a month of the very finest weather, when I had hoped to obtain and preserve a host of fine insects" (S40 1858, 5891). But, uncomfortable though they were, such intervals punctuated Wallace's regular routine with leisure to read and reflect on the hypothesis that had brought him to the field. His theoretical notes and essays intended for publication were written at such times, when he had "nothing to do but look over my books and ponder over the problem which was rarely absent from my thoughts ...": the "Law" paper of 1855 during Borneo's rainy season; "Attempts at a Natural Arrangement of Birds" (S28 1856) while recovering from injured feet; the 1858 Ternate essay following a bout of malarial fever (S729 1906 1:354; S20 1855; S28 1856; Darwin and Wallace 1858). But such intervals were irregular, unpredictable, and, most significantly, continuous with his collecting practice.

This continuity is reflected in Wallace's theoretical notebooks, which are laced with descriptions and counts of species, as well as prosaic details of collecting:

locality labels, methods for treating feathers, descriptions of particular specimens and the difficulties of capturing them.[43] One entry lists biogeological generalizations drawn from Pictet's *Paleontology*, evaluating the cogency of each and ranking their importance for a "Law of Ecological Development." This suggestive argument sketch is flanked by lists of moth and bird species (with species counts of 114 and 161, respectively) from Singapore and Malacca. "Notes from Lyell's Principles" and a "Note for Organic Law of Change" are interspersed with remarks on "name tickets" labeling beetle specimens (MS180 34–42). Notes on an 1855 paper by Richard Owen (read to the Geological Society, 16 May) are given an evolutionary interpretation: "The above is what might be expected, if there has been a constant change of species, by the modifications of their various organs, producing a complicated many branching series" (*ibid.*, 54). The remark is immediately followed by "species counts" of birds, and a detailed description of a Singapore species (*ibid.*, 54–55). The most speculative of Wallace's entries are interspersed with details of his collection and its organization. His theorizing on the origin of species was of a piece with the collecting practice that dominated his life in the Archipelago.

Wallace's first explicitly theoretical essay pertaining to evolution shows this influence. From nine facts of geographical and geological distribution, he derived "the law which has regulated the introduction of new species ... Every species has come into existence coincident both in space and time with a pre-existing closely allied species" (S20 1855, 184–86). Wallace went on to argue that his law "agrees with, explains and illustrates all the facts ..." of natural affinities among species, distribution of animals and plants in space and time (with particular attention to the Galapagos islands), and rudimentary organs. It was, moreover, compatible with what was unknown (extinct forms not preserved in the patchy fossil record), and in this respect superior to Forbes' polarity theory, which presupposed complete knowledge of past and present forms. Though he was silent as to the mechanism by which the pattern of gradual succession of species was produced, using "antitype" to designate the immediate predecessor of a species, Wallace's 1855 view was incipiently evolutionary. Rejecting special creation, he posited an "unbroken and harmonious system" of species related by affinities reflecting gradual modification and co-localization in time and space (S20, 184–86).

Attending to Wallace's practice highlights the evolutionary aspects of his 1855 view. At the beginning of the paper, Wallace alludes to his scientific motivation for going to the field:

> The great increase of our knowledge within the last twenty years, both of the present and past history of the organic world, has accumulated a body of facts which should afford a sufficient foundation for a comprehensive law embracing and explaining them all, and giving a direction to new researches. It is about ten years since the idea of such a law suggested itself to the writer of this paper, and he has since taken every opportunity of testing it by all the

> newly ascertained facts with which he has become acquainted, or has been able to observe himself (S20, 185).

The "idea of such a law" apparently suggested itself to Wallace around the time he read *Vestiges* and encountered the developmental hypothesis that spurred him to become a professional specimen collector. In this capacity, he observed many facts of "Organic Geography" for himself; there is thus a clear, if implicit, connection between Wallace's evolutionary ideas and his 1855 law. The list of facts (four geographical, five geological) from which Wallace's law is deduced immediately follows the quotation above. Though the geological facts were primarily drawn from Wallace's close reading of Lyell's *Principles*, the geographical facts were based at least in part on his own observations from the field: "3. When a group is confined to one district, and is rich in species, it is almost invariably the case that the most closely allied species are found in the same locality or in closely adjoining localities ..." Such facts generalize the results of Wallace's practice, inventorying species (primarily beetles and butterflies) at multiple localities, on two continents. The 1855 paper is Wallace's first attempt to fulfill his original aim of testing the transmutation hypothesis of *Vestiges* with facts assembled in the field, materially represented by his growing collection of specimens.

Wallace's view of "the true Natural System" was also informed by his collecting practice. His branching metaphors are descriptive of evolution: "a complicated branching of the lines of affinity, as intricate as the twigs of a gnarled oak or the vascular system of the human body" (S20, 185, 187). Significantly, he supposed that many of these connections ("the stem and main branches") are represented by extinct species. Working out this system was analogous "to plac[ing] in order ... a vast mass of limbs and boughs and minute twigs and scattered leaves ..." It was, so to speak, the task of a comprehensive collector. Wallace's daily activities aimed at representation and arrangement on a smaller scale: "Nature must be studied in detail, and it is the wonderful variety of the species of a group, their complicated relations and their endless modification of form, size, and colours, which constitute the pre-eminent charm of the entomologist's study."[44]

This implicitly evolutionary conception of the organic world, and its basis in Wallace's collecting practice, are even more apparent in his "Attempts at a Natural Arrangement of Birds," written the following year (S28 1856). In this paper, Wallace proposed arrangements of two Passerine groups (Scansores and Fissirostres) based on affinity in the "essential points of their structure and œconomy" (S28, 195–96, 208).[45] Following Strickland's (1841) inductive method, Wallace related avian groups by similarity of character rather than division into fixed classes, representing their systematic relations in unrooted tree diagrams (S28, 205, 215; O'Hara 1986, 1991). Significantly, he modified Strickland's procedure by making branch lengths proportional to the "distance" between taxa, interpreting "gaps" and internal nodes as extinct species (S28, 206, 214). Rather than assume fixed divisions among groups of any rank, Wallace took as "an article of our zoological faith" that the Natural System as a whole is harmoniously interconnected by gradual relations of modification, of which we can observe only the fragments now existing (S28, 206). Though still lacking a mechanism by which those fragments were whittled into shape, he was evidently inclined toward an evolutionary interpretation of the Natural System.

The transmutation hypothesis, of course, had been with him from the outset. The specific avian arrangements that gave it substance and support were grounded in Wallace's collecting work. He distinguished avian groups on the basis of key adaptations; for example, Fissirostres had "minimized feet and maximized wings, always connected with some modification of structure, adapted to give facilities for seizing the food with the mouth ..." Affinities among members of a group were inferred from variations on these essential features of structure and habit. Wallace's conception of these features and their significance was based on his experience "observing the habits of many tropical birds in a state of nature ...," and his own "constant habit" of skinning them, thereby "obtaining much information on very important parts of their internal structure" (S28 1856, 202, 194). Habits and structures were, so to speak, the foci of the two halves of Wallace's daily collecting routine. Capturing or shooting specimens required careful observation of their habits, especially those pertaining to locomotion. Wallace had detailed knowledge of the habitual motions of hundreds of bird species, acquired over many hours in the field. In the afternoons and evenings, he prepared thousands of avian specimens, in the process observing details of plumage, skin texture and thickness, skeletal structure, and stomach contents. In order for these specimens to contribute to his collection, scientifically or financially, Wallace had to identify them, at least provisionally, as "good" or "new" species. His routine collecting practice thus constructed the relation between individual organisms and species, via his own manual and conceptual labor of capturing and processing specimens. In this sense, the species comprising Wallace's collection, and his evolutionary theory, were constructed from the habits and adaptive structures of individual organisms.

A decisive moment in Wallace's evolutionary theorizing occurred in February 1858, when the connection between the pattern of natural affinities among species and the Malthusian struggle for existence "suddenly flashed upon" him in the wake of a malarial fever.[46] "On the Tendency of Varieties to Depart Indefinitely From the Original Type" (S43) was written over the next two days. The argument is in seven parts:

1. received view of species and varieties in the wild and under domestication
2. struggle for existence and individual selection
3. relative abundance of different species of an allied group
4. modified varieties replace parent species
5. differences between domesticated and wild varieties
6. against Lamarck's hypothesis
7. conclusion and summary

The resemblance to Darwin's theory is striking. (The latter remarked, in some distress, that "If Wallace had my MS sketch written out in 1842 he could not have made a better short abstract!"[47]) Yet Wallace's theory developed independently of Darwin's, and reflects its immediate context: his extensive and intense collecting practice. In particular, the concepts of species abundance and variation, honed in his routine fieldwork, are important components of his theory.

Their importance is brought out by a long-standing historiographic puzzle. Wallace's evolutionary theory, on the face of it, describes selection on varieties or races. The explicit aim of the 1858 essay is to show "that there is a general principle in nature which will cause many *varieties* to survive the parent species, and to give rise to successive variations departing further and further from the individual type" (S43, 54; italics in original). Darwin's theory of evolution by natural selection, in contrast, describes selection as acting on individual organisms, except in special cases (*e.g.*, sterile castes of social insects). A number of historians and philosophers have remarked on this significant theoretical difference.[48] Oddly, however, Wallace and Darwin themselves seem not to have noticed it (Marchant 1916b, 105–262). In their correspondence (1858 to 1881), the co-discoverers consistently described the mechanisms at the core of their theories as the same. And indeed, Wallace does describe the struggle for existence in terms of individual organisms:

> ... so long as a country remains physically unchanged, the numbers of its animal population cannot materially increase. If one species does so, some others requiring the same kind of food must diminish in proportion. The numbers that die annually must be immense; and as the individual existence of each animal depends upon itself, those that die must be the weakest—the very young, the aged, and the diseased,—while those that prolong their existence can only be the most perfect in health and vigour—those who are best able to obtain food regularly, and avoid their numerous enemies. It is ... "a struggle for existence," in which the weakest and least perfectly organized must always succumb (S43, 56–57).

But Wallace immediately seems to equate these selective processes operating on individuals with those operating on groups: "it is clear that what takes place among the individuals of a species must also occur among the several allied species of a group" (S43, 57). He then goes on to describe "variety selection" in some detail:

> Now, let some alteration of physical conditions occur in the district—a long period of drought [*etc.*] ... —any change in fact tending to render existence more difficult to the species in question, and tasking its utmost powers to avoid complete extermination; it is evident that, of all the individuals composing the species, those forming the least numerous and most feebly organized variety would suffer first, and, were the pressure severe, must soon become extinct. The same causes continuing in action, the parent species

> would next suffer, would gradually diminish in numbers, and with a recurrence of similar unfavourable conditions might also become extinct. The superior variety would alone remain, and on a return to favourable circumstances would rapidly increase in numbers and occupy the place of the extinct species and variety (S43, 58).

Viewing Wallace's evolutionary theory as continuous with his collecting practice clarifies this seemingly abrupt move from individuals to varieties and species. Rather than confusion on Wallace's part, it follows naturally from his conceptions of abundance and variation within a species. In his writings from the field, Wallace repeatedly related individual organisms and the groups of which they are members via the concept of abundance: an abundant species is "numerous in individuals." For six hours a day, for hundreds of days, Wallace made estimates of species abundance in the course of capturing specimens. The concept was a necessary component of his daily collecting routine. Wallace was confronted with variation within species on a daily basis as well. One aspect of preparing specimens for his collection was arranging the mass of individual variation in his bottles and nets into distinct species and varieties, which could be counted, compared, and sold. So not only did he routinely estimate variation within species, assembling "good series" for each, but he also attempted to resolve it into discrete groups: varieties or races. The Malthusian struggle for existence provided a mechanistic link between these two concepts: "variations from the typical form of a species must have some definite effect, however slight, on the habits or capacities of the individuals ...," and the "best-adapted... necessarily obtain and preserve a superiority in population ...," while those that exhibit "some defect of power or organization ... must diminish in numbers, and, in extreme cases, become altogether extinct" (S43, 57).

On Wallace's view, "continuance of the species and the keeping up of the average number of individuals ..." come to the same thing, and have the same causes: namely, survival of the fittest *individuals.* This resolves the puzzle over his 1858 theory, clarifying the relation between individual organisms and groups. Furthermore, this account locates Wallace's theory within the broader context of his natural history practice. The 1855 "Law" paper addresses his original scientific goal: testing the transmutation hypothesis. His 1856 "Arrangement of Birds" ties this proto-evolutionary view of the organic world to the details of routine collecting work. The concrete work of making a collection led him to focus on adaptations of individual organisms, as well as variation within species, and the abundance of species and other groups. In 1858, the Malthusian mechanism linking these core concepts "flashed upon" Wallace, animating his evolutionary view with a principle "exactly like that of the centrifugal governor of the steam engine, which checks and corrects any irregularities almost before they become evident" (S43, 62). This completed the core of his theory of the origin of species, which developed (in often surprising directions) over the next several decades.

Conclusion

Many elements of Wallace's natural history practice were established early, during his years as an amateur botanist and entomologist. His land-surveying work set a pattern of itinerant mapping that continued to structure his working life in the Amazon and Malay Archipelago. In collaboration with Bates, Wallace's drive to identify species became focused on key groups, and their traffic in "duplicate" specimens anticipated his dual collections in the field. The key concepts of species abundance and variation were incorporated into Wallace's amateur practice: he kept qualitative records of species abundance at a locality, and represented intra-specific variation with "series" of specimens. The two powerful motivations led him to the field as a professional collector—a drive to test the theory of transmutation of species proposed in *Vestiges*, and the need to support himself financially—drove and constrained Wallace's collecting practice in the Amazon and the Malay Archipelago, shaping its material and conceptual products.

To accommodate his dual aims, Wallace followed an intense daily routine, capturing and arranging thousands of specimens, channeling his fascination with relations among species toward groups that would sell profitably (tropical birds, beetles, and butterflies). In his natural history practice, Wallace was routinely confronted with variation within species, and relative abundance of species within and among localities. His collection represented both: the former by a "good series" of specimens, and the latter with estimates of abundance, elaborated

Figure 15 A selection of beetle species collected by Wallace in the Moluccas.
From S715. Out of copyright.

into "species counts" for particular families, orders, and higher taxa. Placing Wallace's evolutionary theory in the context of this practice illuminates important features of his view and its development, dispelling the unflattering obscurity of Darwin's shadow. This essay, of course, is only a first step. I have attempted to explain the "special turn of mind" of "the collector and the species-man" in terms of Wallace's own extensive collecting practice. Wallace himself favored a phrenological explanation: "The shape of my head shows that I have *form* and *individuality* but moderately developed, while *locality, ideality, colour,* and *comparison* are decidedly stronger ..." (Marchant 1916b, 91-96; S729 1906, 1:25). But the shape of Wallace's life, the routines of his collecting practice, reveals more than the shape of his head.

Notes

1. Charles Lyell and Joseph Hooker, communication of Darwin and Wallace (1858); reprinted in Loewenberg (1959, 41).
2. Hodge and Radick 2003, 149. On Wallace's alleged error or confusion: Bowler (1976), Gould (2002). On the question of priority: McKinney (1972a); Brackman (1980); Browne (1980); Brooks (1984); Beddall (1988a).
3. See Fagan (2007) for a comparison with Darwin focused on field practices.
4. Recent biographies that examine Wallace's fieldwork include Fichman (2004), Slotten (2004). Camerini (1993, 1996, 1997, 2002) has written extensively on Wallace's fieldwork. Beddall (1968) discusses the development of Wallace's theory.
5. The most detailed and authoritative account of Wallace's life is his autobiography, *My Life* (S729 1906); see also Raby (2001).
6. S729 1906, 1:110, 194; WP1/3/14 (letter of 3 October 1845). Archival materials cited in this essay are designated as follows: Wallace family correspondence (Natural History Museum, London, Archives): WP1/3/11–14, WP1/3/19–72; Malay Journals (Linnean Society, Archives): MS178a–d; Zoology notebook with records of consignments to Samuel Stevens, 1855–58 (Linnean Society, Archives): MS179; Natural history notebook, 1855–59 (Linnean Society, Archives): MS180; Notebook on butterflies of the Malay Archipelago, 1854–62 (Linnean Society, Archives): MS181.
7. See Allen (1976, 2001) for more on the British pastime of natural history during this period.
8. See Fichman (2004, 16–17) for details.
9. S729 1906, 1:237; WP1/3/13 (letter of 26 June 1845).
10. WP1/3/13 (letter of 26 June 1845). See also letters of 9 November 1845 and 28 December 1845.
11. *Ibid.*
12. *Ibid.* Wallace did single out "particular families" for more thorough representation (Lamellicornidae, Elateridae, Crioceridae, Galerucidae, Chrysomelidae).
13. *e.g.*, WP1/3/13–14 (letters of 26 June 1845, 3 October 1845).
14. WP1/3/11 (letter of 11 April 1846).
15. WP1/3/14 (letter of 3 October 1845).
16. WP1/3/19 (letter of 11 October 1847).

17. WP1/1/17 (letter of 28 December 1845); see also WP1/3/15 (letter of 9 November 1845).
18. A comprehensive survey of Wallace's travels in the Amazon and Malay Archipelago is beyond the scope of this essay. For more detail see S714 1969; S715 1869b; Camerini (1996, 2002); Shermer (2002); Slotten (2004).
19. His younger brother Herbert briefly joined him in 1850, but did not take to collecting. Wallace and Bates' reasons for separating have not been recorded; their contemporaries speculated they might have quarreled, or chosen to maximize their respective collections. In any case, Wallace never again collected with an equal partner, only assistants (though he valued some of these greatly; see Camerini 1996).
20. The term is borrowed from Browne (1992).
21. His lost species "would have rendered (I had fondly hoped) my cabinet, as far as regards American species, one of the finest in Europe" (letter to Spruce, in S729 1906, 1:303–09).
22. Letter to George Silk (Marchant 1916b, 43). Wallace's early interest in wildflowers never waned—his wife shared this enthusiasm, and Wallace took botanical excursions (in Europe and the United States) throughout his life.
23. MS179 (Aru, July 1857; Ceram, February 1860).
24. Farber and Mayr (1986) discuss the importance of ornithology in evolutionary theory more generally.
25. S48 1859, 129; S67 1862; S715 1869b, 575. See also Camerini (2002, 103–41).
26. MS178a, 18 (July–August 1856).
27. *Ibid.*
28. Wallace questioned the traditional distinction between species and varieties as early as 1845, endorsing the view that "true species" were distinguished from "mere varieties" not by degree of differentiation, but by constancy and ability to propagate from a few individuals throughout an entire race (WP1/3/17; S56 1860, 107).
29. MS179; Marchant (1916b, 68); S715 1869b, vii–viii.
30. S729 1906, 1:337–38. See also S21 1855, 4805-07; S45 1859, 111-13.
31. S729 1906, 1:328–30. See also S21 1855; Slotten (2004). For details of naturalists' equipment in the 1840s to 1860s see Larsen (1996); for an overview Farber (2000).
32. Wallace's principal sources were Boisduval's *Histoire Naturelle des Insectes* (1836) and Bonaparte's *Conspectus Generum Avium* (1850); S729 1906, 1:327–29.
33. S729 1906, 1:338–40, 382–83; Camerini (1996, 1997). Other assistants, such as Baderoon, a young man of Macassar, worked for Wallace for shorter periods than Charles and Ali.
34. MS180, 132–33. What is needed, he argues, is "an approximate law" correcting for biases in partial collections.
35. S13 1853, 254; S9 1853, 3884; S14 1854, 4396; S38 1857, 473; S21 1855, 4803; WP1/3/42 (letter of 2 March 1858); S6 1850, 494–96; S729 1906, 1:269. See also: S35 1857, 91; MJa–b (entries from October 1856, January 1857, March 1857, September 1857). On orang-utans: S24 1856; S26 1856; S715 1869b, 51–53, 57–74.
36. S45 1859, 113; S4 1850, 156-57; also S3 1849, 74–75; S38 1857, 479, 484; S44 1858, 6124; S62 1861, 285; MS178b (entries of January 1857, September 1857).
37. WP1/3/39, 4 (letter of 30 April 1856).
38. WP1/3/47 (letter of 25 November 1859).
39. MS181, 1–2. This comment dates from 1861 or later.
40. WP1/3/46 (letter to Thomas Sims, 25 April 1859).

41. For more on Wallace's evolutionary theorizing in the field and thereafter, see Kottler (1985); Fichman (2004; esp. Chapters 3 and 6); Bulmer (2005).
42. He did, however, have intermittent access to well-stocked libraries, notably that of Sir James Brooke, the first "White Rajah of Sarawak" (1842–68). See S715 1869b, 102–04; S729 1906, 1:327, 341, 346–47.
43. Wallace's theoretical notes are mainly in MS180, a "Natural History" notebook with entries from 1855 to 1859. Additional theoretical notes appear at the back of MS179, a zoology notebook containing entomological species lists (1855–58) and records of consignments to Stevens (1854–61).
44. WP1/3/42 (letter to F. Bates, 2 March 1858, 1).
45. Wallace identified as the essential characteristics of a group the adaptations distinguishing its members from a similar outgroup (though he did not use the latter term); for example, Fissirostres had "minimized feet and maximized wings, always connected with some modification of structure, adapted to give facilities for seizing the food with the mouth" (S28 1856, 202). Affinities among members of a group were inferred from variations on these essential features of structure and habit.
46. S729 1906, 1:360–63; Moore (1997). McKinney (1972a) argues that Wallace's insight occurred on Gilolo rather than Ternate. Wallace's correspondence, however, indicates that his theory was worked out before 2 March 1858 (WP1/3/42).
47. Burkhardt and Smith (1991, 107): letter to Lyell of 18 June 1858.
48. The contrast is typically interpreted in terms rather unfavorable to Wallace: confusion, hasty generalization, or simply missing the essential feature of Darwin's mechanism: Osborn (1894, 245); Beddall (1968); Bowler (1976, 24); Ruse (1980); Gould (1980); Kleiner (1985); Gayon (1998, 19–59): Ruse (1999, 233); Gould (2002, 126–37); Browne (2002, 18). For "neutral theories" see: Kottler (1985); Slotten (2004, 159). For dissenting views, see: Mayr (1982, 494–97); Bulmer (2005, 133). George Beccaloni (pers. commun.) argues, following Kottler (1985), that by "varieties" Wallace meant not subspecies or "permanent varieties," but individuals differing from the norm. He also makes the good point that if Darwin and Wallace really did hold different interpretations in this respect that they would have remarked upon them at some point in their writings.

4

Wallace's Annotated Copy of the Darwin-Wallace Paper on Natural Selection*

George Beccaloni

Introduction

In 2002 the Natural History Museum (NHM), London, purchased an important collection of over 5000 of Wallace's manuscripts, books, photographs and other items, from his grandsons Richard and John Wallace (http://www.nhm.ac.uk/nature-online/collections-at-the-museum/wallace-collection/index.jsp; Beccaloni 2002). They had inherited the collection from their father William, who had inherited it from Wallace after his death in 1913. The collection is especially rich in documents relating to Wallace's family life, and until the NHM acquired it only a handful of scholars had ever accessed it. One of the items in the collection of particular note is an authors' offprint in buff coloured wrappers of Darwin and Wallace's seminal 1858 paper[1] on natural selection (NHM catalogue number WP7/9). It is of interest because it is the only copy of this paper known to have been annotated by Wallace. The annotations, which have not been analysed before, are discussed below.

The Darwin-Wallace paper was read at a meeting of the Linnean Society on 1 July 1858 and published by the society on 20 August 1858. Freeman (1977) noted:

> There are five different forms in which the original edition can be found, but they are all from the same setting of type. Four of these are the results of the publishing customs of the Linnean Society of London and the fifth is the authors' offprints. *The Journal* came out in parts and was available to Fellows of the Society with *Zoology* and *Botany* together in each part, *Zoology* alone, or *Botany* alone. Later it appeared in volume form made up from reserved stock of the parts with new title pages, dated in the year of completion of the volume, and indexes. This again was available complete or as *Zoology* or

* I thank Christopher Wells for first drawing the 1858 offprint to my attention and Richard Wallace for allowing me to photograph it whilst it was still in his possession. I am grateful to Charles Smith and Judith Magee for their helpful comments on an earlier draft of this chapter.

> *Botany* alone. The *Zoology* was signed with numbers and the *Botany* with letters. The Darwin-Wallace paper occurs in the complete part in blue wrappers, or in the *Zoology* part in pink wrappers; the *Botany* parts were in green. The Linnean Society has all the forms in its reference files, although it does not hold the offprint.
>
> The authors' offprints were issued in buff printed wrappers with the original pagination retained. They have "From the Journal of the Proceedings of the Linnean Society for August 1858" on page [45]. They were printed from the standing type but, presumably, after the copies of the number had been run off. The only copies which I have seen have been inscribed personally by Darwin ...

It is unknown how many offprints were produced and to whom they were sent. However, it is clear that Charles Darwin received a number of copies and that he posted some of them to Wallace in South East Asia. In a letter to Hooker dated 12 October 1858 Darwin wrote:

> I have sent 8 copies by post to Wallace, & will keep the others for him, for I could not think of anyone to send any to (Burkhardt and Smith 1991).

In a subsequent letter to Wallace, posted approximately three months later on 25 January 1859, Darwin wrote

> I sent off, by same address as this note, a copy of Journal of Linn. Soc. & subsequently I have sent some ½ dozen copies of the Paper. I have many other copies at your disposal; & I sent two to your friend Dr. Davies(?) author of works on men's skulls (Burkhardt and Smith 1991).

In October 1858, long before he received Darwin's 12 October letter or the copies of the paper that Darwin sent, Wallace asked his agent Samuel Stevens in London to obtain some copies for him. In a letter to Stevens from Batchian (Bacan Island, Indonesia), dated 29 October 1858, Wallace wrote:

> An Essay on Varieties which I sent to Mr. Darwin has been read to the Linnaean [*sic*] Society by Dr. Hooker and Mr. C. Lyell on account of an extraordinary coincidence with some views of Mr. Darwin, long written but not yet published and which were also read at the same meeting. If these are published I dare say Mr. Kippish will let you have a dozen copies for me. If so send me 3 and of the remainder send one to Bates, Spencer and any other of my friends who may be interested in the Matter and who do not attend the Linnaean (transcript in NHM library).

It is unclear whether Wallace was asking for copies of the journal part containing the paper, or for offprints, and it is unknown whether Stevens sent the copies which Wallace requested. Therefore, it is not known whether the annotated copy held in the NHM was originally sent to Wallace by Darwin, Stevens, or someone else. All we can be sure of is that he received it whilst he was in South East Asia before February 1860, as this is the date of an annotation on a blank page at the back of the copy (see below).

The only other extant copy of the offprint known to have been owned by Wallace has an inscription by him on the wrapper which reads "J. Barnard Davis Esq., Shelton, Staffordshire, with Mr. Wallace's Compliments." This copy belonged to the late Quentin Keynes and is now in the possession of The Charles Darwin Trust (Simon Keynes pers. commun.). Joseph Barnard Davis (1801–81) formed a large collection of skulls and skeletons which was sold to the Royal College of Surgeons in 1880. From the letter sent by Darwin to Wallace mentioned above, we know that Darwin also sent two copies of the offprint to Davis.

Wallace's Annotations

All of the annotations on the offprint are written in pencil or ink in Wallace's handwriting and apart from the dated pencil annotation on the blank back page it is unfortunately not possible to determine when they were written. In many cases pencil annotations are overlaid by annotations in ink, so indicating that the pencil ones are earlier. The annotations will be discussed sequentially and referred to by the page numbers of the original paper.

Front wrapper (not illustrated): Pencil annotation at the top of the wrapper reads "Species, Darwin and Wallace."

Page 55 Ink annotation of "*" after the printed text "... nearly ten millions!" Ink annotation in right-hand margin reads "* Really more than two thousand millions!" This has been written over a pencil annotation which reads "1000 millions!" When Wallace reprinted his section of the 1858 paper in his 1870 book *Contributions to the Theory of Natural Selection. A Series of Essays* (S716) he inserted the footnote (on page 29) "* This is underestimated. The number would really amount to more than two thousand millions!"

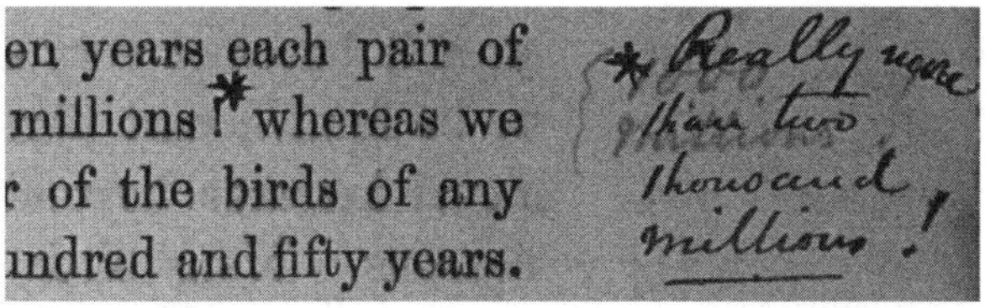
en years each pair of
millions !* whereas we
r of the birds of any
undred and fifty years.
* Really more than two thousand millions !

Page 56 The word "itself" near the bottom of the page has been underlined in pencil.

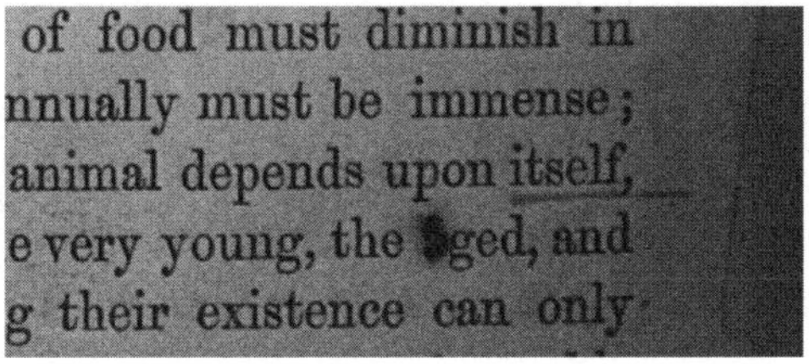
of food must diminish in
nnually must be immense;
animal depends upon itself,
e very young, the aged, and
g their existence can only

Page 57 The pencil and ink annotations in the text and in the right margin are proof correction marks which denote that Wallace wanted a hyphen inserted between "food" and "supply." In S716 this is printed as "food-supply" (on page 33).

organization are the least
es of food supply, &c., must
e cases, become altogether
species will present various

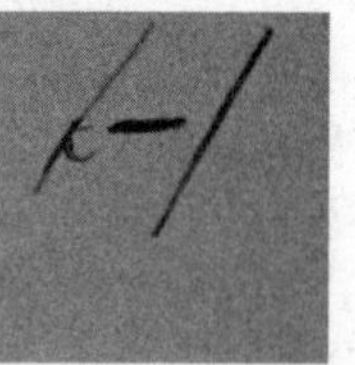

Page 58 Wallace inserted the word "could" between "they" and "inhabit" in the sentence at the top of the page. This annotation is written in ink overlaying the same word in pencil. In S716 "could" has been inserted into the sentence as indicated (on page 35).

58 MESSRS. C. DARWIN AND
could
country which they inhabit. It is
would affect, either favourably or

Page 59 The word "the" 'written in ink in the left-hand margin, is inserted between "and" and "periods." In S716 the text has been altered to read "... and the periods ..." (on page 38).

which nature works is so vast
the periods of time with which she
that any cause, however slight, a
counteracted by accidental circu

Page 60 There are several annotations on this page. They are (from top to bottom of the page) as follows:

(i) The ink annotation "become /" in the left-hand margin indicates that Wallace wanted to substitute this word for "are" in the text, which he crossed out. The sentence would therefore read "Half of its senses and faculties become quite useless ...", which is the form of the text published in S716 (on page 39). The annotation "become" overlies an annotation in pencil which reads "to itself." Wallace has indicated that these words should be inserted after "useless." The sentence would therefore have read "Half of its senses and faculties are quite useless to itself ..."

become /
Half of its senses and faculties ~~are~~ quite useless ; and
half are but occasionally called into feeble exercise, whi
muscular system is only irregularly called into action.

(ii) The pencil annotation "Since man's interference & Selection is left out of consideration" in left-hand margin, appears to qualify the statement in the following sentence of the printed text: "Again, in the domesticated animal all variations have an equal chance of continuance; and those which would decidedly render a wild animal unable to compete with its fellows and continue its existence are no disadvantage whatever in a state of domesticity."

Again, in the domesticated
chance of continuance ; and th
wild animal unable to compet
existence are no disadvantage
Our quickly fattening pigs, s
and poodle dogs could never l

(iii) The ink annotation "gradually" in left-hand margin overlays the same word in pencil. Wallace indicated this should be inserted between "each" and "lose" and this is how the text was published in S716 (on page 40).

gradually /
probably soon
might each lose
into action, an
type, which mu

(iv) The word "would" has been crossed out in ink and in pencil. The mark in the right-hand margin signifies that it should be deleted. This word is absent from this sentence in S716 (on page 40).

eme qualities which would never be called
generations ~~would~~ revert to a common
n which the various powers and faculties

(v) The annotation in ink at the bottom of the page reads "That is, they will vary and the variations which render them best adapted to the wild state and therefore approximate them to wild animals will be preserved. Those that do not vary quickly enough will perish." This has been written over a pencil annotation which reads "That is natural selection will rapidly pick out & accumulate all these [those?] variations [the remainder is illegible]." In S716 a version of this statement appears as a footnote on page 40. It qualifies the sentence which ends "... *or become altogether extinct.*" The footnote in S716 (on page 40) is slightly differently worded and reads "* That is, they will vary, and the variations which tend to adapt them to the wild state, and therefore approximate them to wild animals, will be preserved. Those individuals which do not vary sufficiently will perish."

near the type of the original wild stock, *or become altogether extinct.*

That is, they will vary and the variations which render them best adapted to the wild state and therefore approximate them to wild animals will be preserved. Those that do not vary quickly enough will perish.

Page 61 There are two annotations on this page. They are (from top to bottom of the page) as follows:

(i) Wallace indicated that the ink annotation "the permanence of" at the top of this page (which partly overlies an identical pencil annotation) should be inserted between "to" and "varieties." The text in S716 (on page 40) includes this alteration.

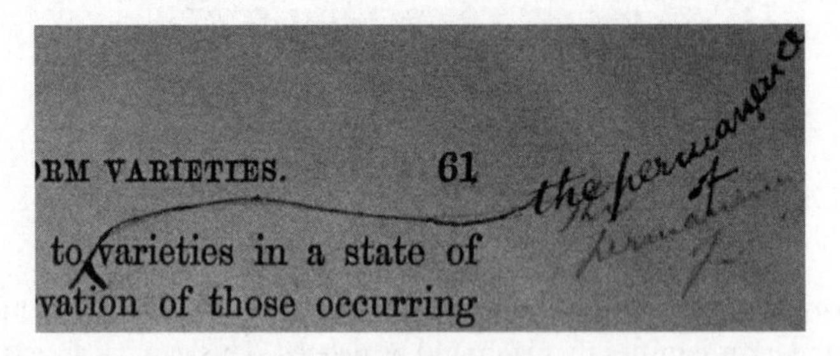

RM VARIETIES. 61

to varieties in a state of

vation of those occurring

the permanence of

(ii) The ink annotation "ations/" in the left-hand margin corrects "varieties" to "variations." It partly overlies an annotation in pencil which reads "variations." The text in S716 (on page 41) includes this correction.

animals are abn

ations/ varieties which

nature: their ve

Page 62 Wallace indicated that the ink annotation "when turned wild" in the right-hand margin should be inserted between "tendency" and "to." In S716 there is a slight difference in that this sentence (on page 43) reads " . . . tendency, when they become wild, to . . .".

state of nature

dency to revert when turned wild,

steps, in various

Blank end page A relatively long pencil annotation on this page reads:

1860. Feb.

> After reading Mr Darwin's admirable work "On the Origin of Species", I find that there is absolutely nothing here that is not in almost perfect agreement with that gentlemans facts & opinions.
> His work however touches upon & explains in detail many points which I had scarcely thought upon,—as the laws of variation, correlation of growth, sexual selection, the origin of instincts and of neuter insects, & the true explanation of Embryological affinities. Many of his facts & explanations in Geographical distribution are also quite new to me & of the highest interest—

ARWallace [signed] . . . Amboina

This pencil note is of considerable historical interest as it gives Wallace's earliest known reaction to his reading of Darwin's *On the Origin of Species* for the first time. What Wallace appears to be saying in the note is that on reading *Origin* he could not find anything in the Darwin-Wallace paper (it is not clear whether he was referring to Darwin's sections, to his, or to both) which was not in almost perfect agreement with what Darwin had written in *Origin*. It is curious that he signed and dated this note—as if it were intended to be read by others in the future.

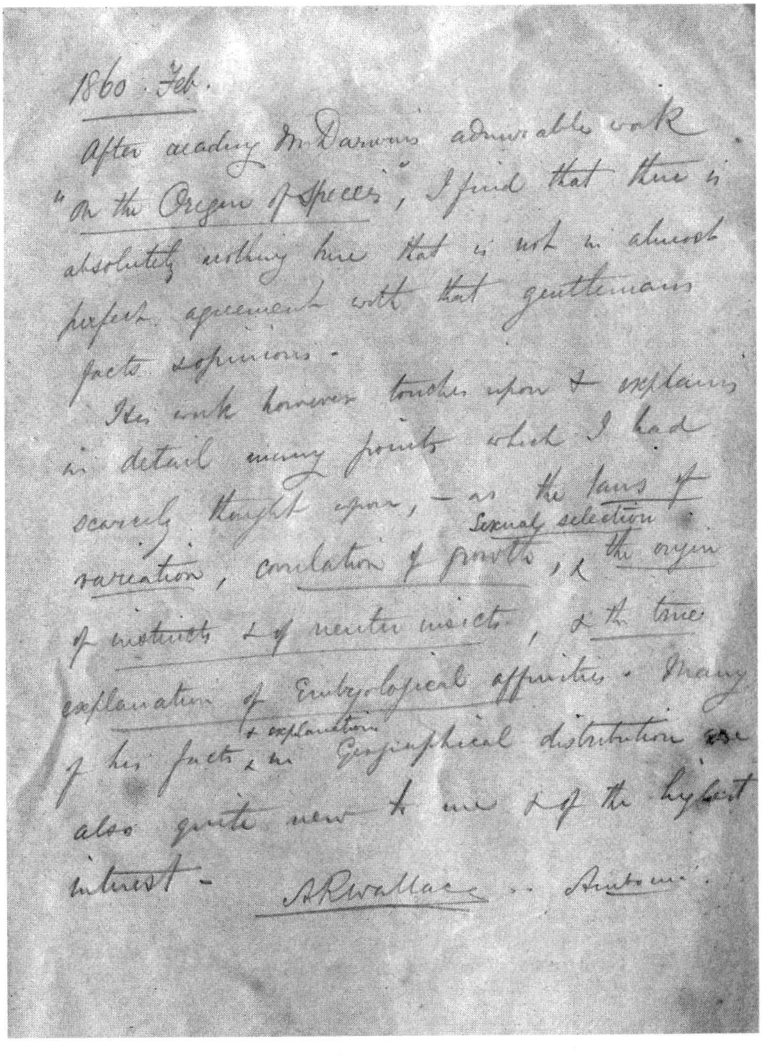
1860 Feb.
After reading Mr Darwins admirable work "On the Origin of Species", I find that there is absolutely nothing here that is not in almost perfect agreement with that gentlemans facts & opinions.
His work however touches upon & explains in detail many points which I had scarcely thought upon,—as the laws of variation, correlation of growth, sexual selection, & the origin of instincts & of neuter insects, & the true explanation of Embryological affinities. Many of his facts & explanation in Geographical distribution are also quite new to me & of the highest interest—
ARWallace .. Amboina

Wallace's Copy of *Origin* and His Appraisal of It

Origin was officially published on 24 November 1859 and in a letter to Wallace dated 13 November 1859 Darwin wrote:

> I have told Murray to send you by Post (if possible) a copy of my Book & I hope that you will receive it at nearly same time with this note. (NB I have got a bad finger which makes me write extra badly—) If you are so inclined, I sh.[d] very much like to hear your general impression of the Book as you have thought so profoundly on subject & in so nearly same channel with myself. I hope there will be some little new to you, but I fear not much. Remember it is only an abstract & very much condensed. God knows what the public will think. No one has read it, except Lyell, with whom I have had much correspondence. Hooker thinks him a complete convert; but he does not seem so in his letters to me; but he is evidently deeply interested in subject.—I do not think your share in the theory will be overlooked by the real judges as Hooker, Lyell, Asa Gray &c. (Burkhardt and Smith 1991).

The book was probably sent to Wallace "care of Hamilton Gray & Co. Singapore" (Burkhardt and Smith 1993, 556).

Wallace possibly received Darwin's letter and the copy of *Origin*[2] whilst he was on Amboyna[3] (Ambon Island, Indonesia). He at least acknowledged receipt of the book from there in a letter to Darwin dated 16 February 1860—the same month as he wrote the pencil note in his copy of the offprint. Darwin replied to this letter on 18 May 1860:

> I received this morning your letter from Amboyna dated Feb. 16th, containing some remarks & your too high approbation of my book. Your letter has pleased me very much, & I most completely agree with you on the parts which are strongest & which are weakest. The imperfection of Geolog. Record is, as you say, the weakest of all ... Before telling you about progress of opinion on subject, you must let me say how I admire the generous manner in which you speak of my Book: most persons would in your position have felt some envy or jealousy. How nobly free you seem to be of this common failing of mankind.—But you speak far too modestly of yourself;—you would, if you had had my leisure done the work just as well, perhaps better, than I have done it (Burkhardt and Smith 1993).

The letter in question from Wallace to Darwin has not been found, and until the discovery of the pencil note in Wallace's offprint, the only extant record of his initial reaction to reading *Origin* is given in annotations he made in his first edition copy of *Origin* (Beddall 1988b), plus the following three letters: one to his childhood friend George Silk dated 1 September 1860; a second to his friend Henry Walter Bates dated 24 December 1860 (Burkhardt and Smith 1993), and a third to his brother-in-law, Thomas Sims, dated 15 March 1861 (Beddall 1988b). Of the three letters, the last gives his most detailed assessment of the contents of Darwin's book:

It is a book in which every page and almost every line has a bearing on the main argument, and it is very difficult to bear in mind such a variety of facts, arguments and indications as are brought forward. It was only on the *fifth* perusal that I fully appreciated the whole strength of the work, and as I had been long before familiar with the same subjects I cannot but think that persons less familiar with them cannot have any clear idea of the accumulated argument by a single perusal.

... It seems to me ... as clear as daylight that the principle of Natural Selection *must* act in nature. It is almost as necessary a truth as any of mathematics. Next, the effects produced by this action *cannot be limited.* It cannot be shown that there is any limit to them in nature. Again, the millions of facts in the numerical relations of organic beings, their geographical distribution, their relations of affinity, the modification of their parts and organs, the phenomena of intercrossing, embryology and morphology—all are in accordance with his theory, and almost all are necessary results from it; while on the other theory they are all isolated facts having no connection with each other and as utterly inexplicable and confusing as fossils are on the theory that they are special creations and are not the remains of animals that have once lived. It is the vast *chaos* of facts, which are explicable and fall into beautiful order on the one theory, which are inexplicable and remain a chaos on the other, which I think must ultimately force Darwin's views on any and every reflecting mind. Isolated difficulties and objections are nothing against this vast cumulative argument. The human mind cannot go on for ever accumulating facts which remain unconnected and without any mutual bearing and bound together by no law. The evidence for the production of the organic world by the simple laws of inheritance is exactly of the same nature as that for the production of the present surface of the earth—hills and valleys, plains, rocks, strata, volcanoes, and all their fossil remains—by the slow and natural action of natural causes now in operation. The mind that will ultimately reject Darwin must (to be consistent) reject Lyell also. The same arguments of apparent stability which are thought to disprove that organic species can change will also disprove any change in the inorganic world, and you must believe with your forefathers that each hill and each river, each inland lake and continent, were created as they stand, with their various strata and their various fossils—all appearances and arguments to the contrary notwithstanding (Marchant 1916a, 1:77–78).

Conclusion

As noted above, Wallace reprinted his section of the 1858 paper in his 1870 book *Contributions to the Theory of Natural Selection. A Series of Essays* (S716). This version was later reprinted with minor changes in his 1891 collection *Natural Selection and Tropical Nature; Essays on Descriptive and Theoretical Biology* (S725). In both of these versions Wallace or the publisher included subheadings in the text which were not present in the original 1858 version, and the corrections and footnotes which Wallace inscribed in ink on his copy of the offprint were also incorporated. It is therefore possible that the ink annotations on Wallace's offprint were intended for the publishers of his book S716. He may have submitted the offprint to the publishers and asked them to reproduce the text in the book, incorporating the corrections he made in ink. The only differences between the annotations in the offprint and the version of the paper reproduced in S716 are in the wording of the statement at the bottom of page 60 of the offprint and page 40 of the book, and the slightly different wording of the statement on page 62 of the offprint and page 43 of S716. These could, however, have been altered when Wallace received the proofs of the book (and note that no manuscript of this book exists, which is what would be expected if the text had been directly transcribed from previously published articles). The pencil annotations on the offprint were not reproduced in S716, S725, or elsewhere and may even have been made whilst Wallace was in the Malay Archipelago, possibly shortly after he received the offprint from Britain. This, however, must remain a speculation.

In several of his publications Wallace complained that he was not given the opportunity to correct the proofs of his 1858 essay. For example, in the earliest known account he gave of his discovery of natural selection (which was conceived during an attack of fever whilst on the island of Gilolo[4] (Halmahera, Indonesia) in February 1858) he stated:

> As soon as my ague fit was over I sat down, wrote out the article, copied it, and sent it off by the next post to Mr. Darwin. It was printed without my knowledge, and of course without any correction of proofs (Letter from Wallace to A. B. Meyer dated 22 November 1869 [Meyer 1895, 415].)

Reprinting his essay in S716 finally gave him the chance to correct the text—albeit twelve years too late!

Notes

1. I treat this publication as a single paper authored by Darwin and Wallace with the title "On the Tendency of Species to Form Varieties; And On the Perpetuation of Varieties and Species by Natural Means of Selection." It contains an introduction authored by Lyell and Hooker, two sections authored by Darwin, and one section authored by Wallace.
2. Wallace annotated this copy of the first edition of *Origin* and eventually gave it to his friend Richard Spruce. It is now part of the Keynes Collections in the library of the University of Cambridge (Beddall 1988b).
3. Wallace was on Amboyna between 25 December 1859 and 24 February 1860 (Baker 2001).
4. Wallace's letter to Darwin, containing his essay on natural selection, was sent from Ternate on 9 March 1858. The essay was marked as being written on Ternate in February 1858, but this cannot have been the case since Wallace's unpublished *Malay Field Journal* in the Linnean Society shows that he was on Gilolo during the whole of February, only returning to the nearby island of Ternate on 1 March. In his book *The Malay Archipelago*

(S715 1869) Wallace records that he was ill whilst on Gilolo, which may have been a reference to his famous "ague fit." It is probable that he wrote "Ternate" on the essay simply because this was the island where he had his base and because it was his postal address. The alternative is that he got the month wrong and should have written "March" instead of "February." Why he never corrected either the date or the place of his discovery in his published accounts of this event is unclear. For more analysis of this situation see McKinney (1972a).

5

Wallace and the Species Concept of the Early Darwinians

James Mallet

••••••

> **... I finished yesterday your paper in the *Linnean Transactions* ["On the Phenomena of Variation and Geographical Distribution as Illustrated by the Papilionidae of the Malayan Region," 1865]. It is admirably done. I cannot conceive that the most firm believer in Species could read it without being staggered. Such papers will make many more converts among naturalists than long-winded books such as I shall write if I have strength.**

(letter from Darwin to Wallace dated 22 January 1866: Marchant 1916a)

Introduction

One of the extraordinary features of modern evolutionary biology is an inability to agree on a common definition of species. This lack of agreement, together with changes in the species concepts of taxonomists, is leading to an unprecedented level of taxonomic instability. In a sense, this is perhaps less of a problem than it at first appears, since almost everyone in the debate agrees that species evolve from populations within species, and that speciation is liable to be gradual: there will thus always be intermediate stages that are hard to classify. But in some cases, disagreements spill over into human affairs and cause practical problems. This is especially true in conservation (Isaac *et al.* 2004; Meiri and Mace 2007).

My own involvement in this debate dates from 1995, when I saw that the roots of the controversy might lie in misinterpretations of Darwin's and the early Darwinians' concept of species of the latter part of the nineteenth century (Mallet 1995). The standard view among evolutionists, which in modern form appears to originate with Ernst Mayr (1942), was that Darwin was confused about species, and that this led to an inability on his part to properly formulate a theory of speciation (Mallet 2008a). Doubting this perceived weakness in Darwinian theory, I argued (Mallet 1995) that modern genetic data were leading us in exactly the opposite direction, towards a "genotypic cluster" definition of species close to Darwin's view of species as morphological clusters of similar individuals, rather than as reproductively isolated communities as in the Mayr view. The reader of this chapter should be advised that my views have not yet been widely accepted. Nonetheless, I do not believe that appreciating the historical conclusions of this chapter depends on the reader adopting my own very pro-Darwinian view of species.

Here, I outline the contribution of Alfred Russel Wallace and that of other Darwinians to the understanding of the nature of species, and their evolution. In particular, I summarize their views about the importance of "reproductive isolation." Since 1995, I have gradually been collecting information on how early Darwinians viewed species. These readings lead me to believe that all of the most important Darwinian evolutionists soon after 1859 carefully read and generally agreed with views expressed about the evolution of species in the *Origin*, often corresponded with Darwin, and appeared to close ranks on a common viewpoint. By 1889, when Darwinism was under a number of threats, Wallace reiterated his support for the Darwinian point of view on species in his book *Darwinism* (S724), which remains today probably the most complete statement of the early Darwinian position. I believe that Wallace's clearly enunciated statements negate the traditional view of the Darwinian species concept as poorly developed and inadequate for the purpose of studying evolution (Mallet 2004; Mallet 2008a).

Wallace and Bates on the Nature of Species *c.*1845

Wallace and Henry Walter Bates (1825–92) had corresponded on the matter of species before their joint trip to Amazon. Wallace first ran into Bates in Leicester, where the younger man sparked his interest in natural history collecting, and beetles in particular. Wallace had already been reading natural history books, but around the time he met Bates he read the anonymously penned *Vestiges of the Natural History of Creation* (Chambers 1844), and was particularly impressed. He wrote to Bates about it (McKinney 1969). A little later, Wallace seems to have decided that the evolution of species should be his life work, and suggested to Bates that they travel together to a tropical location, in order to study natural history and the species question, financing the trip by selling specimens to collectors:

> In the autumn of 1847 Mr. A.R. Wallace, who has since acquired wide fame in connection with the Darwinian theory of Natural Selection, proposed to me a joint expedition to the river Amazons, for the purpose of exploring the Natural History of its banks; the plan being to make for ourselves a collection of objects, dispose of the duplicates in London to pay expenses, and gather facts, as Mr. Wallace expressed it in one of his letters, "towards solving the problem of the origin of species," a subject on which we had conversed and corresponded much together (Bates 1863, 1:iii).

Writing to Bates in the Amazon several years later in January 1858 from the Malay Archipelago, Wallace discussed Darwin's interest in the origin of species question:

> I have been much gratified by a letter from Darwin, in which he says that he agrees with "almost every word" of my paper. He is now preparing his great work on "Species and Varieties," for which he has been collecting materials twenty years. He may save me the trouble of writing more on my hypothesis, by proving that there is no difference in nature between the origin of species and of varieties; or he may give me trouble by arriving at another conclusion; but, at all events, his facts will be given for me to work upon. Your collections and my own will furnish most valuable material to illustrate and prove the universal application of the hypothesis (S729 1905a, 1:358).

Darwin's gratifying letter had been in response to Wallace's own earlier "Sarawak Law" paper on species (S20 1855): the exchange of letters took place just before Wallace's famous bout of fever on the island of Gilolo, in which he suddenly had the revelation of natural selection (Raby 2001). As is now well known, this led to a quick letter and paper dashed off to Darwin explaining his theory, and the ultimate joint publication of both Wallace's and Darwin's ideas on natural selection (Darwin and Wallace 1858).

Darwin's View of Species in *The Origin*

Elsewhere, I have documented in great detail Darwin's (1859) view of species, and have also attempted to refute the idea that Darwin was confused about species (Mallet 2008c), as is commonly believed even today. In brief, Darwin strongly believed that species came to be well defined after a long evolutionary period of divergent evolution:

> To sum up, I believe that species come to be tolerably well-defined objects, and do not at any one period present an inextricable chaos of varying and intermediate links ...
>
> ... if my theory be true, numberless intermediate varieties, linking most closely all the species of the same group together, must assuredly have existed; but the very process of natural selection constantly tends ... to exterminate the parent-forms and the intermediate links (Darwin 1859, 177, 179).

Species thus differed from varieties only in that varieties were still connected together, whereas species did not blend into one another, but were separated by gaps in the distribution of morphologies. At the same time, he argued that species were not "real" in the sense that they did not differ "essentially" from varieties and geographic forms below the species level. The rank at which we define the word "species" is up to us.

> ... it will be seen that I look at the term species, as one arbitrarily given for the sake of convenience to a set of individuals closely resembling each other, and that it does not essentially differ from the term variety, which is given to less distinct and more fluctuating forms.

> In short, we shall have to treat species in the same manner as those naturalists treat genera, who admit that genera are merely artificial combinations made for convenience. This may not be a cheering prospect; but we shall at least be freed from the vain search for the undiscovered and undiscoverable essence of the term species (Darwin 1859, 52, 485).

He recognized that because of the continuous nature of evolution, and the lack of an appropriate essence of species, there would always be difficulties in defining species in the early stages of divergence.

> But cases of great difficulty, which I will not here enumerate, sometimes occur in deciding whether or not to rank one form as a variety of another ... Hence, in determining whether a form should be ranked as a species or a variety, the opinion of naturalists having sound judgment and wide experience seems the only guide to follow (Darwin 1859, 47).

Although "essentially" and "essence" are words that can be used in colloquial English in a somewhat imprecise way, it seems to me clear that Darwin was using these words advisedly; he was evidently referring to species essentialism in a strict, Aristotelian sense, and rejecting it. He knew exactly what he meant by species (summed up in the word "gaps"), but he specifically argued that there is no single "essence" of species true in all cases; in particular he argued in great detail in his chapter "Hybridism" against the idea that species were always isolated by hybrid inviability or sterility. In so doing, he rejected an older "reproductive isolation" notion of species dating back to Ray and Buffon. This view must have been as commonly accepted in his day, and earlier, as it was from the 1940s onwards. Although he did not deny that many species were intersterile, Darwin (1859) in his chapter "Hybridism" strongly argued against sterility of hybrids between forms as a *sine qua non* of species status. The major arguments he used were: that some pairs of species occurring together were largely interfertile; that populations of plants within species were often intersterile; that fertility of hybrids between a particular pair of species varied depending on the populations used in the crosses; and that infertility in one direction of cross (*e.g.* male of species A x female of species B) was accompanied by fertility in the other (male of B x female of A). Darwin argued that post-mating reproductive isolation arose largely as a by-product of changes after separation of the two species, rather than being itself a useful definition of species.

Henry Walter Bates' View of Species

Bates is today most famous for his natural selection-based theory of mimicry, whereby the colour pattern of one species converges for predator defence on the colour pattern of another that is defended against predators. Curiously, in 1860 Wallace had already written to Darwin about the same phenomenon of mimicry in butterflies:

> P.S. "Natural Selection" explains *almost* everything in Nature, but there is one class of phenomena I cannot bring under it,—the repetition of the forms & colours of animals in distinct groups, but the two always occurring in the same country & generally on the *very same spot.* These are most striking in insects, & I am constantly meeting with fresh instances. Moths resemble butterflies of the same country—*Papilios* in the east resemble *Euplœas,* in America *Heliconias* (Darwin Correspondence Project, letter 2627).

Ironically, his friend Henry Walter Bates was to discover the theory that could explain such resemblances, by means of natural selection, the very next year (1861), and publish it a year later (Bates 1862). Wallace had in a sense lost out again! Yet Wallace was as fair with Bates as he was with his admiration of Darwin; he was effusively complimentary about Bates' new theory (S96 1865). Bates had argued that rare species, palatable to predators, gain an advantage in nature if their colour patterns are similar to much commoner, unpalatable species. Predators, such as birds, learn the patterns of the common species, and rarer species without such protection benefit if they have the same colour patterns. Bates' theory chiefly argued that palatable species (dismorphiine pierids) mimicked unpalatable species (ithomiines in the genus *Melinaea* and *Mechanitis*). However, Bates also intuited that rare unpalatable species (for example, in the butterfly genus *Heliconius* (Heliconiinae) benefited by mimicking more common unpalatable species (*e.g. Melinaea*). Thus, Bates was the first to promote the idea that later became known as Müllerian mimicry, after Müller's (1879) paper. Müller's main, and indeed considerable achievement was to develop a mathematical theory to explain why it was mutually advantageous for unpalatable species to mimic one other, and estimate the relative advantage to each (see Chapter 8).

However, Bates' paper was not merely about mimicry. It was largely a systematic treatise, incorporating a somewhat understated theory of speciation by natural selection. Mimicry in particular played a major role in speciation, in Bates' view. To underpin the treatise as well as to define species from among the bewildering array of forms he found among the Amazonian butterflies, Bates had in mind a Darwinian definition of species. In his systematic discussion of ithomiines of the genus *Mechanitis* Bates describes divergent forms living together, but not intergrading, and views them as separate species:

> The new species cannot be proved to be established as such, unless it be found in company with a sister form which has had a similar origin, and maintaining itself perfectly distinct from it. Cases of two extreme varieties of a species being thus brought into contact by redistribution or migration, and not amalgamating, will be found to be numerous (Bates 1862, 530).

In this view, he is closely following Darwin. Although reproductive isolation is clearly an important mechanism of species maintenance, it is the lack of intergradation or intermediacy in the actual specimens he has collected in the wild which leads Bates to characterize these forms as separate species. Bates argued that mimicry was an example of the kind of natural selection that can explain the origin of new species.

Wallace's 1865 Paper "On the Phenomena of Variation and Geographical Distribution as Illustrated by the Papilionidae of the Malayan Region"

In this paper, I think Wallace is the first to lay out the Darwinian definition of species and apply it to geographic and non-geographic contexts clearly. The problem was to define species as distinct from geographical and local varieties, and Wallace, after his travels on the Amazon and in the Malay Archipelago, had unrivalled experience on which to base his opinions:

> What is commonly called variation consists of several distinct phenomena which have been too often confounded. I shall proceed to consider these under the heads of— ... 1. *simple variability* [probably equivalent to quantitative variation] ... 2. *polymorphism or dimorphism* [discrete forms separated by morphological gaps, which nonetheless belong to the same species] ... 3. *local form, or variety* [gradually varying forms differing from place to place] ... 4. *coexisting variety* ... a somewhat doubtful case [reserved for coexisting forms which differ in very few constant characters, but which may be separate species; "sibling species" would perhaps be the modern equivalent] ... 5. *race, or subspecies* ... 6. [true] species (S96 1865, 5–14).

I have argued elsewhere that this is the forerunner of similar and highly influential classifications of geographic and non-geographic variation by the evolutionists E. B. Poulton, Karl Jordan, and Ernst Mayr (Mallet 2004). Wallace gives his definition of species thus: "Species are merely those strongly marked races or local forms which, when in contact, do not intermix, and when inhabiting distinct areas are generally believed to have had a separate origin, and to be incapable of producing a fertile hybrid offspring" (S96, 12). This statement so far approximates the pre-Darwinian understanding of species, but as we shall see is followed by a partial rebuttal. As a Darwinian, Wallace instead believes that all species derive from one another and do not, in fact, have a separate origin. He argues also that the use of the sterility of hybrids as a species definition is tautological. Although he doesn't define here exactly what he means by a species, he clearly agrees with Darwin's view that species cannot easily be distinguished from varieties. He goes on:

> But as the test of hybridity cannot be applied in one case in ten thousand, and even if it could be applied, would prove nothing, since it is founded on an assumption of the very question to be decided—and as the test of separate origin is in every case inapplicable—and as, further, the test of non-intermixture is useless, except in those rare cases where the most closely

> allied species are found inhabiting the same area, it will be evident that we have no means whatever of distinguishing so-called "true species" from the several modes of variation here pointed out, and into which they so often pass by an insensible gradation (S96, 12).

Wallace has already introduced a major difficulty: how to define species when distinct populations are found on different islands:

> The rule ... I have endeavoured to adopt is, that when the difference between two forms inhabiting separate areas seems quite constant, when it can be defined in words, and when it is not confined to a single peculiarity only, I have considered such forms to be species. When, however, the individuals of each locality vary among themselves, so as to cause the distinctions between the two forms to become inconsiderable and indefinite ... I class one of the forms as a variety of the other (S96, 4).

Wallace here attacks the problem of geographic variation, a complication which has blocked agreement on the definition of species even today. There are still "splitters" who would argue that every geographic form with a fixed difference should be defined as separate species, and "lumpers" who in contrast argue that such forms should as far as possible be defined as subspecies within much more widely-distributed species (Isaac *et al.* 2004).

Wallace was also among the first to appreciate that Darwin's idea of species being morphologically different from one other is problematic for a group of special cases, such as mimetic butterflies. *Papilio memnon* males appear to be different species from the females, as they have entirely different colour patterns, and the females themselves are polymorphic, some with tails and some without, each one mimicking a different species of poisonous Papilionidae. Wallace was able to apply Bates' mimicry theory to this situation, citing data showing forms that were reared from eggs laid by mothers with different colour patterns than their own. He gives a graphic illustration of how extraordinary it is that all these markedly different forms of *Papilio memnon* belong to the same species:

> The phenomena of *dimorphism* and *polymorphism* may be well illustrated by supposing that a blue-eyed, flaxen-haired Saxon man had two wives, one a black-haired, red-skinned Indian squaw, the other a woolly-headed, sooty-skinned negress—and that instead of the children being mulattoes of brown or dusky tints ... all the boys should be pure Saxon boys like their father, while the girls should altogether resemble their mothers. ... yet the phenomena ... in the insect-world are still more extraordinary; for each mother is capable not only of producing male offspring like the father, and female like herself, but also of producing other females exactly like her fellow-wife, and altogether differing from herself (S96, 10 n.).

Clearly, these forms have morphological gaps between them, but Wallace does not in any way view them as equivalent to species. In this paper, Wallace built up

Figure 16 Two female morphs of *Papilio memnon.*
From S715. Out of copyright.

perhaps the most complete theory of species and speciation put forward in the early days of Darwinism. His theory was based both on his novel biogeographic ideas, and his knowledge of many details of local natural history, variation, polymorphism, and evidence for natural selection. In the taxonomic sections of this same paper, he tends to assign geographic races as separate species rather more than we would today, but he admits this quite frankly, feeling that the important geographic subspecies in many *Papilios* across the archipelago of South East Asia would be lost if he did not name them as taxonomic species (Mallet 2008b). At this time, it would have been possible to name a local form as a variety (usually written as "var."), but this risked confusion of strongly-marked, geographically informative subspecies with trivial local sports and variants; the subspecies had not yet formally been accepted in the existing trinomial system of nomenclature developed in the period 1890–1910 by ornithologists and lepidopterists, among them David Starr Jordan, Karl Jordan, and Walter Rothschild (Stresemann 1975; Rothschild 1983; Mallet 2004). Essentially, Wallace, in 1865, had laid out the full understanding of what later came to be known as the polytypic or "biological" species concept, even though he had, with Darwin, rejected too great a dependency for his own concept on reproductive isolation. It was no accident that K. Jordan and E. B. Poulton (a good friend of Wallace's in the latter's old age), both acknowledged pioneers of the biological species concept that Mayr (1942) later adopted, were also both experts on the Papilionidae. They had read and absorbed Wallace's paper on the topic (Mallet 2004).

Benjamin D. Walsh in the USA in the 1860s

Walsh was a correspondent of Darwin's, and one of the earliest Americans to appreciate fully and apply the Darwin-Wallace theory of species and speciation, in his case to insects. In over ninety pages, Walsh argues with great verve and verbosity for Darwin's idea of species:

> The only valid practical criterion of specific distinctness is the general non-existence ... of intermediate grades in the distinctive characters, whence we may reasonably conclude that the two supposed species are distinct, i.e. that they do not now in general mix sexually together, or if geographically separated that they would not do so supposing them to be placed in juxtaposition. ... They may even now mix sexually together in some few rare instances [*i.e.*, hybridization between species] and yet if they do not commonly and habitually mix together the species will remain distinct. Hence all experiments on artificial hybridization seem to me to prove nothing as to the distinctness of species unless they are conducted, as they necessarily cannot be, on the same gigantic scale as that upon which Nature works. ... Immediately we assume any other criterion of specific distinctness than the general non-existence in a state of nature of the intermediate grades, either proved by actually examining numerous specimens or inferred from the analogy of allied species, all is chaos and confusion ... (Walsh 1863, 220).

On p. 221 he continues: "I am not ignorant of the existence in the Vegetable Kingdom of what are called Dimorphous species ..."

There follows a long list of polymorphisms and dimorphisms without intermediates in insects. Examples include: neuters in social insects, gynandromorphs in *Dytiscus, Papilio,* and *Colias,* orange females of the dragonfly *Agrion ramburii* [= *Ischnura ramburii*], dimorphism of horns in male *Siagonium* beetles and allies, brachyptery/macroptery in Orthoptera, Hymenoptera, Heteroptera, Homoptera; agamous species of dimorphic forms of a sexual species, trimorphic heterostyly in plants, soldier castes in *Atta* ants and in termites, forms of females in the butterfly *Vanessa interrogationis* [= *Polygonia interrogationis*]. However,

> In the meantime, the general non-existence of intermediate grades between two closely-allied forms may and must be taken as prima facie evidence of their specific distinctness. That "the exception proves the rule" is an old and not very philosophical saying; but that there are exceptions to almost all rules in Natural History is undoubtedly true. Monomorphism is the rule; Dimorphism is the exception (Walsh 1863, 221).

This article was cited by Wallace in his 1865 article, particularly with respect to the case of *Papilio memnon* as described above. Walsh had discovered in Illinois a similar case of sexual dimorphism and female-limited polymorphism in what is now *Papilio glaucus,* which has some females that are black, mimics of *Battus philenor,* and some yellow, and non-mimetic, like the male. The very next year, 1864, Walsh addressed the extraordinary case of the host races of *Rhagoletis pomonella,* and again came up with a reproductive isolation definition of species and mechanism of speciation, similar to Poulton's (Berlocher and Feder 2002).

Wallace's Views on Species in Later Life

Wallace differed with Darwin on a number of issues, such as the evolution of man and sexual selection by female choice (S729 1905), but he never seems to have had strong disagreements with Darwin on species. In 1868, he wrote to Darwin with a suggestion that hybrid sterility might be explained via natural selection (Darwin Correspondence Project, letter 5966). This was perhaps in response to the apparent problem T. H. Huxley had raised: that in order for natural selection to be a complete theory of the origin of species, it must also explain hybrid sterility. Darwin replied to Wallace arguing that he could not entertain this idea (Darwin Correspondence Project, letter 6033): natural selection can never act to reduce fertility of individuals, even if it may eventually be advantageous to species divergence to lack gene flow. As sterility was not a necessary or sufficient characteristic of species, its explanation was not crucial to the theory of the origin of species by natural selection. Wallace seems to have accepted this, and did not publish his theory. Twenty-one years later he described the episode in print (S724 1889).

Wallace also demolished a later theory to explain hybrid sterility between species by natural selection, a phenomenon called "Physiological Selection" (Romanes 1886), by means of a numerical argument (S389 1886, S724 1889). George Romanes had been an earnest and devoted disciple of Darwin's, but as Wallace points out, argued by assertion rather than marshalling any facts in support of his ideas. In *Darwinism* Wallace includes a long chapter "On the Infertility of Crosses ..." (S724, 152–86) in which he is in complete agreement with Darwin's argument that sterility is a by-product of evolution, rather than an "intended" consequence of natural selection. Nonetheless, Wallace himself still seems dissatisfied, and produces a somewhat rambling, five-page theory of his own to explain sterility; yet, as is typical of Wallace's honesty, recognizing his theory's tortuousness, he attempts a brief summary of his argument, in footnotes lasting a good three pages ("As this argument is a rather difficult one to follow ... I add here the following briefer exposition ...", pp. 179–81). The important additional ideas he lays out here are that, provided divergence takes place in different environments, natural selection for ecological divergence may exceed the power of natural selection to prevent the evolution of intersterility, and that selection for a "disinclination to crossed unions" may occur.

> The constant preference of animals for their like, even in the case of slightly different varieties of the same species, is evidently a fact of great importance in considering the origin of species by natural selection, since it shows us that, so soon as a slight differentiation of form or colour has been effected, isolation will at once arise by the selective association of the animals themselves ... (S724, 172–73).

Wallace is here proposing what later became known as "reinforcement," an idea now generally attributed to Dobzhansky (1940). This is somewhat surprising, as the idea has been dubbed the "Wallace effect" by Grant (1966) and Murray (1972). Recent evidence has abundantly proved that the idea is correct, although we don't yet know how common it is in a state of nature (Coyne and Orr 2004; see also Chapter 6).

Much later, in 1900, Wallace was involved in correspondence about species concepts with Henry Bernard, reprinted in Cock (1977). In his letter of reply, late in life, and this time with the aid of a diagram, Wallace again expounds his (and Darwin's) theory of speciation via extinction of intermediates. Interestingly, Mayr (1982) chooses this single, casual, and unpublished example to demonstrate Wallace's primitive and "typological" lack of understanding of species. Mayr apparently does not realize that his own geographic, polytypic, biological species concept stems ultimately from Wallace's original work with Papilionid butterflies, as well as his vast knowledge of birds of South East Asia. I think one can easily argue, based on the 1865 monograph and 1889 book, that Wallace knew exactly what he was talking about: Mayr, who cut his own teeth as a bird collector on a much shorter trip to New Guinea, simply does not do Wallace justice.

Conclusion

Wallace, in 1865 and 1889, developed and supported perhaps the clearest conceptualization of species of all the early Darwinians. In this, he did not in the slightest bit deviate from Darwin's own conception of species, although he clarified and greatly extended its geographic scope. He had always found Darwin's arguments on this topic both convincing and worthy of admiration. Even thirty years after the *Origin*, Wallace sides with him:

> Generally speaking, it may be said that the varieties of any one species, however different they may be in external appearance, are perfectly fertile when crossed, and their mongrel offspring are equally fertile when bred among themselves; while distinct species, on the other hand, however closely they may resemble each other externally, are usually infertile when crossed, and their hybrid offspring absolutely sterile. This used to be considered a fixed law of nature ... [however] ... The elaborate and careful examination of the whole subject by Mr. Darwin, who has brought together a vast mass of evidence from the experience of agriculturists and horticulturists, as well as from scientific experimenters, has demonstrated that there is no such fixed law in nature as was formerly supposed (S724, 152–53).

Species were, to Darwin, Wallace, Bates, and Walsh exactly the same kinds of things as varieties, differing only in the presence of morphological gaps between them. To erect a theory of the transmutation of species, they had to reject the old creationist idea that species were intersterile, while varieties within species were interfertile. They didn't deny that hybrid sterility was a tendency to which species are prone, but there are exceptions to this rule, and the very odd laws of sterility rule out hybrid sterility as a good definition. Wallace in particular enunciated a clear species concept that combined Darwinism with his knowledge of the geography of biodiversity in South America and South East Asia (S96 1865). He saw the logic in Darwin's stance on species, and stuck with it. Furthermore, Wallace's extension of the Darwinian species concept to broader geographic regions, far from being superseded by the polytypic or biological species concept of Mayr in the 1940s, in fact forms a clear forerunner of the geographic parts of that idea.

Subsequent generations of evolutionary biologists ignored these subtleties, and eventually, by the 1960s accepted a new species concept, based on the very essence of reproductive isolation that Darwin and Wallace had recommended discarding. These post-World War II ideas weren't so much wrong, as lacking the depth already explored by the early Darwinists. By the 1980s, species concepts were again becoming a battleground for evolutionary biologists, with the rise of the phylogenetic species concepts. Now we seem again to be on the verge of entering a new age of enlightenment about the complexity of species and speciation. Is the wheel of ideas ready to turn again?

6

Direct Selection for Reproductive Isolation: The Wallace Effect and Reinforcement*

Norman A. Johnson

Introduction

The similarities between Alfred Wallace and Charles Darwin, the pioneering thinkers who separately formulated theories of evolution by natural selection, are striking. Both were ardent students of physical and biological nature whose extensive studies in remote locales left indelible marks on them. Independently influenced by the writings of the economist Thomas Malthus, Wallace and Darwin each recognized that animals and plants were engaged in a "struggle for existence," an insight both men viewed as critical for the formulation of their theories of evolution (Browne 1995, 2002; Shermer 2002; Smith 2003–).

These parallels, however, can sometimes detract from our appreciation of the important differences that also exist between Darwinian and Wallacean evolution. For one, Wallace broke with Darwin over the importance of the female choice aspect of sexual selection (see Cronin 1991; Shermer 2002). Four decades after the original publication of the Wallace/Darwin's theory, Wallace noted, "Sexual selection resulting from the fighting of males is indisputable, but, differing from Darwin, I do not believe there is any selection through the choice of the females, and the drift of scientific opinion is towards my view" (S738 1898, 121). At the time, Wallace was correct; scientific opinion was drifting toward Wallace's point of view. Not until the last third of the twentieth century would sexual selection via female choice become a subject of high interest in evolutionary biology (Cronin 1991; Andersson 1994; Birkhead 2000). Darwin and Wallace also differed in their views on the relevance of artificial selection and domesticated organisms to evolution in the wild. In the *Origin of Species*, Darwin begins his "one long argument" about

* I thank Charles Smith for inviting my participation in this book, and for his patience. Mohamed Noor, Ben Normark, and Michael Wade provided many thoughtful comments on an early draft of this chapter.

evolution by natural selection with analogies to breeding, especially in pigeons. In contrast, Wallace was skeptical about the inferences from domesticated animals and plants that might be applied to organisms in nature (see Shermer 2002, 116).

Another substantial difference between Wallace and Darwin involved whether hybrid sterility can evolve via the direct action of natural selection. That is, can hybrid sterility *per se* ever be adaptive? Darwin and Wallace agreed that in the initial stages of divergence, whatever hybrid sterility that evolved between incipient species would owe not to direct selection for the partial sterility of the hybrids, but instead arise as an incidental by-product of other kinds of character divergence. Darwin thought all hybrid sterility and inviability were by-products. In contrast, Wallace viewed selection as being able in some circumstances to increase directly the sterility of already partially sterile hybrids. Moreover, Wallace saw himself extending natural selection into an area where Darwin had not gone. Late in life and long after Darwin's death, Wallace, while discussing his contributions to evolution by natural selection, listed "maintaining the power of natural selection to increase the infertility of hybrid unions" (S729 1905, cited by Shermer 2002, 149) as an important extension of his to Darwinism.

In general, evolutionary biologists after Darwin and Wallace have not found much evidence that hybrid sterility or hybrid inviability is directly adaptive. In contrast, the increase of mating discrimination occurring between two incipient species (pre-mating reproductive isolation) through the direct action of selection has been an active topic in evolutionary studies both during and after the formulation of the modern synthetic theory of evolution. This latter process is frequently known as reinforcement. Verne Grant (1966), however, suggested using the term "the Wallace effect" to describe all cases of reproductive isolation being driven by the direct action of selection, and not as incidental by-product of other character divergence. In this chapter, I will distinguish between reinforcement as the direct selection for premating reproductive isolation and the Wallace effect for direct selection for post-mating reproductive isolation.

Reproductive isolating barriers are traditionally divided into two categories depending on whether they occur before or after mating: hence, the use of the terms pre-mating and post-mating reproductive isolating barriers (Dobzhansky 1937; Mayr 1963). Sometimes barriers are classified into pre-zygotic and post-zygotic categories, depending on whether they occur before or after zygote formation (see section on conspecific gamete preference). I prefer using the term "barriers" instead of "mechanisms" in describing reproductive isolation because the former does not carry teleological connotations.

As Cronin (1991) has extensively reviewed the debate between Wallace and Darwin over the possible adaptive significance of hybrid sterility and inviability, I will only briefly highlight the rationales for their arguments. I next summarize the history of the early studies of reinforcement and the fluctuations in its acceptance over the last decades of the twentieth century. Following the summary of reinforcement, I consider verbal models and theoretical treatments of the

Wallace effect. I close with a discussion of how the Wallace effect might operate on conspecific gamete precedence, a form of reproductive isolation that occurs after mating but before zygote formation, and suggest avenues for further study.

Wallace and Darwin's Views on Hybrid Sterility

> The view generally entertained by naturalists is that species, when intercrossed, have been specially endowed with the quality of sterility, in order to prevent the confusion of all organic forms. This view certainly seems at first highly probable, for species in the same country could hardly have been kept distinct had they been capable of crossing freely. The importance of the fact that hybrids are very generally sterile has, I think, been much underrated by some later writers. On the theory of natural selection the case is especially important, inasmuch as the sterility of hybrids could not possibly be of any advantage to them, and therefore could not have been acquired by the continual preservation of successive profitable degrees of sterility. I hope, however, to be able to show that sterility is not a specially acquired or endowed quality, but is incidental on other acquired differences (Darwin 1985, 264).

Most contemporary biologists view hybrid sterility and hybrid inviability as a reproductive isolating barrier, and thus as potentially playing a role in speciation. Darwin and Wallace, however, viewed hybrid sterility from a different perspective.

For Darwin, hybrid sterility seemed a challenge to his theory of evolution by natural selection. How could hybrid sterility be adaptive, and how could it evolve by the small, successive steps that Darwin viewed as critical for this theory? As a breeder, Darwin was struck by the frequency with which changes occur in traits that are not favored by the breeder. Again and again in the *Origin of Species*, Darwin stresses the role that correlated responses (what he called "the laws of correlation") play in evolution, and that traits can evolve not by the direct action of natural selection alone. In fact, it seems that correlated responses are Darwin's default explanation when he cannot explain a trait via direct natural selection. Darwin does seem prescient; modern evolutionary genetics has shown that correlated responses do play a major role in evolution (Futuyma 1998) and that much of the accumulation of hybrid sterility between species arises from the incidental effects of other divergence (Dobzhansky 1937; Mayr 1963; Johnson 2000; Coyne and Orr 2004).[1]

In his arguments against special creation, Darwin (1859) emphasized that no strict line separated varieties from species (Mayr 1982; Cronin 1991; Shermer 2002). Darwin repeatedly mentioned partial sterility in hybrids, and the fact that hybrids of reciprocal crosses often vary in degree of sterility. To Darwin, species were just another step in the continuum of divergence.

Wallace, on the other hand, was interested in extending the scope of natural selection. He had corresponded with Darwin during the 1860s on the possibility of hybrid sterility being adaptive, but had put the matter down for a number of years. By the time Wallace returned to this issue in the 1880s, the fact of evolution was no longer a controversy within the scientific community (see Huxley 1880 for a synopsis of the state of evolution at the time). In contrast, the scope and power of natural selection was still a matter of contention (see Gould 2002).

Wallace agreed with Darwin that the initial partial hybrid sterility would evolve as an incidental by-product. He argued that after partial sterility, natural selection could increase the degree of sterility for species that lived in proximity to one another. In his book *Darwinism*, Wallace presents the argument:

> The simplest case to consider, will be that in which two forms or varieties of a species, occupying an extensive area, are in process of adaptation to somewhat different modes of life within the same area. If these two forms freely intercross with each other, and produce mongrel offspring which are quite fertile *inter se*, then the further differentiation of the forms into two distinct species will be retarded, or perhaps entirely prevented; for the offspring of the crossed unions will be, perhaps, more vigorous on account of the cross, although less perfectly adapted to the conditions of existence than either of the pure breeds; and this would certainly establish a powerful antagonistic influence to the further differentiation of the two forms (S724 1889, 174).

Wallace's argument is difficult to follow, and several authors (*e.g.*, Kottler 1985;[2] MacNair 1987) have chastised Wallace for using naïve group selection arguments to explain the evolution of further hybrid sterility via direct natural selection. It is true that Wallace often appealed to the benefit of the species, but it is not clear whether he was viewing hybrid sterility as a species-level or individual-level adaptation. As Cronin (1991) points out, "good for the species" language pervades both Wallace's and Darwin's writings, and may be a short cut in much the same way that contemporary biologists talk about the desires of organisms without implying they are conscious.

Reinforcement

> The grossest blunder in sexual preference, which we can conceive of an animal making, would be to mate with a species different from its own and with which the hybrids are either infertile or, through the mixture of instincts and other attributes appropriate to different courses of life, at so serious a disadvantage as to leave no descendants (Fisher 1930, 130).

Unlike most evolutionary biologists of his era, Sir Ronald Fisher took female mating preferences and their evolution seriously. Fisher (1930) recognized the existence of selective pressures operating on individual females to avoid committing this grossest blunder, and proposed that selection would favor the evolution of female discrimination against males of different species. Thus, pre-mating reproductive isolation could evolve via direct natural selection. This process is currently most commonly known as "reinforcement," although that term wasn't coined until Blair's (1955) study of mating-call evolution in frogs.[3]

The history of reinforcement studies is too vast to detail fully in this chapter (for reviews, see Butlin 1989; Howard 1993; Noor 1999; Marshall *et al.* 2002; Servedio and Noor 2003; Coyne and Orr 2004). Complicating this history are the wild swings in acceptance of reinforcement during the twentieth century that Noor (1999) has likened to stock market fluctuations. A brief historical sketch of this subject, however, is instructive for understanding both reinforcement and the Wallace effect.

Reinforcement would find a champion in the Russian-born American evolutionary biologist Theodosius Dobzhansky, who was struck by the patterns he observed of geographical variation within species for mating preferences with another. Specifically, Dobzhansky (1940) found that *Drosophila pseudoobscura* females from populations that come into contact with the related species *D. miranda* show higher mating discrimination against *D. miranda* than do those from populations that are outside the range of *D. miranda.* He noted, "Any gene that raises an effective barrier to the mingling of incipient species is adaptively valuable, and hence may become the basis of speciation" (Dobzhansky 1940, 320). With time, reinforcement took on increasing importance in Dobzhansky's views on speciation to the extent that he seemed to think that reinforcement was virtually ubiquitous. Given the immense influence that Dobzhansky—through his books, papers, and students—had on the field, it is not surprising that reinforcement was in vogue during the middle of the twentieth century.

After Dobzhansky's death in 1975, reinforcement as a topic for evolutionary studies waned. During the 1970s and 1980s, evolutionary biologists turned increasingly skeptical about its importance, and some even began questioning its likelihood. This "bear market" for reinforcement arose primarily from the accumulation of objections based on verbal scenarios and theoretical models, and not from new empirical studies (see Butlin 1989; Coyne and Orr 2004, 353). One of the main objections against reinforcement is that in each generation recombination erodes beneficial genetic associations between the mating preference alleles, the alleles for the male trait, and the alleles involved in the fitness of hybrids. Unless selection is very strong, the preference loci and the trait loci would have to be tightly linked in order for reinforcement to operate. This argument derives from a highly influential theoretical paper by Joseph Felsenstein (1981) on non-allopatric speciation. Although Felsenstein did not explicitly consider reinforcement in his original paper, his point about the antagonism between recombination and selection influenced subsequent reinforcement models of the 1980s.

When I started graduate school in 1987, a nadir in the acceptance of reinforcement among evolutionary biologists had been reached. That October, I attended a symposium on speciation hosted by the Academy of Natural Sciences in Philadelphia. At this meeting Jerry Coyne presented work he and Allen Orr had carried out to systematically investigate patterns of reproductive isolation in the genus *Drosophila.* Their results were based on a massive compilation of data on both measures of the genetic distance between species (taken from frequencies of protein variants), and indices of their pre-mating and post-mating reproductive isolation. One surprising finding from Coyne and Orr's work was that many pairs of recently diverged pairs of species (as measured by genetic distance) with currently overlapping distributions had high levels of pre-mating reproductive isolation, while none of the comparably recently diverged species whose distributions did not overlap had high levels of pre-mating reproductive isolation. No such pattern was observed for post-mating reproductive isolation. Jerry Coyne, who said that he was surprised himself when he saw the results, offered reinforcement as the best explanation for the observed pattern.[4]

Coyne and Orr's (1989) work, with others', led to a renaissance for reinforcement studies during the 1990s. Case studies from a variety of organisms and meta-analyses like Coyne and Orr's suggested that reinforcement, or something similar, was occurring in nature (reviewed in Noor 1999; Marshall *et al.* 2002; Noor and Servedio 2003; Coyne and Orr 2004). In their book *Speciation*, Coyne and Orr (2004, 354) stated: "By about 1990, reinforcement once again became popular, reflecting the emergence of new data revealing that *something* interesting was happening in sympatry and that, at the least, this something resembled reinforcement."

As Coyne and Orr note, these new data on reinforcement led to a rethinking of reinforcement by a new generation of theorists (see Noor and Servedio 2003). These new theoretical studies showed that far from being unlikely, reinforcement of pre-mating isolation might be easy to evolve. One difference between the old and the new models is that many of the newer models are based on female discrimination, not female preference (see Kelly and Noor 1996). Instead of the female's actual "tastes" changing, her level of discrimination changes. Such models are referred to as "one allele" models, as opposed to the "two allele" models in which female tastes actually change with the male trait. Reinforcement is easier to evolve in the "one allele" models because the antagonism between recombination and selection that Felsenstein (1981) identified is no longer an impediment. The extent to which reinforcement operates via a "one allele" versus a "two allele" process in nature is not yet known, but one study (Oritiz-Barrientos and Noor 2005) provides some support for "one allele" reinforcement between closely related species of *Drosophila.*

The "Wallace Effect"

> Isolating mechanisms which block or restrict interbreeding between species can originate in two ways: as by-products of evolutionary divergence, and as products of selection for reproductive isolation per se (Grant 1966, 99).

In his 1966 paper, Verne Grant dubbed the latter process of direct selection for reproductive isolation the "Wallace effect." In Grant's usage, the Wallace effect includes selection for pre-mating isolation (reinforcement) in addition to selection for post-mating reproductive isolation. He cites the literature on what was then known as reinforcement (*e.g.*, Fisher 1930; Dobzhansky 1940; Blair 1955) as examples of the Wallace effect. Here, however, I will consider only direct selection for post-mating reproductive isolation to be the Wallace effect, reserving reinforcement as the name for pre-mating reproductive isolation evolving via direct selection.

Grant's interest in direct selection for reproductive isolating barriers was prompted by his extensive studies of these barriers between species of *Gila*, a genus of annual herbaceous plants. In this genus, a combination of pre-fertilization (mating) and post-fertilization barriers prevents the exchange of genes between species. Summarizing decades of study, Grant (1966) presented evidence that both types of these isolating barriers evolve by direct selection, in addition to being by-products of other divergence.

In proposing a mechanism for the direct selection of post-fertilization reproductive barriers, Grant (1966) noted that early in development, lost flowers or ovules could easily be replaced by new flowers. So, if hybrid flowers that are sterile are destroyed early on, the parents can compensate for the lost flowers with fertile flowers produced by pure-species matings. Grant (1966, 104) concluded, "Therefore a post-fertilization block which prevents the formation of mature hybrid seeds would have a high selective advantage over no such block in annual plants." A few years later, Coyne (1974) constructed a scenario for the evolution of hybrid inviability via the Wallace effect that is explicitly based on parental investment. If the costs associated in raising offspring that are sterile reduce the level of investment that the parent can provide for future offspring, then inviability of the sterile hybrids would be favored.

John Maynard Smith (1975) and Bruce Wallace (1988) came up with Wallace effect explanations to explain observations seen in some interspecific crosses of cotton. Ordinarily, crosses between Bourbon cotton (*Gossypium hirsutum var. marie-galante*) and Sea Island cotton (*G. barbadense*) yield sterile hybrids. These sterile hybrids are usually luxuriant, showing what is often known as hybrid vigor. Certain crosses of *G. h. marie-galante* and *G. barbadense*, in contrast, yield dwarf hybrids with corky outgrowth (Stephens 1950). This "corky" phenotype is not observed in pure species, and appears to be the result of alleles at two different complementary (sometimes known as synthetic) loci. Bruce Wallace and others present two lines of evidence that this corky dwarf phenotype is an example of the Wallace effect: first, that the carrier alleles for the synthetic corky-causing loci are fairly common, and second, that these alleles are much more frequently found in areas where the two species of cotton meet in sympatry than when they are in allopatry.

In a textbook, Maynard Smith (1975) proposed a kin-selection argument for the Wallace effect. He noted that the conspecific half-siblings of sterile hybrid cotton could be in competition with the hybrids. Thus, the presence of the hybrids could reduce the fitness of their relatives. Given that the fitness of the hybrids is zero, they would pay no additional cost if a mutation decreased their viability. Maynard Smith noted that this scenario would require the hybrid and the conspecific plants to be living in sufficient proximity to be competing. He added: "If this argument is correct, then selection for hybrid inviability will be more effective in plants and sessile animals with relatively poor dispersive powers, since in such groups the offspring of a given individual compete with one another" (Maynard Smith 1975, 256).

Leibowitz (1994) constructed a formal theoretical treatment of Maynard Smith's argument in the context of Hamilton's (1964) kin selection equation wherein "altruism" would be favored given the product of relatedness (R) times the benefit to the recipients (B) is greater than the cost to the altruist (C). Here, altruism is decreasing the viability of hybrids. In Leibowitz's model, hybrid inviability was able to evolve by the Wallace effect but the conditions were generally restrictive, and the hybrids already had to be largely sterile. The mating system affected the likelihood of this process: hybrid inviability evolved more readily under polygamy than under monogamy (Leibowitz 1994).

A critical assumption of Leibowitz's model is that underdominance at a single locus, where heterozygotes are reduced in fitness, is the basis for hybrid sterility. Both theoretical understandings and empirical data show that hybrid sterility via underdominance is at the least exceedingly unlikely (Johnson 2000; Coyne and Orr 2004). Instead, hybrid sterility is usually due to negative interactions of at least a pair of loci. It isn't clear how readily hybrid inviability would evolve via the Wallace effect in the Leibowitz model with a more realistic genetic basis for hybrid sterility, but as Leibowitz himself noted, increasing the loci involved would tend to diminish the likelihood of the Wallace effect.

Johnson and Wade (1995) also modeled the Wallace effect for hybrid inviability, entertaining a somewhat different set of conditions. Instead of a kin-selection model, their approach, following Wallace (1988), was based on soft selection (a form of frequency and density-dependent selection), and considered density regulation to be operating within families. (Wade [1985] has shown that this form of soft selection bears a formal resemblance to kin selection.) Unlike the other models, Johnson and Wade (1995) also considered cases where the alleles that decrease hybrid inviability also have negative effects in pure-species individuals. Thus, they were testing whether soft selection could overcome individual selection and yield hybrid inviability via the Wallace effect.

Hybrid inviability did evolve by the Wallace effect in Johnson and Wade's (1995) model, sometimes regardless of the fitness costs to the pure-species individuals, but the conditions were fairly restrictive. Johnson and Wade (1995, 498) reach the conclusion:

> ... that direct selection for hybrid inviability via soft selection could be an important phenomenon in taxa in which there is: (i) regulation of population size by density-dependent factors on the local scale of a family; (ii) frequent contact with heterospecifics, such as in long persisting hybrid zones; and (iii) little opportunity for the evolution of prezygotic reproductive isolation due to the genetics and mating behavior of the system. We believe that the first criteria [*sic*] will often occur and thus be the least restrictive. We do not know how often the second criterion will apply. The third criterion will probably be the most restrictive: prezygotic reproductive isolation will probably evolve faster than hybrid inviability in most taxa.

Studies of reproductive isolating barriers since 1995 suggest that the third criterion may be more restrictive than Johnson and Wade (1995) thought. Prezygotic reproductive isolation not only includes mating discrimination, but also preferential fertilization of gametes from members of the same species over that of members of another species. We turn to this post-mating, prezygotic reproductive barrier next.

Conspecific Gamete Precedence and the Wallace Effect

Before the middle 1990s, most evolutionary biologists studying reproductive isolation focused either on pre-mating isolation (including behavioral isolation due to mate preferences and/or habitat differences) or hybrid sterility and inviability. Although evolutionary biologists realized that barriers between mating and zygote formation could impede gene flow between incipient species, such pre-mating related to post-zygotic barriers received little attention. During this time, Gregory and Howard (1993) working with two species of ground crickets (*Allonemobius* spp.) and Wade *et al.* (1994), working with two species of flour beetles (*Tribolium* spp.) independently discovered the same phenomenon. In both instances, the offspring from a female that mated with both a male from her own species and a male from the other species were almost always pure species; few hybrids were produced. Because females would readily mate with the other species and interspecific crosses did produce offspring, a logical inference is that females preferentially used the sperm from males of their own species over the sperm from males of the other species. Subsequent studies confirmed this inference (reviewed in Howard 1999). Furthermore, this phenomenon has since been observed in numerous and diverse groups of plants and animals, and been named conspecific gamete precedence[5] (see reviews in Howard 1999; Marshall *et al.* 2002; Johnson 2006).

Can conspecific gamete precedence evolve via the direct action of natural selection? If it did, it would be considered a Wallace effect by the definition used in this chapter because the isolating barrier occurs after mating. In a recent theoretical paper, Patrick Lorch and Maria Servedio (2007) found that conspecific gamete precedence can indeed evolve readily under a number of different conditions.

Marshall *et al.* (2002) noted that in groups where conspecific gamete precedence is common, reinforcement is not (and vice versa). This negative correlation suggests that each process may impede the other. Lorch and Servedio (2007) tested this hypothesis in their models, and found support for it. When two nascent species displayed strong conspecific gamete precedence, pre-mating isolation evolved much more slowly by reinforcement than it did in the absence of conspecific gamete precedence. Likewise, strong pre-mating reproductive isolation generated by reinforcement inhibited the evolution of conspecific gamete precedence by the Wallace effect. Lorch and Servedio (2007) also modeled the simultaneous evolution of pre-mating isolation by reinforcement and conspecific gamete precedence by the Wallace effect. They found that while each process inhibited the other, more reproductive isolation evolved when these processes were paired than when they were separate.

The presence of conspecific gamete precedence does not imply that it evolved via the Wallace effect. Male–female co-evolution occurring independently in the two nascent species could also lead to conspecific gamete precedence as an incidental by-product. The finding of greater mating discrimination when species overlap—in contrast with when they are geographically isolated—has provided support for reinforcement for pre-mating reproductive isolation (Servedio and Noor 2003; Coyne and Orr 2004). Similar tests examining whether conspecific gamete precedence is greater in species pairs that overlap as opposed to those that don't would support the hypothesis that conspecific gamete precedence arose via direct selection (Wallace effect). As of early 2008, such studies are still lacking.

Conclusions and Future Directions

> If we are seeking selection's cut-off point, the distinction between pre- and postmating is irrelevant and misleading. The correct distinction is pre- and post "weaning," where "weaning" stands for any parental investment. It is this that is the great divide. On the one side, natural selection could act to save costs to the parents; on the other side, it stands powerless (Cronin 1991, 393).

As it is the action of direct natural selection for increased pre-mating reproductive isolation that often manifests itself as altered female preferences, reinforcement is the intellectual hybrid descendant of Darwinian and Wallacean views. Despite disagreements over the relative importance of reinforcement over other modes of speciation, most evolutionary biologists at the start of the twenty-first century agree that reinforcement is theoretically plausible, occurs some of the time in nature, and is worthy of active study (Marshall *et al.* 2002; Serevido and Noor 2003; Coyne and Orr 2004).

In contrast, there is far less support that the Wallace effect as defined as post-mating reproductive isolation evolves by the direct action of natural selection. Lorch and Servedio (2007), however, demonstrate that the Wallace effect can readily evolve for conspecific gamete precedence. What is now desired is a search for more empirical examples of conspecific gamete precedence evolving via direct selection. Effort should also be put into further comparative work testing Marshall *et al.*'s (2002) finding of a negative correlation between the evolution of pre-mating reproductive isolation via reinforcement, and conspecific gamete precedence.

Thus, support for hybrid inviability evolving via the Wallace effect appears to be weak, at best. Theory suggests it is unlikely, and empirical examples are lacking. Moreover, latter-acting barriers to gene flow likely allow for less compensation and more wastage than earlier acting ones. Still, Cronin (1991) is correct; we should not dismiss out of hand the possibility of pre-weaning reproductive isolation evolving by the Wallace effect. We should also remember that the new theory indicating pre-mating isolation should be relatively easy to evolve via reinforcement emerged largely through empirical studies identifying reinforcement. If several cases of hybrid inviability should be found that appear to be due to the Wallace effect, a similar burst of activity on theoretical models of hybrid inviability via the Wallace effect could ensue.

Notes

1. Cronin (1991, 410) notes that Darwin's investigations of *Primula* led him to the phenomenon of self-incompatibility, an inbreeding avoidance mechanism wherein matings between close relatives are rendered infertile. (We now know that this self-incompatibility is not exceptionally rare in plants, and that analogous processes may be occurring in a few animals.) Darwin for a brief period considered that an analogous phenomenon might allow natural selection to increase the sterility of partially sterile hybrids.
2. Kottler (1985, 416) states, "Wallace was not proposing the selective origin of reproductive isolation mechanisms in general, but rather the selective origin of the particular post-mating mechanisms of cross- and hybrid sterility. Since, according to current theory, these forms of sterility are precisely the types of reproductive isolation that cannot be produced by selection, the Darwin-Wallace debate provides little historical justification for the term 'Wallace effect'."
3. Some authors (*e.g.*, Butlin 1989) have objected to using the term "reinforcement" where the hybrids are already completely sterile or inviable. They argue that if hybrids have zero fitness, then speciation is complete and thus does not require reinforcement. Others (Coyne and Orr 2004) use the term more expansively to cover any case in which natural selection has led to a decrease of mating between different nascent or complete species. I use the latter definition throughout this chapter because my interest is on situations where reproductive isolating barriers can be adaptive in themselves.
4. The symposium volume from this meeting is Otte and Endler (1989). The patterns of speciation study presented by Coyne is not in that volume, but was instead published in *Evolution* (Coyne and Orr 1989).
5. Also known as conspecific sperm precedence in animals, and conspecific pollen precedence in plants.

The Colours of Animals: From Wallace to the Present Day I. Cryptic Coloration

Tim Caro,* Sami Merilaita, Martin Stevens

• • • • • •

To the ordinary observer the colours of the various kinds of molluscs, insects, reptiles, birds, and mammals, appear to have no use, and to be distributed pretty much at random. There is a general notion that in the tropics everything—insects, birds, and flowers especially—is much more brilliantly coloured than with us; but the idea that we should ever be able to give a satisfactory reason why one creature is white and another black, why this caterpillar is green and that one brown, and a third adorned with stripes and spots of the most gaudy colours, would seem to most persons both presumptuous and absurd. We propose to show, however, that in a large number of cases the colours of animals are of the greatest importance to them, and that sometimes even their very existence depends upon their peculiar tints.

(S318 1879, 128)

Introduction

Alfred Russel Wallace had an abiding interest in animal and plant coloration, not only because he saw it as a way to advance the theory of evolution through natural selection, but because he was a superb naturalist. Wallace viewed coloration as one of the most important features of natural selection; in his book *Darwinism* he stated: "Among the numerous applications of the Darwinian theory in the interpretation of the complex phenomena presented by the organic world, none have been more successful, or are more interesting, than those which deal with the colours of animals and plants" (S724 1889, 187). He wanted to demonstrate the *utility* of the colours and patterns for their bearers, not as creations to please

* Authors are listed in alphabetic order. Correspondence to Tim Caro. We thank George Beccaloni, Leena Lindström, Charles Smith, and Mike Speed for encouragement.

humans, or as artefacts resulting from exposure to heat or light, but as adaptations increasing their bearers' chances of survival. "We are thus compelled to look upon colour not merely as a physical but also as a biological characteristic, which has been differentiated and specialised by natural selection, and must, therefore, find its explanation in the principle of adaptation and utility" (S724, 189).

In addition, Wallace was an acute observer sensitive to the importance of coloration in species identification and behaviour. He recognized the fundamental importance of considering markings in their natural habitat when interpreting the role and function of visual signals correctly, and was aware of the risk in examining specimens under artificial conditions: "... concealment is effected by colours and markings which are so striking and peculiar that no one who had not seen the creature in its native haunts would imagine them to be protective" (S724, 200). For example, he thought that the Taiwanese bat *Kerivoula picta* had an orange body and orange-yellow and black wings to blend in with decaying orange and black longan *Nephelium longanum* (= *Dimocarpus longan*?) leaves among which it rested. If it had been described only from the hand it might be thought of as warningly coloured. Similarly, the orange spot between the shoulders of a sloth *Bradypus tridactylus* resembles a branch snapped from a tree trunk (S724, 201), something impossible to determine from museum specimens alone. Similar arguments were frequently presented by Edward Bagnall Poulton, his influential contemporary: "We cannot appreciate the meaning of the colours of many animals apart from their surroundings, because we do not comprehend the artistic effect of the latter" (Poulton 1890, 25; see also Thayer 1909). Students of coloration still have to grapple with this issue.

Wallace classified the coloration of animals and plants into five categories. For animals, the subject of this and the next chapter, they are: 1. Protective colours; 2. Warning colours of, (a) creatures specially protected, and (b) defenceless creatures, mimicking 'a'; 3. Sexual colours; 4. Normal colours; and for plants, 5. Attractive colours (S725 1998 [1891]). Broadly, contemporary understanding still follows this classification. Nonetheless, biologists are currently making great strides in the field of protective coloration, the subject of this chapter, and have made giant strides in the fields of warning colours, mimicry, and sexually selected coloration (see Chapter 8) all of which have occurred after a considerable lull in interest during the first two thirds of the twentieth century. The purpose of these two chapters is to view Wallace's work on coloration in the light of these conceptual and empirical advances.

Colour and Visual Perception

> What we term colour is a subjective phenomenon, due to the constitution of our mind and nervous system; while, objectively, it consists of light-vibrations of different wavelengths emitted by, or reflected from, various objects (S724 1889, 188).

Wallace's comprehension that colour is something subjective is just one example of an understanding which pervaded much of his work on animal coloration. Accurate theories of colour perception stemmed from before Wallace's time; as Newton (1718) realized, it is the brain that "colours" rays of light. Furthermore, by the 1880s, Sir John Lubbock (1882) had shown that some animals are sensitive to ultraviolet light, illustrating that vision between species is variable. However, Wallace's comprehension of the details of visual perception and colour was nevertheless remarkable, and he was well aware that there was a perceptual world unavailable to us: radiation from the sun "consists of sets of waves ... of which the middle portion only is capable of exciting in us sensations of light and colour" (S725 1998 [1891], 355). Wallace here is describing the "electromagnetic spectrum," which includes light that humans can see ("visible light"), plus gamma rays, X-rays, microwaves, radio waves, and so on. Wallace understood that in humans, visible light of longer wavelengths led to a perception of red, then orange, yellow, green, and blue as the wavelengths of light decrease. A vast array of colours could be produced simply by differing the composition of the light reflected from, or transmitted by, an object. Wallace knew that there are three sets of nerve-fibres in the retina each of which is primarily sensitive to different wavebands of light, thus describing the basics of trichromatic theory which was proposed by Thomas Young around the start of the 1800s. Here, a sensation of colour is produced by a comparison of the outputs of each receptor type. This stage of colour perception means that the same light spectra will produce different colours to animals that differ in the properties or number of photoreceptors (Wyszecki and Stiles 1982; Zeki 1993).

Wallace (S725 1998 [1891], 410) was also aware of the other main theory of colour perception: "If we look at pure tints of red, green, blue, and yellow, they appear so absolutely contrasted and unlike each other, that it is almost impossible to believe... that the rays of light producing these very distinct sensations differ only in wave-length and rate of vibration ... " This is the opponent-processing theory which describes antagonistic colour pairs being processed in separate neural channels and, in human vision, results in yellow and blue, plus red and green, representing opponent neural signals (Wandell 1995). As Wallace noted, this explains why some colours never appear in the same colour patch (*e.g.* no colour ever appears reddish-green or bluish-yellow). Simplistically, the trichromatic theory explains colour vision at the photoreceptor level, whereas the opponent-processing theory explains the way in which photoreceptors are interconnected neuronally.

Far from simply understanding some of the basics of colour vision in humans, Wallace (thanks in part to the works of John Lubbock and others) realized that visual perception differs among animals. Not only was he aware of the widespread presence of colour vision in many animals, but he also correctly speculated (along with Grant Allen) that colour vision in birds is more acute than in humans, possibly because of their need to find coloured fruits and insects from a distance

(Allen 1879; S304 1879). This understanding of Wallace and Allen was at times quite detailed; for example, stretching to an awareness of the presence of retinal oil droplets in birds: "in addition their cones [are] furnished with variously-coloured globules, which are supposed to give a still more perfect perception of colour" (S304 1879, 503). These oil droplets, found in various vertebrates, are lipid-based globules located in the cone inner segments, which selectively absorb light below a certain value before it reaches the photopigments, thus changing the spectral sensitivity of the cones and modifying colour vision (Bowmaker 1980; Goldsmith 1990), potentially enhancing colour discrimination (Vorobyev 2003). Wallace was not always spot on, however. For example, in reviewing another of Allen's books (S348 1882), he questioned the proposal that humans' colour sense stems from an advantage of being able to detect bright fruits in tropical forests, because the colours of fruits did not necessarily indicate that they were edible. While this is true, current evidence indicates that the evolution of trichromacy in humans and Old World apes does indeed stem, at least in part, from an advantage in being able to detect ripe yellow or red fruit or young leaves against a dappled background of mature green leaves (*e.g.* Regan *et al.* 2001; Párraga *et al.* 2002).

Physics of Coloration

Wallace also understood much about the physics of colour production in general, and of the processes of absorption and interference which are involved (S724, S725). This was certainly not unique to Wallace, as most contemporary naturalists studying colour signals discussed these mechanisms (*e.g.* Poulton 1890; Beddard 1895). It was broadly known during Wallace's era that colour production stems either from pigments, or by the interactions of certain light waves with biological structures (Prum 2006). Structural colours are produced by the physical interaction of light with fine-scale structures, whereas pigments are chemicals that selectively absorb and reflect different wavelengths of light. Wallace indicated that the colours produced by pigments were more frequent, and this, he reasoned, was due to the molecular structure of the body and the range of different substances which are found within animals for reasons originally unrelated to colour production. For example, many substances used for adaptive purposes (such as in colouring birds' eggs) are apparently similar to non-adaptive colours (such as that of blood). In truth, the coloration of many animal markings can be complex, and may stem from a range of chemical pigments, the interactions of certain light waves, or a combination of the two (Prum 2006). Structural colours are often responsible for the blues and greens of the feathers of many birds, as well as certain butterfly wings and beetle shells, among other signals. Variation in the pattern's structure can also give rise to iridescent effects, as seen in peacock *Pavo cristatus* or starling feathers, because the perceived colour depends on the viewing angle. Structural mechanisms generally also lead to white "colours" in otherwise unpigmented structures and ultraviolet signals (Osorio and Ham 2002; Prum 2006).

In contrast to structural mechanisms, pigments differentially absorb and reflect specific wavelengths of light, depending on the structure and concentration of the pigment molecules. A range of pigments is known to be involved, but two in particular are especially common. Carotenoids absorb more short wavelengths of light, generally producing red, orange, and yellow colours, and are found in many invertebrates and in all groups of vertebrates (McGraw 2006a). The specific colour that carotenoids produce depends on the exact form and concentration of the substance, in addition to the material in which the carotenoid is presented (McGraw 2006a). In addition, melanins, largely producing blacks, browns, and greys, are the most abundant and widespread of all pigments (McGraw 2006b). Overall, many of the most striking colours in nature result from a combination of structural and pigmental mechanisms, such as non-iridescent greens and certain yellows (Prum 2006).

The Special Problem of Tropical Versus Temperate Animals

During Wallace's era, naturalists frequently attributed the colour of various animals as arising from the direct action of heat and light from the sun, thus accounting for the large number of brightly coloured species found in the tropics. Much of this debate devolved from the amount of emphasis different naturalists attributed to the action of natural selection. Frank Beddard (1895), for example, unlike Poulton and Wallace, generally ascribed less influence to natural selection in producing animal colours (Pycraft 1925; Kingsland 1978). While Beddard did accept a role for protective markings (*e.g.* camouflage, warning colours, and mimicry) and sexual selection, he attributed a much greater direct effect to the environment in producing animal colours. Many effects, Beddard argued, could be attributed to the direct result of temperature, and not natural (or presumably sexual) selection. For example, the colours of desert animals need not be a case of protective general resemblance, as Wallace and most others argued, but rather due to the effects of temperature and a lack of moisture.

On this subject, Wallace (S725 1998 [1891]) was as astute as ever in realizing that what really needed to be considered was not the number of bright tropical and bright temperate species *per se*, but the proportion of brightly coloured animals in each region. While the number of vivid tropical species was higher than anywhere else, the proportion of bright species in the tropics compared to other regions apparently did not differ greatly, and the presence of a great number of dull coloured tropical species indicated that there must be other important factors at work. Wallace pointed out that many tropical birds are actually dull in colour, and that those species occurring with diverse geographical distributions are often no more brightly coloured in the tropics than elsewhere. In fact, many groups of birds and insects are gaudier in temperate regions than in the tropics. Wallace recognized that the direct effect of sunlight was not important, otherwise, why should species found on the Galapagos Islands, situated right on the equator, be so dull in colour? Likewise, in the case of desert animals, Wallace argued that if light and temperature were important in producing the brightness and intensity of colours, then animals living in these locations should be brightest and most gaudy; the fact that they are not renders the theory incorrect. What mattered was natural selection. The direct action of light too was of little significance in most cases. For instance, many deep sea fauna, living where light does not penetrate, are brightly coloured red when viewed at the surface. However, because longwave light does not reach the ocean depths, a red animal is not seen in the deep ocean because there is no red light (except for red bioluminescence, which is very rare). While it is now also known that light and the environment can cause changes in colour, this is adaptive and a result of natural selection. For example, some African grasshoppers are found in different coloured morphs, and it was suggested by both Bacot (1912) and Poulton (1926) that the dark morph is protected by resembling the scorched substrate and vegetation after a fire: "fire melanism." Later, it was shown that those grasshoppers that develop on burnt black earth develop into a black morph, and those that develop on green vegetation develop into a green morph; *i.e.*, both are camouflaged on their respective backgrounds (Burtt 1951).

Background Matching

The first category in Wallace's functional and biological classification of the colours of living organisms was protective coloration, which he also termed useful or protective resemblances (S725 1998 [1891]). According to his definition this category includes animal coloration that conceals the animal from its enemies or its prey: "The fact that first strikes us in our examination of the colours of animals as a whole, is the close relation that exists between these colours and the general environment. ... The obvious explanation of this style of coloration is, that it is protective, serving to conceal the herbivorous species from their enemies, and enabling carnivorous animals to approach their prey unperceived" (S724 1889, 190). Visual similarities between the coloration of animals and their habitats are very common in nature (Plate 7). For naturalists of today the evident aim of this is the need of concealment, or crypsis (here defined simply as a strategy to reduce the probability of detection) either from enemies or from prey but, before the theory of natural selection, the reason was not so obvious. Other competing ideas were that such similarities were caused by the action of light, health, soil, or food on the coloration of animals. Wallace presented several logical arguments against these other reasons as discussed above.

The principle of concealment through coloration that Wallace recognized was visual similarity between the animal and its background (now often referred to as "background matching" or "resemblance to background"). He listed numerous examples of "protective colours which serve to harmonize animals with their general environment" (S724, 199), and "general assimilation of colour to the surroundings" (S724, 193) in many different taxa and habitats as examples of

Figure 17 The Indian leaf-butterfly (*Kallima inachis*) resembles a dead leaf when at rest with its wings closed.
From S318 1879. Out of copyright.

prey adaptations to avoid predation, or sometimes to predator adaptations to approach prey unnoticed (S389 1886). As a further evidence for the concealing function of coloration Wallace mentioned the ability of some animals to change their colour to harmonize with their environment through a rapid adjustment, as in many flatfish, or through developmental change, as in many Lepidoptera (S724). In addition to the appearances of the background and the animal, Wallace also pointed out the importance of the "powers of vision" and "faculties of perception" in successful concealment (S725). The appearance of animal coloration and its habitat, together with the visual system of the viewer, are nowadays considered the main factors influencing the evolution of camouflage. Furthermore, in addition to the appearance of an animal, including both coloration and morphology, Wallace gave several examples of the importance of behaviour that completes the disguise (S725).

Wallace gave seven main examples of background matching (S134 1867, S724 1889, S725 1998 [1891]): white animals living in the arctic and animals changing colour seasonally in lower latitudes; light brown animals living in deserts; the commonness of green ground colour in birds living in the tropics in contrast to the commonness of brown ground colour in birds living in the temperate region with deciduous leaves; dusky grey animals that are nocturnal; the colour of birds' eggs; transparency in oceans (see below); and, finally, dark dorsal but light ventral coloration in larger marine animals for hiding from enemies above and below.

Considering first the white of arctic animals, Wallace dismissed animal coloration as assimilating climatic variables because some arctic animals are not white (see also S425 1890). He also dismissed thermoregulation, arguing that species that are not white have alternative antipredator defences to crypsis (musk ox *Ovibos moschatus* that may need to be brown to see each other for grouping, sable *Martes zibellina* that hunt in trees, and pugnacious ravens *Corvus corax* that do not need defences [S318 1879, S378 1885, S424 1890, S724 1889]). These exceptions are still the strongest argument for white colour being an antipredator rather than a thermoregulatory device in the arctic. Debate over the importance of protection and thermoregulation in driving camouflage occurred in the 1920s over the coloration of mammals living in variously coloured soil surfaces in North America (Sumner and Swarth 1924; Benson 1936) but was eventually resolved in favour of crypsis. Subsequent phylogenetic analyses have confirmed these background matching associations in arctic and desert regions in carnivores and artiodactyls (Caro 2005a). On the other hand, Wallace thought that in butterflies and some other insects it was more probable that melanism could also be driven by thermoregulatory factors (S378 1885). This has subsequently been shown to be important; for example, in influencing mobility when escaping threats (*e.g.* Forsman *et al.* 2002).

Another example was bird egg coloration. It had been noticed that birds nesting in cavities have white eggs. Wallace hypothesized that a white egg without patterning was the ancestral type (S724). He also pointed out that some other birds, which have a nest or a cryptically coloured parent that conceals the eggs or which are able to effectively defend their eggs, have white eggs too (S724). Wallace suggested that the function of egg coloration was concealment to decrease the risk of egg predation, which also would explain the lack of colours and patterns in cavity-nesting species or in eggs that were otherwise well protected. In a thorough account of bird egg coloration, Kilner (2006) reviewed previous studies and used a comparative analysis to investigate the evolution of egg coloration across species. In accordance to the hypotheses put forward by Wallace, Kilner found that ancestral bird egg coloration was non-patterned and white, and she concluded that predation was the most plausible general explanation for deviations from the ancestral type.

Related to this, Wallace also thought that in birds the female is drab when she incubates the eggs (S134 1867, S139 1868; see next chapter). He regarded colour as being a default characteristic because he thought that the surface of animals automatically produces colours—assuming, for instance, that gaudy male plumage is the norm but drab female plumage the result of active selection. As the emphasis of modern sexual selection theory has been on explaining the evolution of bright ornaments in males, the issue of dowdy female plumage has been all but disregarded. Recently, however, bird species with brighter plumage have been shown to suffer greater predation than those with dull plumage (Martin and Badyaev 1996; Huhta *et al.* 2003). Moreover, in species where females sit on the nest, females are duller in shrub nesting species where nest predation peaks, than in ground and canopy nesting species where nest predation is lower (Martin and Badyaev 1996); in males there is no such effect. These findings show that there is direct selection on female drab coloration that is different from that on males, just as Wallace surmised (Plate 8)!

Wallace's arguments on protective coloration centred on qualitative associations between coloured integuments, feathers, and pelages and the assumed colours of particular environments. Subsequently, many studies have shown that cryptically coloured prey survive better than conspicuously coloured prey. Evidence has accumulated slowly over a long period, first through observations in the field and simple tests, and later with more extensive and systematic experiments (*e.g.* Popham 1941; Kettlewell 1955). Today, investigation of similarity in colour between animals and their habitats is not necessarily based on the eye of the researcher only but confirmed by directly measuring the reflectance spectrum (Norris and Lowe 1964; Endler 1984); and, further, this information can be combined with a colour vision model of the viewing species (*e.g.* Stuart-Fox *et al.* 2004; Endler and Mielke 2005).

A parallel idea is that certain colour changes that occur during an individual's development closely match its microenvironment (Booth 1990). Wallace described how young sphingid caterpillars have longitudinal stripes when young and living on grass stems, but acquire diagonal stripes as they grow matching the oblique veins of leaves on which they later feed. Some adult larvae have red spots that resemble small red flowers on which they browse (S338 1881). Remarkably little work has been done on ontogenetic colour in the intervening period but the phenomenon is widespread including many insects and nidifugous birds (Caro 2005b). In bovids and cervids, for instance, species with spotted young are found in lightly wooded forest and in grassland habitats respectively (Stoner *et al.* 2003).

One of Wallace's chief contributions was to link behaviour to coloration. For example, nocturnal caterpillars are green, feeding at night but resting motionless on foliage of the same colour as themselves during the day. By contrast, diurnal caterpillars are bright and hairy (aposematic) (S134 1867, S318 1879). In addition there are behavioural-morphological complexes (*e.g.*, slow, noisy, aposemes: S272 1887). He described, for example, how he had observed the leaf-mimicking *Kallima* butterflies land only on those parts of the environment where their disguise is effective, and how stick insects (Phasmida) improve their (behavioural-morphological) disguise by stretching out their legs asymmetrically (S725 1998 [1891]). Today we know that such coupling of morphology and behaviour may take place at the genetic level, resulting in interdependent evolution of these two types of traits. As an illustration, in the garter snake, *Thamnophis ordinoides*, anti-predator behaviour has been found to be genetically coupled with colour patterning such that striped individuals are more likely to flee when threatened, whereas non-striped and spotted individuals tend to rely on more secretive behaviour (Brodie 1989). Also, it has been experimentally established that in the pygmy grasshoppers, *Tetrix subulata*, coloration and behaviour jointly determine an individual's susceptibility to predation (Forsman and Appelquist 1998, 1999).

Pattern Blending

The coloration of numerous animals cannot be described in terms of colour only, but through geometry as well: the pattern and distribution of colours (or visual textures, such as mottling) on the surface of the animal. Wallace thought that the evolution of concealing coloration was a gradual process. At the first and most general stage of adaptation an animal reproduces the general tint of the environment (S134 1867). Next, Wallace considered special resemblances, *i.e.*, local adaptations to a smaller spatial scale. He wrote that "This form of colour adaptation is generally manifested by markings rather than by colour alone" (S724, 199). According to him, a patterning represents a more specialized and elaborated stage of adaptation. Wallace wrote that the protection gained through these general and special protective resemblances "varies in degree, from the mere absence of conspicuous colour or general harmony with the prevailing tints of nature, up to such a minute and detailed resemblance to inorganic or vegetable structures as to realise the talisman of the fairy tale, and to give its possessor the power of rendering itself invisible" (S725 1998 [1891], 47).

Comparative data show that pattern blending in the form of spotting in artiodactyls and felids is associated with forest habitats where light is likely dappled, that spots are found in young ungulates that hide their young after birth, and that striped carnivores are found in grasslands (Ortolani and Caro 1996; Stoner *et al.* 1993). Moreover, experimental evidence shows patterning influences survival of prey: both in terms of the spatial distribution of pattern elements, and the level of matching between the elements of prey patterning and the background, affecting the detection of the prey by predators (Merilaita *et al.* 2001; Merilaita and Lind 2005). Thus, currently, we expect pattern to be selected for resemblance as much as for colour.

Wallace mentioned symmetry in prey as a factor that increases its risk of being detected (S134 1867, S318 1879, S725 1998 [1891]). It has been experimentally confirmed that symmetry tends to increase the probability of a cryptic pattern being detected (Cuthill *et al.* 2006; Merilaita and Lind 2006). However, the detrimental influence of pattern symmetry on concealment varies between colour patterns, such that for some patterns it is very low or even non-existing (Merilaita and Lind 2006). This is likely to influence the evolution of the appearance of cryptic colour patterns.

Masquerade

Wallace thought that at the most specialized and advanced level of adaptation, morphology is also modified to promote concealment through protective imitation of particular objects (S724) or special modifications of colour (S725). These imitations of objects are seen but appear uninteresting to the predator. Examples include the sand-coloured nighthawk *Caprimulgus rupestris* (= *Chordeiles rupestris*) on rocks, seahorses and pipefish looking like seaweed, buprestid beetles resembling dung, and the *Kallima* butterfly that looks like a leaf (S725). Masquerade is considered mechanistically distinct from concealment through cryptic coloration, because it aims to obstruct recognition rather than avoid detection (Ruxton *et al.* 2004a; Stevens 2007). Nonetheless, the concept is difficult to define

(Endler 1981), as Wallace realized (S725), because resembling innocuous objects (such as a phasmid impersonating a stick) is called masquerade, but markings on moths resembling patches of lichen would be called pattern blending. Additionally, a sphingid moth caterpillar inflating its body to resemble the head of a venomous snake might be called masquerade, whereas a hoverfly resembling a wasp would be termed mimicry.

We now think that although an evolutionary succession of the kind envisaged by Wallace may occur in many cases, there seems no *a priori* reason to assume that a non-patterned, uniform coloration would always be a less specialized or less effective adaptation than a patterned coloration. The adaptive value of a concealing coloration is case-specific, depending on the animal bearing the coloration and its behaviour, the visual environment, and the visual system of the viewer. Indeed, Wallace recognized the importance of these three factors for the evolution of concealment.

General Concerns

In a very original departure, Wallace also outlined situations in which animals do not resemble their background because (i) they have alternative methods of defence, as in armoured animals or those with pungent smells (S134 1867), or (ii) because they do not need to be cryptic. Examples are gaudy female birds nesting in holes in trees (S725) or birds' eggs resembling the background only in those species that do not sit on them (S724). Contemporary scientists often fail to try to explain absence of traits, but they can be helpful in explaining patterns of evolution.

Although background colour matching, pattern blending, and even masquerade may appear as quite simple topics at a first glance, a closer scrutiny reveals that they are actually challenging subjects for evolutionary ecology. There are still many gaps in our knowledge of how these phenomena interfere with the visual processes of object detection or recognition, and how natural selection shapes the appearance of animal coloration to minimize the risk of detection.

For example, one previously dominant hypothesis argued that the highest level of camouflage is produced by a coloration that represents a random sample of the visual habitat (Endler 1978). This idea, although it has progressed the field and may still have some heuristic value, is now seen as problematic in several ways. For one thing, experimental evidence shows that if the habitat is not visually very simple and homogeneous (which is seldom the case in nature), there is no reason to expect that a random sample would always produce the best crypsis or that different random samples would be equally cryptic (Merilaita and Lind 2005). So how should an animal gain best protection through background matching in heterogeneous habitats consisting of visually different patches (Merilaita *et al.* 1999, 2001; Houston *et al.* 2007; Sherratt *et al.* 2007)? These studies provide both theoretical and experimental evidence that the solution is a compromise that combines features from two or more different patches. Thus, these results contradict the idea of Wallace (as well as the idea of crypsis through random sample matching) that the most specialized coloration would always represent the most advanced outcome of evolution. The next obvious questions for this line of research are, when will compromise and specialist strategies be favoured, and how common are these two strategies in nature?

Also, visual heterogeneity on a finer spatial scale, such as visual complexity on a scale smaller than the animal, may be important for the evolution of camouflage. It has been suggested that such background visual complexity could influence how close the resemblance has to be to produce a given level of concealment (Merilaita 2003). This is because in complex backgrounds the viewer has to process more visual information than in simple backgrounds in order to obtain some information useful for detection.

One still largely unanswered question about the evolution of background matching coloration in animals is the role of different aspects of colour patterns, namely colours, brightness, and pattern geometry. The sensitivity of visual systems to detect deviations in these various aspects will often differ between species. Therefore, all aspects are not necessarily equally important for effectual background matching, and, further, their relative importance may vary in relation to other factors, such as viewing distance (Hailman 1977; Merilaita and Lind 2005; Stevens 2007; Plate 9).

Disruptive Coloration

Generally, Wallace, Poulton, Beddard, and other nineteenth-century naturalists interested in protective coloration seem not to have anticipated or realized an important component of camouflage which may be widespread in animals. At this time, discussions of concealment were firmly centred on the idea of simply matching the background environment as accurately as possible. By the end of the nineteenth century, however, the American painter Abbott Thayer realized that such descriptions of camouflage were not wholly adequate. The natural history of most animals meant that they would often be mobile and found on a range of backgrounds, and specialist background matching, as discussed by Wallace and others, would be ineffective in concealing an animal from its predators. Something else was needed. One of the key problems with simple background matching was that the outline of an animal's body, if left unmodified, would reveal its presence due to discontinuities between the body and the background (Thayer 1909).

Thayer's solution: disruption and dazzle

Gerald Thayer (1909), Abbott Thayer's son, was one of the first to argue that camouflage consisted both of blending (background matching), and disruption (G. Thayer called it "ruptive")—the latter being where the animal's appearance is broken up by strongly contrasting patterns that mask the outline of the body (see also Behrens 1988; Stevens *et al.* 2006a). Similarly, disruptive patterns may disguise otherwise conspicuous or vulnerable body parts such as the legs or eyes. In addition to disruptive coloration, Thayer also pioneered the related idea of "dazzle markings" (the term "dazzle" stemming from the American term "razzle dazzle," to confuse). While disruptive coloration and dazzle markings have often been considered synonymously, it seems they are functionally distinct, or target distinct perceptual processes; disruptive markings are peripheral and break up the appearance of an object, whereas dazzle markings may either draw the attention of a viewer towards the markings or away from the outline of the object (see Stevens 2007).

Although the far-fetched ideas of G. Thayer were clearly fallacious (for example he eventually argued that *all* animal coloration was involved in concealment, most famously depicting flamingoes as concealed at sunset against the pink sky), his ideas were crucial in illustrating the different methods by which animals can achieve concealment. Renewed interest in Thayer's ideas has recently led to a resurgence in work investigating disruptive camouflage in nature (see below).

Cott's formalization

Gerald Thayer pioneered the idea of disruptive camouflage but he did not manage to formalize his ideas. Interest waned for thirty years until 1940, when the British zoologist Hugh Cott's published his classic and highly influential book *Adaptive Coloration in Animals.* Cott's book (1940) set out an extensive formalization of Thayer's theory, based on several key ideas. These included (1) *differential blending,* where adjacent peripherally placed disruptive markings both blend in and stand out from the background, (2) *maximum disruptive contrast,* where those peripheral markings are highly contrasting, and (3) *coincident disruptive coloration,* where different parts of the body are linked to hide their form (such as appendages and limbs). Cott's formulation was crucial because it put Thayer's ideas into a testable framework—one that has only recently been rigorously addressed.

Wallace and disruptive camouflage

It is clear that the idea of disruptive coloration was pioneered by the Thayers, but did Wallace, or his contemporaries have any input on the subject? The Thayers' main arguments were published towards the end of Wallace's life; whether Wallace was aware of Abbott Thayer and his ideas seems not to be known, but Thayer's work was highly controversial at the time, at least in the USA, and it would be surprising to find that Wallace had not heard of it. In fact, Abbott Thayer's first writings on disruption were published in 1903 (Thayer 1903), in an English society's transactions, and it also seems that his work did attract the attention of Poulton, who often corresponded with Wallace on matters regarding animal coloration. Finally, Abbott Thayer was also known for his independent (from Poulton) formalization of the ideas of countershading, which were published as early as 1896 (Thayer 1896).

It therefore seems likely that Wallace did come across Thayer's ideas in some form, but we may never know whether his silence regarding disruptive coloration stems from him not being fully aware of the theory, or because he did not merit it—though the former would seem more likely because Wallace rarely shied away from criticizing a theory he disagreed with.

Whatever the situation with Wallace, Poulton had recognized the principle of disruptive markings as early as 1890. With regards to privet hawk moth larvae, he wrote: "Although the caterpillar looks so conspicuous, it harmonises very well with its food plant, and is sometimes difficult to find. The purple stripes increase the protection by breaking up the large green surface of the caterpillar into smaller areas ..." (Poulton 1890, 42). Wallace certainly would have read this, so it seems he did not appreciate its broader importance. On various occasions he did, however, consider the markings of caterpillars. For example, in a review of Weismann's book *Studies in the Theory of Descent* (S338 1881), Wallace discusses Weismann's thoughts on the markings of caterpillars, which commonly consist of longitudinal stripes on species which feed on plants with straight line features. Wallace essentially referred these markings to background matching whereas Poulton hinted that they may do more than simply resemble features of the environment. Yet he too developed this idea no further—despite the fact that by the time of his death much of Cott's ideas were published, and the use of disruptive and dazzle markings had become widespread in military applications for several decades (Behrens 1999, 2002).

Despite apparently being unaware of disruptive coloration, some of Wallace's abundant writings on camouflage do illustrate similarities between his ideas and those of Thayer. One interesting issue raised by Wallace was the commonness of spots found on forest or tree-dwelling animals, particularly cats of large size (S725 1998 [1891]), and that such spots seem to harmonize in the dappled light and shadows created by the forest canopy. For instance, small peripheral spots on butterflies would harmonize with gleams of sunlight (S724 1889). This is interesting because Thayer (1909) clearly believed that such markings were disruptive, and helped to break up the form of the animal against the background (Thayer 1903). There are associations between spots and dappled habitats in some mammal taxa (see above) but the mechanism by which these species might enjoy crypsis is not yet clear. Others have argued that eyespots on some butterflies are very noticeable in patches of forest floor directly exposed to sunlight (Young 1979), so they may serve a dual purpose (see below).

Wallace also had differences in opinion with his contemporaries (from Thayer as well as Poulton and Darwin) on other matters. For instance, Wallace noted that stripes are common on mammals which are found in reeds and grasses, illustrating that he often thought of such markings as background matching (S725). However, while some of these markings would seem to be background matching (*e.g.*, tigers *Panthera tigris*: Godfrey *et al.* 1987; Ortolani and Caro 1996), others may be disruptive. Thayer suggested that this was the case for zebra *Equus burchelli* markings, especially at low light levels. Wallace also discussed zebra markings and while he mentions the idea that they may be concealed at low light levels (not originally proposed by him), he preferred his theory that they were a form of recognition mark (S724). There is no strong positive or negative evidence for any of these ideas at present (Ruxton 2002). Recognition was Wallace's primary explanation for a range of high contrast markings, such as the bars on the breast of plovers. Today, these are frequently discussed as being disruptive, or more commonly have not received the attention that they deserve (see next chapter).

The power of disruptive coloration

Following Cott's (1940) book, disruptive coloration gained widespread acceptance, and rapidly became standard textbook material in explaining animal markings. Surprisingly, this was despite any empirical evidence demonstrating a survival advantage in nature for disruptive markings, over and above any benefit provided by simple background matching. In fact, it was not until a study by Cuthill and colleagues (2005) (see also Sherratt *et al.* 2005) that the theory gained its first significant experimental support. Cuthill *et al.* used artificial "moth prey" designed to resemble oak bark, as perceived by a foraging bird, with background matching and disruptive markings. They found that differentially blending disruptive targets survived the longest, because they were most effectively concealed from the avian predators. A second experiment showed that these disruptive markings were most effective when they were highly contrasting, supporting Cott's maximum disruptive contrast theory (Cuthill *et al.* 2005). This study stimulated a range of related and follow-up work, which not only illustrated the power of disruptive camouflage over simple background matching, but also illustrated that it may be highly valuable to animals that cannot utilize background matching alone (Cuthill *et al.* 2006; Schaefer and Stobbe 2006; Stevens *et al.* 2006b), just as Thayer had proposed. Concurrently, experiments with avian foragers in laboratory environments (Merilaita and Lind 2005), and studies with human foragers (Fraser *et al.* 2007) have also demonstrated the importance of disruptive camouflage. In addition, Stevens and Cuthill (2006) showed, by modelling the visual perception of birds, that disruptive coloration seemingly works by exploiting edge detection mechanisms that operate in early visual processing, creating "false" edges within the body, inhibiting successful detection of the true body outline (see also Stevens 2007). Therefore, almost a century after Thayer's main work, the power of disruptive coloration as a concealment strategy in natural systems finally has significant empirical support.

Disruptive coloration in nature

Many animals from a vast range of taxonomic groups have been pointed to as possessing disruptive markings. For instance, studies spanning over twenty years have investigated the expression of disruptive markings in cuttlefish *Sepia officinalis* (see *e.g.* Hanlon and Messenger 1988; Chiao *et al.* 2005; Kelman *et al.* 2007). Yet proving the existence of disruptive camouflage on real animals has been difficult, stemming from the problem of isolating disruptive markings, over-and-above simple background matching. Only one study, for example, has rigorously tested (and found support) for the presence of marginally placed disruptive markings in an animal (a species of marine isopod), as opposed to purely background matching (Merilaita 1998). In addition, tests of a survival advantage for disruptive markings in real animals have been few (in butterflies, Silberglied *et al.* 1980, but see Merilaita and Lind 2005; Stevens *et al.* 2006a). Despite lack of rigorous empirical support, disruptive coloration, subjectively at least, appears widespread in animals as diverse as insects, arachnids, fish, cephalopods, mammals, birds, and reptiles, in both terrestrial and aquatic systems (Stevens *et al.* 2006a). Given the strong survival advantage it has provided in experimental systems, it would be surprising were this not the case!

Countershading

Wallace recognized that most fish are "protectively coloured by the back being dark and the belly light, so that, whether looked at from above on the dark background, or from below on the light one, they are equally difficult to see" (S176 1870, 85). While these markings in fish and aquatic mammals could be clear cases of background matching (and indeed Wallace gave an example of this in porpoises: S272 1877), light bellies are more difficult to explain in terrestrial species and nocturnal species where they are also common. One possibility first described by Poulton (1888, 1890) and also elucidated independently and most extensively by the Thayers (1896, 1909) is that the light area below counteracts the dark shadow produced from the sun shining overhead (Kiltie 1998; Ruxton *et al.* 2004b). Using coloured pastries set out on lawns as bait for songbirds, Edmunds and Dewhirst (1994) showed that countershading provides "survival benefits" (for pastries!) over and above background matching; later, Speed and colleagues (2004) showed that this was specific only to certain bird predators. More recently, Rowland *et al.* (2007b) modelled countershaded pastry prey with respect to avian vision, and concurrently showed that countershaded prey had significantly higher survival than background matching and control treatments against a range of predator species. Nonetheless, rather little work has been achieved in this area in the last 150 years and it is unclear whether countershading is really a way to minimize shadow, or help camouflage in real animals.

Transparency

Wallace also made cursory acknowledgement of animal transparency, pointing to its prevalence in aquatic lower organisms (S176 1870) and in butterfly wings (S725). It seems self-evident that transparency is a way of reducing the probability of being spotted by predators but this is by no means certain: it could simply be a way to reduce costs of pigmentation. Little work has been done on transparency (Ruxton *et al.* 2004a).

Lures and Baits

Another interesting set of ideas to which Wallace contributed was the use of lures and baits by animals. Here, lures are loosely defined as something that draws an animal towards a specific object or feature, and as such they can be used by both predators ("aggressive lures") and prey ("defensive lures").

Aggressive Lures

Aggressive lures seem to be particularly common in many spiders and mantids, and may be utilized in disguise to attract prey within reach (cf. aggressive resemblance/mimicry). Wallace discussed two particularly interesting examples in his *Darwinism* (S724) in 1889. First, a spider *Thomisus citreus* (= *Misumena vatia*), which closely resembles the flower buds of trees on which it waits to catch insects that come too close; and second, a mantis *Hymenopus bicornis* (= *Hymenopus coronatus*), that elaborately resembles the colour and structure of a pink orchid flower. Nineteenth-century naturalists like Wallace considered these to be examples of extreme concealment or disguise; subsequently, there have been some fascinating discoveries in crab and orb spiders. Comparisons of the perception and detectability of certain spiders to both their prey (insects) and their predators (birds), has shown that some spiders may use their coloration to lure prey, rather than for camouflage, as previously thought. For example, instead of relying on camouflage to remain undetected by potential prey (*e.g.* bees) and predators (*e.g.* birds), some crab spiders are conspicuous (in the ultraviolet) on the flowers on which they wait seemingly to attract their victims (Heiling *et al.* 2003, 2005). Humans cannot see this but birds and many insects can. Some species of orb spiders also seem to use their body markings and features on their webs to attract prey (Tso *et al.* 2006), but this comes at the cost of increasing the risk of attack from predatory wasps (Cheng and Tso 2007).

Defensive Lures

Lures need not always attract prey to a waiting predator. Instead, they may divert the attention or actions of predators away from a potential prey animal, or from important features of the prey. It has been hypothesized for almost two hundred years that wing spots (termed "ocelli" in Wallace's era) on butterflies may, in addition to intimidating predators, deflect the attacks of predators to less important regions of the body—for example, they may deflect the pecks of birds away from the main body and towards the outer edges of the wings that may be able to cope with some level of damage (Stevens 2005). Darwin spent several pages on the function of ocelli in *The Descent of Man and Selection in Relation to Sex* (Darwin 1871), generally referring to their use in sexual selection in birds. With regard to conspicuous butterfly coloration, Darwin was clear: if not to advertise unpalatability (or mimicry), then such bright colours are usually used in sexual selection. Wallace (S725), however, remarked how so many ocelli and bright patches of colour are on the wing tips or margins; this, he argued, was because many insects are quite visible when flying, and therefore susceptible to attacks by birds. The distance that the spots are from the main body may afford some protection (so that predators attack the less vulnerable regions). Wallace (S725) was aware of the potential implications of wing damage relating to the use of lures or deflective spots in insects. He realized that individuals with broken wings may have escaped capture from birds, and that their large wings may have also allowed some protection for the main body. After Wallace, there have been a number of studies that have derived evidence for a deflective function of butterfly markings by assessing wing damage (Stevens 2005). However, there is a key problem with such an interpretation, pointed out by Edmunds (1974b), that it is difficult to determine whether a high frequency of predation marks indicates a high rate of escape or a high rate of attack.

On this subject we can see a genuine difference in the thoughts of Darwin and Wallace, and today, both scientists are viewed as partly right. Evidence for a deflective function of spots in butterflies is exceedingly poor, although there is some indirect support relating to differences in wing strength around deflective and non-deflective wing regions in related butterfly species (Hill and Vaca 2004), and some limited support from studies with captive predators (Lyytinen *et al.* 2004). There is also growing evidence that wing spots are used in sexual selection, at least in one species of butterfly (Breuker and Brakefield 2002; Robertson and Monteiro 2005) as Darwin might have argued. If support is scarce in butterflies, deflective markings do seem to function in other animal groups. For example, deflective tail markings seem to play a role in protection of young lizards from birds (Hawlena *et al.* 2006), and in deflecting raptor attacks from weasels (Powell 1982). Further, spots on the tails of tadpoles seem to deflect the attacks of dragonfly larvae (Van Buskirk *et al.* 2004). And many butterflies possess elaborate "false-heads" at the posterior end of the body, at least some of which appear to mimic appendages and eyes; here, it is difficult to argue against some form of manipulative influence on predator attack behaviour. Wallace was well aware of the intricate forms found on many lycaenid butterflies, and noted that such morphologies are often linked with specific behavioural displays (S725).

Conclusion

Wallace's grasp of crypsis was impressive but his focus was almost exclusively on protective resemblance. In this area he used astute observations of the natural world, especially from the tropics, to develop general examples of how animals blend in with their background. Although he recognized other routes to crypsis, including transparency, he developed these little. After a long period of inactivity, in the last decade there has been a considerable growth in interest in camouflage, driven by many conceptual and empirical studies. Scientists now view both protective resemblance and disruptive coloration as possible routes to crypsis, although the relative importance of these and other mechanisms has yet to be investigated.

8

The Colours of Animals: From Wallace to the Present Day II. Conspicuous Coloration

Tim Caro,* Geoffrey Hill, Leena Lindström, Michael Speed

• • • • • •

The second class—the warning colours—are exceedingly interesting, because the object and effect of these is, not to conceal the object, but to make it conspicuous. To these creatures it is useful to be seen and recognised.

(S725 1998 [1891], 350)

I have long held this portion of Mr. Darwin's theory to be erroneous, and have argued that the primary cause of sexual diversity of colour was the need of protection, repressing the female whose bright colours which are normally produced in both sexes by general laws.

(S725 1998 [1891], 364–65)

There is also, I believe, a very important purpose and use of the varied colours of the higher animals in the facility it affords for recognition by the sexes or by the young of the same species; and it is this use which probably fixes and determines the coloration in many cases.

(S725 1998 [1891], 367)

Introduction

In taking on the task of explaining all aspects of coloration in animals and plants, Wallace tackled conspicuous coloration in animals in three ways. First, he recognized that conspicuousness in nature, especially when found in juveniles of both sexes, signalled distastefulness and was therefore an anti-predator adaptation. He cited numerous examples across taxa to show this. Building on this idea, and in conjunction with his fellow fieldworker Henry Walter Bates, he propounded and

* Authors are listed in alphabetic order. Correspondence to Tim Caro. We thank George Beccaloni, Sami Merilaita, Charles Smith, and Martin Stevens for encouragement.

developed one theory of mimicry; later he expounded on a second theory that came to his attention sixteen years later as advanced by the German naturalist Fritz Müller. Second, for avian species in which the colour of the sexes differ, he singled out lack of conspicuous coloration (in females) as being the phenomenon to explain. In this respect he differed strongly from Charles Darwin, who saw the change as explaining gaudy coloration in (usually) male birds. Third, Wallace pointed to individual recognition within members of the same species as being another driving force for the evolution of conspicuous coloration in mammals. These three categories of animal coloration are the subject of this chapter.

Warning Colours

> But there are other caterpillars which seem coloured on purpose to be conspicuous, and it is very important to know whether they have another kind of protection, altogether independent of disguise, such as a disagreeable odour and taste. If they are thus protected, so that the majority of birds will never eat them, we can understand that to get the full benefit of this protection they should be easily recognised, should have some outward character by which birds would soon learn to know them and thus let them alone; because if birds could not tell the eatable from the uneatable till they had seized and tasted them, the protection would be of no avail, a growing caterpillar being so delicate that a wound is certain death (S130 1867, 206).

The idea that Wallace introduces here is that being distasteful is not sufficient for a caterpillar to survive, because if predators had to sample the caterpillars to find them distasteful they might injure them fatally during this process. "Not believing that any animal could have acquired a character actually hurtful to it without some more than counterbalancing advantages" (S389 1886, 305), he suggested that were a caterpillar conspicuously coloured and distinct from palatable prey (*i.e.*, cryptic prey), the predator would "soon learn to distinguish them at a long distance, and never waste any time in pursuit of them" (S134 1867, 20), thereby increasing the "survival value" of these caterpillars. Later Edward Bagnall Poulton (1890) would elaborate on animal coloration, introducing the term "aposematism" ($\alpha'\pi'o'$: away; $\sigma\eta\mu\alpha$: sign) to describe Wallace's theory.

Wallace's insight in the development of the theory of warning coloration was pivotal but, additionally, it shows the power of scientific collaboration. While formulating the theory of sexual selection (see below), Darwin was convinced that the beautiful colours of male birds and butterflies (Plate 10) were frequently a result of sexual selection and female preference (Darwin 1871). However, he could not understand why some lepidopteran larvae exhibit conspicuous colours. Larvae are not sexually active and thus coloration could not be attributed to selection by the other sex. So Darwin asked first whether Bates (see below) had an explanation for why larvae are aesthetically coloured. As Bates could not provide an

Figure 18 Caterpillars of the cinnabar moth (*Tyria jacobaeae*) gregariously feeding on ragwort (*Senecio jacobaea*). Their black and orange coloration advertises the fact that they are toxic.
From S318 1879. Out of copyright.

explanation, he suggested Darwin ask Wallace (Marchant 1916a). Although Wallace's letter of reply is missing, he wrote in his autobiography (S729 1905) that he immediately saw the answer to Darwin's question. His answer was a "theory of gaudy colours" and at the next meeting of the Entomological Society in London he asked for observations to support his theory (S130 1867). Many subsequent observations and experiments at the time proved his theory right (Poulton 1887, 1890). Further corroborating evidence emerged over the following century (Cott 1940; Edmunds 1974a; Ruxton *et al.* 2004a) and the colours of many animals, ranging from beetles to frogs (Plate 11), are now recognized as being aposematic.

Wallace, like all naturalists of his time, defended his theory through the use of numerous examples. His writings feature scores of examples of animals and plants in different habitats (deserts, tropics, sea, temperate forests, and arctic regions) that illustrate his various categories of animal coloration. To prove his theory of conspicuous coloration acting as a defence against predation, he also needed proof that birds avoid conspicuously coloured insects, and that insects have a distasteful character or odour to begin with (S129 1867). So he asked for examples and data from colleagues (S130 1867). But once he had such, those who did not believe his theory employed the same evidence as disproof, since, they argued, if predators do not eat these conspicuously coloured prey, how can the phenomenon have been selected in the first place? Wallace replied that there are differences in the predators' habits in acting against different prey species, and that the aposematism of caterpillars was primarily targeted against birds and lizards, and not, for instance, toads (S535 1897). But he also understood that every defence strategy is not entirely effective; *e.g.* predators might have adaptations to counter those defences—bees, for example, are not protected against bee-eaters (S724 1889). Similarly, hungry or young birds and lizards might be more willing to attack conspicuously coloured prey (S319 1879, S724). Wallace also was aware that the theory might have some exceptions. Some animals are large and so cannot hide, rendering them conspicuous, and some deadly poisonous snakes are cryptically coloured in order to ambush their prey (S134 1867).

In hindsight, the theory of Batesian mimicry, introduced a few years earlier (Bates 1862; see below) has features similar to Wallace's ideas on warning coloration. Bates understood that some butterflies use colourful patterns to communicate that they are distasteful to their predators, birds (the very same argument that Wallace used). Bates also observed that those patterns are mimicked (or imitated) by other species for their own protection even though they are palatable (Plate 12). Therefore, it is interesting to speculate on why it was Wallace rather than Bates who came up with the answer to Darwin's question. It could be related to the fact that Wallace's view of conspicuousness was set in the context of his theories about protective coloration (see Chapter 7); *i.e.*, warningly coloured caterpillars need to differ from the camouflaged and edible prey. The butterflies that Bates studied (*e.g.* the Ithomiinae) are not very brightly coloured (they are brown, black, white, yellow, or even transparent), so perhaps he did not realize the significance of bright coloration in signalling distastefulness. He may have thought that it was simply that a colour pattern was remembered by predators because the insect bearing it was distasteful. Alternately, Wallace may just have connected different observations (S725 1998 [1891]).

Wallace's genius lay in finding the explanation that cryptic and warning coloration evolved by the same mechanism: differential predation. Later, Sir Ronald Fisher (1930) pointed out that conspicuousness or distinctiveness from cryptic prey makes it difficult to explain how this anti-predator strategy originally evolved if the predators are naïve, and particularly if every predator generation has to learn the association between coloration and unprofitability (Edmunds 1974a). Thus Fisher saw warning coloration as a form of evolutionary paradox. If the conspicuousness was good against educated predators, how did it initially evolve if the ancestral form was cryptic and rare conspicuous mutants faced increased risk of being killed by naïve predators (see Lindström *et al.* 2001)? The latter part of the last century's research has centred on this apparent paradox (Gittelman and Harvey 1980; Alatalo and Mappes 1996; Lindström *et al.* 1999; see Ruxton *et al.* 2004a).

Wallace did not touch directly on the origin of conspicuousness but he repeatedly argued that insects reproduce in great numbers (S319 1879). He certainly knew the apparent cost of being conspicuous as he remarked that domestication and consequent reduction of predation pressure increased the number of white varieties in swallows and blackbirds (S389 1886), so it is strange that he did not perceive the origin of conspicuous warning coloration as a problem. Recall, genetics was not well understood during Wallace's life and therefore an understanding of frequency dependent selection was not available. Fisher, who was a statistician, soon realized that conspicuousness (and increased detectability) can create a very strong barrier for the evolution of this anti-predator defence because conspicuous signalling is beneficial only when it is relatively common (Sword 1999; Lindström *et al.* 2001).

Wallace's copious writings created a basis for an active research field (see reviews in Cott 1940; Edmunds 1974a; Ruxton *et al.* 2004a). His explanation for the underlying mechanism of conspicuous coloration, predation, was followed for the next century and it now integrates behavioural ecology (Endler 1991), psychology (Guilford 1990), and phylogeny (Sillén-Tullberg 1988) in making rigorous experimental tests. We now know that predators indeed learn more rapidly to avoid conspicuous than cryptic unpalatable prey (Gittleman and Harvey 1980; Sillén-Tullberg 1985; Riipi *et al.* 2001) and we know that conspicuousness carries costs which can be partially balanced by prey grouping (Alatalo and Mappes 1996; Gamberale and Tullberg 1998; Sword 1999; Riipi *et al.* 2001) or by rapid predator learning (Lindström *et al.* 1999; Riipi *et al.* 2001). Thus we know, exactly as Wallace suggested, that the conspicuous colour pattern makes it easier for predators to learn to avoid aposematic prey and to remember it for longer (Roper 1994; Speed 2000). But we also know that predator species (Exnerova *et al.* 2007) and individuals are variable, that they have innate biases to avoid warningly coloured prey (Shuler and Hesse 1985), and that hunger does indeed make predators select toxic prey strategically (Barnett *et al.* 2007). Recently some theoretical work addressed Wallace's notion that learning arises from conspicuousness being more easily *separated* from cryptic prey (Sherratt and Beatty 2003).

Wallace overlooked the effects of developing toxicity/unpalatability in warningly coloured prey species; this has been an active field of research in plant-animal interactions since the 1960s (*e.g.*, Brower *et al.* 1968). Insects are known chemists of the animal phyla, and they sequester plant secondary substances in special glands either directly, or metabolize these into new chemical compounds (Blum 1981; Bowers 1990; Rothschild 1993). They are adapted to use plants as food but simultaneously employ plants to provide the raw materials for their defence against predators. It has been argued that the costs of dealing with plant toxins are outweighed by the benefits of escaping predation through conspicuous signalling (Rothschild 1993). In other words, predation has maintained the adaptation to utilize host plant toxins (Berenbaum 2001; Dobler 2001). We also know from phylogenetic analysis that in some groups, conspicuousness and advertising co-evolves with the toxicity of the host plant (Farrell and Mitter 1998; Summers and Clough 2001). Instead Wallace focused on species that were palatable but mimicked unpalatable conspicuous species.

Mimicry

When read in chronological order, Wallace's comments and writings on mimicry make for fascinating reading because they allow us to trace the historical challenges to which the ultimately successful theories of mimicry were subject after the publication of Bates' original formulation (Bates 1862). Initially, Wallace had recognized the existence of shared resemblances between members of distantly related species; he had found much evidence for this during his fieldwork in Malaysia (S715 1869a) but he had no evolutionary explanation for the phenomenon. In a letter to Darwin (December 1860), Wallace wrote "Natural Selection explains *almost* everything in Nature, but there is one class of phenomena I cannot bring under it,—the repetition of the forms & colours of animals in distinct groups, but the two always occurring in the same country & generally on the *very same spot.* These are most striking in insects, & I am constantly meeting with fresh instances" (Burkhardt and Smith 1993, 504). Shortly afterwards, Bates began communicating his theory of mimicry via personal letter (*e.g.* H. W. Bates to C. R. Darwin, 30 September 1861); a public announcement of the idea was published the next year (Bates 1862). Bates' theory explained the resemblance between insect prey that are defended by virtue of being "unpalatable" and those which lack such a defence. Batesian mimicry (Plate 12) is therefore a form of deceptive mimicry because "palatable" prey deceive predators by their resemblance to undesirable species.

Bates combined his insight into the benefits of mimicry with the newly announced Darwin/Wallace theory of natural selection, so that not only did he articulate the ecological functions of mimicry, he also deduced a selective mechanism which would explain its existence without recourse to special creation. Thus Bates described how bird predators may cause mimicry to evolve, "the selective agents being insectivorous animals, which gradually destroy those sports or varieties that are not sufficiently like [the distasteful models] to deceive them" (Bates 1862, 512). It is fair to say that such "Batesian" mimicry became a champion cause for proponents of natural selection, and that a major campaigner was Wallace, who knew Bates well, having travelled with him in the 1840s (Bates 1864).

Wallace and Batesian mimicry

Wallace's first, and arguably most important, contribution to the study of mimicry was presented to the Linnean Society in March 1864, and summarized shortly thereafter in *The Reader* of 16 April 1864 (S96 1865). In this work Wallace examined "Batesian" mimicry and demonstrated the remarkable phenomenon that mimetic resemblance could be limited to the female sex (now termed sex-limited mimicry). Wallace described data which showed that within (edible) species from the *Papilio* genus (Papilionidae) of butterflies (such as *P. polytes*), mimicry of noxious model species occurs in the female but not in the male. Prior to this, certain males and females in the *Papilio* genus had often been considered separate species, and the complete absence of one sex in each "species" was an enigma. By demonstrating that mimetic females and non-mimetic males could emerge from a single brood, Wallace solved a problem for tropical lepidopterists "that no male of *P. polytes* has ever yet been found, although the species is very common" (S96, 491).

More importantly for the nascent field of evolutionary biology, Wallace later explained sex-limited mimicry in terms of differences in the force of selection on male and female conspecifics. In 1867 he argued that males experience relatively lower threats from predation than females, and hence face little or no evolutionary pressure to evolve mimicry; he wrote "In insects the case is very different [to higher vertebrates]; they pair but once in their lives, and the prolonged existence of the male is in most cases quite unnecessary for the continuance of the race" (S134 1867, 36). In contrast, the female needs to survive for longer and to expose herself to threats of predation during oviposition, so selection for mimicry as a form of anti-predator defence is much stronger: "The female, however, must continue to exist long enough to deposit her eggs in a place adapted for the development and growth of the progeny. Hence there is a wide difference in the need for protection in the two sexes; and we should, therefore, expect to find that in some cases the special protection given to the female was in the male less in amount or altogether wanting" (S134, 36–37). This explanation for sex-limited mimicry survives in current literature (Mallet and Joron 1999; Ohsaki 2005). Mallet and Joron (1999) argue, for example, that females may need to engage in mimicry because selection of oviposition sites requires slow flight and one way to be protected while flying slowly is to copy the colour patterns of typically slow-flying and chemically defended aposematic species. There is some evidence that in some butterflies with sex-limited mimicry the female may be subject to heightened predation risk during flight (Ohsaki 1995).

Now, in addition, we know that biases in female mate choice may also explain sex-limited mimicry by imposing excessive costs on males if changes to their colour patterns lead to sufficient reduction in mating opportunities (*e.g.* Turner 1978; Krebs and West 1988). This explanation appears to originate with the naturalist Thomas Belt (1874), who argued that the costs of mimicry were too great because females may prefer males of the "primordial" colour.

Wallace did not propound a sexual selection argument for sex-limitation (see below) presumably because he viewed costs of predation for animals in general to be very large, and to outweigh other functions of animal coloration such as female choice-based sexual selection (S724 1889). Nonetheless, the modern view suggests that both sex-differences in exposure to predation and sexual selection can combine to cause sex-limited mimicry (Mallet and Joron 1999). This dual approach has been used to explain male-limited Batesian mimicry in the beetle *Chrysobothris humilis*; Hespenheide (1975) argued that males of this species require mimicry because they are highly exposed to predation when displaying to females.

In addition to sex-limitation, Wallace revealed the equally remarkable existence of genetic polymorphism within mimicry (S96 1865, S134 1867); *i.e.*, females from a single brood may have radically different colour patterns, because they mimic alternative, distinctive noxious model species. Wallace used these observations to formulate distinctions between concepts of monomorphism, dimorphism, polymorphism, variation, and species. Mimetic polymorphisms in *Papilio* species (especially *P. polytes*) have been repeatedly used as a case study to explain how density dependence and frequency dependence can explain the maintenance of genetic variation within a breeding population (*e.g.* Turner 1977). In essence, edible Batesian mimics have some "burden of discovery," in that predators may start to attack mimics and their models if predators learn that the valuable mimics are sufficiently common. Splitting the mimetic appearances of a locally abundant edible species into discrete mimetic forms which copy several different models lowers this burden of discovery to the benefit of all individuals.

Wallace had no ready explanation for these mimetic polymorphisms. However the phenomenon itself was used to argue against the Darwin/Wallace theory of natural selection by members of the "mutationist school," headed by Reginald Punnett. In fact J. C. F. Fryer, a colleague of Punnett's, compared the relative frequencies of contemporary mimetic forms of *Papilio polytes* with those found in older records, from 50 to 150 years previously, and found no substantial difference between them (Fryer 1914). For the mutationists, and presumably many other biologists of the time, natural selection was evidenced by change in the relative frequency of alternative forms across generations. Fryer's finding of evolutionary equilibrium between mimetic forms of *P. polytes* led Punnett to conclude that in respect of mimetic resemblances natural selection does not exist for *P. polytes* in Ceylon (Punnett 1915; Gerould 1916). Some years later Fisher decisively refuted Punnett's conclusion, when he explained stable polymorphisms in Batesian mimics on the basis of frequency dependent natural selection (Fisher 1927, 1930) and put mimicry at the heart of the modern synthesis on which contemporary evolutionary biology now rests.

Two years after the presentation of his landmark paper on mimicry (S96) Wallace published the first of several vigorous refutations of the sceptic's arguments against Bates' mimicry theory. Entomologists such as John Westwood and David Sharp argued that mimicry between species may exist for reasons other than those proposed by Bates. For example, species may show mimicry because of accidental resemblance, or because of adaptation to similar ecological conditions, similar conditions of life, or because of shared recent phylogeny, in Wallace's words "reversion to a common ancestral type" (S123 1866). Wallace refuted the arguments of the sceptics and went on to formulate a set of criteria through which Batesian mimicry could be validated in the natural world (*e.g.*, S123 1866, S724 1889). Thus, for example, in 1867 (S134) Wallace proposed three "mimicry laws" consistent with the law of survival of the fittest.

The first law is, that in an overwhelming majority of cases of mimicry, the animals (or the groups) which resemble each other inhabit the same country, the same district, and in most cases are to be found together on the very same spot. This is still taken to be true in modern mimicry studies. However, seasonal polyphenism has been investigated in relation to mimicry and it is sometimes argued on theoretical and empirical grounds that, in seasonal species, individual Batesian mimics may have greatest survival if they emerge later than their models, in order that they minimize their "burden of discovery" (Huheey 1980). Similar arguments have recently been made in relation to asynchrony in daily activity patterns of models and mimics in butterflies (Pinheiro 2007).

The second law is, that these resemblances are not indiscriminate; but are limited to certain groups [models], which in every case are abundant in species and individuals, and can often be ascertained to have some special protection. This contention is certainly true; there are no known cases of "pointless" mimicry, in which a prey mimics another for no clear adaptive reason.

Wallace's third law was *that the species which resemble or "mimic" these dominant groups, are comparatively less abundant in individuals, and are often very rare.* The view that examples of Batesian mimicry are rare persists to the present (*e.g.* see review in Huheey 1988). Furthermore it is still widely accepted that individuals do best in populations of mimics that are rare relative to their models; here the risk that mimicry will be discovered by predators is low (and as described above, mimetic polymorphism may be a mechanism by which members of an abundant population gain high levels of protection from mimicry). However this rarity rule is not necessarily true (an early critique is in Punnett 1915): very highly noxious models, which pose significant danger to predators, need not be more numerous than their mimics.

Although the origins of mimicry theory, and the initial data used to support it, came from observations of butterflies and moths (and especially the mimetic assemblages centred on Amazonian nymphalids [Nymphalidae] and the East Asian *Papilios*), Wallace and others rapidly expanded their application of mimicry theory to other groups, such as beetles, crickets, and dipteran flies, and to some vertebrates—birds, snakes, and frogs. Nonetheless, Wallace considered that there would prove to be good reasons why mimicry was less common in vertebrates, arguing that modification of vertebrate skeletons would prove a significant evolutionary constraint: "the skeleton being internal the external form depends almost entirely on the proportions and arrangement of that skeleton, which again is strictly adapted to the functions necessary for the well-being of the animal. The form cannot therefore be rapidly modified by variation" (S134 1867, 31). Furthermore, Wallace realized that limited opportunity would lead to reduced rates of mimetic evolution in the vertebrates; he argued that "The number of [vertebrate] species of each group in the same country is also comparatively small, and thus the chances of that first accidental resemblance which is necessary for natural selection to work upon are much diminished" (S134, 31). Wallace allowed an exception to this rule in the snakes, where an abundance of venomous species as potential models made the evolution of mimicry more likely.

Current research on Batesian mimicry is still keenly focused on insects, especially Lepidoptera, beetles, ants, and dipteran flies as well as jumping spiders, with less work being done on coral snakes, fishes, and amphibians (Eagle and

Jones 2004). It is hard to know how much of the continued emphasis on Batesian mimicry in the invertebrates reflects a greater frequency in this group or the historical legacy left by the great Victorian naturalists and the early evolutionary biologists. Certainly Batesian mimicry in mammals seems very scarce and its distribution in the birds is poorly understood (Caro 2005); in marine fishes Batesian mimicry may be more common, although under-appreciated (Randall 2005).

Though much modern work on Batesian mimicry assumes that models will be chemically defended, such as with the venoms of snakes and wasps, Wallace had no difficulty extending the idea of deceptive mimicry to putative models protected by physical (even social) defences: "a genus of honeysuckers called *Tropidorhynchus* [= *Philemon*], good sized birds, very strong and active, having powerful grasping claws and long, curved, sharp beaks. They assemble together in groups and small flocks, and they have a very loud bawling note, which can be heard at a great distance, and serves to collect a number together in time of danger. They are very plentiful and very pugnacious, frequently driving away crows and even hawks" (S134, 32). Wallace argued that a much less well protected species of oriole (genus *Mimeta* (= *Oriolus*)) were in fact Batesian mimics of the *Tropidorhynchus* (= *Philemon*) friarbirds, stating that "on a superficial examination the birds are identical, although they have important structural differences, and cannot be placed near each other in any natural arrangement" (S134, 33). To this day, the idea that Batesian mimicry can be sustained on the basis of some non-toxic defence is still understudied and controversial (Srygley 1994; Brower 1995; Ruxton *et al.* 2004c; Sherratt *et al.* 2004).

Aggressive mimicry

Wallace's interest was in mimicry in general, so perhaps he followed Bates' lead (H. W. Bates to C. R. Darwin, 30 September 1861) in devising and popularizing an alternative use for mimetic resemblance, now known as "aggressive mimicry." Wallace wrote that "There are a number of parasitic flies whose larvæ feed upon the larvæ of bees, such as the British genus Volucella and many of the tropical Bombylii, and most of these are exactly like the particular species of bee they prey upon, so that they can enter their nests unsuspected to deposit their eggs" (S134, 29–30). Wallace used his extensive knowledge of tropical natural history to add examples of aggressive mimicry, for example, writing that "There is a genus of small spiders in the tropics which feed on ants, and they are exactly like ants themselves, which no doubt gives them more opportunity of seizing their prey" (S134, 30). Furthermore, Wallace considered that mimicry of another species for the purposes of enhanced predation may extend to mammals, citing the example of an insectivorous tree shrew (genus *Cladobates* (= *Tupaia*)) which resembles a local squirrel: "the use of the resemblance must be to enable the Cladobates to approach the insects or small birds on which it feeds, under the disguise of the harmless fruit-eating squirrel" (S134, 34). Unsupported anecdotes about cheetah cubs *Acinonyx jubatus* resembling pugnacious ratels *Mellivora capensis* and aardwolves *Proteles cristatus* resembling striped hyaenas *Hyaena hyaena* are still found in the literature (reviewed in Caro 2005).

Wallace and Müllerian mimicry

The idea of Batesian mimicry (in which an edible prey copies the appearance of one with strong defences) was easy to grasp, and quickly accepted. In contrast, the explanation for mimicry between defended species proved elusive for more than sixteen years after Bates published his original theory. Neither Wallace nor Bates had an explanation for mimicry between defended prey, so that Wallace later recorded that "In these cases both the imitating and the imitated species are protected by distastefulness, and it was not therefore clear how the one could derive any benefit by resembling the other. Accordingly, Mr. Bates did not consider these to be true cases of mimicry, but to be due, either to identical parallel variations of externally similar form, or 'to the similar adaptation of all to the same local, probably inorganic, conditions'" (S353 1882, 86). There is some irony that Bates had recourse to the argument of shared environment to explain these cases of mimicry, which Wallace had refuted in clear cases of Batesian mimicry. Indeed Wallace himself had attributed this puzzling form of mimicry to *unknown local causes* as late as 1876 (S257). In 1878, Fritz Müller published an explanation for this form of mimicry—now known, of course, as Müllerian mimicry (Müller 1878). This account, however, was not published in English, and it appears that Wallace and his British contemporaries remained ignorant of the idea until a translation of Müller's second statement of his theory was published (Müller 1879).

Müller had argued that if insectivorous birds learn to avoid unpalatable prey, and take a fixed number of given appearance during their education, then mimicry between unpalatable prey would be beneficial to individuals because the mortality costs of predator education would be partitioned out between members of the mimetic species. Mimicry then evolves not to deceive a predator, but because it is mutually beneficial to members of all co-mimic species. Müller had presented what is effectively the first mathematical model of selection, showing that "If both species are equally common [and equally well defended], then both will derive the same benefit from their resemblance—each will save half the number of victims which it has to furnish to the inexperience of its foes. But if one species is commoner than the other, then the benefit is unequally divided, and the proportional advantage for each of the two species which arises from their resemblance is as the square of their relative numbers" (Müller 1879, xxvii).

Wallace saw the value in the new mimicry theory, and readily conceded that his earlier statements about non-predatory, localized causation of shared resemblances were incorrect. Wallace therefore wrote that "Dr. F. Müller's theory appears to me to afford a clue (with some slight modifications) to most of the cases of close individual resemblance of not-nearly-related species of butterflies yet observed"

(S353 1882, 86). Wallace's contribution was primarily to support Müller's theory, and characteristically, to consider how the theory might be applied across a general range of ecological conditions. Nevertheless, Wallace rejected the significance of Müller's model, pointing out that what we now call "fitness" is not measured as a relative term between two species, but as a term relative to individuals within a species. Wallace wrote:

> I am, however, not quite sure that this way of estimating the *proportionate* gain has any bearing on the problem. When the numbers are very unequal, the species having the smaller number of individuals will presumably be less flourishing, and perhaps on the road to extinction. By coming to be mistaken for a flourishing species it will gain an amount of advantage which may long preserve it as a species; but the advantage will be measured solely by the fraction of *its own numbers* saved from destruction, not by the proportion this saving bears to that of the other species. I am inclined to think, therefore, that the benefit derived by a species resembling another more numerous in individuals is really in inverse proportion to their respective numbers, and that the proportion of the squares adduced by Dr. Müller, although it undoubtedly exists, has no bearing on the difficulty to be explained (S359 1882, 482).

Wallace also expanded the scope of the theory to include situations in which levels of defence were unequal between mimetic species. One reading of Wallace's papers on Müllerian mimicry is that he could not reconcile Müller's theory (which focused on the effects of varied abundance in equally defended co-mimetic species) with his knowledge that unpalatability of insect prey is variable and may often be "partial." Although the text is not absolutely clear, it is not unreasonable to argue that Wallace appears to have fused some ideas from Batesian mimicry (that a less unpalatable species may gain from the greater protection of a more unpalatable species) with Müller's argument for mutual gain between defended species, in order to explain *in general* how mimicry can come about in the many cases where inequality in defence pertained (Rowland *et al.* 2007a).

Specifically, Wallace argued that a combination of factors (shared recent ancestry facilitating mimicry, unequal levels of unpalatability, and/or unequal levels of abundance) would combine to cause "Müllerian" mimicry. He even hinted that this form of mimicry, aided by shared ancestry between co-mimics, may more typically occur than Batesian mimicry:

> But it is evident, that, if these differences [in unpalatability] exist, it will be advantageous for the less protected to mimic the more completely protected species, and the fact of the affinity between the different genera, with perhaps some tendency to revert to a common style of coloration or marking, will afford facilities for the development of this class of mimicry even greater than occur in the case of the distinct and often remote families of completely unprotected butterflies. We need not, therefore, be surprised

> to find whole series of species of distinct genera of Heliconoid butterflies apparently mimicking each other; for such mimicry is antecedently probable on account of the greater need of protection of some of these species than others, arising either from some species being less distasteful to certain enemies, or less numerous, and therefore likely to suffer to a serious extent by the attacks of inexperienced birds. When these two conditions are combined, as they often would be, we have everything necessary for the production of mimicry (S353 1882, 87).

Despite the fact that Wallace considered it his synthesis, this whole view is now generally referred to as "Müllerian mimicry" (*e.g.* Turner 1977). Thus in a preamble to the above text, Wallace referred to his explanation of mimicry with unequal levels of defence: "There is however yet another cause which may have led to mimicry in these cases, and one which does not appear to have been discussed by Dr. Müller." Perhaps, then, we should consider mimicry between unequally defended species to be "Wallacean" rather than Müllerian!

Wallace's insight was not universally noticed by other researchers in the field. Thus in his 1908 paper on Müllerian mimicry, Guy Marshall added his own discussion about inequalities in defence and mimicry stating, "yet in practice, the application of the Mullerian interpretation involves the assumption of a uniform standard of inedibility, and the complications which would be introduced by inequality in this respect have not been taken into account" (Marshall 1908). Marshall then began what has become a century-long argument about whether or not less well-defended forms should be considered Müllerian, Batesian ("due to the simple operation of the principle enunciated by Bates"), or some combination of Batesian and Müllerian mimicry (Fisher 1927; Mallet 1999). More importantly perhaps, some vagueness in Wallace's writing left open the question of coevolution between Müllerian co-mimics and a rancorous exchange over this question between Marshall and Frederick followed (see Marshall 1908; Dixey 1908a, 1908b) that has only recently been readdressed at a theoretical level (Mallet 1999).

Recent debate about the nature of Müllerian mimicry has focused on the actual selective mechanisms by which convergence of colour patterns may come about. One problem with the mechanism suggested by Müller and favoured by Wallace and others, is that the cost of educating young naïve birds about distinct colour patterns may not be sufficiently large to cause the evolution of mimicry, especially very precise forms of Müllerian mimicry (see Rowe *et al.* 2004). In fact, Wallace himself proposed an alternative mechanism which may cause selection for mimicry even after learning is complete. Writing about mimicry rings (where several unpalatable species resemble each other) in Heliconidae (= Ithomiinae and Heliconiinae: Heliconiini), he stated that "the types of coloration are few and very well marked, and thus it becomes easier for a bird or other animal to learn that all belonging to such types are uneatable. This must be a decided advantage to the family in question, because not only do fewer individuals of each species need to be sacrificed in order that their enemies may learn the lesson of their inedibility, *but they are more easily recognized at a distance, and thus escape even pursuit*" [italics added] (S724 1889, 255). About a century later, a similar idea was used to explain aposematic signalling (the "distance-detection hypothesis": see Guilford 1986; Gamberale-Stille 2000). In articulating the idea, Wallace was effectively extending his own original view that aposematism functions by making unprofitable prey distinctive and easy to recognize. The idea that enhancement of recognition accuracy is actually the major cause of mimetic convergence between defended prey species is currently gaining some ground (MacDougall and Dawkins 1998; Beatty *et al.* 2004).

Sexual Dimorphism and Monochromism in Birds

Many birds have beautifully coloured feathers and patches of skin (Hill and McGraw 2006a, 2006b). In some species there is extreme divergence in appearance of males and females (sexual dichromatism); in others, the sexes are exactly alike. One of the greatest triumphs of Wallace's application of natural selection to explanations of the evolution of feather coloration was his hypothesis that exposure of incubating females on nests drove the evolution of dichromatism (S139 1868). Wallace proposed that the basic condition of both males and females of most species was gaudy coloration—a contention that lies at the centre of Wallace's views on sexual selection (see below). According to Wallace, it was predation on females in species that incubated on open cup nests that caused natural selection to lead to the evolution of drab plumage in these females. As males of most species did not incubate, they were relieved of such natural selection against bright coloration and hence dichromatism, with brightly coloured males and drably coloured females, was a result of natural (survival) selection on females.

Wallace tested his theory for the evolution of sexual dichromatism using an *ad hoc* comparative study and without publishing the number of species that he examined he found almost perfect support for his theory in his survey of bird species (S139 1868, S724 1889). There were a few exceptions to his predicted pattern, but in his view, these cases actually further supported his hypothesis when considered in more detail (S139). For instance, the pitta family lays eggs in open cup nests on the ground, a nesting behavior that should lead to drab female coloration; but both sexes are brightly coloured, an apparent exception to the pattern predicted by Wallace. Wallace pointed out, however, that the ornamental coloration of pittas is restricted to the ventral feathers that are concealed when the female is incubating (S139).

The debate between Wallace and Darwin

In *Descent of Man*, Darwin (1871) rejected Wallace's hypothesis that natural selection acting through nesting behaviour shapes the evolution of plumage dichromatism. Much has been made of the debate between Darwin and Wallace regarding the role of female choice in the process of sexual selection and evolution of colourful feathers (Cronin 1991; Blaisdell 1992). But in his review of *Descent of Man* (Darwin's definitive statement on female choice as a key selective agent in sexual selection), Wallace spent many more words defending his explanation for sexual dichromatism in plumage and the role of natural selection in shaping such dichromatism than he did criticizing Darwin's view on mate choice (S186 1871).

In *Descent of Man*, Darwin argued it was unlikely that natural selection could exert independent selective influences on the two sexes. Wallace appears to have been especially wounded by Darwin's failure to accept what he saw as incontrovertible evidence supporting the role of nest predation and hence natural selection in shaping dichromatism. Particularly troubling to Wallace was Darwin's willingness to recognize a conspicuous portion of avian morphology, dichromatism, as being beyond the influence of natural selection. To Wallace, such a view was inexplicable given Darwin's willingness to attribute difference between the sexes in primary sexual characteristics to natural selection. Why arbitrarily exclude sexually selected traits? Wallace convincingly asked—"he [Darwin] appears to be unnecessarily depreciating the efficacy of his own first principle when he places limited sexual transmission beyond the range of its power" (S186, 181).

Wallace looked always to natural selection as the architect of the natural world. In his insistence on the primacy of natural selection, he was more of a staunch Darwinian than Darwin himself. And to Wallace, there was no greater triumph of the theory of evolution via natural selection than the colours of animals. Before the theory of natural selection, there was no rational or logical explanation for colour displays and no hope that such an explanation could be achieved. With natural selection theory, however, most coloration of most species of animals could be explained:

> Among the numerous applications of the Darwinian theory in the interpretation of the complex phenomena presented by the organic world, none have been more successful, or are more interesting, than those which deal with the colours of animals and plants. To the older school of naturalists colour was a trivial character, eminently unstable and untrustworthy in the determination of species; and it appeared to have, in most cases, no use or meaning to the objects which displayed it. The bright and often gorgeous coloration of insect, bird, or flower, was either looked upon as having been created for the enjoyment of mankind, or as due to unknown and perhaps undiscoverable laws of nature. But the researches of Mr. Darwin totally changed our point of view in this matter (S724 1889, 187).

Given Wallace's reluctance to admit that any trait of any animal fell beyond the working of natural selection, it is understandable why he was so dismayed to have Darwin identify such a large part of avian colour diversity as beyond the control of selection: referring to natural selection, he wrote, "With this principle as our guide, let us see how far we can account both for the general and special colours of the animal world" (S724, 190). Wallace saw these views of Darwin as clearly flawed.

Only recently and more than a century after he proposed the idea has Wallace's hypothesis that nest conspicuousness drives the evolution of female plumage crypsis been tested with comparative studies that statistically account for phylogeny and other potentially confounding variables (Johnson 1991; Martin and Badyaev 1996). These comparative analyses support Wallace's hypothesis insofar as they find a significant association between nest concealment and female conspicuousness. That said, modern studies of sexual dichromatism still do not recognize nest concealment as the primary explanation for sexual dichromatism (Badyaev and Hill 2003). The generally accepted view of sexual dimorphism in bird feathers, as well as of sexual dimorphism in the ornamental traits of all organisms, is that it is shaped by the intensity of sexual selection acting on both males and females (see Amundsen and Parn 2006; Hill 2006). Natural selection, such as the death of females while incubating, seemingly serves as a force that counteracts sexual selection that leads to trait elaboration and in this way also influences dimorphism (Irwin 1994; Badyaev and Hill 2003).

Recent studies of the more general phenomenon of sexual dichromatism have found that patterns of dichromatism across taxa are driven by change in female coloration rather than change in male coloration and in several taxa, strong dichromatism is the ancestral state and sexual monochromatism is the derived state (Badyaev and Hill 2003; Omland and Hofmann 2006). Moreover, the discovery of sex chromosomes and sex-specific hormone profiles (Kimball 2006) have shown that dichromatism can be shaped as readily by natural selection as any trait (Badyaev and Hill 2003). These studies indicate that Wallace was correct in emphasizing selection on female coloration as a key to understanding the evolution of sexual dichromatism.

Sexual selection and mate choice

Darwin and Wallace came to explanations of ornamental coloration from very different perspectives. For Darwin, ornamental traits posed a serious challenge to his theory of evolution by natural selection: "The sight of a feather in a peacock's tail, whenever I gaze at it, makes me sick!" (Darwin 1887a, 296). Explaining ornamental traits and particularly brilliantly coloured feathers of birds was a hurdle to Darwin's theory of natural selection as a universal explanation for the traits of animals. In contrast, Wallace never seemed to have viewed brilliant animal coloration as a challenge to natural selection. In his early writing, Wallace accepted Darwin's explanations of sexual selection including female mate choice for why "higher" animals, including especially birds, were colourful (S186 1871) although from his earliest writing he expressed doubt that there was mate choice among "lower animals": "Passing now to the lower animals—fishes, and especially insects—the evidence for sexual selection becomes comparatively very weak" (S186, 181). This challenge went unheeded for well over a century. It was only in the last year that the first empirical support for female choice of colour display in any species of butterfly or moth was published. Kemp (2007) showed in a series of carefully controlled experiments that females of the butterfly *Hypolimnas bolina* prefer to mate with more colourful males.

Wallace focused more on explaining the differences between males and females rather than on why one of the sexes was brightly coloured. But in his later writing, when he more directly took on the challenge of explaining bright animal coloration, Wallace rejected not only Darwin's ideas regarding a role for female mate choice in sexual selection—"Amid the copious mass of facts and opinions collected by Mr. Darwin ... there is a total absence of any evidence that the females admire or even notice this display" (S272 1877, 400)—but even the need for an adaptive explanation for most bright coloration. Wallace proposed that bright coloration is simply a by-product of the chemical activity of the body: "Colour may be looked upon as a necessary result of the highly complex chemical constitution of animal tissues and fluids" (S724 1889, 297).

So as Darwin struggled for an explanation for extravagant colour displays like bright red feathers, Wallace saw such coloration as no greater challenge to natural selection theory than the crimson coloration of blood or the yellow hue of fat: "... as differences of colour depend upon minute chemical or structural differences in the organism, increasing vigour acting unequally on different portions of the integument ... would almost necessarily lead also to variable distribution of colour ..." (S272 1877, 399). To invoke female mate choice to explain such phenomena seemed to Wallace unnecessary and indeed unrealistic. The thorough understanding of the biochemical basis for animal coloration that has developed since the mid-twentieth century (Needham 1974; Fox 1976; Fox and Vevers 1960; Hill and McGraw 2006a), however, clearly shows that Wallace was wrong in this hypothesis. Feather coloration can result either from the microstructure of the feathers interacting with ambient light (Prum 1999, 2006) or from the deposition of colourful pigments in feathers (McGraw 2006a, 2006b, 2006c). Structural coloration, as the latter form is called, is demonstrably not simply a "necessary result of the highly complex chemical constitution of animal tissues and fluids." Such coloration is the result of the precise arrangement of the molecular components of the feather and is certainly the result of selection for bright colour displays (Prum 1999, 2006). Among pigment-based coloration, carotenoid pigments create some of the brightest colour displays of vertebrates (Goodwin 1984; McGraw 2006a). Carotenoid-based yellow and red colour displays are the forms of external coloration that most closely match the brilliant yellow and red coloration of internal organs. But carotenoids do not incidentally get into feathers in the high concentrations needed for bright colour displays. Birds have evolved sophisticated mechanisms of absorption, transport, modification, and deposition of carotenoids to achieve bright colours (Hill 2002; McGraw *et al.* 2005). These mechanisms also certainly evolved in response to selection for bright coloration. Finally, melanin pigmentation, which creates the bold black patches and patterns on feathers, is the result of activation of melanocytes (McGraw 2006b). Such melanocyte activation seems to be under tight genetic control (Mundy 2006) and shaped by selection (Griffith and Pryke 2006; Plate 13).

With regard to the evolution of ornamental plumage coloration, Darwin was right. Such colours cannot be dismissed as the default state of biological systems; they do pose a challenge to evolution by natural selection, and they do require a functional explanation. And Darwin was correct that female mate choice is the primary selective force in the evolution of such colour traits (Houde 1997; Hill 2006) although such colour traits can also function in intrasexual signalling (Senar 2006). Since Wallace commented on the absence of evidence for female mate choice for ornamental traits, hundreds of such studies on traits ranging from the eye stalks of flies to the combs of roosters have been published (Andersson 1994). There have now been several dozen studies showing evidence for female mate choice for coloration in birds (Hill 2006). Wallace failed to see such a prominent role for female mate choice in the evolution of colourful plumage, but his assessments were reasonable given the empirical evidence that he had to assess.

The indicator model of sexual selection

Wallace is sometimes credited with foreshadowing what is termed the indicator model of sexual selection in current literature (Andersson 1994). The forerunner of the indicator model, the handicap model, was first proposed by Zahavi in the early 1970s (Zahavi 1975, 1977) and developed into indicator models in the 1980s and 1990s. Modern indicator models propose that ornamental traits evolve as reliable signals of individual quality, where the quality in question can be either body condition or genetic quality (Andersson 1994). Repeatedly in his writings Wallace proposed that colour expression was proportional to the health or vigour of a male: "In as far as these peculiarities [color displays] show a great vital power, they point out to us the finest and strongest individuals of the sex" (S724 1889, 296). But Wallace never directly asserted that females assess such traits to gain information about the quality of potential mates. He came closest to such a suggestion in a response to a book review of Romanes' *Darwin, and After Darwin* (S459 1892), but here he most clearly rejects the idea of female choice for signals of quality. First Wallace concludes that general vigour and ornament expression are linked such that more vigorous males have greater colour display, a fundamental principal of indicator models. He then considers whether females would base their choice on vigour or ornament expression, concluding that it would make no sense to take a weak male with a large ornament over a vigorous male with a small ornament. He then concludes with "I further admit that the display of ornament by the male is one of the means of exciting this desire; but mainly because it is an indication of sex, of sexual maturity, and of sexual vigour, probably not at all on account of details of color or pattern" (S459, 749). So Wallace flirted with the idea that vigour would enhance ornamentation and that more ornamentation would attract females, but he never saw signalling quality as a reason for the evolution of ornamentation.

When sexual selection was rediscovered by evolutionary biologists in the late twentieth century after decades of neglect, there was an initial rejection of Zahavi's handicap model by theoreticians (Maynard Smith 1976, 1978). It was not until indicator models were shown to be theoretically viable in genetic models (Grafen 1990) that they received broad acceptance as likely explanations for some ornamental traits (Maynard Smith 1991; Andersson 1994). The development of theories of runaway or Fisherian sexual selection (Fisher 1958; Maynard Smith 1991; Andersson 1994) and indicator models of sexual selection represent the greatest advances in the study of secondary sexual traits since the writings of Wallace and Darwin.

The intricacies of coloration

Wallace thought that it was very unlikely that female choice could lead to the evolution of colour patterns as intricate and precise as those seen in the feathers of birds and the wings of butterflies: "Successive generations of female birds choosing any little variety of colour that occurred among their suitors would necessarily lead to a speckled or piebald and unstable result, not to the beautifully definite colours and markings we see" (S186 1871, 182). To Wallace, the only explanation for the evolution of such intricate and precise pattern was some underlying organizing force that was yet to be discovered. In the involved and exact detail of colour patterns in the feathers of birds and the wings of butterflies, Wallace saw evidence for "some such law of development, due probably to progressive local segregation in the tissues of identical chemical or organic molecules, and dependent on laws of growth yet to be investigated" (S724, 298). With the publication of new models for the ontogeny of feathers and the generation of within-feather colour pattern, Wallace's ruminations about unknown laws of growth controlling colour patterns are strikingly prophetic. Prum and Williamson (2002) simulated feather growth and pigmentation using a six-parameter reaction-diffusion model. They were able to generate feathers with intricate bands, spots, and bars that match almost exactly the diversity of within-feather patterns observed in real feathers. Moreover, they showed that some feather patterns were epiphenomena, a by-product of selection for other feather patterns. While Wallace showed keen insight in deducing that intricate feather patterns would be manifestations of changes in growth parameters and not selection on each intricate part of the whole, he was wrong in surmising a selective pressure, such as female choice, could not generate such intricate patterns. The work of Prum and Williamson (2002) strongly suggests that selection acts on parameters of feather growth that will affect general features of the colour pattern.

At the time of writing, the evolutionary forces that shape the extremely precise and complex patterns of coloration, especially those in the feathers of birds or the wings of butterflies, remain poorly explained by theory or empirical results. Virtually all studies of mate choice in birds to date address either gross patterns of coloration, for instance how much of the body surface has red or black coloration, or the hue or brightness of the coloration (Hill 2006). The few studies that have assessed intricate patterns of coloration have focused on preference for symmetry, not preference for the pattern *per se* (Swaddle and Cuthill 1994). Some have tested for female mate choice relative to the "immaculateness" of feather coloration, which is defined as the sharpness of the transition between two colour patches (Ferns and Lang 2003; Ferns and Hinsley 2004). Such studies of immaculateness come closer to testing the aspects of coloration that Wallace proposed could not result from female choice, but still no studies have assessed mate choice in relation to the complex colour patterns that are so common in feathers (Plate 13) and butterfly wings (Plate 10).

Normal or Typical Colours

The fourth category of animal coloration envisaged by Wallace was recognition. "I am inclined to believe that its [color functioning in recognition] necessity has had a more widespread influence in determining the diversities of animal coloration than any other cause" (S724 1889, 217). Wallace had a category for these examples called normal or typical colours (S725 1998 [1891]). In essence this was a catch-all category (S272 1877) that included recognition of conspecifics, of group members, or of family members (S724).

First, Wallace thought that individuals needed to recognize members of their own species to avoid hybridization and that coloration would facilitate this (S527 1896). For instance, in considering the coloration of butterfly wings, Wallace noted that in some species groups, males of the different species were similar in coloration while females differed markedly. He suggested that the different colour displays of females were an adaptation to enable rapidly flying males to locate mates of the appropriate species (S272). Thus Wallace thought the sexes needed species-specific markers to recognize each other. He saw coloration, particularly coloration marks (S527 1896), as a means of attracting sexes to each other rather than allowing each sex to make subtle discriminations among mating partners (S432 1891, S527 1896). From the beginning of the twentieth century and for a further seventy years, this portion of Wallace's recognition hypothesis, proposing that colour displays functioned in species recognition in choice of mates, became the nearly universal explanation for ornamental coloration in animals. Yet throughout this whole stretch of time and more recently, when the species recognition hypothesis has been relegated to a much lesser role in the sexual selection literature (*e.g.* Andersson 1994), there have been very few empirical tests of the idea that plumage coloration evolves for species recognition, especially within the reproductive context that Wallace proposed. In a classic study on colour as a species isolating mechanism, N. G. Smith (1966) manipulated the eye-ring coloration of gulls and showed a strong effect on formation of interspecific pairs (Plate 14).

Second, Wallace thought group-living individuals needed to keep in contact with each other and that this would be facilitated by characteristic coloration

(S724). He saw the need to rapidly recognize group members as being responsible for the evolution of bold and striking colour displays like the flashing white tails of fleeing deer or the bold breast bands of plovers. He considered this important for birds congregating in flocks or migrating. As an illustration, Wallace noted that group-living artiodactyls had white rump patches and attributed it to grouping facilitation whereas solitary forest species did not (S724). Others have discussed this since (Kingdon 1982) and comparative analyses support the idea of white rumps being found in species living in intermediate sized groups, but these markings still remain largely enigmatic (Stoner *et al.* 2003a). Wallace even thought the shape of ungulate horns would serve the same purpose (S724), but other explanations involving fighting explain this better (Lundrigan 1996; Caro *et al.* 2003). Wallace (S499 1891) saw pattern symmetry on each side of the body as an additional aid to recognition; the functional significance of pattern symmetry outside mate choice is still unclear.

Third, Wallace thought that young animals might need to recognize their mothers (S272 1877) and this would be facilitated by mothers having a characteristic external coloration. White patches behind the ears of forest carnivores and black behind the ears of open habitat carnivores perhaps supports this (Ortolani and Caro 1996) since young animals follow their mothers, but no other systematic data are available on this point.

Wallace never discussed coloration as signalling age class, although this occurs in many dichromatic birds (Rowher *et al.* 1980) and primates (Treves 1977). Neither did he ever discuss colour patches as signalling fighting ability (Senar 2006), as signalling to predators (Caro 1995), as signal amplifiers (Maynard Smith and Harper 2003), or as thermoregulatory devices (Burtt 1979). That said, for one person, he engaged an extraordinary breadth of functional hypotheses, and most in considerable depth.

Conclusion

Wallace spearheaded predator avoidance as being the selective force driving the evolution of conspicuous warning colours in the animal kingdom; he recognized the significance of both sorts of mimicry that were discovered in his lifetime and expanded upon them; and he carried out a running debate on the significance of colour dichromatism in birds with his colleague and intellectual equal, Darwin. He also recognized that there were many aspects of coloration that do not fit into these categories, or that of protective coloration, and, in contrast to the other categories, we are little further on in understanding these today. More generally, and incredibly, after more than a century since Wallace established coloration as being a subject of enormous biological significance, "The varied ways in which the colouring and form of animals serves for their protection, their strange disguises as vegetable or mineral substances, their wonderful mimicry of other beings, offer an almost unworked and inexhaustible field of discovery for the zoologist, and will assuredly throw much light on the laws and conditions which have resulted in the wonderful variety of colour, shade, and marking which constitutes one of the most pleasing characteristics of the animal world, but the immediate causes of which it has hitherto been most difficult to explain" (S134 1867, 42), we still know remarkably little about animal coloration. For example, we still do not understand why European rabbits *Oryctolagus cuniculus* lift their white tails when they bolt for refuges (S724 1889; Stoner *et al.* 2003b), why many species flaunt a bright patch of colour upon fleeing (S724; Stevens 2007); why disparate species living on islands are similarly coloured, often white (S724, S725); or why zebras *Equus burchelli* have black and white stripes (S725; Kingdon 2006). As Wallace wrote in *Natural Selection and Tropical Nature* (S725 1998 [1891], 340) "... colour is by no means so unimportant or inconstant a character as at first sight it appears to be; and the more we examine it the more convinced we shall become that it must serve some purpose in nature, and that, besides charming us by its diversity and beauty, it must be well worthy of our attentive study, and have many secrets to unfold to us." There is still work to do.

9

Alfred Russel Wallace, Biogeographer*

Bernard Michaux

• • • • • •

Biogeographer. noun. One who studies the branch of biology concerned with the geographical distribution of animals and plants.

Introduction

> The writings of these revolutionaries, like their life histories, advertise their Magellanic nature. Wallace's "On the Law Which Has Regulated the Introduction of New Species," Darwin's *Origin of Species*, Wegener's *The Origin of Continents and Oceans*, du Toit's *Our Wandering Continents*, do not smell of the classroom; they smell of swamps, jungles, rivers and beaches. Such risk-takers are not likely to be awed by professors or cowed by textbooks (McCarthy 2005).

Wallace, whilst not strictly the founder of the science of biogeography, was without doubt the most important figure in its genesis. His reputation is based on numerous technical publications and four books: *A Narrative of Travels on the Amazon and Rio Negro* (S714 1853), *The Malay Archipelago* (S715 1869), *The Geographical Distribution of Animals* (S718 1876), and *Island Life* (S721 1880). Wallace was a man of little formal education and no social standing in a class-ridden society, and how he rose to such a pre-eminent intellectual position within the scientific and wider community, is one of the truly great stories of nineteenth-century science. That he should fall into relative obscurity after his death makes his story all the more poignant.

* I would like to thank the editors, Charles Smith and George Beccaloni, for inviting me to contribute to this volume, and Charles Smith and Rich Leschen for their valued suggestions for improving an earlier draft of the paper.

Shermer (2002) suggested that Wallace's working-class origins predisposed him to develop "heretical" theories, such as his teleological explanation for the evolution of the human brain. He argued that Wallace's restless intellect, unencumbered as it was by received wisdom and nurtured by exposure to the educational programmes and opportunities afforded to working-class men at Mechanics' Institutes, became receptive to radical ideas. Wallace's working-class background was also important in the development of Wallace as a self-reliant, resourceful, and practical man. In my view it is doubtful he would have succeeded as a collector in the tropics without these attributes. In the summer of 1837, at the age of fourteen, Wallace was apprenticed to his eldest brother William as a trainee surveyor. For most of the next six years the brothers travelled extensively through rural Britain. Wallace enjoyed the life and became increasingly interested in the natural history of the areas he worked in. As a future zoologist of renown, it might surprise people to know that his first systematic scientific interest was in botany. The fact that common plants of the fields and hedgerows had scientific names and could be identified came as something of a revelation to him. He bought himself his first identification guide, Lindley's *Elements of Botany*, which he had to annotate extensively from Loudon's *Encyclopaedia of Plants* to make it more useful for identifying what he was collecting. He also wrote out passages from Darwin's *Journal of the Voyage of the Beagle* in the margins of his copy of *Elements* (Raby 2001). This work by Darwin was to be an early influence on Wallace, inspiring a desire to travel to exotic places himself.

Wallace did not confine his growing scientific interest to botany, but also started to read widely on geology, including the influential *Principles of Geology* by Charles Lyell. This work was to be important in the development of Wallace's ideas because it gave him a good general background in the subject, which he made an integral part of his biogeography, and because it espoused the principle of uniformitarianism. This principle stated that past events could be interpreted in terms of present-day, observable processes, implying that geological changes were continuous over long time periods, and dispensing with the need for catastrophic changes such as the biblical flood. Uniformitarianism provided an essential foundation for the theory of evolution by means of natural selection. It reinforced the idea that change is gradual and gave natural selection the time needed to bring about new species. Wallace had read Swainson's *Treatise on the Geography and Classification of Animals* during this early period, a book combining the two major preoccupations of Wallace's future intellectual life, but he was not impressed, writing: "To what ridiculous theories will men of science be led by attempting to reconcile science with scripture" (quoted in McKinney 1972a).

Wallace regarded the period 1840 to 1843 as one of the turning points in his life, a period during which he had set the course for his future. He had become fascinated with the natural world, had approached his self-education in a systematic way, and had acquired a broad, practical skill base that complemented his growing theoretical outlook. What was needed now was the spark to ignite him into action. That spark was to be provided by Henry Walter Bates.

The two young men met when Wallace took up a teaching position in Leicester following the collapse of his brother's business during "the hungry forties." Bates was an avid collector of insects, particularly beetles, and it was he who introduced Wallace to entomology in general and beetle-collecting in particular. Wallace was fascinated by the shapes, markings, coloration, and above all, by the diversity shown by beetles. He was astounded to learn that there were probably a thousand different species to be found within ten miles of Leicester. If botany piqued his interest in the "species problem" then beetle diversity required its solution. In Bates Wallace found a kindred spirit, someone with similar interests with whom he could discuss ideas, especially about evolution. There was an excellent subscription library in Leicester and it was here that Wallace read three very influential works. Humboldt's *Personal Narrative of Travels in South America* described his expedition along the Orinoco River which inspired Wallace to experience tropical nature himself. Malthus's *Essay on the Principle of Population* was to prove a key to his and Darwin's formulation of natural selection. Lastly, Robert Chambers's anonymously published *Vestiges of the Natural History of Creation*, a popular best-seller, brought the idea of evolution (the "species problem") into general debate and fired the imagination of these two young naturalists.

Wallace left Leicester in 1845 to resume surveying work but kept up a correspondence with Bates. Wallace wrote to him from Wales: "I begin to feel rather dissatisfied with a mere local collection, little is to be learned by it. I should like to take some one family to study thoroughly, principally with a view to the theory of the origin of species" (quoted in Raby 2001). As most families contain geographically dispersed genera and species, it is clear that Wallace was already thinking about the evolution of species in terms of space as well as time. It is probable that their plan to go to the Amazon was first hatched in 1846 when Bates was visiting Wallace in Wales. This "rash adventure," as Williams-Ellis (1966) called it, was to be financed by the collection and sale of specimens.

Travels on the Amazon, 1853

Wallace and Bates arrived in South America in 1848 and set up base at Pará, near the mouth of the Amazon (see Fig. 19). Bates (1910, 1–2) described the environs of Pará thus:

> To the eastward the country was not remarkable in appearance, being slightly undulating, with bare sand-hills and scattered trees; but to the westward, stretching towards the mouth of the river, we could see through the captain's glass a long line of forest, rising apparently out of the water; a densely-packed mass of tall trees, broken into groups, and finally into single trees, as it dwindled away in the distance.

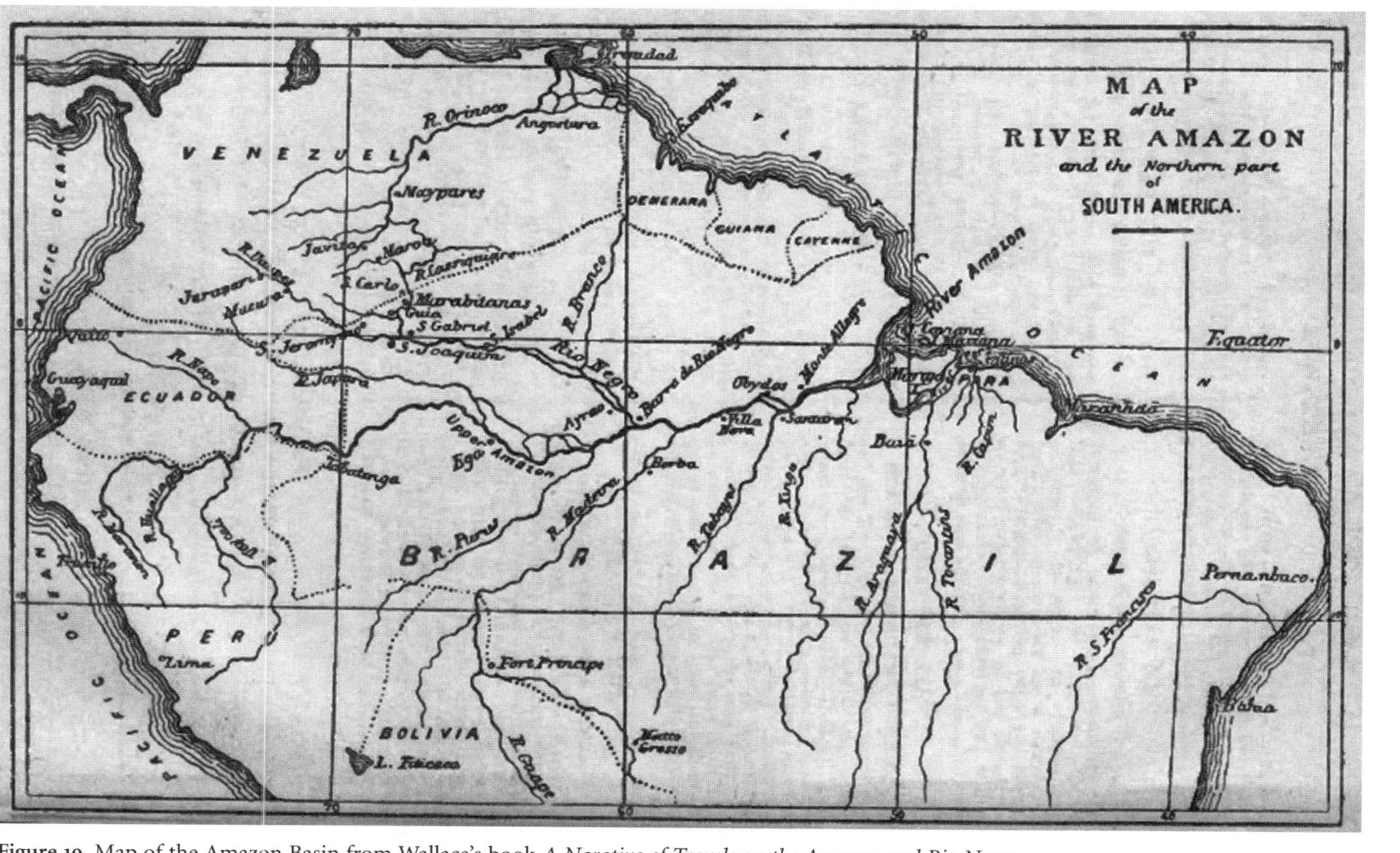

Figure 19 Map of the Amazon Basin from Wallace's book *A Narative of Travels on the Amazon and Rio Negro.*

Primeval forest was what they had come to explore, but it was just beyond reach from their base in Pará, and it wasn't until they moved upstream to Santarem that they experienced it in all its glory. Wallace was to comment: "Perhaps no country in the world contains such an amount of vegetable matter on its surface as the valley of the Amazon. Its entire extent, with the exception of some very small portions, is covered with one dense and lofty primeval forest, the most extensive and unbroken which exists upon the earth" (S714 1911, 300). The Amazonian rain forest was unlike any temperate forest he was familiar with because of the extraordinary diversity of the vegetation: "Instead of extensive tracts covered with pines, or oaks, or beeches, we scarcely ever see two of the same species together, except in certain cases, principally among the Palms" (S714 1911, 302).

Wallace was to offer an explanation for the increased diversity of plants and animals at lower latitudes in another important book, *Tropical Nature and Other Essays* (S719), in 1878. He suggested that tropical climates have been stable over geological timescales, giving greater time for diversity to develop, and that extinction rates in such equitable environments are lower than in the more physically demanding temperate regions. For example, Pleistocene glaciations would have caused the extinction of many species in the northern temperate regions (but see Coope 2004). Later he also suggested that the tropics were "museums," full of species that had migrated from the north under the influence of deteriorating climatic conditions during the Ice Ages, and now preserved there (S302 1879). Tropical diversity could thus be explained by a combination of habitat persistence, lower extinction rates, and immigration. Since Wallace's time there has been a considerable research effort to understand this most global of biogeographic patterns. Mittelbach *et al.* (2007) provide an excellent review of the present state of play in the subject, which is beyond the scope of this work. However, it is worthwhile to note that Wallace's original suggestions continue to contribute to this understanding.

Wallace was to spend four and a half years collecting in the Amazon basin; Bates would remain eleven years and go on to achieve renown as a tropical entomologist and originator of the hypothesis of Batesian mimicry. Wallace's greatest achievement was to explore the upper reaches of the Rio Negro and the headwaters of the Orinoco, in so doing fulfilling his ambition of emulating his great hero Humboldt. Collecting specimens in such rugged conditions was arduous and dangerous. Wallace suffered recurring bouts of fever, attacks by biting and stinging insects, and periods of near starvation. Wallace did not mount costly expeditions in the manner typical of European explorers of the time, but travelled lightly, without a great retinue of porters. He lived as the locals did because he knew how superior this was when collecting in isolated regions. For example, while exploring the Rio Negro and its tributaries he used existing trading and communication networks where possible, or hired local guides and travelled by dugout when not. A major food source for travellers was fish, and one of the camp chores at the end of the day would be to go fishing. This could be done either by netting or using *timbo*, a fish poison containing the alkaloid rotenone. *Timbo* was extracted from the liana *Paullinia pinnata*, by pounding the roots with a mixture of water and clay and extracting a juice which, when released into a stream, shuts down fish respiration and causes them to float to the surface. The collection would be inspected for new specimens before the rest went into the cooking pot. Wallace's fish drawings were among the few documents to survive his shipwrecking on his return journey to England and have recently been published (S734a 2002; reviewed by Harold 2005).

The beginnings of a biogeographer

Wallace's scientific output from these four years was not great. He did publish a few well-received technical papers and two books—a small ethnobotanical work on palms (S713 1853) and *A Narrative of Travels on the Amazon and Rio Negro* (S714 1853). Unfortunately neither book sold well, nor added much to Wallace's kudos. Nevertheless, collecting in the Amazon did lead to considerable personal and professional growth (Raby 2001). The experience had transformed him from a naïve enthusiast into a confident and skilled field biologist, professional traveller, and writer. There was certainly little left of the rather clumsy young man who had almost shot off his own hand early on during his Amazonian adventure. And he had made some significant advances in the practice and theoretical development of biogeography.

Wallace was an evolutionist in a pre-evolutionary world who, faced with an extraordinary array of diversity, had to develop the intellectual tools needed to understand how it had originated. The task of tracing Wallace's thought processes at this time is made more difficult by his theoretical understandings having been in advance of the little he wrote (Colwell 2000). His major biogeographical theme was, as George (1964) put it, the correlation between animals and plants with locality. Wallace showed that while diversity in tropical rain forests is high, occurrence need not be random. Occurrences often seemed to be localized, with clusters of endemic species occupying restricted localities (what we would now term "areas of endemism"). Wallace showed how the collection of accurate locality data could lead to the mapping of biology, and that interesting avenues for research—were, for example, the localization of distributions correlated with physical features?—could thus be opened up.

Wallace's first biogeography paper (S8) was read at the December meeting of the Zoological Society of London, and published in their *Proceedings* for 1852. His mapping of Amazonian monkey distributions showed that species' range boundaries were often major rivers. These also appeared to restrict the distributions of some birds and insects. On the basis of these data Wallace recognized four "districts" within Amazonia as a whole: Guiana and Ecuador, both north of the Amazon and separated by the Rio Negro, and Brazil and Peru, both south of the Amazon and separated by the Madeira. It is a testament to Wallace's powers of observation that his districts are still apparent within currently recognized Amazonian areas of endemism (Cracraft 1985; Brown *et al.* 1995; da Silva and Oren 1996; Hall and Harvey 2002; Racheli and Racheli 2004). Racheli and Racheli (2004) investigated the relationship between Amazonian interfluvial areas based on papilionid butterflies. Their study recognized some sixteen areas of endemism which grouped into three clades—Guyana, South East Amazonia, and West Amazonia. Guyana and South East Amazonia are equivalent to Wallace's Guiana and Brazil respectively, while West Amazonia is a combination of Wallace's Peru and Ecuador districts. Within West Amazonia a clade of two endemic areas (Imeri 1 and 2) is broadly equivalent to Wallace's Ecuador district, while the sister clade, composed of the remaining endemic areas, is broadly equivalent to Wallace's Peru district.

Wallace expanded on the theme of interaction of species and locality in *Travels*. His observation that rivers can act as barriers to the dispersal of Amazonian monkeys (S8 1852) was part of a general discussion about the means by which animal ranges are restricted. In some cases it was easy to imagine how physical features such as mountain ranges or large rivers could confine a distribution, or where specific ecological requirements, such as the granite outcrops necessary for nesting by cock-of-the-rocks (*Rupicola rupicola*), could restrict a species range. But other examples were less straightforward to interpret. For example species' boundaries could run across major rivers dividing the upper from the lower reaches of a river, or where extreme diversity in composition had been achieved in a continuous medium (as in the case of Amazonian fishes).

Wallace's study of Amazonian monkeys eventually led to the "Riverine Barrier Hypothesis," the oldest hypothesis for the origin of Amazonian diversity (Colwell 2000). While this hypothesis is still influential enough to continue to excite research interest (Gascon *et al.* 2000), it also provides insight into Wallace's development as a biogeographer. George (1964) argued that his work on monkeys was formative in developing Wallace's understanding of the role that barriers to dispersal have in the speciation process. Colwell (2000) went further to suggest that Wallace already saw these barriers as causes of speciation, even if he didn't say so publicly. Whilst it is quite possible that Wallace had understood the relationship between barriers and speciation while still in the Amazon, I suspect it didn't become clear to him until he reached Sarawak and wrote "On the Law Which Has Regulated the Introduction of New Species" (S20 1855).

The Malay Archipelago, 1869

Wallace's return to Britain was brief, about sixteen months in total, during which time he arranged his affairs and prepared for his next expedition. There was little doubt that he would return to the field because he had no available alternatives. As Huxley said in a letter to his sister: "Science in England does everything—but pay. You may earn praise but not pudding" (quoted in George 1964).

For a while Wallace briefly considered East Africa as a possibility, but eventually concluded the Malay Archipelago was safer, relatively uncollected, and at least as ripe for biological exploration. The eight years Wallace spent there would provide the raw material to feed his scientific theorizing, as well as much of the subject matter for his biogeography books. There is no doubt that he had come to the archipelago to achieve some ambitious goals, as he made clear in a letter to his family, possibly responding to their pleas to return to civilization:

> But I am engaged here in a wider & more general study—that of the relations of animals to time and space, or in other words their Geographical & Geological distribution & its causes. I have set myself to work out this problem in the Indo-Australian Archipelago & I must visit & explore the largest number of islands possible & collect animals from the greatest number of localities in order to arrive at any definite results (Raby 2001, 144).

At first Wallace used Singapore as a base from which to explore the Greater Sunda Islands (Borneo, Sumatra, and Java), but then moved to the more centrally located town of Makassar on Sulawesi to investigate the smaller islands to the east and on to New Guinea. Once again collecting expeditions were financed through the sale of specimens, so Wallace was obliged to keep a constant flow going back to his agent Samuel Stevens in London.

Wallace was to publish some forty-three scientific papers during his time in the Archipelago, on subjects as varied as descriptive lists, avian higher level systematics, and biogeography and evolution, as well as keep up a tremendous correspondence with peers, friends, and Stevens. This output included the famous Ternate paper of 1858 (S43) in which he independently proposed the theory of evolution by natural selection, and "On the Law Which Has Regulated the Introduction of New Species" (S20) written in Sarawak at the beginning of 1855, not long after his arrival in the East. He had spent a few months in Singapore and Malacca collecting and learning Malay before moving to Sarawak under the patronage of the "White Rajah" Sir James Brooke (Williams-Ellis 1966). When the rainy season brought collecting to a halt, Wallace retired to a small house at the mouth of the Sarawak River. There, nestled at the base of Santubong mountain, he wrote "Every species has come into existence coincident both in space and time with a pre-existing closely allied species" (S20, 186). The "law" clearly looks back to Wallace's Amazonian studies, particularly of the monkeys, and is significant in its linkage of evolution with space as well as time. He shows what a dynamic world view he had developed when reminding the reader that "the present state of the earth, and the organisms now inhabiting it, are but the last stage of a long and uninterrupted series of changes which it has undergone, and consequently, that to endeavour to explain and account for its present condition without any reference to those changes ... must lead to very imperfect and erroneous conclusions" (S20, 184). And later, speaking specifically of geographical distribution: "... the present geographical distribution of life upon the earth must be the result of all the previous changes, both of the surface of the earth itself and of its inhabitants" (S20, 185).

Thus, geographical distributions were the result of the dynamic between biology and geography. Species evolve and have an inherent capacity to increase their numbers and occupy new areas. At some times geographical change isolated species, promoting differentiation; at other times barriers were removed leading to dispersal and mixing of faunas. Changes in climate could also promote dispersal or cause isolation. Organisms tracking their preferred habitats (Coope 2004) would be an example of climate change promoting dispersal, while fragmentation of once-continuous ranges would exemplify its role in promoting isolation and possible speciation.

In Wallace's view the presence of endemic species indicated isolation because barriers to dispersal promoted speciation. In modern terminology, a population can become isolated at the periphery of a species' range, or the range may become divided by a geological, ecological, or climatic barrier, allowing sister-species to evolve through vicariance. Differences required time to evolve and so the degree of endemism or uniqueness (as indicated by the taxonomic rank of the endemics) became a function of the length of isolation. Thus the occurrence of endemic families or genera would indicate longer periods of isolation than would endemism at the species level. When animal and plant species showed little or no taxonomic differentiation there were no barriers to dispersal, and the area was not biologically isolated from its surroundings. Wallace was to use these conceptual tools to reconstruct past geographies in his attempt to understand the origin of the Archipelago and its animals.

The Malay Archipelago (S715), published in 1869 seven years after his return to England, was Wallace's most commercially successful book. He was able to bring into Victorian sitting rooms vivid accounts of the life, peoples, and landscapes of a region that might have been straight out of a fable. His clever interweaving of story lines, the wealth of detail so clearly presented, and his obvious love for what he was writing about ensured a continuing public demand. Detailed and often amusing descriptions of incidents from his everyday life, such as staying in Dyak long houses amongst the famed head-hunters of Borneo, or evicting giant pythons from his roof, made the ordinary extraordinary. What is often overlooked, however, is that *Malay Archipelago* is also a first-hand description of a biodiversity hotspot, by an experienced tropical biologist, before any large-scale habitat destruction had taken place. As such it is a historically important document concerning one of the most globally important centres of biodiversity and biological uniqueness (Wilson *et al.* 2006).

The Spice Islands revisited

Adventurer Tim Severin understood the historical importance of Wallace's work when he and his companions were inspired by *The Malay Archipelago* to sail a traditionally constructed prahu in Wallace's footsteps among the Spice Islands of eastern Indonesia to compare his descriptions of places with what they found nearly 150 years later. *The Spice Islands Voyage* (1997) is a homage to Wallace, who Severin regarded as an early environmental advocate:

> He was acutely aware that the magical places he described were under threat from the moment when he brought them to the attention of the outside world. Passionately he asked his readers to remember that the rare creatures he had discovered, particularly the rich bird-life, had lived undisturbed for countless generations. He pleaded that careful thought should be given to the preservation of these marvellous creatures; he strenuously urged the protection of the tropical environment for the benefit of its people and animals (Severin 1997, 12).

But fashioning Wallace as an environmental advocate in this manner is somewhat problematic. In a recent play written by Nick Drake—*Dr. Buller's Birds*—the life of this eminent New Zealand ornithologist and collector is played out to highlight the contradiction between his study of the New Zealand avifauna and his role in its destruction. Buller amassed sizeable personal collections of bird skins during the course of his studies and, whilst not strictly a dealer, was not averse to supplying them to museums or overseas collectors for either profit or the promotion of his own self-interest. He not only collected rare and endangered species himself, but also commissioned collectors throughout New Zealand to obtain specimens on his behalf. The extinction of the magnificent huia (*Heteralocha acutirostris*) by 1907, a year after Buller's death, can in part be put down to his rapacious collecting of this species. Buller gave accounts of hunting huia in his seminal work *A History of the Birds of New Zealand* first published in 1873. He apparently held the view that the bird was doomed anyway, so that specimens should be collected to preserve them for posterity (and personal gain no doubt). By the time of the publication of his final work in 1905—*Supplement to the Birds of New Zealand*—he had changed his mind and supported conservation policies which were then gaining tentative acceptance. Of course his own collecting days were over and his role in bringing about the circumstances that called for such policies was ignored. Nick Drake's play addressed the complex question of whether Buller can be regarded as an environmentalist, and while he leaves the audience to come to their own conclusions, it is clear that the answer cannot be a simple yes or no. I would suggest that there are some striking parallels when assessing Wallace as an environmental advocate.

Wallace estimated his own personal collection from the Archipelago consisted of 3,000 bird skins, 20,000 pinned beetles and butterflies, and sundry other mammalian and molluscan specimens (S715 1883). According to Shermer (2002, 126) his total collection from this period was 125,660 specimens. Wallace gave a clear insight into his views on collecting in a paper published in 1863:

> It is for such inquiries the modern naturalist collects his materials; it is for this that he still wants to add to the apparently boundless treasures of our national museums, and will never rest satisfied as long as the native country,

> the geographical distribution, and the amount of variation of any living thing remains imperfectly known. He looks upon every species of animal and plant now living as the individual letters which go to make up one of the volumes of our earth's history; and, as a few lost letters may make a sentence unintelligible, so the extinction of the numerous forms of life which the progress of cultivation invariably entails will necessarily obscure this invaluable record of the past. It is, therefore, an important object, which governments and scientific institutions should immediately take steps to secure, that in all tropical countries colonised by Europeans the most perfect collections possible in every branch of natural history should be made and deposited in national museums, where they may be available for study and interpretation (S78, 234).

This is clearly a version of the "posterity" argument that Buller used. Note how Wallace seems more concerned about the scientific value of endangered species than their ultimate survival. And his argument about the inevitability of extinction of some species echoes Buller's own pessimistic view. However, unlike Buller, Wallace's own collecting efforts probably didn't threaten the survival of individual species, but persistent collecting at that level most certainly did. Birds of paradise feathers have always been traded locally (like the huia), but their popularity in Europe and Asia saw harvesting reach unsustainable levels (Gilliard 1969). Mayr (1942) estimated that between 1870 and 1924 more than a million birds of paradise skins had been exported from New Guinea. One might expect that Wallace of all people would have spoken up about this had he really been an environmental advocate. Whilst his collecting may not have led to any extinctions, it may well have been detrimental to local populations. His description of shooting mias (orang-utan) is probably an example. Not only did he take large numbers of animals from restricted locations, but he also indiscriminately shot nursing females. Judged by the standards of the day, this otherwise remarkably likeable man was simply executing his trade in an efficient way for a noble purpose. In this respect Wallace was very much a man of his times. And just as Buller came to unequivocally support measures to protect the environment in later life, so Wallace was to articulate a clear conservation message (see for example S732 1910, 278–80). His masterful observation and study of nature, however, did not necessarily make him an environmental advocate; in my view the issue is far more complex. Still, I'm sure he would be pleased to know that his field writings retain the power to inspire modern environmental advocacy.

The results from Severin's study were mixed. In 1857 Wallace described Ambon's harbour:

> Passing up the harbour, in appearance like a fine river, the clearness of the water afforded me one of the most astonishing and beautiful sights I have ever beheld. The bottom was absolutely hidden by a continuous series of corals, sponges, actiniae, and other marine productions, of magnificent dimensions, varied forms, and brilliant colours. The depth varied from

> about twenty to fifty feet, and the bottom was very uneven, rocks and chasms and little hills and valleys, offering a variety of stations for the growth of these animal forests. In and out among them, moved numbers of blue and red and yellow fishes, spotted and banded and striped in the most striking manner, while great orange or rosy transparent medusae floated along near the surface. It was a sight to gaze at for hours, and no description can do justice to its surpassing beauty and interest (S715 1883, 294–95).

In the 1990s Severin and his companions "encountered a slimy yellow slick of plastic bags, old bottles and raw sewage floating out on the tide." They noted that degraded environmental conditions occurred where population growth had been poorly managed, or where government agencies set up to protect biodiversity were poorly funded and equipped. Still, examples were found where isolation and local control of resources resulted in an ecology as healthy or even healthier than Wallace experienced. Habitat loss continues to pose a serious threat to Indonesian wildlife (Benjamin 2007).

The Geographical Distribution of Animals, 1876

This two-volume opus, Wallace's first book devoted exclusively to biogeography and the concept of global regions, had its origins in Ternate (in the Spice Islands) where Wallace found himself based during 1858 and the early part of 1859. It was here he read Philip Sclater's paper "The General Geographical Distribution of the Members of the Class Aves" (Sclater 1858) in which global faunal regions were first described. Wallace responded immediately: "My Dear Mr. Sclater, Your paper on 'The Geographical Distribution of Birds' has particularly interested me, and I hope that a few remarks and criticisms thereon may not be unacceptable to you" (S52 1859, 449). In this work Wallace accepted Sclater's broad arrangement of six regions, with some amendments, and argued that the boundary between the Indian and Australian regions lay between Bali and Lombok in the south and Borneo and Sulawesi in the north. He was undecided about on which side of the boundary the Philippines lay. A year later (S53 1860) he presented additional data, mainly mammalian distributions, to further support Sclater's regions, and placed the Philippines in the Indian Region.

The Geographical Distribution of Animals (S718) was published sixteen years later and represents a synthesis of Wallace's long investigation into biogeography. By this time he was an established figure within British scientific circles, and certainly the person most eminently qualified to produce such an overview. The work was universally praised and became enormously influential, but I have to disagree with George's (1964) assessment that the book was revolutionary. True, it was novel to incorporate fossil evidence and treat it like any other data, but the evolutionary interpretation of Sclater's creation-inspired regions was not new, and conceptually it is rather confused. Wallace touched directly on the reason for this confusion in his final paper on the subject. Regions are conceptually valid only if they are "natural" and they are natural:

> ... because during the more recent geological periods they have formed single more or less continuous areas, while separated either by geographical, climatal, or biological barriers from the adjacent areas (S494 1894, 612).

But how would we know that a region is natural, or in modern terms how would we know a locality is an area of endemism? While the answer to this seemingly simple question is complex (Platnick 1991), such areas can only be recognized by the overlap of endemic species' ranges. Sclater's regions were originally defined in terms of bird distributions, and were certainly not natural entities as Wallace understood the term. Wallace knew that Sclater's regions were inspired by a creationist world view and were permanent features, so to speak. As a good evolutionist Wallace knew that there could be no "absolute character of independence"—the earth was constantly changing too. Global regions, therefore, weren't really natural entities but were conventions. By this he meant they were convenient. Distributional data have to be systematized to be useful, for example for data retrieval or making comparisons. And regions provided a type of pigeonholing into which these data could be ordered. In *Geographical Distribution* Wallace provided a current checklist of species for each region to demonstrate this utility. The popularity and longevity of faunal regions, one might argue, can be put down to their utility rather than to any real insights into the natural world they provided. The concept of regions is now devoid of any theoretical content, but the names are still widely in use for the same utilitarian reasons that appealed to Wallace.

A whole theoretical edifice was built on these shaky foundations. Animals and plants evolved at centres of origin and then dispersed outwards. Wallace developed and promoted this model to help explain disjunct southern distributions (such as the rhea, ostrich, emu, and cassowary, and kiwi being confined to South America, Africa, Australia, and New Zealand respectively). He did not believe in land bridges as solutions to the puzzle of disjunct distributions, but proposed that populations isolated on southern continents were originally derived from northern ancestors, who migrated southwards to displace "less well adapted forms." When the northern forms became extinct the southern forms were left isolated on their respective land masses. While this explanation is ingenious, we know in the light of plate tectonic theory and evidence of southern origin for many important groups that he was wrong on this matter.

Wallace's association with an idea that outlived its utilitarian function and stifled new developments in the subject may well have disproportionately damaged his scientific reputation. Croizat (1962), for example, roundly criticized Wallace for what he regarded as his unwarranted speculations, while ignoring his pioneering methodological and theoretical developments. One can only feel that Wallace's decision to ignore the conceptual difficulties in applying endemic areas at a global scale was a mistake and led nowhere. Only an association of endemic species found at a specific locality can be "natural" in the sense Wallace meant—locality and life share a history—which continental-scale associations can never be.

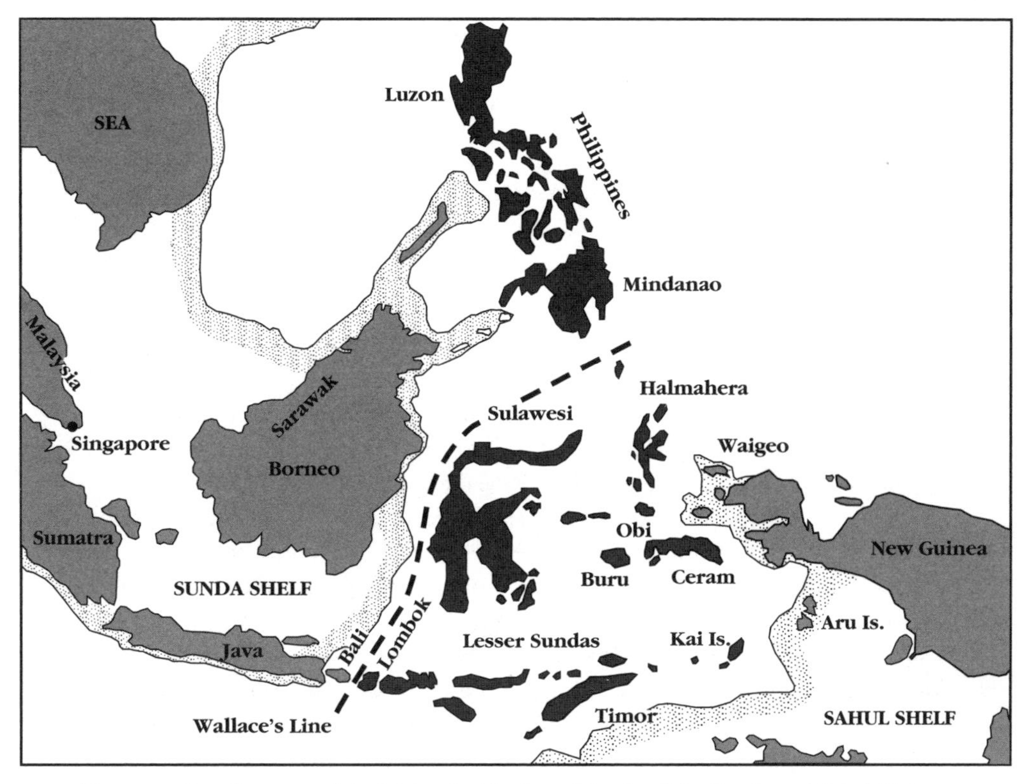

Figure 20 Locality map of the Malay Archipelago. Islands of the Sunda and Sahul Shelves are lighter shaded. Wallacea is darker shaded. Wallace's Line is indicated with a dashed line. Copyright Bernard Michaux.

Wallace's Line

Regions have boundaries and the most famous of these is Wallace's Line, the boundary of the Indian and Australian regions (Fig. 20). It is one of the ironies of his story that Wallace had never really intended to visit Bali and Lombok, but was forced to go to Sulawesi (Celebes) a roundabout way because of missed connections and delays in Singapore. His real intention was to use the port of Makassar as a base to explore eastward after birds of paradise, but he was first diverted to Bali, and then to Lombok, before reaching Sulawesi late in 1856. His early descriptions are very low key. In a letter written at the time he comments: "The Islands of Baly and Lombock, for instance, though of nearly the same size, of the same soil, aspect, elevation and climate, and within sight of each other, yet differ considerably in their productions, and, in fact, belong to two quite distinct zoological provinces, of which they form the extreme limits" (S31 1857, 5415).

By 1859 Wallace had reconstructed a remarkably modern model of the geological history of this complex plate collision zone:

> Here then is the key to the problem: Sumatra, Java, Borneo, and the Philippines are parts of Asia broken up at no distant period (an elevation of 50 fathoms would in fact join them all again); Celebes [Sulawesi], Timor, the Moluccas [Spice Islands], New Guinea, and Australia are remnants of a vast Pacific continent ... Celebes is in some respects peculiar, and distinct from both regions, and I am inclined to think it represents a very ancient land which may have been connected at distant intervals with both regions ... (S52 1859, 453).

To Wallace the line itself was of little interest. It was the history behind the juxtaposition of Asian and Australian faunas that he wanted to understand. He was not surprised that many groups of insects and plants ignored the line, because he believed these groups were older than the discontinuity. What did surprise him was the abruptness of the change and how far west of Australia it was. Interestingly, groups that ignore Wallace's Line have the potential to illuminate its nature. In a molecular-based study of Asian and Australian agamid and varanid lizards, Schulte *et al.* (2003) demonstrated the existence of Asian and Australasian sister groups. Furthermore, they were able to estimate divergence dates for the node connecting these sister groups, and hence to estimate the time of origination of Wallace's Line. Their dating of Wallace's Line to between 150 and 112 million years ago is consistent with the derivation of east Gondwanan terranes and their accretion onto Asia (Metcalfe *et al.* 2001). As these authors were to comment, Gondwanan fragmentation and subsequent juxtaposition provide the primary explanation for the phenomenon of Wallace's Line.

It is perhaps the biggest irony of all that Wallace's name should be most remembered for a term coined by Huxley in 1868 which, to add insult to injury, he put in the wrong place. Wallace's, or more accurately Huxley's line, generated plenty of discussion and controversy, even up to the present day (Mayr 1944; Van Oosterzee 1997; Metcalfe *et al.* 2001). Its reality has been questioned, its placement changed so many times George Gaylord Simpson was led to cry "too many lines" (Simpson 1977), and its nature dissected in detail: abrupt or diffuse, transitional or filtered, lottery or stepping stone. Wallace's original question—what was the relationship of this faunal discontinuity with geology?—had largely been forgotten until more recent times (see for example Ladiges *et al.* 1991; Hall and Holloway 1998; Metcalf *et al.* 2001; Schulte *et al.* 2003).

Island Life, 1880

The last of Wallace's biogeography books is devoted to the problem of understanding the history of insular floras and faunas. Wallace regarded this work (S721) as a companion volume to *The Geographical Distribution of Animals.* The first part deals with special subjects within the realm of biogeographical studies. In particular it covers Wallace's ideas on continental glaciation and how this influences the distribution of animals and plants (see Chapter 10). The second part concerns the classification of islands and contains a comprehensive review and interpretation of a number of island faunas and floras. Wallace used biological and geological criteria to distinguish between oceanic and continental islands, terms previously introduced by Darwin in *The Origin of Species.* Oceanic islands had impoverished mammalian or amphibian faunas and were composed of basalt or other volcanic rocks, often of no great age. Continental islands were geologically more complex, often contained very old rocks, and were considered detached continental fragments. Wallace subdivided continental islands into recent and ancient. Islands recently attached to a continent have a flora and fauna that is virtually indistinguishable from the mainland, and contain few unique (endemic) species. Ancient continental islands have been isolated biologically for a long time, their floras and faunas being characterized as much by what is missing as by what is present (tending to make them highly unique). They are usually distant (but not always so) and are separated from the closest continent by deep seas.

In *Island Life* Sumatra, Java, and Borneo are used to exemplify recent continental islands. Their biological similarity, to each other and to mainland South East Asia, shows they were not originally islands but were, with the surrounding Sunda Shelf, part of mainland Asia. His analysis of their biological differences allowed Wallace to propose a very testable historical hypothesis. He suggested that the first Greater Sunda Island to become physically isolated was Java, because its fauna is the most biologically isolated—both in terms of endemic species and species absence. Absences could, as he had already suggested, be evidence of physical barriers that were interrupting dispersal. Encroaching seas eventually isolated Borneo and finally cut Sumatra's connection to the mainland. This analysis shows how sophisticated Wallace's thinking had become. He had switched from interpreting biology (species' ranges) in terms of geographical change (barriers to dispersal), as he was doing in South America, to reconstructing geological change by using biology. The conceptual tools were simple: biological similarity meant connectedness, difference meant barriers to dispersal and the promotion of speciation, the degree of difference was proportional to time, with absence possibly indicating the operation of physical barriers.

Sulawesi he classified as an ancient continental island, albeit an anomalous one. Sulawesi lies at the heart of the Archipelago and is surrounded both by large and small islands (Fig. 20). Wallace assumed that Sulawesian species would represent a selection of those from the different source areas, perhaps with some development of endemism, because there were no obvious barriers to dispersal:

> As so often happens in nature, however, the fact turns out to be just the reverse of what we should have expected; and an examination of its animal productions, shows Celebes to be at once the poorest in the number of its species, and the most isolated in the character of its productions, of all the great islands in the Archipelago (S715 1883, 270).

Sulawesi appears anomalous because its isolated and impoverished fauna is out of character with respect to its location. This proved a sore test for Wallace's method of historical reconstruction because he lived in a pre-Wegenerian world where continents did not move laterally. Despite this limitation, his reconstructions are remarkably modern and show the power of his theoretical and methodological approaches to understanding biogeography. Here is what he had to say in 1860, while he was still exploring eastern Indonesia:

> Facts such as these can only be explained by a bold acceptance of vast changes in the surface of the earth. They teach us that this island of Celebes is more ancient than most of the islands now surrounding it, and obtained some part of its fauna before they came into existence. They point to the time when a great continent occupied a portion at least of what is now the Indian Ocean, of which the islands of Mauritius, Bourbon [Reunion], &c. may be fragments, while the Chagos Bank and the Keeling Atolls indicate its former extension eastward to the vicinity of what is now the Malayan Archipelago. The Celebes group remains the last eastern fragment of this now submerged land, or of some of its adjacent islands, indicating its peculiar origin by its zoological isolation, and by still retaining a marked affinity with the African fauna (S53 1860, 177–78).

Wallace's flirtation with lost continents was only fleeting, and underlines just how committed he was to explain Sulawesi's geological history in a way consistent with his analysis of its faunal relationships. He was sure the Sunda Shelf marked the edge of the Asian continent and the Sahul Self the edge of the Australian continent. In between were a host of islands which he thought were part of the Australian region, and which he interpreted as remnants of a now submerged continent. The Island of Sulawesi was something else, and not related to its surroundings.

Island Life is now mainly of historical interest. The term oceanic island is still in use, but the book's classification of islands and its pre-Wegenerian world view have become victims of history. It is all the more remarkable therefore, that by 1860 Wallace should have developed a model of Indonesia's geological history that modern geotectonic models and animations seem to confirm are broadly correct.

Wallacea

Wallacea was the name given by Dickerson *et al.* (1923) to those islands located in the present-day collision zone between the Sunda and Sahul Shelves (Fig. 20). The name honoured Wallace for his pioneering work in the region, and no memorial is more suitable. The name has remained in usage (as of this writing 1,310 hits on *Google Scholar*) but the modern boundary of Wallacea has changed with the exclusion of the Philippines (*e.g.* White and Bruce 1986). From the very beginning there has been uncertainty about the positioning of the Philippines (Diamond and Gilpin 1983). Wallace decided the Philippines, together with the Greater Sunda

Islands, were part of Asia. He could just as easily have grouped them with Sulawesi and the Spice Islands, as Huxley clearly favoured in 1868. Wallace never really resolved the biogeographical position of the Philippines—he simply made a decision. Resolving this issue is important because Wallacea should be a natural region—in Wallace's terms it should have a distinct fauna and be bounded geographically through time—if it is to have any real biogeographical meaning.

Sulawesi is the key island of Wallacea because its biological characteristics—distinctness, endemism, and unclear relationships—mirror that of Wallacea as a whole. The distinctiveness of the Sulawesian fauna is as much defined by what's missing as by what's present. It was as though invisible barriers were stopping colonization, and that these barriers had been biologically isolating Sulawesi and promoting speciation for a long time. The biological relationship of Sulawesian endemics to Asian relatives appears to be only distant (Anderson 1991; Turner *et al.* 2001). Sulawesian species' relationships to Australasian species are more complex. In Indo-Australian groups (those whose distributions straddle Wallace's line), Sulawesian species appear to be basal to the Australasian clade; that is, they are older and more primitive than Australasian species (Schuh and Stonedahl 1986; Muona 1991; Duffles and Turner 2002). Sulawesian species that are members of Australasian clades, such as the cockatoo *Cacatua sulphurea*, are derived species within the clade (Brown and Toft 1999).

Wallace was in the twilight of his life when a young German scientist named Alfred Wegener gave a lecture entitled "The Formation of the Major Features of the Earth's Crust (Continents and Oceans)." Wegener used his lecture to introduce the idea of continental drift. But Wallace had been dead for two years when *Die Entstehung der Kontinente und Ozeane (The Origin of Continents and Oceans)* (Wegener 1915) was published; moreover it wasn't until 1924 that an English translation became available (Miller 1983). I'm sure Wallace would have welcomed the concept, but at the time Wegener and his ideas were ignored (or ridiculed) in most quarters and the idea languished until the advent of plate tectonics in the mid-1960s.

Wallacea is as complex geologically as it is biologically, and the dynamics and history of this collision zone have become clearer in the past two decades. The literature is extensive and expanding, but Ladiges *et al.* (1991), Hall and Holloway (1998), and Metcalfe *et al.* (2001) are useful starting points. Penny Van Oosterzee's excellent *Where Worlds Collide* provides the reader with a very comprehensive and readable account (Van Oosterzee 1997). Figure 21 is a summary terrane map of Wallacea and its surroundings based on Michaux (1991, 1994, 1995, 1996). Terranes are continental fragments, detached rift structures, extinct island-arcs, slices of oceanic crust, or a mixture thereof. While their geology is diverse their locations are all exotic, that is they didn't originate where they are now found. The terranes of Sundaland and South East Asia had amalgamated by the end of the Mesozoic, long before Australasia's northward drift had started. The terranes in the collision zone are predominantly of Australian origin. Intensive shearing along the collision margin conveyed terranes westwards during the later part of the Tertiary. One of these terranes is now embedded in Sulawesi. Australia/New Guinea's interaction with a clockwise rotating Pacific plate also resulted in a shearing component that moved Melanesian margin and island-arc fragments west and north along the Pacific plate margin. The Wallacea terrane is composed of fragments of continental crust (Sulawesi, north Borneo and southern Philippines) bordering old oceanic crust (Sulu and Celebes Basins). Where this composite terrane originated is unclear, but was likely to have been closer to the Greater India sector of Gondwana than to the Australian craton, and probably offshore and biologically isolated. How it came to be incorporated into the collision zone is unknown.

Including the Philippines makes Wallacea a natural (Wallacean) region—a geographical location bounded in time. Although terranes originally come from distinct source areas, they were collected together in the collision zone where they even amalgamated in places. The other Wallacean criterion of naturalness is biological distinctness from surrounding areas. The fauna of Wallacea is distinct from its surroundings and not just different because it is impoverished. What unites the faunas of these diverse islands is an intermixing of three different

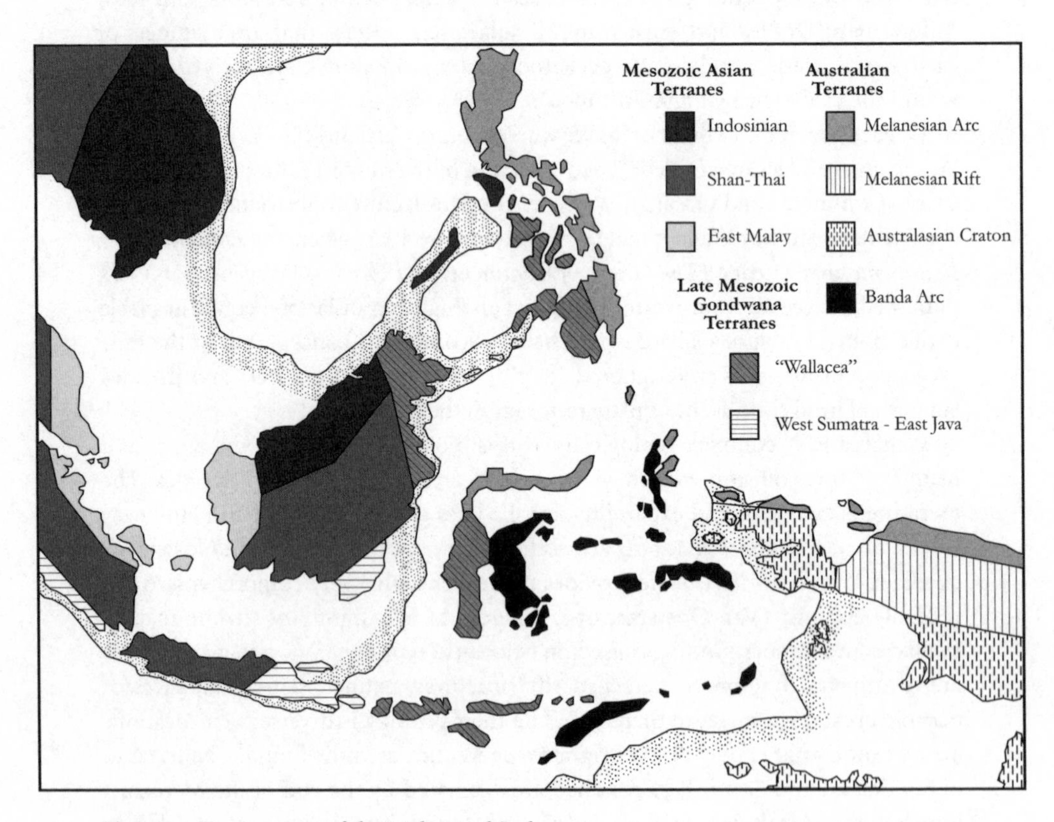

Figure 21 Terrane map of the Malay Archipelago.

phylogenetic strands; an ancient Asian fauna associated with the Wallacea terrane, an equally ancient Australian fauna associated with the Banda Arc terranes, and a more recent and speciose Papuan fauna associated with the Melanesian Arc terranes (see Fig. 21). Individual islands have their own mixture of these elements, which makes comparison between islands difficult. Nevertheless, I would argue that Wallacea (*sensu* Dickerson *et al.* 1923) is indeed a natural region as Wallace defined it.

Alfred Russel Wallace: A Personal Appreciation

It must have been at the start of the 1990s that I came across a copy of the 1891 tenth edition of *The Malay Archipelago* in the Kaukapakapa, New Zealand library (established by the early European settlers of the district in 1865). I had already become interested in biogeography, but although Wallace's name was known to me I had scant knowledge of who he was and what he'd achieved. While never a panbiogeographer, I have to admit that my early views on Wallace were probably unduly influenced by my reading of the writings of Leon Croizat, who dismissed Wallace as a tired old dispersalist with nothing relevant to contribute to modern biogeography. Remember, this was a time of renaissance when both panbiogeography and vicariance biogeography were revolutionizing the subject. Imagine my surprise then when I started to read what Wallace actually said, rather than relying on second-hand commentaries.

Malay Archipelago remains a very good read to this day—I can recommend it to anybody interested in biogeography, natural history, Indonesia, or the history of science. Wallace's insistence that present-day distributions can only be understood in the light of past geological events certainly resonated strongly with me. And his bold geological model for the development of the Archipelago was stimulating enough to send me in search of what modern geological studies had to say on the subject. My researches into Indonesian biogeography eventually led, via New Guinea and the south-west Pacific, to my present interest in the biogeography of New Zealand. Wallace's personal qualities are also apparent in his writings. His energy and enthusiasm, his steadfastness in the face of hardship and difficulty, his openness to the experiences of life, and above all his honesty and open mindedness, shine through his writing. And the more I found out about him, the more there was to admire. In my view Wallace was certainly the most interesting and possibly the most important of the Victorian biologists. I hope that this volume goes some way to restoring his reputation among the wider scientific community, and that his remarkable story becomes more widely known.

10

Wallace and the Great Ice Age

Keith Tinkler

Introduction

We can say without the chance of a serious argument that geology was the pre-eminent natural science of the first half of the nineteenth century, and that its primary achievement in that century was the elucidation of the geological column. This enterprise was fraught with difficulty because of the spatial fragmentation of rock exposures or "outcrops," and because of the frequently contorted and dislocated nature of the primary evidence. William Smith famously made a start on this enormous project, using both the subtle clues of the landscape itself and the distinctive identities of organic forms to almost single-handedly map England and Wales by 1816. Smith—a civil engineer—forced the hand of the geological establishment, and it was soon clear that meticulous field mapping had to be the basis of progress in extending his column of strata beyond the spatial confines of the relative orderly and superposed strata to be found in central, southern, and eastern England.

Nobody formally set out to build the geological column: it emerged piecemeal as fragments of it were found to overlap and correlate, and not without substantial controversies as Rudwick (1985) and Secord (1986) have shown, to select just two well-known cases. The fortuitous discovery of identical faunas, separated by thousands of miles of ocean, seemed to justify the principle of uniformity in space as well as time because Cambrian and Silurian faunas matched across the Atlantic just as well as Tertiary forms did.

Underlying geological exploration were assumptions not uniformly held by all parties—particularly ones relating to the rates of operation of geological processes. Although the concept of geological time as consonant with biblical and genealogical chronologies effectively had been banished by all serious geologists by the early decades of the century, it was far less clear what was to replace it. Charles Lyell's uniformitarianism (1830–33, with editions to 1875) (I shall use the simpler term "uniformity") was the most evangelized conception: that processes in the past had operated in the same manner and with the same force and frequency as they do now. Still, it was subject to vehement debate by Murchison (1839),

Sedgwick (1843), and many lesser luminaries, the lineal descendants of the catastrophists of the previous century.

As the century progressed, philosophical objections to uniformity arose on the assumption that the earth's thermal history was one of systematic cooling. In principle, if not in any computational fact, this implied a systematic change in the rate at which geological processes might operate, bringing into question the rates of seismic and volcanic activity in remote periods, and with it the rates of other surface processes. On these bases Lyellian uniformity was opposed in principle. There is no sign though that philosophical differences over the matter of processes, their rates of operation, and the timeframe into which they were thought to fit, ever inhibited anyone in their fieldwork (Rudwick 2005).

As evidence for the sheer magnitude and complexity of the geological column accumulated, however, it forced the profession—as it was becoming—to consider the time consumed in its construction. Matching the observed sediments with: (1) anecdotal evidence about the erosion of cliffs at the coast, and (2) the building of deltas and floodplains by rivers and floods, led inevitably to a realization of what historians of the natural sciences now term "deep time." Eventually, Archibald Geikie, in his 1892 Presidential Address to Section C of the British Association for the Advancement of Science Meetings at Edinburgh (Geikie 1905b), concluded that sedimentological data indicated a period of at least 400 million years as the time required to accumulate the stratified sediments of the geological column, and allowing for breaks and gaps he thought 450 million was reasonable.

If uniformity, and its subsequent corollary "deep" geological time was one underlying concept of the geological column, it cannot be denied that the nature and meaning of organic forms contained within the rocks was another. Although different from minerals, and increasingly recognized for the fossilized forms they really were (instead of being "sports of nature" made of "plastic virtue"), they were essentially used as "markers" either alone or in assemblages to distinguish stratigraphic units. Even early in the nineteenth century it was evident that as one progressed up the geological column, the organic forms became more complex. By the 1840s one of Lyell's primary objectives in visiting the east coast of the United States (Lyell 1845) was to study the very complete Tertiary sequences of gradually changing mollusks there: using these data the period could be classified into distinguishable subdivisions based on the proportions of archaic and modern forms.

Just as the evident scale of geological time was not in itself dependant on the conception of how processes operated through that time, so the procession of increasing organic complexity—whether by successive "creations" and "extinctions" à la Cuvier, or by gradual change à la Lamarck, or in the middle years of the century according to *Vestiges of the Natural History of Creation* (Chambers 1844, published anonymously)—did not undermine the usefulness of organic forms as stratigraphical markers. The advent of Charles Darwin and Alfred Wallace in 1858 was important for starting the process welding a theory of organic change onto geological time, whose prime metric was uniformity, but it cannot be said to have seriously affected how geological mapping proceeded.

Squatly straddling the century—almost as its own metaphor—was the Ice Age controversy. For many in the early years the very concept was a "wild dream" (Curwen 1940, Mantell diary entry from 1851). Starting as a one-stop explanation for the loose drift deposits scattered over northern Europe (Agassiz 1840) and North America, it gradually was recognized as a multi-stage process that, by the end of the century, had grown into a Great Ice Age (Geikie 1874, 1877, 1894). It was then understood to comprise several clear glacial stages, with fully expanded ice caps capable of significant erosion, separated by interglacial periods such as the present. It was understood to be the last significant, but very short, episode in geological time as it was known from the geological column then extant. The post-glacial period was seen to be relatively insignificant with respect to the rest of geological time.

Given this grand perspective why should Wallace, whose knowledge of world flora and fauna was probably more extensive than any other nineteenth-century naturalist, have been so interested in the Ice Age as to write his own accounts of it, and why did he go to considerable lengths to attack one of the most frequent arguments used against it: that ice was incapable of erosion of actual rock? In particular why should he write so late (his most important works being from 1879 and 1893), when for professionals and academics the Ice Age debate was all over bar the shouting, and James Geikie's *The Great Ice Age* was into its third edition?

Geological Time

To explain Wallace's interest in these matters we must examine the nineteenth century's battle with geological time—not so much within geology's own ranks, but against the physicists. Early geologists refused to speculate on precise numbers for the age of the earth. That was still a lively topic in nineteenth-century social circles if not scientific ones, but it was generally accepted that it numbered in the many millions. The details have been well covered elsewhere (Burchfield 1975), but roughly stated, the lengthy timescales eventually envisaged as a corollary of the observably slow rates of erosion and sedimentation caused no particular problem in the first half of the nineteenth century. The full scale of the geological column, especially the vast tract of time involved in the Pre-Cambrian, was not yet as evident as it would become. Assuming millions, tens of millions, or even a hundred million years really had little impact on thinking, because there was no inbuilt metric within uniformity against which to gauge it. The geological column was, as said, piecemeal even in its most complete sequences (Geikie 1905b).

However, the advent of natural selection changed all that (S43 1858; Darwin 1859). Whatever the underlying process by which change was effected, it was clearly linked to generations whose time span changed with the species under discussion. It was not precise, but one could imagine the fabric of deep time measured with the metric of generational units. The principle in itself did not help to estimate the span of time necessary to see the speciation of disparate species, but because such change was not readily noticeable in many species, or took substantial generations

Figure 22 Portrait of James Croll.
From James Campbell Irons' book *Autobiographical Sketch of James Croll* (1896). Out of copyright.

even with the breeding of domestic stock, one could sense the immensity of the time, and feel if not its length, at least that it had to exceed some numbers that were just beginning to be imaginable. Even Darwin admitted when reading the explanatory material by James Croll (1868) (Fig. 22), in the aftermath of his own unfortunate attempt to estimate geological time from the denudation of the Weald (Darwin 1859; Burchfield 1974), that he had little comprehension what a million years meant. Poulton (1896), while the controversy was still festering thirty years later, documents in considerable detail both the primary physical arguments for the age of the earth and the biological and geological evidence pointing to huge spans of time, while remaining respectful towards the physicists. John Perry, in a letter quoted by Lord Kelvin (1895) summarized the matter thus: "The biologists have no independent scale of time; they go by geological time."

The difficulties originally arose, however, when the physicist William Thomson, later Lord Kelvin, began to apply thermal calculations to models of the earth's evolution (Thomson 1862, 1863). The physicists' numbers would reduce from 100 million years (Thomson 1862) to nearer 20 million (Tait 1869 [anonymous in original]), whereas the naturalists' sense of time expanded and the twain were not to meet. Tait's view was certainly severe and by no means universally accepted, even by physicists. By 1895 Kelvin (1895, and quoted by Poulton 1896), was remarking that he would be "exceedingly frightened to meet him [*i.e.* Archibald Geikie, the geologist] now with only twenty million in my mouth." The 1895 note in *Nature* aptly summarizes, because it includes a series of letters between Perry and Tait, the controversies on assumptions that existed within the community of physicists. The thermal conductivity of the interior was one important element in the computations.

Needless to say, the naturalists were not unanimous in their attitude to the disparity. Uniformity in its purest form preached the steady monotonous tread of erosional processes and sedimentary deposition through almost endless time. But, apart from the catastrophic leanings in some professionals earlier in the century, the application of physics to deducing the time available brought some respectability to the already existing view that perhaps the tread had slowed, or quickened, in phase with the thermal history of the globe. Thus, if in some way biological change went faster in the past, the metric of time was nonlinear, and the required evolutionary changes would fit within the computed period. Darwin and Wallace had different reactions to the developing problem of fitting the requisite change into time (Burchfield 1974). If the time could not be changed, how might change itself be modified?

Wallace on the Ice Age

Once the physicists were in ascendancy after about 1870, Wallace and Darwin responded differently to how biological change might mesh with the physical picture. Wallace was anxious to establish the Ice Age, and then to fit its shifting climatic peaks and troughs within a chronology based on James Croll's astronomical theories (Croll 1864, 1868, 1875) connected to changes over time in the eccentricity of the earth's orbit. Basically, he wanted to use the environmental stress of oscillating climate to drive natural selection faster so that it might conflict less with the physicist's estimates. He took the hint from T. H. Huxley's 19 February 1869 lecture to the Geological Society of London (Huxley 1909, 308–42), meshing points made in this with Croll's astronomical calculations so in the longer term he could show that the astronomical variations so graphically illustrated by Croll—who had plotted them for three millions years back, and a million forward (see Fig. 23)—likely drove variations in climate, and concomitantly selection, throughout geological time. Not unnaturally, he found abundant fossil evidence of climatic changes within the Ice Ages from innumerable sources and travellers' accounts. Darwin's response was less direct, as he tended toward allowing that external forces could influence morphological change (à la Lamarck), and to a quiet state of resigned puzzlement that something was wrong with the physical arguments, but he did not know what.

Wallace's two primary essays on the subject were from 1879 and 1893, that is to say long after Croll had written his fundamental works, and at about the time that the younger of the two Geikie brothers, James, was progressively shifting toward a full-fledged multiple Ice Age hypothesis posing between four and six major glacial peaks similar to the most recent one (Geikie 1874, 1877, 1894). Wallace's works also appeared long after he himself had tried to rescue natural selection from the physicists (S159 1870). If Wallace was to use the astronomical driver to natural

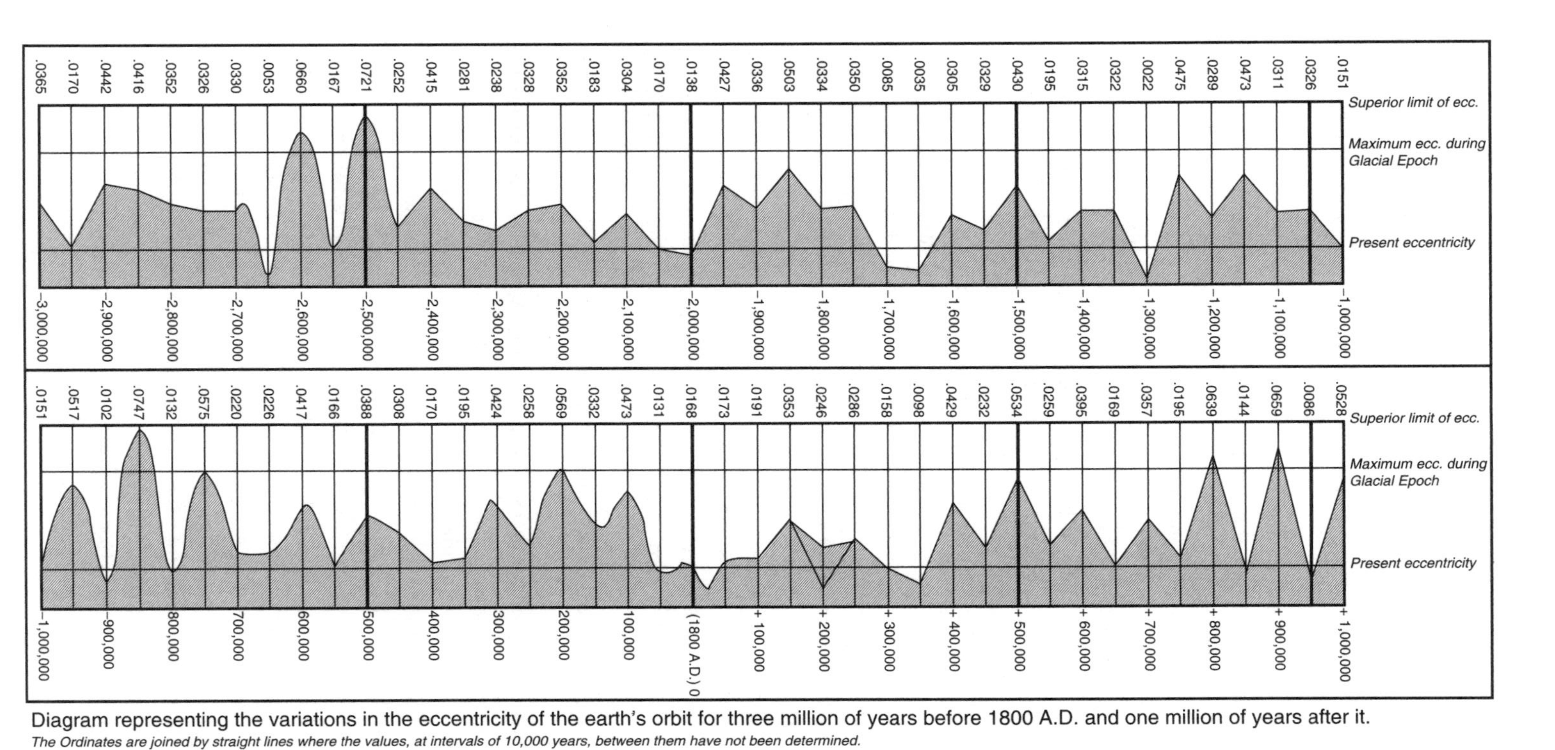

Figure 23 Croll's diagram (redrawn) showing variations of the Earth's orbital eccentricity, three million years into the past, and one million years forward.

selection that he thought Croll's theory provided, he had to vanquish any existing professional scepticism of glaciation, especially amongst naturalists for whom glaciation was perhaps seen merely as a recent dramatic landscape of limited import to theories of biological change. Thus, they might see it only as a backdrop to change, instead of realizing it was the very vehicle of change that Wallace envisaged.

Wallace's two glacial essays are masterpieces of accurate keen description of the known facts about the Ice Age and careful reasoning about the deductions drawn from them. He was thoroughly versed in the works of James Geikie, his primary source. In addition he had experience as a surveyor as a very young man travelling and working in England and Wales. While still a fairly young man he had had a brief vacation in Switzerland, and he went again in 1868 when the age of the earth controversy was peaking. These at least would have enabled him to see some of the evidence of which he wrote, even if he had not interpreted the land in that manner on every visit. The first paper, based on a review of books on glaciology and polar travel, is a view of the whole, but it devotes almost half the available space to Croll's theories and its import for the Ice Age. The second, and much later paper, is a reasoned attack on the primary objection to the Ice Age: the existence of lake basins, and its corollary, the ineffectiveness of ice as an erosive agent. The latter was the theory of glacial protection that still had limited but vigorous support, in Europe from people like Bonney (summarized in Bonney 1893b and Garwood 1910), and in North America from J. W. W. Spencer, who was attempting to reconstruct the pre-glacial drainage of Great Lakes basin (Spencer 1881, 1890a, 1890b; Middleton 2004). However, Wallace himself regarded this paper as addressing a topic that fascinated him in its own right, and that was still, on his own statement, an active controversy—with professionals balanced in a fifty: fifty ratio between active glacial erosion, and more or less complete glacial protection of landscapes.

His agenda in the first paper (S313 1879) becomes clear as he proceeds. His emphasis is on the overwhelming and compelling picture of the Ice Age as a whole to be drawn from the distribution of erratics and till. He is not to be distracted by local disputes and minutiae and his target too is clear: not so much the anti-glacialists, but the "old theory of the cooling earth" which he judges "totally inadequate" to explain the "wonderful series of phenomena" exhibited by glaciation. A simplistic view would have recent glacial climates as the inevitable result of systematic long-term cooling. In claiming he has not the space to discuss the matter (and puzzle) of glacial submergence, he states (p. 123) that "our special object is the reality and magnitude of that wonderful and comparatively recent change of climate termed the glacial epoch." It becomes clear that his argument is embedded in a longer vision of the earth. Substantial high latitude evidence of warm flora and fauna in the Miocene and Pliocene are cited (p. 131 *et seq.*) across the northern hemisphere, although not until Croll's theory has been thoroughly expounded.

His summary of Croll is exemplary and he makes a number of telling points: the earth can accumulate cold as ice, but not heat; gravitational shifts due to fluctuating ice sheets might go some way to explain submergences, and intercalated fossiliferous terrestrial beds and glacial tills show that radical climatic shifts took place within the Ice Age itself. From the well-recorded high Arctic evidence of limited present land ice, he infers (incorrectly in light of modern knowledge) that the interior of Antarctica is likely bare rock too, or at least not occupied by an ice sheet on the scale of Greenland.

Having thus summarized both Croll, and the abundant evidence for warm climates in high northern latitudes across the globe in the Pliocene and Miocene, Wallace makes the additional inference that the Ice Age was simply the latest manifestation of systematic change in the eccentricity. Noting that the evidence of warm climates is most evident in currently cold regions, he then asks what evidence there is for cold climates being revealed in currently warm regions. Although less abundant than the former case, he manages to document a substantial number of glacial episodes through erratic blocks and polished and striated surfaces in Palaeozoic sediments in Scotland, Nova Scotia, Ohio, India, and Australia. The plot has become clear. The Ice Age is a convenient cartoon for the general process of organic change whose intensity was, in his view, driven by the endlessly variable astronomical cycles. Wallace may have been the first (S159 1870) to illustrate Croll's calculations, which he certainly exhibited at the 1870 Liverpool British Association Meetings (S171 1871) (Fig. 23). Astronomically driven organic change was not a concept that Darwin was happy with, for in no way did it undermine the physical arguments about the earth's age based on thermal cooling. It did however add a quite different twist to physical arguments about the earth's history.

I shall treat Wallace's second paper (S481 1893) initially on the same basis that he himself did, as an independent entity of abiding interest, but it is well to note he began writing on glacial topics in 1867 (S124), and was commenting on the physics of glacial motion soon afterwards. Wallace's professional distance from glacial work gave him a perspective those closer to the subject lacked. Although Wallace may have been interested in the matter in its own right, it has a tangential bearing on his general reasoning. If organic development is in synchrony with the astronomical theory, and the physical behaviour of the landscape is in synchrony with the Ice Age, a stasis in the physical processes of glaciation might have some correlative spin-off in the organic world. Nevertheless he tackles the problem of lake basins as a geologist, not as a biologist.

The glacial erosion hypothesis had been established by Sir Andrew Ramsay (1862) after a twenty-year period during which the glacial theory had primarily been related to the distribution of surficial "drifts" explained by marine submergence and localized, but unexplained, uplift. Ramsay had drawn on a vast array of evidence from well-known glaciated regions, including North America. Wallace's paper (S481) is a masterly demolition of the glacial protection, or "no erosion" position, and especially as it related to lake basins. A full appreciation demands reading Wallace's paper, not to mention Ramsay's. A skirmish in letters to *Nature* throughout 1893 almost certainly began with a review by T. G. Bonney (1892) of Forel's book on Lake Geneva (Forel 1892) in which he approved of Forel's rejection of the glacial excavation hypothesis for the lake. A few months later Bonney (1893a) was commenting on a recent *Atlas des Lacs Français* issued by the French Ministry of Public Works (Delabecque, 1892). He summarized the bathymetric data made available, concluding they did not "lend themselves readily to the glacial excavation hypothesis; but to be more favourable to that which regards the larger Alpine lakes as mainly formed by movements of the earth's crusts after the erosion of the valleys in which they lie."

The response to these reviews by Bonney included an approving letter from the Duke of Argyll (1893), disclaiming the glacier theory. This letter prompted Wallace into action (S462 1893) on account of the Duke's complete misconception of the glacial theory, and his ignoring any evidence in favour of it. This note is essentially an abstract of the later paper (S481), and public interest doubtless led to Bonney being invited to address the Royal Geographical Society, a talk subsequently published in June (Bonney 1893b). Wallace's formal "reply" (S481) was in the *Fortnightly Review* in December, and because Bonney had discussed Lake Geneva at length it forms a prominent part of his paper as an example.

Few disputed the size and scale of glaciation, the issue was whether ice eroded rock, or did not. Crucial to this matter was the origin of lakes that in a "deep" time perspective were granted by all to be young landscape features (otherwise they would either have been filled with sediment, or filled with water that spilt over an outlet col to cut a gorge which drained them). He distinguishes between plateau lakes that are barely maintained by their inflow, and deep valley lakes characteristic of the recently and still glaciated terrains of Europe. Then he calls in the negative evidence, tracts of mountains in tropical regions (for example in Mexico and the West Indies, and Australia's ranges from Victoria up to Cape York Peninsula) that lack any such incised rock basins and valley lakes. Finally, Wallace points out that "excessive glaciation" (S481, 754) leads to "comparatively large and deep valley lakes" whereas less severely glacially eroded areas exhibit only small high lakes, "tarns," and/or smaller valley lakes. Thus he identifies a gradation of process. He draws on bathymetric data for larger glacial lakes to provide further fuel for the argument. He points out that if the lakes came about through ponding of existing valley systems (he illustrates rias from southwestern England), then the system's side valleys should be flooded, and the bathymetric maps should display the expected contours of a flooded valley system. The evidence was quite the contrary: submerged scoured troughs seen in bathymetry mimicked the contours of glacial troughs—often to be found further up the same valleys, and with no indication that the flooding was from "accidental" causes.

What were the alternative theories? In fact there were none that were satisfactory or comprehensive. The arguments for subsidence due to solution, or to localized faulting, popular in the decades *before* Ramsay wrote had largely been discounted

Figure 24 An Alpine glacier.
From the third edition of James Geikie's book *The Great Ice Age* (1894). Out of copyright.

thanks to the detailed geological fieldwork that characterized the century. Wallace, writing three decades after Ramsay, could quite rightly expect to find a comprehensive tectonic theory to explain lakes in glaciated regions, if glaciation were not responsible. There was no such theory worth the name. Bonney (1893b) called upon a complex series of "earth movements" happening just before glaciation began, which were imagined to cause flexes in the surface which trapped the lakes. Wallace points out, using the example of the Grand Canyon and a recent case cited by Lyell from Sicily (S481 1893, 762), that rivers usually have little difficulty maintaining their courses in the face of earth movements that all were agreed would be very slow. He points out too how extraordinary is the coincidence between the strength of the imagined flexures needed to produce lakes, and independent evidence—such as proven ice thicknesses—for the vigour of glaciation. Bonney's theory had to produce very strong tectonic flexuring to induce lakes exactly where glacial ice was thickest and had the strongest surface slope, and the flexuring had to grade in strength to be small in regions where only small "tarns" were to be found, such as the Pyrenees. In addition it had to act specifically on "soon to be glaciated areas" because a clinching argument, from negative evidence, is that there are many equally large mountain ranges that lack valley lakes. Why have these been neglected by flexuring? Because ice does not erode lake basins on the Bonney argument, the flexuring must just pre-date glaciation.

There is perhaps however a little more to this interchange than one might notice at first glance. Bonney was probably invited to give his paper to the Royal Geographical Society (RGS) on the basis of the letters in *Nature*, and he admitted that his primary interest dated back twenty years to papers he had written between 1871 and 1877. I have located three in 1871, 1873, and 1874 so far, and the abstracts, at least, make no mention of flexuring. Bonney did not cite them, and admitted that since then he had been primarily a petrologist. The RGS welcomed him warmly and recommended to themselves they should have more speakers from the geological realm. But one might wonder whether Bonney was picking a potentially receptive audience that would be relatively mute in its criticisms. The paper was long, detailed, and avuncular in tone. Bonney claims his alternative flexuring argument was posited in the 1870s and ignored. If that is so, it is odd that in his 1893 paper the only real example he uses is Spencer's map (1890a, 1890b) of the Great Lakes, where Bonney calls upon the differential movements documented by the shoreline of Lake Iroquois (which had occupied the Ontario basin in the immediate post-glacial). The fact that differential movements are needed to produce lakes pre-glacially, not post-glacially, is not mentioned. Bonney was present when Spencer (1890b) spoke to the Geological Society of London. He responded favourably to Spencer's paper and argued against Dr Hinde, who had local knowledge, and who may have been the same Hind(e) who advised Ramsay on the Great Lakes basins in the 1850s. Very conveniently Bonney argues that the complex stratigraphy typical of rocks over large areas would hide any sign they might themselves display of the mild warping that would lead to lakes that would be lightly modified by ice. It is then a specious argument unless one can use the evidence of former lake shorelines. Whatever the example he cited in the 1870s it cannot have been based on Spencer, whose first paper reconstructing the pre-glacial rivers in the Great Lakes region was in 1881. The only comparable area would be Scandinavia where similar earth movements had long been known to exist. The idea may pre-date Bonney, as Geikie (1877, 277 n.) explains that he ignored Charles Lyell's explanation of the rock basins of Switzerland through a huge depression of the central Alps (as "unequal movements of upheaval and depression") in his first edition (1874), but that in his second he mentioned it reluctantly, if only to negate it. Presumably this was because Lyell or others such as Bonney had resurrected it.

The flurry of letters on lake basins and erosion to *Nature* in 1893 demonstrates a continuing public interest in the topic, but it is noteworthy that apart from Bonney no establishment geologist commented, and both Geikie brothers held strong opinions on the matter. In 1896 Bonney wrote a book titled *Ice Work* for the popularly intended "International Scientific Series." Claiming in his Preface to take a fair stance on controversies he provides a curiously "arm's length" review of the debate in which, apart from Ramsay (who had died in 1891), Wallace is the only one named. Bonney again utilizes Spencer's material on the Great Lakes and is careful to admit to a small amount of polishing and scratching and the possibility of small rock basins being eroded by ice. Subsequently, serious Quaternary geologists had little or no time for Bonney. In later works James Geikie makes no mention of him, Hobbs (1911) only very briefly outlines, then dismisses, his 1871 paper, and there is no mention of him in the second edition of Wright's *The Quaternary Ice Age* (1937).

With the viewpoint of hindsight, Wallace's comprehensive paper appears to have closed the argument, but the protectionist argument rumbled on in mute tones for decades more (Garwood 1910). It may be said that most of the vocal protagonists were independent and socially well-placed amateurs (*e.g.* the Duke of Argyll, Sir Henry Howorth, and G. F. Wright, a Congregational minister and Professor at Oberlin College). As is often the case, resolution required the deaths of the protagonists, and the replacement of that controversy with another.

In the late nineteenth century theories of mountain building and earth movements had not been adequately linked to cadastral surveys. A belief in slow warping was encouraged by a century of uniformity and widespread field evidence in Scandinavia and the Great Lakes region—both of which were revealing, as we now know, evidence for isostatic uplift following the removal of the load of an Ice Sheet. A proper understanding of mountains required the structural insights that geodesy on the one hand, and radical remapping of the Alps on the other (Suess 1883–1904) was revealing. In the modern era the glacial erosion issue has still not entirely vanished, with the rival merits of strongly streamlined erosion by debris-rich ice and debris-rich sub-glacial melt water still being hotly debated. It has simply shifted its ground. The original glacial erosion hypothesis is much easier to accept now in the light of current understanding from ocean cores that there were about twenty-four glacial episodes, than it might have been with the belief contemporary with Wallace that there were just one or two, before Geikie (1894) promoted the notion of between four and six.

The Croll / Wallace Correspondence

An autobiographical sketch of Croll's life (Irons 1896) contains a long series of letters between Croll and Wallace. Most of these deal with their mutual interest in glaciation, and in the astronomical controls Croll had devised to explain glacial and climate cycling. In an early exchange, Croll had clearly sent Wallace some off-prints and Wallace replied expressing his great interest in them (letter of 14 March 1870), admitting that although he has a "very scanty acquaintance with practical geology, [I] am exceedingly interested in all the wider problems with which it deals" (Irons 1896, 247). Croll's replies to these first letters are not preserved and Wallace's next deals with the movement of glacial ice discussed in Croll (1870) on the cause of motion in glaciers, and takes issue with the nature of shearing within the ice, and the role of meltwater in summer glacial motion (later summarized in S184 1871). At the end of the letter he mentions that he exhibited at the recent British Association for the Advancement of Science Meetings at Liverpool (S171

1871) a large diagram constructed from Croll's published tables (reproduced in S159 1870), for its bearing "on climate and organic change." It excited discussion, although an unnamed Russian mathematician had doubts about extrapolating the calculations beyond plus or minus 10,000 years.

The next exchange that has been preserved took place in 1879 following the unsigned review paper by Wallace (S313) on "Glacial Epochs" in the *Quarterly Review.* The paper gave Croll "intense pleasure" (p. 334) and Croll had had a note from James Geikie who also approved. Wallace's reply highlights the insecurity he felt: "I can only hope I have fallen into no serious errors or misstatements such as dabblers in subjects that don't belong to them are always liable to." Wallace is curious to obtain measurements of solar radiation near the equator and asks for Croll's help with possible meteorological contacts who might have such information, then touches on the distribution of gravel and loess in Europe in relation to glacial damming of the North Sea and the English Channel. Croll's reply finds no fault with Wallace's summary although he differs over the possible nature of the Antarctic Ice Sheet, whilst sharing Wallace's opinion that there had never been a substantial Arctic Polar ice-cap.

The next year, after a letter from Croll to Wallace (19 July 1880) about "Glacial Epochs," Wallace reiterates his debt to Croll's work and outlines his idea that the transfer of heat by currents to polar regions could greatly modify the effect that the "changing phases of the perihelion" could have on the progress of glaciation during the late Secondary and early Tertiary. Croll's reply is keen to underplay any appeal to changing physical geography since he notes that such changes operate irregularly, and with "extreme slowness," and therefore in no way gel with the evidence of rapid and extreme climate change that glacial evidence, shortly to be outlined in Geikie's *Prehistoric Europe* (1881), could show. Croll seemed to agree with Wallace that Antarctica is a low plateau upon which ice is relatively negligible, apart from marginal mountain systems. Yet he continued to maintain that the ice surface must rise inland, while the underlying rock surface does not. This led eventually to an estimate of a twelve-mile ice thickness, though by making the profile more convex he reduced this estimate by half. In all, it is an interesting interchange only a generation before the reality of the matter was revealed by inland exploration.

A letter from Wallace some months later (2 December 1880) suggests missing correspondence, for it is brief and regrets Dr Croll's "continued ill health, as far as regards mental work," and notes again his reliance "almost wholly on your own researches." Some final letters follow the appearance in the *Saturday Review* (*SR*), for October 1887, of reviews of a reissue of Croll's *Climate and Time* together with the new volume *Discussions on Climate and Cosmology* (Croll 1885). Copies had been sent to Wallace (who possibly could have been the anonymous *SR* reviewer), for on 12 December 1885 he had already written to Croll to take up points in the last two chapters bearing on computations—not so much on the age of the earth, but upon the time elapsed since it was cool enough to permit the development of life.

This as we have seen was an issue already hotly debated, and about to emerge again in the 1890s. Croll's reply raises the issue, later debated by the physicists in the 1890s, about the conducting properties of the earth with respect to computations about its age, seeing the geological evidence as requiring more than twice the twenty million years commonly allowed by the more extreme physicists.

In all, the letters show that the men shared much common ground and had great mutual respect, with Wallace admitting his heavy reliance on Croll's theories to help him understand biological change at what we would now term its "macro-evolutionary" level. Croll in his turn was pleased to find someone who clearly understood his theoretical work and applied it to geology in the whole. The Antarctic issue was more important than it might appear since Croll saw that his theory required alternating ice ages in the northern and southern hemispheres, and the two men were equally in agreement that the Arctic had not been glaciated recently and showed evidence of hosting "warm climate" flora and fauna at least in the Tertiary. With the general lack of large landmasses far enough south in the southern hemisphere to show substantial glaciation, and the unknown state of the Antarctic interior, both men saw the need to find evidence of full glacial conditions in high latitudes in the northern hemisphere in order to establish the truth of Croll's astronomical theory of climatic change. As each was long past his prime for undertaking fieldwork, they were both reliant on what evidence Croll could gather from his contacts at the Geological Survey.

Concluding Remarks

Wallace's interest in glaciation was an important element of his lifelong work of explaining organic change. Prior to 1858 the issue of geological time, and indeed glaciation, likely was of little concern to Wallace, but with a systematic process on hand to explain the structure of the entire organic sector of the planet, time became a central issue. He and Darwin had established the nature of sequential organic change, largely with reference to living examples, but it was immediately clear that an immense time was needed to lead to the enormous diversity of life on earth, with a large proportion of that time being consumed with the development of the most elemental forms. All the evidence for that was embedded in rocks. Almost as soon as the need of "deep" time became evident, the physicists seemed to work to negate its existence.

Wallace, via Huxley (1909), and using already existing computations by Croll (1868), thought he saw a way to weld organic change onto an index of time in a manner that might drive organic change. It was a bold stroke. It would be ironic if eventually DNA could show that there was some truth in his hypothesis, but leaving that aside, it created an additional line of argument against the physicists. Geikie's attack on the physicists (1905b, 1905c) was based largely on the twin pillars of measured denudation rates and its corollary, the immensity of the geological column. The biologists showed the immense complexity and diversity of the organic world was not to be squeezed into a few million years, work well summarized by Poulton (1896). Wallace used Croll and a vast knowledge of published material, to show how they might all be related to systematic climate change driven by astronomical properties. Appreciation of Croll's work has undergone a renaissance thanks to the general acceptance of Milankovitch's work, but one could argue that Wallace kept Croll's work before the public in the closing decades of the nineteenth century, providing it with a strong biological endorsement adding to its support from the Geikie brothers. Whatever the merits of the arguments Wallace and Croll presented, Wallace drew attention to the complexity of the fossil biological data, to the way it was frequently inconsistent with both existing climate and theories of historical climate, and to the need for explanations to be both global in scope and rooted in deep time. One can only marvel at the way Wallace mastered the glacial literature to pursue his naturalist's agenda.

It was James' brother Archibald who had the satisfaction of hammering home the nails that put the lid on the time controversy. In the reprint (Geikie 1905c) of his 1899 address he was able to add a lengthy footnote (dated October 1904) citing the work and opinions of George Darwin (son of Charles) (Darwin 1903), Joly (1903), and Rutherford (1904, 346), all of whom saw that radioactivity provided a new heat source that argued for a greatly prolonged age of the earth. Thus in a puff of radiation the time controversy vanished, and with it so did the rationale for trying to tie organic change to astronomical properties of the earth's orbit.

11

Wallace, Conservation, and Sustainable Development*

Sandra Knapp

Introduction

From a twenty-first-century conservation biology perspective it is difficult to reconcile the images of a man who shot and killed eighteen orang-utans (the nineteenth he shot got away), with that of the man who advocated the establishment of botanical reserves in the tropics (S732 1910) and wrote "He [the naturalist, meaning himself] looks upon every species of animal and plant now living as one of the individual letters which go to make up one of the volumes of our earth's history; and, as a few lost letters may make a sentence unintelligible, so the extinction of numerous forms of life which the progress of cultivation invariably entails will necessarily obscure this invaluable record of the past" (S78 1863, 234). Alfred Russel Wallace did all of these things, and his relationship with nature and the environment is both typically Victorian and astonishingly modern.

Perhaps more than any other nineteenth-century British naturalist, Alfred Russel Wallace was a true field biologist (Camerini 2002). His collections, despite the tragic loss of many of them in the fire aboard the *Helen* as he returned from his first major tropical foray to South America (Knapp 1999), truly expanded knowledge of the biological diversity of both the New and Old World tropics. Wallace's interest in natural history began when he was an apprentice surveyor with his brother, grew while he taught school in Leicester and made the acquaintance of Henry Walter Bates, and blossomed during his twelve years collecting in the field in

* My thanks to Charles Smith and George Beccaloni for inviting me to put down my thoughts about Alfred Russel Wallace and his conservation views on paper; Jim Mallet (University College, London), David Williams (NHM, London), and Kate Jones (Zoological Society of London) for conversations about Wallace and conservation; and Gina Douglas and the Linnean Society of London for loans of books and materials. My taxonomic and conservation work, from which my ideas in this paper stem, has been funded by the National Science Foundation Planetary Biodiversity Inventory (PBI *Solanum*, DEB0316614) and the UK government's Darwin Initiative (Defra).

the tropics (Camerini 2001; Berry 2002; Raby 2002). An interest in natural history, however, does not automatically make a person interested in its conservation.

Ideas about the conservation of the environment were developing in both Britain and the United States in the late nineteenth century, spurred by three main concerns, (1) the decrease in natural resources, (2) the fate of "sublime" wilderness, and (3) increasing pollution (*e.g.* Worster 1994). Historians of the conservation movement outside the United States emphasize its colonial roots; Britain's Society for the Preservation of the Wild Fauna of the Empire was founded in 1903, and today is one of the oldest international conservation organizations, Fauna and Flora International (FFI). The men who founded the Society were concerned not only with the sustainability of hunting big game in Africa, but in the preservation of such game in perpetuity by means of the establishment of reserves and protected areas. Various acts for the preservation of game were in place in British colonies in Africa by the end of the nineteenth century, and the 1900 Convention for the Preservation of Animals, Birds, and Fish in Africa attempted to regulate this effort on a continental scale (Prendergast and Adams 2003). Edward North Buxton wrote a book outlining the principles behind his big-game centred view of conservation (Buxton 1902) in which he set out a clear agenda not only for the avidly reading Victorian public (certainly including Wallace), but also for the British government. The Society for the Preservation of the Wild Fauna of the Empire became an effective pressure group for game reserves and other measures now seen as the beginnings of the world conservation movement, and built on earlier concerns about deforestation and its effect on climate derived from experiences in the British colonies in the Caribbean (Grove 1995). In the United States, the end of the nineteenth century brought the origins of the National Park Service—first with the preservation of the Yosemite Valley and the Mariposa Big Tree Grove in the state of California in 1864, then the establishment of the first federally managed national park in Yellowstone (Wyoming) in 1872, and eventually the establishment of the US National Park Service in 1916 by an Act of Congress to "... conserve the scenery and the natural and historic objects and the wildlife therein and to provide for the enjoyment of the same in such manner and by such means as will leave them unimpaired for the enjoyment of future generations" (Albright 1971). Although wildlife was important to the conservation movement in the United States, the base laid by influential American naturalists such as Henry David Thoreau, John Muir (founder of the Sierra Club), and Aldo Leopold meant that the preservation of wilderness and scenic beauty was of primary concern. Conservation in its early days therefore entailed concern both for beauty and for particular species; this then, combined with emergent concerns over pollution of air and water (as captured in Rachel Carson's immensely popular book *Silent Spring*; Carson 1962), led to a twentieth-century environmental movement concerned with overuse of the environment and extinction of charismatic species, mostly large animals (epitomized by the WWF's panda logo, see **http://www.wwf.org.uk/core/about/aboutwwf.asp**).

Today's conservation movement is quite different, though it faces many of the same challenges as did the Society for the Preservation of the Wild Fauna of the Empire and the US National Park Service (Prendergast and Adams 2003). In the late twentieth and early twenty-first centuries conservation also became international, no longer the province of colonial powers or local interests, but governed by international treaties and agreed upon sets of targets and goals. Much of the emphasis in the modern interpretation of conservation is on sustainable development and the equitable sharing of the benefits derived from the use of natural resources. The objectives of the Convention on Biological Diversity (CBD 1992; see **http://www.cbd.int**) are "conservation of biological diversity, the sustainable use of its components and the fair and equitable sharing of the benefits arising out of the utilization of genetic resources." Conservation of biological diversity is defined by the Convention as "the variability of living organisms from all sources, including, *inter alia*, terrestrial, marine and other aquatic ecosystems and the ecological complexes of which they are part; this includes diversity within species, between species and of ecosystems" (CBD 1992). The signing of CBD by more than 180 countries around the globe following the first Earth Summit in Rio de Janeiro in 1992 has inextricably linked the conservation of biodiversity (as defined in the Convention) with its managed use by humans, thus creating the concept of sustainable development.

These concepts—conservation and sustainable development—are big, confusing, and almost indefinable (Knapp 2003). At the World Summit for Sustainable Development in Johannesburg in 2002, the assembled governments endorsed the 2010 target (earlier agreed on by the CBD) to "achieve by 2010 a significant reduction in the current rate of biodiversity loss at global, national and regional levels as a contribution to poverty alleviation and to the benefit of all life on earth." This challenge explicitly links people with the rest of the diversity of life on Earth, thus moving conservation away from a charismatic species and beautiful landscape focus to one involving human well-being. This contributes directly to the UN Millennium Development Goal of ensuring environmental stability, by promising the delivery of a reduction in the rate of loss of environmental resources. Whether we achieve this challenging goal in the face of accelerating environmental change is in the balance (Butchart *et al.* 2005), but more important to the discussion of Wallace's view on conservation here are the rest of the UN Millennium Development Goals (Table 11.1, also see **http://www.un.org/millenniumgoals/**), all of which are concerned with the betterment of the human condition. The focus away from nature for its own sake and instead on its role in promoting and sustaining human well-being is at the heart of modern thinking on sustainable development. Wallace's concern with social justice and equity, coupled with his deep love for nature means that his approach and views about conservation, while deeply rooted in the nineteenth century, were consonant with this modern perspective.

Alfred Russel Wallace lived at a time of excitement and establishment of the international conservation movement at the cusp of the twentieth century, but

Table 11.1. The United Nations Millennium Development Goals[a]

1	Eradicate poverty and extreme hunger
2	Achieve universal primary education
3	Promote gender equality and empower women
4	Reduce child mortality
5	Improve maternal health
6	Combat HIV/AIDS, malaria, and other diseases
7	Ensure environmental stability
8	Develop a global partnership for development

[a] See http://www.un.org/millenniumgoals/for reports on progress and other related targets.

appears to have played little direct part in it. In Britain, he lacked the African colonial establishment credentials of many of the key players, and he only visited America in later life. In California he met John Muir, who took him to see the "Big Trees," but in his account of his American travels Wallace (S729 1905a, vol. 2) mentions nothing about him other than his "beautiful volume" (Muir's 1894 book entitled *The Mountains of California*)! His collecting experiences in some of the most diverse places on Earth gave him an intense, almost spiritual, appreciation for the grandeur and beauty of natural places and the sheer variety of their denizens, and his acute observation of the habits and ranges of many species of plants and animals gave him the data needed to analyse the impact of humans on the environment. Wallace, however, was not really a "proto-conservationist" in a traditional (if there is such a thing) sense. His deep sense of the importance of human beings and his passionately held social reformist views meant his take on conservation was less wildlife and beauty-oriented than that of his contemporaries. Instead, he saw much of nature in terms of its relationship to human beings. In this paper, I will use Wallace's own words from his many published works to trace some of his thinking that might be related to concepts of conservation and sustainable development. Wallace never wrote a single, definitive article or book that can be said to be "about" conservation, but the theme of man's relationship with the environment permeates his writing, particularly in his later years. It is altogether too easy to raise figures from the past onto a pedestal as thinkers ahead of their times; in the case of Wallace and conservation, I think a more accurate assessment would be of a man who would be completely in sympathy with the modern view of conservation as tightly linked with sustainable development and improvement of the human condition for the poorest on Earth.

Interest in Nature

Alfred Russel Wallace was not a born naturalist as are some of today's great natural historians (*e.g.* Professor E. O. Wilson; Wilson 1994). His interest in nature only blossomed gradually, through seeing other's enjoyment of it, and knowledge about its elements:

> At that time I hardly realized that there was such a science as systematic botany, that every flower and every meanest and most insignificant weed had been accurately described and classified ... This wish to know the names of wild plants, to be able even to speak of them, and to learn anything that was known about them, had arisen from a chance remark I had overheard about a year before [*c*.1836 in Hertford]. A lady, who was governess in a Quaker family we knew at Hertford, was talking to some friends in the street when I and my father met them, and stayed a few moments to greet them. I then heard the lady say, "We found quite a rarity the other day—the Monotropa; it had not been found here before." This I pondered over, and wondered what the Monotropa was. All my father could tell me was that it was a rare plant ... (S729 1905a, 1:111).

Wallace as a young man had always loved the out-of-doors; this familiarly British love of walking and fresh air, combined with his nascent interest in a nature composed of things one could learn about and understand, opened up a new world for him. His purchase of a flora of Britain allowed him to learn the names of plants for himself, something he practised throughout the period he was apprenticed to his older brother William in the late 1830s and early 1840s, and beyond: "At such times I experienced the joy which every discovery of a new form of life gives to the lover of nature, almost equal to those raptures which I afterward felt at every capture of new butterflies on the Amazon, or at the constant stream of new species of birds, beetles, and butterflies in Borneo, the Molucas [*sic*], and the Aru Islands" (S729, 1:195). William was a land surveyor, and working with him in Bedfordshire and Wales, Wallace experienced first hand the effects of the enclosure of common land:

> ... at the time ... I certainly thought it a pity to enclose a wild, picturesque, boggy, and barren moor, but I took it for granted that there was *some* right and reason in it, instead of being, as it certainly was, both unjust, unwise, and cruel (S729, 1:158).
>
> The "General Inclosure Act" states in its preamble, "Whereas it is expedient to facilitate the inclosure and improvement of commons and other lands now subject to the rights of property which obstruct cultivation and the productive employment of labour, be it enacted," etc. But in hundreds of cases, when the commons, heaths, and mountains have been partitioned out among the landowners, the land remains as little cultivated as before. It is either thrown into adjacent farms as rough pasture at a nominal rent, or is used for game-coverts, and often continues in this waste and unproductive state for half a century or more, till any portions of it are required for railroads, or for building upon, when a price equal to that of the best land in the district is often demanded and obtained ... and if this is not obtaining land under false pretences—a legalized robbery of the poor for the aggrandisement of the rich, who were the law-makers—words have no meaning (S729, 1:151).

Wallace did not, it seems to me, rail against land enclosure for its effect on biodiversity, but rather for its effect of putting land beyond the reach of those who might cultivate or improve it. His indignation over this wastefulness of cutting people off from the land is palpable. He was not thinking about the effect of land enclosure on the landscape or the creatures it contained, but on the impoverished local residents who depended upon it. His was not a conservationist's view, in a way Britain had been too long cultivated and intensively managed by human beings for him to see the "wild, picturesque, boggy, and barren moor" as a landscape worth preserving in its own right. Wallace's word pictures of British habitats always focus on the cultivation of the land and its use by people:

> Another thing that should be attended to in all such inclosures of *waste land* [my italics] is the preservation for the people at large of rights of way over it in various directions, both to afford ample means of enjoying the beauties of nature and also to given pedestrians short cuts to villages, hamlets, or railway stations. One of the greatest blessings that might be easily attained if the land were resumed by the people to be held for the common good, would be the establishment of ample footpaths along every railway in the kingdom ... (S729, 1:155).

Much of Wallace's writing about his early experiences was done late in his life, thus seen through the lens of his experiences in the tropics. Nevertheless, his experience with surveying and outdoor work and his early interest in the components of the natural world meant that his ideas about the relationships between landscapes and the species they contained were complex. To a certain extent, it seems as if for Wallace, species were things to be collected and "understood," while landscapes were intrinsically important for their potential in bettering the human condition.

Tropical Experiences

In his travels to the Amazon (1848–52) and South East Asia (1854–62), Wallace combined his acute observational skills with his ability to find, identify, and catalogue collections of the most marvellous specimens of natural history, mostly birds, butterflies, and beetles. His writing, both from the field and after he returned, gave to the late Victorian public lyrical, sometimes slightly hyperbolic for some modern tastes, descriptions of the plants and animals of those regions. Camerini (2002, 3) has described this perfectly—"His writing reveals an independence of thought, which though shaped by his time and place, bears the stamp of his thoughtfulness, curiosity, and openness, along with his penchant for being opinionated and outspoken ... [he] maintained a lack of pretension, an optimism, and an enormous compassion, which come through in his writing and capture the heart of many a reader." Wallace's books about his tropical travels were written for the entertainment of the general public. He imparted his enthusiasm for the sights and sounds he experienced with no embarrassment. In a letter to his brother-in-law Thomas Sims, sent from Ternate, he says:

> So far from being angry at being called an enthusiast (as you seem to suppose), it is my pride and glory to be worthy to be so called. Who ever did anything good or great who was not an enthusiast? The majority of mankind are enthusiasts only in one thing—in money-getting; and these call others enthusiasts as a term of reproach because they think there is something in the world better than money-getting. It strikes me that the power or capability of a man in getting rich is in an *inverse* proportion to his reflective powers and in *direct* proportion to his impudence (S729, 1:368).

His descriptions of the plants and animals of the tropics are almost poetic; in a letter to his friends back at the Mechanic Institute at Neath he describes the tropical forests of the Amazon in a way that would make anyone want to down tools and join him straight away, and his description of orchids in the flooded forests of the Rio Negro makes them spring to life instantly:

> There is, however, one natural feature of this country, the interest and grandeur of which may be fully appreciated in a single walk: it is the "virgin forest." Here no one who has any feeling of the magnificent and the sublime can be disappointed; the sombre shade, scarce illumined by a single direct ray even of the tropical sun, the enormous size and height of the trees, most of which rise like huge columns a hundred feet or more without throwing out a single branch, the strange buttresses around the base of some, the spiny or furrowed stems of others, the curious and even extraordinary creepers and climbers which wind around them, hanging in long festoons from branch to branch, sometimes curling and twisting on the ground like great serpents, then mounting to the very tops of the trees, thence throwing down roots and fibres which hang waving in the air, or twisting round each other form ropes and cables of every variety of size and often of the most perfect regularity. These, and many other novel features—the parasitic plants growing on the trunks and branches, the wonderful variety of the foliage, the strange fruits and seeds that lie rotting on the ground—taken altogether surpass description, and produce feelings in the beholder of admiration and awe. It is here, too, that the rarest birds, the most lovely insects, and the most interesting mammals and reptiles are to be found. Here lurk the jaguar and the boa-constrictor, and here amid the densest shade the bell-bird tolls his peal ... (S729, 1:270–71).

> But what lovely yellow flower is that suspended in the air between two trunks, yet far from either? It shines in the gloom as if its petals were of gold. Now we pass close by it, and see its stalk, like a slender wire a yard and a half long, springing from a cluster of thick leaves on the bark of a tree. It is an *Oncidium*, one of the lovely orchis tribe, making these gloomy shades gay with its airy and brilliant flowers. Presently there are more of them, and then others appear, with white and spotted and purple blossoms, some growing

> on rotten logs floating in the water, but most on moss and decaying bark just above it. There is one magnificent species, four inches across, called by the natives St. Ann's flower (Flor de Santa Anna), of a brilliant purple colour, and emitting the most wonderful odour; it is a new species, and the most magnificent flower of its kind in these regions; even the natives will sometimes deign to admire it, and to wonder how such a beautiful flower grows "atóa" (uselessly) in the Gapó (S714 1853, 178).

Although he wrote poetically about the forests in letters, his overall impressions were of solemnity and gloom: "There is a weird gloom and a solemn silence, which combine to produce a sense of the vast—the primeval—almost of the infinite. It is a world in which man seems an intruder, and where he feels overwhelmed by the contemplation of the ever-acting forces which, from the simple elements of the atmosphere, build up the great mass of vegetation which overshadows and almost seems to oppress the earth" (S725 1891, 240). Wallace's idea of "virgin forest" was a forest in which there were no Europeans, and where cultivation had not yet reached. Today we know that the forests of the tropics have harboured human beings for millennia, and that the idea of completely untouched or undisturbed forest is more of a romantic myth than reality (Willis *et al.* 2004; see papers in Willis *et al.* 2007). Wallace's impressions of the diversity of species of trees and their subsequent overall rarity and sparseness in the Amazon led him to conclude that "[t]his peculiarity of distribution must prevent a great trade in timber for any particular purpose being carried on here" (S714, 437)—something that has changed utterly in the intervening century, even though particular tree species in the Amazon are in fact as locally rare as Wallace thought (see Pitman *et al.* 2001).

His accounts of the many animals he saw were equally descriptive, in the Amazon he saw "a little fish, peculiar to the Amazon, which inflates the fore part of the body into a complete ball, and when stamped upon explodes with a noise similar to that produced by the bursting of an inflated paper bag" (S714, 190–91), but also quite matter of fact: "Several jaguars were killed, as Mr. C pays about eight shillings each for their skins: one day we had some steaks at the table, and found the meat very white, and without any bad taste" (S714, 106). Today we would be horrified at the killing of jaguars, top predators, for bounty, although this was still practised until only very recently in parts of the New World tropics. The collections Wallace made involved much shooting of animals—particularly birds—and his first instinct was often to shoulder his gun.

> All the time we kept a sharp look-out, but saw no birds. At length, however, an old Indian caught hold of my arm, and whispering gently, "Gallo!" pointed into a dense thicket. After looking intently a little while, I caught a glimpse of the magnificent bird sitting amidst the gloom, shining out like a mass of brilliant flame. I took a step to get a clear view of it, and raised my gun, when it took alarm and flew off before I had time to fire. We followed, and soon it was again pointed out to me. This time I had better luck, fired with a steady aim, and brought it down. The Indians rushed forward, but it

> had fallen into a deep gully between steep rocks, and a considerable circuit had to be made to get it. In a few minutes, however, it was brought to me, and I was lost in admiration of the dazzling brilliancy of its soft downy feathers. Not a spot of blood was visible, not a feather was ruffled, and the soft, warm, flexible body set off the fresh swelling plumage in a manner which no stuffed specimen can approach (S714, 221–22).

Wallace's obvious joy in the living organism indicates that he was concerned not only with getting specimens for sale back in London, but also with observing their habits. It must be remembered that much of what we know now about animals that are today rare such as cocks-of-the-rock or orang-utans or birds of paradise comes from the initial impressions and collections made by naturalists like Wallace.

> As I was walking quietly along I saw a large jet-black animal come out of the forest about twenty yards before me, which took me so much by surprise that I did not at first imagine what it was. As it moved slowly on, and its whole body and long curving tail came into full view in the middle of the road, I saw that it was a fine black jaguar. I involuntarily raised my gun to my shoulder, but remembering that both barrels were loaded with small shot, and that to fire would exasperate without killing him, I stood silently gazing. In the middle of the road he turned his head, and for an instant paused and gazed on me, but having, I suppose, other business of his own to attend to, walked steadily on, and disappeared in the thicket. ... This encounter pleased me much. I was too much surprised, and occupied too much with admiration, to feel fear. I had at length had a full view, in his native wilds, of the rarest variety of the most powerful and dangerous animal inhabiting the American continent (S714, 241–42).

Sometimes feelings got the better of Wallace. The sheer excitement of finding something new or exceptionally beautiful can be overwhelming—as he found when he caught species of the genus *Ornithoptera*, the huge birdwing butterflies. He had captured many butterflies in the Amazon, and in South East Asia, all of which he described in his writing as "beautiful," "handsome," or "elegant," but these exceptionally large and strikingly coloured insects were special.

> I trembled with excitement as I saw it [*Ornithoptera poseidon* = *Ornithoptera priamus poseidon*] coming majestically towards me, and could hardly believe I had really succeeded in my stroke till I had taken it out of the net and was gazing, lost in admiration, at the velvet black and brilliant green of its wings, seven inches across, its golden body, and crimson breast. It is true I had seen similar insects in cabinets at home, but it is quite another thing to capture such one's self—to feel it struggling between one's fingers, and to gaze upon its fresh and living beauty, a bright gem shining out amid the silent gloom of a dark and tangled forest (S715 1962, 328–29).

> The beauty and brilliancy of this insect [*Ornithoptera croesus*] are indescribable, and none but a naturalist can understand the intense excitement I experienced when I at length captured it. On taking it out of my net and opening the glorious wings, my heart began to beat violently, the blood rushed to my head, and I felt much more like fainting than I have done when in apprehension of immediate death. I had a headache the rest of the day, so great was the excitement produced by what will appear to most people a very inadequate cause (S715 1962, 257–58).

When, in the Aru Islands, he was brought the prized king bird of paradise by his assistant Ali, Wallace was excited to see this wonderful bird at last, but also philosophical:

> I thought of the long ages of the past, during which the successive generations of this little creature had run their course—year by year of being born, and living and dying amid these dark and gloomy woods, with no intelligent eye to gaze upon their loveliness; to all appearance such a wanton waste of beauty. Such ideas excite a feeling of melancholy. It seems sad that on the one hand such exquisite creatures should live out their lives and exhibit their charms only in these wild, inhospitable regions, doomed for ages yet to come to hopeless barbarism; while on the other hand, should civilized man ever reach these distant lands, and bring moral, intellectual, and physical light into the recesses of these virgin forests, we may be sure that he will so disturb the nicely-balanced relations of organic and inorganic nature as to cause the disappearance, and finally the extinction, of these very beings whose wonderful structure he alone is fitted to appreciate and enjoy. This consideration must surely tell us that all living things were *not* made for man. Many of them have no relation to him. The cycle of their existence has gone on independently of his, and is disturbed or broken by every advance in man's intellectual development ... (S715 1962, 340).

Primates, our closest living evolutionary relatives, always excite in human beings particular feelings of sympathy or concern, perhaps because they look so like us. Conservation NGOs devoted to primates abound (just do the simple Google search "conservation AND primate"!) and as most primates are now rare, they can have a disproportionate effect on conservation policy under some circumstances (see Isaac *et al.* 2004). Wallace first saw monkeys in the Amazon:

> But to me the greatest treat was making my first acquaintance with the monkeys. One morning, when walking alone in the forest, I heard a rustling of the leaves and branches, as if a man were walking quickly among them, and expected every minute to see some Indian hunter make his appearance, when all at once the sounds appeared to be in the branches above, and turning up my eyes there, I saw a large monkey looking down at me, and seeming as much astonished as I was myself. I should have liked to have had a good look at him, but he thought it safer to retreat. ... At last one approached too near for its safety. Mr. Leavens fired, and it fell, the rest

Figure 25 Portrait of a young orang-utan.
From Beddard's 'Mammalia' in *The Cambridge Natural History* (1902). Out of copyright.

> making off with all possible speed. The poor little animal was not quite dead, and its cries, its innocent-looking countenance, and delicate little hands were quite childlike (S714, 41–42).

Later, in Borneo, he kept a baby orang-utan whose mother he had shot—a real encounter with an animal that seemed to him almost human: "The Mias, like a very young baby, lying on its back quite helpless, rolling lazily from side to side, stretching out all four hands into the air, wishing to grasp something, but hardly able to guide its fingers to any definite object; and when dissatisfied, opening wide its almost toothless mouth, and expressing its wants by a most infantine scream" (S715 1962, 35). His sadness when the baby "Mias" eventually died is clear, but ever practical, he prepared the skin and skeleton for sale back in London.

Despite all his vivid descriptions of the life of the tropical forests in which he travelled, Wallace very rarely in his travel books wrote about man's effect on the environments in which he was collecting and travelling. This is not to say he did not notice the effects human beings, both native and colonial, were having on natural habitats in the Amazon and South East Asia, but his opinions on these were confined to matter of fact statements, or to the effects of humans on his collecting.

> The most interesting and useful reptiles of the Amazon are, however, the various species of fresh-water turtles, which supply an abundance of wholesome food, and from whose eggs an excellent oil is made. ... There are such numbers of them, that some beaches are almost one mass of eggs beneath the surface, and here the Indians come to make oil. ... Millions of eggs are thus annually destroyed, and the turtles have already become scarce in consequence. There are some extensive beaches which yield two thousand

> pots of oil annually; each pot contains five gallons, and requires about two thousand five hundred eggs, which would give five millions of eggs destroyed in one locality (S714, 464–65).

> The other great Mammalia of Sumatra, the elephant and the rhinoceros, are more widely distributed; but the former is much more scarce than it was a few years ago, and seems to retire rapidly before the spread of cultivation (S715 1962, 104).

While attempting to collect near the village of Djilolo (on Gilolo, near Ternate) he remarked: "The place was then no doubt much more populous, as is indicated by the wide extent of cleared land in the neighbourhood, now covered with coarse high grass, very disagreeable to walk through, and utterly barren to the naturalist" (S715 1962, 242).

He calculated that dodos had become extinct in less than two hundred years since human occupation of the Mascarene islands, because they "were quite defenceless, and were rapidly exterminated when man introduced dogs, pigs, and cats into the island, and himself sought them for food" (S721 1892, 436). Here he definitely attributes the extinction of a particular species to human intervention, but in general he wrote about extinction as a phenomenon of the past, of the changing climates over geological time.

In contrast, his opinions on the social and moral milieus in which he was travelling were long and pointed—he compared cultures extensively, and not always to the favour of Europeans: "There is in fact as much difference between the various races of savage as of civilized peoples, and we may safely affirm that the better specimens of the former are much superior to the lower examples of the latter class" (S715 1962, 282).

> Before bidding my readers farewell, I wish to make a few observations on a subject of yet higher interest [than the scenery, vegetation and animals of the Malay Archipelago one supposes] and deeper importance, which the contemplation of savage life has suggested, and on which I believe that the civilized can learn something from the savage man. ... Now it is very remarkable that among people in a very low stage of civilization we find some approach to such a perfect social state. ... although we have progressed vastly beyond the savage state in intellectual achievements, we have not advanced equally in morals. ... We should now clearly recognize the fact, that the wealth and knowledge and culture of *the few* do not constitute civilization, and do not of themselves advance us towards the "perfect social state" ... until there is a more general recognition of this failure of our civilization ... we shall never, as regards the whole community, attain any real or important superiority over the better class of savages (S715 1962, 455–57).

In his descriptions of "savage" life Wallace does not outline people's use of the environment, but instead their social state. He does carefully describe the use of particular species such as palms or bamboos, but he does not comment on use of the environment as a whole by local peoples. While he was in the field, he was not an environmentalist in the modern sense, his accounts of his travels do not point out the destruction of the habitats in which he travelled (which was certainly going on at the time) nor do they recount sustainable practices of the local indigenous people. This, perhaps, is due to the commonly held belief in the nineteenth century that the tropical forests were infinitely large, and not because he did not see or think about these things. In 1853 he wrote about the Amazon: "Its entire extent, with the exception of some very small portions, is covered with one dense and lofty primeval forest, the most extensive and unbroken which exists upon the earth. ... Here we may travel for weeks and months inland, in any direction, and find scarcely an acre of ground unoccupied by trees" (S714, 432). Satellite photographs and ground studies today show that the forests of the Amazon, though still large, are disappearing at an alarming rate (Kirby *et al.* 2006), and that the extent of the world's tropical forests may already be too fragmented to maintain diversity as Wallace saw it (MEA 2005).

Synthesis

Once he returned from his tropical travels and established himself as a member of the scientific "establishment" of London, Wallace's writings began to take a synthetic turn, and here, in the relative comfort of Britain he began to voice opinions that to us sound more environmentally radical—he put together all his observations, not for entertainment, but to use as instruction for those in power. He was not directly involved in conservation lobbying groups like the Society for the Preservation of the Wild Fauna of the Empire (see above), but he did write popular books, designed to be read by those who voted, and whose opinions, he felt, could change the way politicians behaved. In his later writings he synthesized information from places he himself had not visited, recounting in particular the destruction of forests on islands, setting such events in a larger biological context. His experiences whilst collecting allowed him to see the complexity of life's interactions, and how simple decisions by those in power could change things utterly, often for the worse.

> ... the clearing of the forests on steep hill slopes, to make coffee plantations, produced permanent injury ... of a very serious kind. The rich soil, the product of thousands of years of slow decomposition of the rock, fertilized by the humus formed from decaying forest trees, being no longer protected by the covering of dense vegetation, was quickly washed away by the tropical rains, leaving great areas of bare rock or furrowed clay, absolutely sterile, and which will probably not regain its former fertility for hundreds, perhaps thousands, of years (S726 1901, 373).

> Every change becomes the centre of an ever-widening circle of effects. The different members of the organic world are so bound together by complex

> relations, that any one change generally involves numerous other changes, often of the most unexpected kind. We know comparatively little of the way in which one animal or plant is bound up with others, but we know enough to assure us that groups the most apparently disconnected are often dependent on each other. We know, for example, that the introduction of goats into St. Helena utterly destroyed a whole flora of forest trees; and with them all the insects, mollusca, and perhaps birds directly or indirectly dependent on them (S718 1876, 1:44).

> Thus, through the gross ignorance of those in power, the last opportunity of preserving the peculiar vegetation of St. Helena, and preventing the island from becoming the comparatively rocky desert it now is, was allowed to pass away (S721 1892, 296).

Despite his romantic feelings about the habitats in which he had collected—"There is a vastness, a solemnity, a gloom, a sense of solitude and of human insignificance, which for a time overwhelm him; and it is only when the novelty of these feelings have passed away that he is able to turn his attention to the separate constituents that combine to produce these emotions, and examine the varied and beautiful forms of life which, in inexhaustible profusion, are spread around him" (S725 1891, 269)—Wallace in later life was full of quite detailed suggestions and ideas for how to manage such habitats and regions:

> It is really deplorable that in so many of our tropical dependencies no attempt has been made to preserve for posterity any *adequate* portions of the native vegetation, especially of the virgin forests. . . . before it is too late our Minister for the Colonies should be urged without delay to give stringent orders that in *all* the protected Malay States, in British Guiana, Trinidad, Jamaica, Ceylon, Burma, etc., a suitable provision shall be made of forest or mountain "reserves," not for the purpose of forestry and timber-cutting only, but in order to preserve adequate and even abundant examples of those most glorious and entrancing features of our earth . . . It is not only our duty to posterity that such reserves should be made for the purpose of enjoyment and study by future generations, but it is absolutely necessary in order to prevent further deterioration of climate and destruction of the fertility of the soil, which has already taken place in Ceylon and some parts of India to a most deplorable extent. . . . I would also strongly urge that, in all countries where there are still vast areas of tropical forests, as in British Guiana, Burma, etc., all future sales or concessions of land for any purpose should be limited to belts of moderate breadth, say half a mile or less, to be followed by a belt of forest of the same width; and further, that at every mile or half-mile, and especially where streams cross the belts, transverse patches of forest, form one to two furlongs wide, shall be reserved, to remain public property and to be utilised in the public interest. Thus only can the salubrity and general amenity of such countries be handed on to our successors. Of course the general position of these belts and clearings should be determined by local conditions; but there should be no exception to the rule that all

> rivers and streams except the very smallest should be preserved as public property and absolutely secured against pollution; while all natural features of especial interest or beauty should also be maintained for public use and enjoyment ... (S732 1910, 77–78).

His ideas sound timely, but they were not really new—he was basically reiterating the advice being given by many others during that time, both in Britain and the United States. Forestry practices that included the concepts of sustainable use were already well established in British India in the mid-nineteenth century (Barton 2002). His plea for preservation of tracts of redwood forests (S441 1891) had already been addressed through the establishment of the state protection of the coastal redwoods in California (see above). His words were added to others working to conserve animals and habitat such as the Society for the Preservation of the Wild Fauna of the Empire, and given weight by his wealth of experience in the tropical forest regions of the world. There is little evidence, however, that he actively lobbied those in power to implement his ideas. His suggestions were rather aimed at a reading public whose opinions influenced those for whom they voted (a strategy consistent with advice he once offered to John Stuart Mill in an early letter published in the magazine *Reader*; see S110 1865).

Some of his ideas, however, were novel and were not taken up during his lifetime, but their eventual implementation has truly had a profound effect on our understanding of tropical nature. In describing a system of young collectors documenting the flora of the islands of South East Asia, Wallace anticipated the practice of today exemplified by the Missouri Botanical Garden's stationing of young botanists in diverse tropical regions of the world (Missouri Botanical Garden, St Louis, Missouri, USA; see **http://www.mobot.org**):

> There must be hundreds of young botanists in Europe and America who would be glad to go to collect, say for three years, in any of the islands [South East Asia] if their expenses were paid. ... And if each of these collectors had a moderate salary for another three years in order to describe and publish the results of their combined work on a uniform plan, and in a cheap form, the total expense for all the nations of Europe combined would be a mere trifle. Here is a great opportunity for some of our millionaires to carry out this important scientific exploration before these glorious forests are recklessly diminished or destroyed—a work which would be sure to lead to the discovery of great numbers of plants of utility or beauty, and would besides form a basis of knowledge from which it would be possible to approach the various great governments urging the establishment, as a permanent possession for humanity, of an adequate number of such botanical, or rather biological, "reserves" as I have here suggested in every part of the world (S732 1910, 79–80).

This practice, very different from the expeditions undertaken by many studying the diversity of the tropics, has promoted cooperation and scientific collaboration, as well as a deeper understanding of the areas and their people.

In the 1880s Wallace became the first president of the Land Nationalization Society, albeit somewhat reluctantly (Raby 2001), and much of his thought on the environment over the latter decades of the nineteenth century is framed in the context of the iniquity of private ownership of land. His lecture tour to the United States in 1886 at the invitation of the Lowell Institute of Boston was not a huge financial success (Raby 2001), but it allowed him to experience the vastness and diversity of the North American continent as he travelled from Boston to California via the Rocky Mountains. He tried to see as much of the country as he could, all the while lecturing on a wide variety of topics from Darwinism to the certainty of life after death and spiritualism. He was immensely impressed with the scenery of the United States, but not so impressed with the people's misuse of their environment nor with their uptake of all he thought the worst in British society:

> Over the greater part of America everything is raw and bare and ugly, with the same kind of ugliness with which we also are defacing our land and destroying its rural beauty. The ugliness of new rows of cottages built to let to the poor, the ugliness of the main streets of our towns, the ugliness of our "black countries" and our polluted streams. Both countries are creating ugliness, both are destroying beauty; but in America it is done on a larger scale and with a more hideous monotony. ... What a terrible object-lesson is this as to the fundamental wrong in modern societies which leads to such a result! Here is a country more than twenty-five times the area of the British Islands, with a vast extent of fertile soil, grand navigable waterways, enormous forests, a superabounding wealth of minerals—everything necessary for the support of a population twenty-five times that of ours ... which has yet, in little more than a century, destroyed nearly all its forests, is rapidly exhausting its marvellous stores of natural oil and gas, as well as those of the precious metals; and as the result of all this reckless exploiting of nature's accumulated treasures has brought about overcrowded cities reeking with disease and vice, and a population which, though only one-half greater than our own, exhibits all the pitiable phenomena of women and children working long hours in factories and workshops, garrets and cellars, for a wage which will not give them the essentials of mere healthy animal existence ... (S729 1905a, 2:193–96).

> But even more insidious and more widespread in its evil results ... [is] ... our bad and iniquitous feudal land system; first by enormous grants from the Crown to individuals or to companies, but also—which has produced even worse effects—the ingrained belief that *land*—the first essential of life, the source of all things necessary or useful to mankind, by labour upon which all wealth arises—may yet, justly and equitably, be owned by individuals, be monopolized by capitalists or by companies, leaving the great bulk of the people as absolutely dependent on these monopolists for permission to work and to live as ever were the negro slaves of the south before emancipation (S729, 2:195).

Although many of Wallace's statements still ring true today, he was not a prophet, nor was he particularly out of step with the zeitgeist of his time. His intensely personal views meant that often he did not really set his opinions in the context of what was going on elsewhere, but instead seems to depict himself as the only one with views on these topics. This dissociation of his opinions from the background of his time can make him seem a lone thinker, perhaps even possessing "devastatingly accurate foresight" (Raby 1996):

> The struggle for wealth ... ha[s] been accompanied by a reckless destruction of the stored-up products of nature, which is even more deplorable because more irretrievable. Not only have forest-growths of many hundreds of years been cleared away, often with disastrous consequences, but the whole of the mineral treasures of the earth's surface, the slow products of long-past eons of time and geological change, have been and are still being exhausted, and probably not equalled in amount during the whole preceding period of human history (S726 1901, 369).

We know that today the rate of exploitation of resources is increasing rapidly (MEA 2005), and if Wallace thought it was bad in 1900, just imagine how he might have felt today!

Pollution of the air and water are contributing factors to two of the principal drivers of biodiversity loss identified by the Millennium Ecosystem Assessment (climate change and pollution with phosphorus and nitrogen; see MEA 2005) which should hardly surprise us, as pollution was also one of the factors in the rise of the environmental movement in the mid-twentieth century, and in fact, had been an issue for those concerned with nature the century before (see Melosi 1980). For Wallace, pollution was an issue intricately tied up with the land, its use and misuse, and with people. His philosophical book *Man's Place in the Universe* (S728 1903a), has chapters entitled "The Earth is the Only Habitable Planet in the Solar System" and "The Air in Relation to Life"—where he lays out his arguments for a just and equitable society based on the proper use of natural resources. Nowhere in the book, however, does he comment on man's destruction of that only habitable planet, instead suggesting that it is the system, not people themselves, who are responsible for misuse of resources, and framing the argument entirely on the betterment of the *human* condition:

> Yet is among those nations that claim to be the most civilised, those that profess to be guided by a knowledge of the laws of nature, those that most glory in the advance of science, that we find the greatest apathy, the greatest recklessness, in continually rendering impure this all-important necessity of life, to such a degree that the health of the larger portion of their populations is injured and their vitality lowered, by conditions which compel them to breathe more or less foul and impure air for the greater part of their lives. The huge and ever-increasing cities, the vast manufacturing towns belching forth smoke and poisonous gases, with the crowded dwellings,

> where millions are forced to live under the most terrible unsanitary conditions, are the witness to this criminal apathy, this incredible recklessness and inhumanity. ... Remember! We claim to be a people of high civilisation, of advanced science, of great humanity, of enormous wealth! For very shame do not let us say "We *cannot* arrange matters so that our people may all breathe unpolluted, unpoisoned air!" (S728 1903a, 259–61).

Wallace's later years were marked by a settling of opinion as to man's central place in nature, and to the specialness of the human species. In his book *Darwinism* (S724 1889) he expounded on the necessity for a "higher power" in the creation of man, and this view led to an increasing distance between him and the Darwinians such as Huxley and Hooker. The apparent "uselessness" of the variety of life he had seen and written about bothered him, but his was a typically Wallacean solution...

> For the great majority of these entities *we* can see no use whatever, either of the enormous variety of species, or the vast hordes of individuals. Of beetles alone there are at least a hundred thousand distinct species now living, while in some parts of sub-arctic America mosquitoes are sometimes so excessively abundant that they obscure the sun. And when we think of the myriads that have existed through the vast ages of geological time, the mind reels under the immensity of, to us, apparently useless life. All nature tells us the same strange, mysterious story, of the exuberance of life, of endless variety, of unimaginable quantity. All this life upon our earth has led up to and culminated in that of man. It has been, I believe, a common and not unpopular idea that during the whole process of the rise and growth and extinction of past forms, the earth has been preparing for the ultimate—Man. Much of the wealth and luxuriance of living things, the infinite variety of form and structure, the exquisite grace and beauty in bird and insect, in foliage and flower, may have been mere by-products of the grand mechanism we call nature—the one and only method of developing humanity (S728 1903a, 320–21).

For Alfred Russel Wallace, human beings were at the centre of his thoughts about nature. Nature existed for human beings, either for rational and equitable use in the present or as a long chain of being developed by evolution culminating in man—for Wallace, it was man that mattered. This I think puts him in a unique position in his thinking about conservation of biodiversity for his time. While some of his thinking seems idiosyncratic or even patently false to a twenty-first-century biologist, it nevertheless is in tune with the twenty-first century's broad, challenging goals set by the international community to put an end to poverty, educate all, empower women, and achieve a stable environment (see Table 11.1). These goals too are centred on human beings, but for a different reason. Today we understand that the human species is one of many, one that has had a huge and ever-increasing impact on the world around us, but still a species of animal like all others. If we are to maintain a dynamic, evolving planet—"the only habitable planet in the solar system"—then human needs must be taken into account. The balance is difficult, and engenders much argument over how and what to conserve (see Royal Society 2003). Conservation thinking has gone a long way from its early beginnings in colonial game reserves and national parks, through species-centric campaigns to conserve charismatic animals, but has come full circle to be framed as sustainable development, broadly in a landscape view that takes into account human beings and their needs. The current focus on sustainable development would have been entirely understandable to Wallace; he too put our own species at the centre of what the world was all about.

Summary

Wallace was not a proto-conservation biologist, although he did glean a lot about nature and the environment from his experience. Many of his statements can be read as astonishingly prescient, but in fact, his views were based not only on personal experience but on ideas "in the air" swirling around during his lifetime. His deeply held beliefs on social justice and equity shaped his views on man's impact on the environment, and how this environment should be used for the benefit of all. These views are consistent with the modern focus on sustainable development, and I feel that Wallace would have been entirely comfortable with many of the international targets set out over the last decade. He recognized that knowledge of nature was a prerequisite to caring about nature, and nowhere is this view more eloquently stated than in his essay on the geography of the Malay Archipelago (S78 1863). This is still true today: "A major obstacle for knowing (and therefore valuing), preserving, sustainably using and sharing benefits equitably from the biodiversity of a region is the human and institutional capacity to research a country's biota" (MEA 2005, 14). I wonder whether he would think we, with all our investment in biodiversity conservation and scientific study of nature, had even moved part way toward achieving what he considered necessary.

> It is for such inquiries that the modern naturalist collects his materials; it is for this that he still wants to add to the apparently boundless treasures of our national museums, and will never rest satisfied as long as the native country, the geographical distribution, and the amount of variation of any living thing remains imperfectly known. He looks upon every species of animal and plant now living as the individual letters which go to make up one of the volumes of our earth's history; and, as a few lost letters may make a sentence unintelligible, so the extinction of numerous forms of life which the progress of cultivation invariably entails will necessarily render obscure this invaluable record of the past. It is, therefore, an important object, which governments and scientific institutions should immediately take steps to secure, that in all tropical countries colonised by Europeans the most perfect collections possible in every branch of natural history should be made and deposited in national museums, where they may be available for study and interpretation.

If this is not done, future ages will certainly look back upon us as a people so immersed in the pursuit of wealth as to be blind to higher considerations. They will also charge us with having culpably allowed the destruction of some of those records of Creation which we had it in our power to preserve; and while professing to regard every living thing as the direct handiwork and best evidence of a Creator, yet, with a strange inconsistency, seeing many of them perish irrecoverably from the face of the earth, uncared for and unknown (S78, 234).

II

In the World of Man, and Worlds Beyond

12

The "Finest Butterfly in the World?": Wallace and His Literary Legacy

Peter Raby

Wallace offered a characteristically self-critical report of his own literary powers in his autobiography, *My Life* (S729 1908), noting that he had a serious defect in verbal memory, which, together with his "imperfect school training" and his "shyness and want of confidence," put him at a great disadvantage as a public speaker. "I can rarely find the right word or expression to enforce and illustrate my argument, and constantly feel the same difficulty in private conversation." However, he continued, in writing this was not so injurious, as he could generally express himself with "tolerable clearness and accuracy" when he had time for deliberate thought. In fact, he concluded, equally characteristically turning his perceived defect into a strength, "the absence of the flow of words which so many writers possess has caused me to avoid that extreme diffuseness and verbosity which is so great a fault in many scientific and philosophical works." But he then proceeded to define another supposed defect, his inability to see "analogies or hidden resemblances and incongruities," which, he claimed, in combination with his linguistic defect, "has produced the total absence of wit or humour, paradox or brilliancy" in his writings (S729 1908, Chapter 8 "Self-Education in Science and Literature," 116–17). One might take issue with Wallace about the validity of this judgement, especially with regard to the last term, but he has many other strengths as a writer, including clarity and accessibility. His 1858 Linnean paper, "On the Tendency of Varieties to Depart Indefinitely from the Original Type" (S43) certainly does not suffer by comparison with Darwin's extracts. By the time of his 1905 assessment, Wallace had a remarkably long list of publications to his name—books, papers for learned journals, articles for more popular newspapers and magazines—on a wide variety of subjects.

If some of his own writings do not obviously and at first sight exhibit much wit and brilliancy, Wallace fully appreciated these qualities in others. He cited Thomas Hood as an early influence, relishing the puns and conundrums of Hood's *Comic Annual*, and later enjoying the humour of Mark Twain and Lewis Carroll.

But he was, from an early age, an eager reader. At home, there were the "good old standard" works, Swift's *Gulliver's Travels*, Bunyan's *The Pilgrim's Progress*, Goldsmith's *The Vicar of Wakefield*, which he returned to repeatedly, and his father would read aloud in the evenings from Mungo Park's travels, or Defoe's *Journal of the Plague Year.* At one point his father held a modest post as a librarian in Hertford, and Wallace would join him there after school for an hour, and on wet Saturday afternoons he would squat on the floor in a corner and make his way through the fiction: Fenimore Cooper, Harrison Ainsworth, Captain Marryat, Bulwer Lytton, as well as classics such as *Don Quixote, Roderick Random*, and *Tom Jones.* By the time he was fourteen, Wallace had absorbed enough literature to be wholly familiar with the idea that life can be seen as a journey, a series of adventures. In addition, he read a great deal of poetry, including *Paradise Lost* (together with *The Pilgrim's Progress*, this was permitted Sunday reading), Dante's *Inferno*, and Pope's translation of *The Iliad* (S729 1908, Chapter 3 "My School Life at Hertford," 39–41). He retained his affinity for romantic fiction, and romantic poetry, throughout his life. He loved Cowper, especially "The Task," and, more surprisingly, Byron. The eccentric humour of Sterne's *Tristram Shandy* particularly appealed to him. He continued to take an interest in new work, quoting Oscar Wilde's "The Ballad of Reading Gaol" as an epigraph to a chapter of his own *The Wonderful Century* (S726 1898a) in the year of the poem's publication.

The habit of reading, and of close reading, stayed with him. His autobiography records his many encounters with serious books of ideas that were, either immediately or retrospectively, significant to him. Among the highlights were Robert Owen's writings, associated with his visits to the "Hall of Science," off Tottenham Court Road, in 1837, and Thomas Paine's *Age of Reason*, when he was only fourteen (S729 1908, 45). When he went to Leicester in 1844, he had access to the town subscription library, and read Humboldt's *Personal Narrative of Travels in South America*, and Malthus's *Principles of Population*, each crucial for him in different ways. (He also acquired Stephen's *Manual of British Coleoptera*, to accompany his earlier *vade mecum*, Lindley's *Elements of Botany.*) Back in Neath in 1845, he was much impressed by Chambers's *Vestiges of the Natural History of Creation*, and began to correspond with Henry Walter Bates about its theories. Among later works which influenced and impressed him were Herbert Spencer's *Social Statics*, Edward Bellamy's Utopian *Looking Backward*, and, inescapably, Darwin's *The Origin of Species*, which he read whilst still on his travels through the Malay Archipelago (S729 1908).

Wallace acquired the habit of writing letters when he left home at the age of fourteen, to live in a succession of lodgings, mostly with his brothers John and William. He wrote to his parents, to his sister Fanny, and, more expansively and humorously, and sometimes sentimentally, to his old schoolfriend George Silk. He wrote clearly structured, factual lectures for the courses he undertook to teach at the Mechanics' Institute, Neath. He learned to record in meticulous detail his botanical and zoological findings. And on his travels up the Amazon and Rio

Negro, he combined all these skills, so that letters to his family could serve to some extent as a record, or so that suitable extracts from letters to his agent Samuel Stevens could also be printed in a specialist journal such as *The Annals and Magazine of Natural History*.

Wallace possessed the qualities of lucidity, and simplicity, in his writing. He liked facts, and was at pains to record them. But he also had a feeling for atmosphere, and for more expansive description. He was conscious of his English readers, or potential readers, aware, from his own reading in classic travel literature, that they might have a picture in their minds of what the equatorial forest might be like, and being prepared either to endorse or to correct this. He was also adept at shaping his narrative. *Travels on the Amazon* (S714 1889), that "absurd" book as he once described it, has a simple chronological structure. "It was on the morning of the 26th of May, 1848," it begins as accurately, even dully, as possible, "that after a short passage of twenty-nine days from Liverpool, we came to anchor opposite the southern entrance to the River Amazon, and obtained our first view of South America." The narrative closes, effectively, at the end of Chapter 13: "On the 1st of October the pilot came on board, and Captain Turner and myself landed at Deal, after an eighty days' voyage from Para; thankful for having escaped so many dangers, and glad to tread once more on English ground."

In between, the story unfolds in an apparently traditional manner. A young man arrives in an exotic country, with a companion. He explores the immediate neighbourhood and begins to learn his trade of collecting. He travels up the mighty Amazon, pushes into increasingly remote territory, suffers a number of accidents and close encounters with dangerous animals, and finally sets sail for home with countless treasures, only to watch helplessly as his specimens, dead and alive, and most of his journals and records go up in flames in mid-Atlantic. The story is gripping enough, though Wallace found himself severely restricted in its writing because of the loss of so many of his records. Nevertheless, his style is crisp and vivid, and conveys his fresh enthusiasm for the country, and for its peoples; and his natural reserve and sense of privacy protects him from indulging in too much personal commentary, so that for the most part his terrible privations are uninflected, and are all the more arresting because of that:

> All this time the Indians went on with the canoe as they liked; for during two days and nights I hardly cared if we sank or swam. While in that apathetic state I was constantly half-thinking, half-dreaming, of all my past life and future hopes, and that they were perhaps all doomed to end here on the Rio Negro. And then I thought of the dark uncertainty of the fate of my brother Herbert, and of my only remaining brother in California, who might perhaps ere this have fallen a victim to the cholera, which according to the latest accounts was raging there. But with returning health these gloomy thoughts passed away, and I again went on, rejoicing in this my last voyage, and looking forward with firm hope to home, sweet home! (S714 1889, 226).

The "dark uncertainty" was indeed dark for Wallace, who by the time he wrote those words had visited his brother's grave in Pará. The "firm hope" of "home, sweet home" may seem something of a commonplace; but it is a commonplace that is sustained by the authority of experience. *Travels on the Amazon* holds the reader, for the most part, by the immediacy and authenticity conveyed by Wallace's plain style, though from time to time he varies this. There is, for example, his breaking out into verse in Chapter 9, nearly two hundred lines of blank verse in the style of Cowper entitled "A Description of Javita," written "in a state of excited indignation against civilised life in general"—a meditation on the contrast between the "civilised" and the "wild." Or the inclusion of a lengthy story about "Compadre Death," which he heard on his return trip down the Amazon. Both these inclusions seem more like random insertions, or perhaps disingenuous attempts to bulk out the narrative. Another, more conscious, technique is the selection of a specific incident for elaboration, and retrospective comment:

> As I was walking quietly along I saw a large jet-black animal come out of the forest about twenty yards before me, which took me so much by surprise that I did not at first imagine what it was. As it moved slowly on, and its whole body and long curving tail came into full view in the middle of the road, I saw that it was a fine black jaguar. I involuntarily raised my gun to my shoulder, but remembering that both barrels were loaded with small shot, and that to fire would exasperate without killing him, I stood silently gazing. In the middle of the road he turned his head, and for an instant paused and gazed at me, but having, I suppose, other business of his own to attend to, walked steadily on, and disappeared in the thicket. As he advanced, I heard the scampering of small animals, and the whizzing flight of ground birds, clearing the path for their dreaded enemy. This encounter pleased me much. I was too much surprised, and occupied too much with admiration, to feel fear. I had at length had a full view, in his native wilds, of the rarest variety of the most powerful and dangerous animal inhabiting the American continent (S714 1889, 166).

Such epiphanies punctuate the texts of both *Travels on the Amazon* and *The Malay Archipelago*, and serve to convey a sense of the unity of nature, and the relationship between the wild and the human. Wallace gazes at the jaguar, and the jaguar gazes at him. Admiration dispels fear. Animal and man share the space and the moment; and, for once, no shot is fired. The sense of wonder at the extraordinary, the rare, the beautiful is even more present in *The Malay Archipelago*, whose lengthy title glosses the region as "The Land of the Orang-Utan, and the Bird of Paradise" (S715 1989), and to some extent reflects that title as it moves from the relatively early encounters with orang-utans in Borneo to the more extended sequences on birds of paradise towards the close. The second travel book, so long in gestation, has a more subtle and finely planned structure. Wallace explains in his Preface that he adopted "a geographical, zoological, and ethnological arrangement" rather than a strictly chronological one. Significantly, he chose to adapt an earlier article to form a closing chapter entitled "The Races of Man in the Malay Archipelago"; the arc of his narrative brings the reader, both in specific episodes, but even more powerfully as the book draws to its close, to a consideration of the overall meaning of his discoveries for man, and the future of civilization.

As a historical figure, Wallace appears in a number of novels. He is referred to as the author of *The Malay Archipelago* in Conrad's *The Secret Agent*, and in Somerset Maugham's story "Neil Macadam." In "Neil Macadam" the young man of the title, a naturalist, reads *The Malay Archipelago* as he sails from Singapore to Borneo to take up a post as assistant curator of a museum; and then finds himself the unwilling object of the sexual attentions of the curator's Russian wife. Macadam confesses to Darya that he read a lot of Conrad on his voyage east, and admired him awfully. "How can you English ever have let yourselves be taken in by that wordy mountebank?" she replies. "That stream of words, those involved sentences, the showy rhetoric, that affectation of profundity: when you get through all that to the thought at the bottom, what do you find but a trivial commonplace?" "There's no one who got atmosphere like Conrad," is Neil's defence. "I can smell and see and feel the East when I read him." Later in the conversation he continues: "I don't know why fiction should be hampered by fact. I don't think it's a mean achievement to have created a country, a dark, sinister, romantic and heroic country of the soul." Conrad's reliance on second-hand sources seems at issue here, but as Maugham's story progresses, it begins to bear a marked resemblance to Wallace's account of his own journey into the interior of Borneo, recorded in Chapter 5 of *The Malay Archipelago.* The smell, sight, and feel of the East, at least, owe much to Wallace's original writing (Maugham 1933, 286–87). Wallace also features as a historical character in A. S. Byatt's novella "Morpho Eugenia," whose naturalist hero William Adamson sets out for the Amazon one year after Wallace and Bates. Many of the details of William's Amazon experience are drawn from the writings of Richard Spruce, as well as from the accounts of Wallace and Bates. Like Wallace, Adamson sees his journals and collections burn at sea on his journey home; like Wallace, he records his sensations on acquiring some rare specimens of tropical butterflies: "When they were brought to me, in such perfect condition, I felt the blood rush to my head, truly felt I might faint with excitement" (Byatt 1992, 18).

Another, earlier, Amazon-centred novel in which Wallace appears is Arthur Conan Doyle's 1912 adventure story, *The Lost World.* As in "Morpho Eugenia," Wallace appears as himself, when the formidable Professor Challenger—a most un-Wallace-like figure, apart from his qualities of courage, resilience, and intellectual curiosity—expounds to Edward Malone, the "straight" journalist, about his previous classical journey to South America, whose object was to verify some conclusions of Wallace and of Bates. Having introduced them as scientific travellers, Conan Doyle draws relatively freely on their published accounts, incorporating details gleaned from Henry Walter Bates's *The Naturalist on the River Amazons* and from Wallace's *Travels on the Amazon* (S714 1889), in addition to more recent sources such as Roger Casement and Colonel Percy Fawcett. The route taken by the expedition to the lost world follows initially that of Wallace and Spruce, as they moved further up the Rio Negro and its tributaries. Challenger leads his small band of intrepid explorers through a forest throbbing with danger, as the drums beat out "We will kill you if we can." "I learned, however, that day once for all that both Summerlee and Challenger possessed that highest type of bravery, the bravery of the scientific mind. Theirs was the spirit which upheld Darwin among the gauchos of the Argentine or Wallace among the head-hunters of Malaya."[1] When the group finally reaches the plateau, they find themselves in a time-warp, inhabited by dinosaurs and a race of ape-men, as well as colonizing Indians. The ape-men, missing links, are bloodily slaughtered. When the members of the expedition make it back to England, they fail to convince a scientific public meeting of their claims, until Professor Challenger removes the lid of a large square packing-case, and unleashes a pterodactyl. The beast certainly convinces the sceptics, but, like King Kong, escapes from its minders, perches on the roof of the Queen's Hall like a diabolical statue for some hours, and is last seen heading out into the wastes of the Atlantic. This wonderful blend of adventure story, science fiction, and parody veers away from its factual basis as it develops; but behind the extravagances of the narrative lies the pattern of Wallace's own scientific travels, and his search for the living species which would help to fill out the picture of the natural world, and its origins. In the Amazon, he hoped to find the white umbrella-bird; from the Malay Archipelago, he brought back, not a pterodactyl, but birds of paradise, as well as the most beautiful butterfly in the world.

While Wallace's shadow or influence can be found in the work of a number of writers, he is most strikingly present in the novels and stories of Conrad. *The Malay Archipelago*, according to Richard Curle, was Conrad's "favourite bed-side book." "He had an intense admiration for those pioneer explorers—'profoundly inspired men' as he called them—who have left us a record of their work; and of Wallace, above all, he never ceased to speak in terms of enthusiasm. Even in conversation he would amplify some remark by observing, 'Wallace says so-and-so,' and *The Malay Archipelago* had been his intimate friend for many years" (Curle 1934, 431). In fact, Conrad's copy of *Travels on the Amazon*, now with *The Malay Archipelago* in the collection of the Canterbury Museum, was also regular bedside reading for Conrad. Jessie Conrad's inscription to Richard Curle names it "one of Joseph Conrad's bedside books which he must have known from one corner to the other." Wallace's experiences on each of his great journeys permeate Conrad's stories and novels in a number of ways.

Conrad began writing *Almayer's Folly*, his first novel, in 1889, soon after his four voyages to the Malay Archipelago, including four visits to the Berau trading post in Borneo. Although Conrad was a sharp observer, his time in port would have been limited, and it is hardly surprising that he drew, in varying degrees, on detailed accounts of Malay life such as Wallace's, in addition to books by or about Rajah Brooke, and other books of travels and memoirs. Conrad was a little defensive about his methods. Taken to task by Sir Hugh Clifford for inaccuracy, he wrote to William Blackwood: "Curiously enough all the details about the little characteristic acts and customs which they hold up as proof I have taken out (to be safe) from undoubted sources—dull, wise books."[2] He clearly regarded *The Malay Archipelago* as both wise and far from dull. Detail after detail in *An Outcast of the Islands* can be traced to its source in Wallace; and in *Lord Jim*, especially, there is a more powerful and resonant presence.

Conrad's dependence on Wallace is especially notable in the central section of the novel, where Marlow recounts his dealings with Stein. The name Stein was probably suggested by Wallace's reference to Bernstein, a German naturalist collecting for the Leiden museum, who appears twice in *The Malay Archipelago*, on each occasion in close conjunction with key passages used by Conrad. It is tempting to see something of Wallace's own physical characteristics (beard excepted), as well as his character, in Conrad's depiction of Stein:

> The gentle light of a simple unwearied, as it were, and intelligent good-nature illumined his long and hairless face. It had deep downward folds, and was pale as of a man who had always led a sedentary life—which was indeed very far from being the case ... It was a student's face; only the eyebrows nearly all white, thick and bushy, together with the resolute searching glance that came from under them, were not in accord with his, I may say, learned appearance. He was tall and loose-jointed; his slight stoop, together with an innocent smile, made him appear benevolently ready to lend you his ear; his long arms with pale big hands had rare deliberate gestures of a pointing out, demonstrating kind. I speak of him at length, because under this exterior, and in conjunction with an upright and indulgent nature, this man possessed an intrepidity of spirit and a physical courage that could have been called reckless had it not been like a natural function of the body—say good digestion, for instance—completely unconscious of itself (Conrad 1920, 202–03).

Conrad was writing before Wallace's autobiography was published, so it is in reality unlikely that many of these attributes actually stem from an attempted portrait; nevertheless, the simple good-nature, sense of benevolence, intrepidity of spirit, and physical courage (the last would have certainly been denied by Wallace himself) strongly invoke the Wallace of *The Malay Archipelago* and *Travels on the Amazon*. Stein is presented, not as a scientific traveller *per se*, but as, originally, a naturalist's assistant, who has remained out in the East—a pattern that suggests Charles Allen, Wallace's young assistant, who later found employment and made his home in Singapore. The routine of Stein's life is drawn from Wallace's description of his Dutch friend, Mesman, who farmed near Macassar. Mesman

> lived in a spacious house near the town, situated in the midst of a grove of fruit-trees, and surrounded by a perfect labyrinth of offices, stables, and native

> cottages, occupied by his numerous servants, slaves, or dependants.... Putting on a clean white linen suit, he then drove to town in his buggy, where he had an office, with two or three Chinese clerks, who looked after his affairs (S715 1989, 233–34).

> At first he had travelled a good deal amongst the islands, but age had stolen upon him, and of late he seldom left his spacious house three miles out of town, with an extensive garden, and surrounded by stables, offices, and bamboo cottages for his servants and dependants, of whom he had many. He drove in his buggy every morning to town, where he had an office with white and Chinese clerks (Conrad 1920, 207).

Conrad drops the slaves, and, whereas Mesman was a coffee and opium merchant, with a prau which traded for mother-of-pearl and tortoise-shell, makes Stein the owner of a small fleet of schooners and native craft, dealing in "island produce on a large scale."

These factual details, while establishing Conrad's use of Wallace as a source, are interesting, but of limited significance. However, more tellingly, he also drew imaginatively on Wallace's description of his emotions when capturing a female specimen of a "new" bird-winged butterfly, which he subsequently named *Ornithoptera croesus.* Wallace records seeing, on his very first walk into the forest at Batchian, an "immense butterfly of a dark colour marked with white and yellow spots." He saw at once that it was "a female of a new species of Ornithoptera, or, 'bird-winged butterfly', the pride of the Eastern tropics":

> I had begun to despair of ever getting a specimen, as it seemed so rare and wild; till one day, about the beginning of January, I found a beautiful shrub with large white leafy bracts and yellow flowers, a species of Mussaenda, and saw one of these noble insects hovering over it, but it was too quick for me, and flew away. The next day I went again to the same shrub and succeeded in catching a female, and the day after a fine male. I found it to be as I had expected, a perfectly new and most magnificent species, and one of the most gorgeously coloured butterflies in the world. Fine specimens of the male are more than seven inches across the wings, which are velvety black and fiery orange, the latter colour replacing the green of the allied species. The beauty and brilliancy of this insect are indescribable, and none but a naturalist can understand the intense excitement I experienced when I at length captured it. On taking it out of my net and opening the glorious wings, my heart began to beat violently, the blood rushed to my head, and I felt much more like fainting than I have done when in apprehension of immediate death. I had a headache the rest of the day. So great was the excitement produced by what will appear to most people a very inadequate cause (S715 1989, 341–42).

This passage is representative of the more heightened style Wallace adopts for incidents of special moment, for example when he secures birds of paradise. He maintains the objective details—the shrub which became his collecting station, a species of *Mussaenda,* with large white leafy bracts and yellow flowers, the fine

1 Watercolour of Wallace's flying frog *(Rhacophorus nigropalmatus)* from the Simunjon coalworks, Sarawak, Borneo, painted by Wallace in 1855 (private collection). Copyright Richard Wallace.

2 Kensington Cottage (now Kensington House), Usk, Wales, as it is today. Wallace was born here in 1823. Copyright George Beccaloni.

3 Watercolour painting of Treeps, Hurstpierpoint, Sussex, by Wallace's wife Annie. This was Annie's family home. Wallace stayed here in 1867 and 1868, during which time he wrote his book *The Malay Archipelago* (private collection). Copyright Richard Wallace.

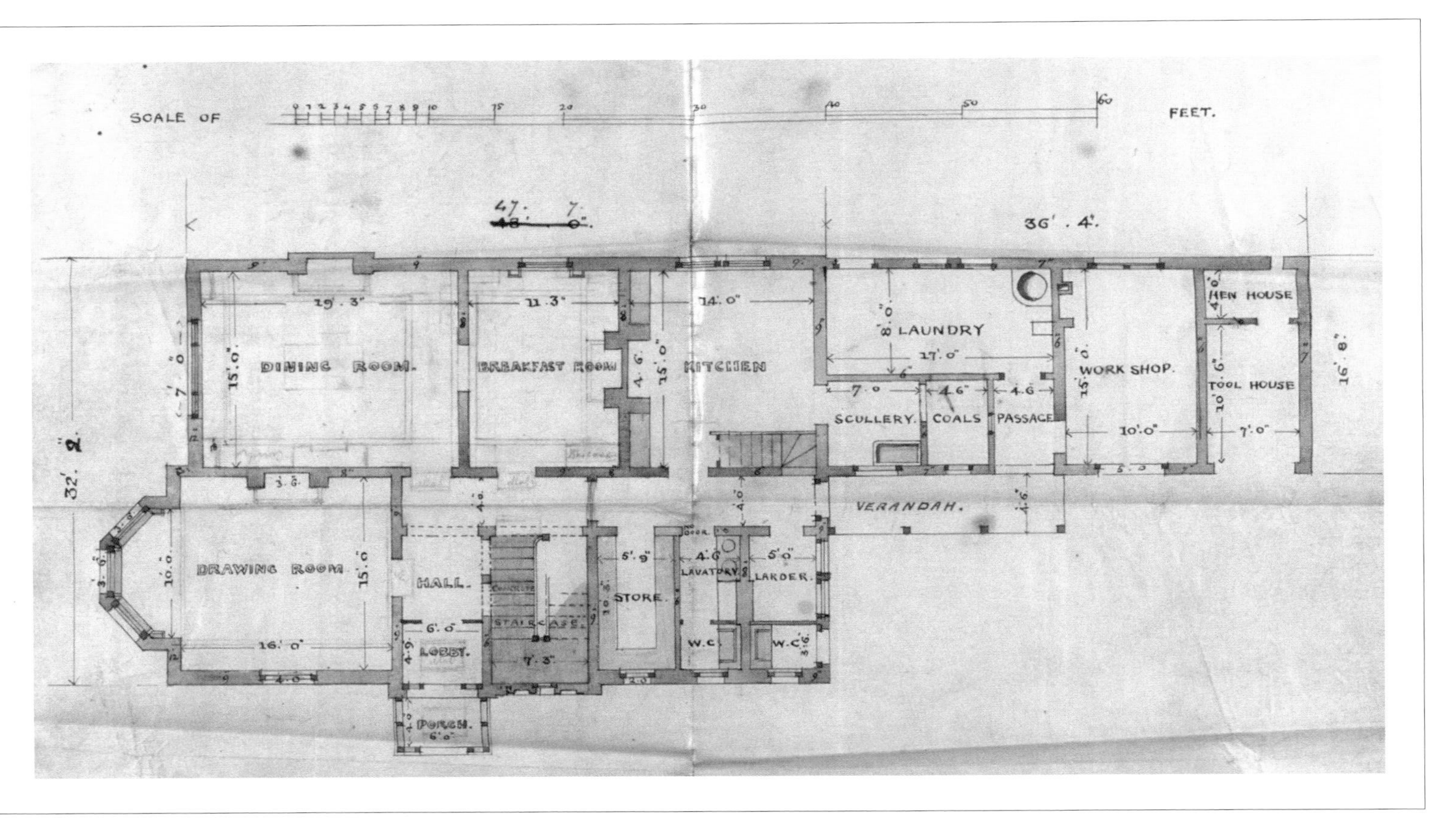

4 Architect's drawing showing the ground floor of The Dell, the house that Wallace built in Grays, Essex (original now in the NHM, London). Copyright Richard Wallace.

5 Watercolour painting of Corfe View, Parkstone, Dorset, by Wallace's wife Annie. Wallace lived here from 1889 to 1902 (private collection). Copyright Richard Wallace.

6 Terracotta monogram of Wallace's initials from his house Old Orchard, Broadstone, Dorset. Copyright George Beccaloni.

7 Nymph of a bush-cricket (Orthoptera: Tettigoniidae) from Ecuador which is camouflaged to resemble the moss on which it rests. Copyright George Beccaloni.

8 While Darwin focused his discussions of feather colouration almost exclusively on bold and bright colouration, Wallace spent as much time explaining cryptic colouration, such as the brown and camouflaged plumage of this sedge wren *(Cistothorus platensis)*, as he did ornamental colouration. Copyright Geoffrey Hill.

9 Lions *(Panthera leo)* feeding on a Burchell's zebra *(Equus burchelli)* they have killed. Even within the same habitat, sympatric mammals have remarkably different coat colours; here, lions presumably have tawny coats to match dry season yellow-brown leaf litter and grassland habitats, whereas the function of the zebra's black and white stripes remains unknown. Copyright Tim Caro.

10 A drawer from Wallace's personal insect collection (now in the NHM, London) showing a display he produced to illustrate sexual dimorphism in butterflies. Males are on the left and females on the right. Copyright George Beccaloni.

11 An aposematic poison arrow frog *(Dendrobates duellmani)* from eastern Ecuador. The skin secretes toxic alkaloids which may sometimes even prove fatal to predators. Copyright George Beccaloni.

12 A drawer from Wallace's personal insect collection (now in the NHM, London) showing a display he produced to illustrate Batesian mimicry in butterflies. The presumed inedible model species are arranged in the far right column and the presumed edible mimics are to the left of them. Note that two of the mimic species are sexually dimorphic and have non mimetic males. Copyright George Beccaloni.

13 Wallace never accepted Darwin's idea that female mate choice was a primary force in the evolution of colourful plumage. In particular, Wallace thought it was implausible that female mate choice could lead to the evolution of complex and intricate colour displays, like the tail of this peacock *(Pavo cristatus)*. Recent models for feather growth and pigmentation suggest that relatively simple developmental programs can create such complex colour patterns. Whether these developmental programs are shaped by female preferences remains to be determined. Copyright George Beccaloni.

14 Wallace proposed that many of the distinctive colour patterns of animals serve as signals of species identity that are used to avoid hybrid pairing, and for cohesion of social groups. One of the few field tests of Wallace's hypothesis focused on eye and eye-ring colouration in gulls, including the yellow eye and yellow eye ring of this western gull *(Larus occidentalis)*, and this study supported Wallace's idea. Copyright Geoffrey Hill.

male specimens "more than seven inches across the wings," "velvety black and fiery orange"—but systematically piles on the adjectives, rare, wild, noble, magnificent—and then proceeds to describe the indescribable, in suggesting the emotion of the capture itself, before bringing the reader down to earth with the admission of the headache, and the slightly self-deprecatory comment about the "inadequate" cause. The heightened language and the suggestion of a trance evoke the tone of poems by Coleridge or Keats—the figure of Porphyro, perhaps, gazing at fair Madeline in Keats's "The Eve of St Agnes":

> She seem'd a splendid angel, newly drest,
> Save wings, for heaven:—Porphyro grew faint:
> She knelt, so pure a thing, so free from mortal taint.

One might assume that Wallace's description was the result of much later, *post facto* revision, when he was working up his journals in 1868 for the publication of *The Malay Archipelago.* The actual journal entry provides evidence of the vividness of the immediate experience:

> The brilliancy of this colour is indescribable, & none but a naturalist can appreciate the intense excitement I experienced on at length capturing it. On taking it from my net & opening the glorious wings my heart beat violently my blood rushed to my head & I have never been so near fainting when in apprehension of instant death, as from the excitement produced by what will to most people appear a very absurd & inadequate cause.

The light punctuation even increases the sense of ecstasy; and the reference to "apprehension of instant death" comes across not as a flight of romantic imagination, but as a fact often experienced by this adventurous and persistent traveller. There is, too, a marked contrast between this passage and the conclusion of the journal entry, as Wallace reverts to a more pragmatic style: "Capture of this insect decided me to stay 3 months longer. Devoted one of my men to it. Rocky stream. Good collecting ground this side—*Buprestidae, Longicorns, Curculionidae* etc. etc."[3]

Conrad draws from this incident, and from Wallace's way of portraying its emotional significance, in an extended passage which, it could be claimed, forms the heart of *Lord Jim.* Stein is an entomologist, and lives surrounded by his collection of Coleoptera. He points out to Marlow the "other things" he sees in the butterfly, which "spread out dark bronze wings, seven inches or more across, with exquisite white veinings and a gorgeous border of yellow spots." "Look! The beauty—but that is nothing—look at the accuracy, the harmony. And so fragile! And so strong! And so exact! This is Nature—the balance of colossal forces. Every star is so—and every blade of grass stands so—and the mighty Kosmos in perfect equilibrium produces—this. This wonder; this masterpiece of Nature—the great artist." Marlow, the plain questioner, comments "cheerfully": "Never heard an entomologist go on like this," and adds, "Masterpiece! And what of man?"

(Conrad 1920, 208). (We now hear, perhaps, an echo of Wallace's unanswerable question to Darwin.) Stein's response, "Man is amazing, but he is not a masterpiece," a judgement that still lies near the centre of contemporary debate, is the preface to his story of capturing the prize specimen of his collection, on a day when he had been ambushed by seven rascals in the forest, and had had to shoot three of them to survive. As he looked at the face of one of them for some sign of life, a faint shadow passes over his forehead: "It was the shadow of this butterfly." He finds the butterfly sitting on a small heap of dirt. Revolver in hand, he stalks forward with his soft felt hat. "One step. Steady. Another step. Flop! I got him! When I got up I shook like a leaf with excitement, and when I opened these beautiful wings and made sure what a rare and so extraordinary perfect specimen I had, my head went round and my legs became so weak with emotion that I had to sit on the ground" (Conrad 1920, 210).

The general correspondence between the feelings evoked in both Wallace and Stein is clear; but Conrad also appears to draw upon another extended passage from *The Malay Archipelago*, when Wallace describes his reaction to the small bird his boy Baderoon brought to him in the Aru Islands, a specimen of the king bird of paradise. "I knew how few Europeans had ever beheld the perfect little organism I now gazed upon, and how very imperfectly it was still known in Europe. The emotions excited in the minds of a naturalist who has long desired to see the actual thing which he has hitherto known only by description, drawing, or badly-preserved external covering, especially when that thing is of surpassing rarity and beauty, require the poetic faculty fully to express them." Wallace goes on to explore two further dimensions, both of which Conrad articulates within the complex texture of *Lord Jim*. Wallace "thought of the long ages of the past, during which the successive generations of this little creature had run their course—year by year being born, and living and dying amid these dark and gloomy woods, with no intelligent eye to gaze upon their loveliness—to all appearance such a wanton waste of beauty. Such ideas excite a feeling of melancholy" (S715 1989, 448–49). Melancholy invades Stein's reverie too, as he strikes a match, which flares violently, and is then blown out. Wallace then proceeds to speculate on man, civilized and uncivilized, and to predict that: "should civilized man" ever reach the Aru islands, he would "so disturb the nicely-balanced relations of organic and inorganic nature as to cause the disappearance, and finally the extinction, of these very beings whose wonderful structure and beauty he alone is fitted to appreciate and enjoy. This consideration must surely tell us that all living things were *not* made for man." In Conrad's story, Marlow takes up the thread, telling Stein that he came to him to describe a specimen: " 'Butterfly?' he asked, with an unbelieving and humourous eagerness. 'Nothing so perfect,' I answered, feeling suddenly dispirited with all sorts of doubts. 'A man!' "

The probing of Jim's nature proceeds, and becomes the search for how to live a life, "how to be!": " 'This magnificent butterfly finds a little heap of dirt and sits still on it; but man will never on his heap of mud keep still. He want to be so, and again he want to be so . . .' He moved his hand up, then down . . . 'He wants to be a saint, and he wants to be a devil—and every time he shuts his eyes he sees himself as a very fine fellow—so fine as he can never be . . . In a dream . . . ' " (Conrad 1920, 213). Then Stein lowers the glass lid, the automatic lock clicks sharply, "and taking up the case in both hands he bore it religiously away to its place, passing out of the bright circle of the lamp into the ring of fainter light—into shapeless dusk at last." Conrad expands with great subtlety Wallace's parallels between specimen of butterfly or bird of paradise and man, between light and darkness, between stillness and struggle, between perpetuation and death. Wallace's own example, as delineated in his travel writing, can be traced both in the figure of the naturalist and philosopher Stein, and in some aspects of "Lord" Jim. The profounder quest for truth about the nature of man, a central strand of the "tale" that Conrad unfolds with such complexity, is also something that he might have found in those bedside books.

Wallace was unusual in the way he infused his skill as a field naturalist and his perception as a scientific observer with a sense of the beauty and rarity of his subjects, jaguar, bird of paradise, beetle, butterfly. He brought an imaginative dimension, a poetic faculty, to the act of seeing and recording, to the interaction between seer and seen, that allies him within his own time to romantic poetry, and that might find its counterpart today in the mediation of a great television naturalist such as David Attenborough. Wallace's letter of 28 January 1859 to Stevens, published in the *Proceedings of the Entomological Society of London* on 3 October of that year, features the first description of the spectacular butterfly *Ornithoptera croesus*. Remarkably, it retains the full range of the sensations—fainting with delight, heart beating violently, even the headache—of his journal entry, but adds some telling detail, as well as a distinctly undispassionate claim: "It is a fiery golden orange, changing when viewed obliquely, to opaline-yellow and green. It is, I think, the finest of the Ornithoptera, and consequently the *finest butterfly in the world?*" (S50 1859, 70).[4] Only Wallace, surely, would combine the emphatic underlining of "finest butterfly in the world" with the speculative question-mark.

This particular butterfly specimen is in the collection of the Natural History Museum, and Dr George Beccaloni has identified the mark of Wallace's fingers on it, as he removed it from the net in 1859 and opened its glorious wings.[5] Fittingly, it has itself been captured in verse, in Anne Cluysenaar's (2001) poem "Stilled":

As I catch a trace
  Of Wallace's fine-tipped quill
    On the tiny round of the label
And the dull glint of the pin
  Through that wizened thorax,
I think of a mind's movement
  Stilled between pages,
    As dead, as rich—
Ready in another mind
  To fly, and settle.[6]

Wallace's writing about nature certainly belies that bleak self-judgement about his total absence of brilliancy. In creating for his readers the wonder of *Ornithoptera croesus*, he is aided by the essence of the butterfly itself, the epitome of transient beauty. But his questioning sensitivity permeates everything he describes in his masterpiece, *The Malay Archipelago*: a sense of the inexpressible beauty and complexity of the natural world, but also of its acute fragility, a fragility shared equally by the races of man.

Notes

1. Conan Doyle (1998) has an extremely informative introduction and notes by Ian Duncan. The reference to Wallace and Darwin is on page 68. Duncan suggests that Conan Doyle's vivid description of the expedition's journey into a natural fairyland (pp. 70–71) is embellished from Wallace's equally evocative description of his voyage to the lake on pages 66–67 of *Travels on the Amazon* (S714 1889). Wallace, unlike Challenger, tends to downplay his encounters with serpents and apes, though the image of the orang-utan attacked by Dyaks in *The Malay Archipelago* might generate sensational responses.
2. Letter to William Blackwood, 13 December 1898, in Karl and Davies (1986, 2:130). There is extensive commentary on the links between Wallace and Conrad. I am particularly indebted to Houston (1997). See also Sherry (1966). Florence Clemens was the first to trace the connections between *The Malay Archipelago* and Conrad (Clemens 1937). For extremely detailed notes on correspondences between Conrad's *An Outcast of the Islands* and *The Malay Archipelago*, see Conrad (2002). For Darwin's influence on Conrad (with shadows of Wallace), see O'Hanlon (1984).
3. Wallace, manuscript journal, Volume 1, in the collection of the Linnean Society, London. See entries 159, p. 74, and 164, pp. 75–76.
4. The letter was read to the meeting of 6 June 1859, and the butterfly was exhibited at the following meeting on 4 July (and at the Zoological Society of London on 28 June). At the June meeting, John Westwood cast doubts on the butterfly being new, commenting that he had "little doubt" it was the *O. tithonus* of De Haan, and subsequently suggested that *O. croesus* might be a local variety of *O. priamus.*
5. Personal communication from Dr George Beccaloni, Natural History Museum, London. This is the specimen figured on the cover of this book.
6. This poem is to appear, as part of an extended sequence on Wallace, in Cluysenaar's book *Batu-Angas* (published by Seren in June 2008).

13
Wallace and Owenism*

Gregory Claeys

Alfred Russel Wallace in his later years came to view the socialist movement as the most marked proof of the nineteenth century's self-reckoning with its flaws and mistakes. As he put it in 1898, in *The Wonderful Century; Its Successes and Failures*:

> ... although this century has given us so many examples of failure, it has also given us hope for the future. True humanity, the determination that the crying social evils of our time shall not continue; the certainty that they can be abolished; an unwavering faith in human nature, have never been so strong, so vigorous, so rapidly growing as they are to-day. The movement towards socialism during the last ten years, in all the chief countries of Europe as well as in America, is the proof of this (S726 1898a, 378).

The socialist component in Wallace's thought, however, has rarely been scrutinized with much care. Nonetheless the early and abiding influence of Owenism upon Wallace is well known and readily documented. Readers may recall Wallace's recollection that "Although later in life my very scanty knowledge of his work was not sufficient to prevent my adopting the individualist views of Herbert Spencer and of the political economists, I have always looked upon Owen as my first teacher in the philosophy of human nature and my first guide through the labyrinth of social science" (S729 1908, 57). In 1904 he wrote that "I am just now reading Robert Owen's Autobiography. What a marvellous man he was! A most clear-seeing socialist & educator ages before his time," adding two days later that: "I go even further & consider Owen one of the *first* as well as one of the *greatest* men of the 19th century; an almost ideally perfect character but too far in advance of his time" (Shermer 2002, 239). Elsewhere, too, he described Owen as "one of the most wonderful men of the nineteenth century" (Marchant 1916a, 2:225).

In what we may term the standard view of the role played by Owenism in Wallace's thought, the secondary literature has broadly construed this influence to entail an early sympathy for socialist ideas on Wallace's part, begun at a tender age in London in the late 1830s. This was mitigated in midlife by the impact of classical political economy and liberal philosophy, which made Wallace sceptical,

* I am grateful to Charles Smith for bibliographic assistance with this chapter.

in particular, about any proposals to abolish economic competition as a motivating force. Wallace's commitment to land nationalization thereafter has been recognized as leading to a re-embracing of socialist ideas in 1890 (*e.g.* Clements 1983, 91-99; Shermer 2002, 238–49). In addition, a somewhat stronger case has been made recently by Greta Jones for seeing the impact of Owenism on Wallace in rather broader and deeper terms (Jones 2002, 73–96). This chapter will assess Wallace's own account of the role played by Owenism in his ideas. It will further detail the context of Wallace's first encounter with early socialism, in London in 1837. It will then weigh the evidence for looking at the stronger as opposed to the standard case for the influence of Owenism, over the longer run of Wallace's development. It will suggest that the Owenite component in Wallace's later socialist thought can be interpreted as much more substantial than any preceding account has suggested, but for reasons different from those previously proposed.

Wallace's Introduction to Owenism

A connection with Robert Owen and Owenism runs through much of Wallace's long life, starting in early adolescence. In *My Life* Wallace states that he was in London from "early" 1837 until "early in the summer" of that year, when, having only turned fourteen in January, he moved to Bedfordshire to commence a career as a land surveyor under his brother William (S729 1908, 58). In London he was living at Robert Street, Hampstead Road, with a London builder, Mr Webster, to whom his brother John was apprenticed. He spent most of his time in the builders' shop, where he may first have encountered Owenites, since many had been involved in Owen's ill-fated effort in the mid-1830s to found a single union of all trades, the Grand National Consolidated Trades Union.[1] Both he and his brother spent most of their evenings in Owenite company, at one of their meeting places at John Street, near Tottenham Court Road. The Owenites he met there shared a commitment to Owen's vision of resettling the poor and unemployed, and eventually the entire population, in self-supporting "co-operative communities" of a few thousand residents in the countryside, where labour and profits would be shared justly and communally. Wallace tells us that

> Here we sometimes heard lectures on Owen's doctrines, or on the principles of secularism or agnosticism, as it is now called; at other times we read papers or books, or played draughts, dominoes, or bagatelle, and coffee was also supplied to any who wished for it. It was here that I first made acquaintance with some of Owen's writings, and especially with the wonderful and beneficent work he had carried on for many years at New Lanark. I also received my first knowledge of the arguments of sceptics, and read among other books Paine's "Age of Reason" (S729 1908, 45).

The initial effect of his encounter with Owenism was in the first instance to challenge, and then destroy, Wallace's religious beliefs. He was puzzled by queries that have disconcerted so many generations of believers: "Is God able to prevent evil but not willing? Then he is not benevolent. Is he willing but not able? Then he is not omnipotent. Is he both able and willing? Whence then is evil?" His father was unable to answer these questions to Wallace's satisfaction. Then he read in London a tract entitled "Consistency," by Robert Dale Owen, eldest son of Robert Owen. This rejected any idea of eternal punishment, and Wallace later wrote that he "thoroughly agreed with Mr. Dale Owen's conclusion, that the orthodox religion of the day was degrading and hideous, and that the only true and wholly beneficial religion was that which inculcated the service of humanity, and whose only dogma was the brotherhood of man. Thus was laid the foundation of my religious scepticism" (S729 1908, 46).

Wallace was also profoundly influenced by the central principle of the Owenite system, the notion that human character was, as Owen put it, formed for man rather than by him, in other words was the product of the environment surrounding the individual rather than any pre-existing nature, much less "Original Sin." As Wallace put it,

> my introduction to advanced political views, founded on the philosophy of human nature, was due to the writings and teachings of Robert Owen and some of his disciples. His great fundamental principle, on which all his teaching and all his practice were founded, was that the character of every individual is formed *for* and not by himself, first by heredity, which gives him his natural disposition with all its powers and tendencies, its good and bad qualities; and, secondly, by environment, including education and surroundings from earliest infancy, which always modifies the original character for better or for worse. Of course, this was a theory of pure determinism, and was wholly opposed to the ordinary views, both of religious teachers and of governments, that, whatever the natural character, whatever the environment during childhood and youth, whatever the direct teaching, all men *could* be good if they liked, all *could* act virtuously, all *could* obey the laws, and if they wilfully transgressed any of these laws or customs of their rulers and teachers, the only way to deal with them was to punish them, again and again, under the idea that they could thus be *deterred* from future transgression. The utter failure of this doctrine, which has been followed in practice during the whole period of human history, seems to have produced hardly any effect on our systems of criminal law or of general education; and though other writers have exposed the error, and are still exposing it, yet no one saw so clearly as Owen how to put his views into practice; no one, perhaps, in private life has ever had such opportunities of carrying out his principles; no one has ever shown so much ingenuity, so much insight into character, so much organizing power; and no one has ever produced such striking results in the face of enormous difficulties as he produced during the twenty-six years of his management of New Lanark (S729 1908, 46–47).

Readers of Wallace's *Life* will also recall its extended discussion of Owen's mill and schools at New Lanark, much of which was taken from Owen's own *Life* (1858), and W. L. Sargant's *Robert Owen and His Social Philosophy* (1860), both of which he rightly assumed were little known to later readers. Wallace commented extensively on the kind treatment of the workers in Owen's New Lanark mills south of Glasgow, and Owen's success in reducing hours of labour, improving conditions of work, and fostering child education. He was deeply impressed by the fact that Owen had managed to

> transform a discontented, unhealthy, vicious, and wholly antagonistic population of 2500 persons to an enthusiastically favourable, contented, happy, healthy, and comparatively moral community, without ever having recourse to any legal punishment what ever, and without, so far as appears, discharging any individual for robbery, idleness, or neglect of duty; and all this was effected while increasing the efficiency of the whole manufacturing establishment, paying a liberal interest on the capital invested, and even producing a large annual surplus of profits which, in the four years 1809–13, averaged £40,000 a year ... (S729 1908, 48).

He later wrote, too, of New Lanark, that he knew "of no more wonderful example in history" (S655 1909, 22). This application of socialist principles, however, Wallace recalled, was in reality only the

> *partial* application of Owen's principles of human nature, most patiently and skilfully applied by himself. They were necessarily only a partial application, because a large number of the adults had not received the education and training from infancy which was essential for producing their full beneficial results. Again, the whole establishment was a manufactory, the property of private capitalists, and the adult population suffered all the disadvantages of having to work for long hours at a monotonous employment and at low rates of wages, circumstances wholly antagonistic to any full and healthy and elevated existence. Owen used always to declare that the beneficial results at which all visitors were so much astonished were only one-tenth part of what *could* and *would* be produced if his principles were fully applied. If the labour of such a community, or of groups of such communities, had been directed with equal skill to produce primarily the necessaries and comforts of life for its own inhabitants, with a surplus of such goods as they could produce most economically, in order by their sale in the surrounding district to be able to supply themselves with such native or foreign products as they required, then each worker would have been able to enjoy the benefits of change of occupation, always having some alternation of out-door as well as indoor work; the hours of labour might be greatly reduced, and all the refinements of life might have been procured and enjoyed by them (S729 1908, 55–56).

His main criticism of Owen was that he had changed the manner in which his principles were applied: "The one great error Owen committed was giving up the New Lanark property and management, and spending his large fortune in the

Figure 26 Portrait of Robert Owen.
From Volume 2 of *Robert Owen, A Biography* by Frank Podmore (1906). Out of copyright.

endeavour to found communities in various countries of chance assemblages of adults, which his own principles should have shown him were doomed to failure" (S729 1908, 56).

In the second phase of Wallace's intellectual development, which occupied most of his middle life, such scepticism about the possibility of great moral improvement of "chance assemblages" of people produced an evident rejection of socialist solutions to poverty. Fortuitously, such doubts coincided with an extended period of economic growth and prosperity, lasting from the late 1840s until the middle and later 1870s. But then his views underwent profound alteration once again.

> Although I had, since my earliest youth, looked to some form of socialistic organization of society, especially in the form advocated by Robert Owen, as the ideal of the future, I was yet so much influenced by the individualistic teachings of Mill and Spencer, and the loudly proclaimed dogma, that without the constant spur of individual competition men would inevitably become idle and fall back into universal poverty, that I did not bestow much

> attention upon the subject, having, in fact, as much literary work on hand as I could manage. But at length, in 1889, my views were changed once for all, and I have ever since been absolutely convinced, not only that socialism is thoroughly practicable, but that it is the only form of society worthy of civilized beings, and that it alone can secure for mankind continuous mental and moral advancement, together with that true happiness which arises from the full exercise of all their faculties for the purpose of satisfying all their rational needs, desires, and aspirations (S729 1908, 326–27).[2]

This did not however mean that Wallace absolutely abandoned this "individualist" perspective even after his conversion to socialism. In 1894 he described as a form of "Social Economy," a scheme "which, while securing many of the beneficial results of Socialism, will preserve all the advantages of individual self-dependence and healthy rivalry, and will so educate and develop social feelings, that if any advance in the direction of Socialism is then desired, it will no longer be impracticable" (S507 1894, 185). The future system, "some form of socialism, which may be briefly defined as *the organization of labour for the good of all*" (S727 1900, 2:512) had thus necessarily for Wallace to be a voluntary system, and might well be preceded by "a period of true individualism—of competition under strictly equal conditions—to develop all the forces and all the best qualities of humanity, in order to prepare us for that voluntary organization which will be adopted when we are ready for it, but which cannot be profitably forced on before we are thus prepared." The creation of a system of equality of opportunity was thus a halfway house on the road to creating a co-operative society:

> Under such a system of society as is here suggested, when all were well educated and well trained and were all given an equal start in life, and when every one knew that however great an amount of wealth he might accumulate he would not be allowed to give or bequeath it to others in order that they might be free to live lives of idleness or pleasure, the mad race for wealth and luxury would be greatly diminished in intensity, and most men would be content with such a competence as would secure to them an enjoyable old age. And as work of every kind would have to be done by men who were as well educated and as refined as their employers, while only a small minority could possibly become employers, the greatest incentive would exist towards the voluntary association of workers for their common good, thus leading by a gradual transition to various forms of co-operation adapted to the conditions of each case. With such equality of education and endowment none would consent to engage in unhealthy occupations which were not absolutely necessary for the well-being of the community, and when such work was necessary they would see that every possible precautions were taken against injury. All the most difficult labour-problems of our day would thus receive an easy solution (S727, 2:519–20).

Indeed "a voluntary co-operation and organization of labour which would produce most of the best results of Socialism itself."

Let us now consider these phases of development in greater detail, with a view to determining whether the schema of early Owenism meeting up with various forms of later socialist sympathy can be fleshed out. To do this we require greater detail about the relationship between his earlier and later socialist sympathies than has been offered elsewhere.

Wallace and London Owenism in the mid-1830s

Owenism in this period had just begun the last and greatest phase of its development. Its first flowering was in the early 1820s, when experimental communities like Orbiston in Scotland were founded, and Owen acquired the Harmony site in Indiana in 1824, which he renamed New Harmony.[3] The effort collapsed, however, and no major further communities were established until the late 1830s, when Owen again acquired land, this time at Tytherly, Hampshire, on which a building was commenced, and a settlement founded, which became generally known as Queenwood or Harmony. Following the failure of the Grand National Consolidated Trades Union, Owenite organization in London recommenced in late 1834 with Owen's assumption of the management of the "Institution" at 14 Charlotte Street, Fitzroy Square, where weekly lectures on Sunday evenings were inaugurated. Discussions of the "new principles" occurred every Thursday evening, while "Social Festivals" were held on Monday evenings.[4] A "Community Coffee House," secretary H. Rose, was also open at the same time a short distance away at 92 John Street, where "Conversation Parties" were held every Sunday afternoon.[5] At this time a "Halfpenny-A-Week Land Fund," also known as the "Social Land Community of the Rational System," existed, founded around April 1834, which transferred operations from 14 Charlotte Street to 92 John Street in late 1834, the premises of "Mr. Presley's Coffee House."[6] Larger Social Festivals were also held from the end of 1834 not far away at the Burton Chapel, 4 Crescent Place, Burton Crescent, Burton Street, near Tavistock Square.[7] Owen's Sunday evening lectures were transferred to the Burton Street chapel, later described as "for some years in the possession of Mr. Owen,"[8] in early 1835, while morning meetings continued at Charlotte Street.[9] About 150 members and candidates were reported in late 1835.[10] The 92 John Street premises, while less important to the Owenite organization, seem nonetheless to have been a daily meeting place to some socialists, who probably attended larger meetings and lectures at the other two nearby addresses, as Wallace likely did. From 1835 a weekly journal was commenced which provided an extensive account of the growth of the branches of the movement (over fifty eventually), called *The New Moral World*, which was at this time printed in Lincoln's Inn Fields. Owen's new organization, The Association of All Classes of All Nations (AACAN), was established in May 1835, chiefly to raise funds to buy land for the newly-planned community. It held its third annual congress in May 1837. This annual congress was the great date on the Owenite calendar, and Wallace would certainly have been privy to plans for its preparation, which included the founding of the National Community Friendly Society, another fund-raising body.

After the AACAN was founded in May 1835, membership increased steadily. In 1835 Owen lectured chiefly at the Institution at Charlotte Street.[11] By early 1836 he spoke mainly at the Burton Rooms, where he was referred to as the "Social Father," and assisted by a "Senior Council." The first annual meeting of the AACAN was held here in May 1836. At this time a Mr Presley, who was the treasurer of the Community Friendly Society, an Owenite benefit and fund-raising organization, was described as operating the "Community Coffee House" from premises which were reported as being located at "49" (probably a misprint for 94), John Street.[12] (The Community Friendly Society was located at this time at 94 John Street.)[13] (This should not be confused with the National Community Friendly Society, founded in June 1837, and run from Salford.[14]) However, the John Street premises were not referred to as the "John Street Institution" or (the term Wallace himself uses in the *Life*), a "Hall of Science" in 1837. In fact, a quite different location used the latter title at this time.[15]

Owen himself returned to London after a tour of Scotland and the Midlands at the end of 1836.[16] He seems to have lectured Sundays at Burton Street during most of January and February.[17] He again left London on 23 February, lectured several weeks in Manchester, once to an audience of two thousand, attended festivals at the Social Institutions at Salford and Bolton, argued with political economists and factory reformers, and met with the Chartist and factory reform leaders Richard Oastler, Joseph Rayner Stephens and others.[18] He had returned to London by mid-June, where his return was celebrated on 18 June by a tea at Ealing.[19] He then embarked for France, arriving in Paris on 7 July, and returning again in late August. Owen himself, when in London, commonly lectured on Sunday; Wallace met him once (but probably not at John Street), and later recalled "his tall spare figure, very lofty head, and highly benevolent countenance and mode of speaking" (S729 1908, 57). If Wallace's chronology as given in the *Life* is correct, thus, he would have seen Owen speak either from late January to mid-February, or in the first two weeks of June, shortly before his own departure for Bedfordshire. If the latter, the subject of the talk may well have been an address to the young princess Victoria, shortly to ascend the throne.[20]

During Wallace's stay in London, Owenism was developing very rapidly. In early 1836 two further branches had been founded, at Manchester (opened 3 January 1836), and Bolton (opened 3 April 1836), but plans were afoot to reorganize London Owenite activities. To this end a "Social Missionary and Tract Society" was established at the "Social Institution," Curtain Road, Shoreditch, with F. Wilby, as secretary, which aimed to "diffuse a correct knowledge of the means of arriving at a Community, by opening rooms for lecturing on the principles, and diffusing explanatory tracts."[21] Some time later its festival was reported as taking place, which indicates rooms had been rented.[22] At this time the Owenites continued to hold balls, soirées, etc., at the Burton Rooms, where a former Dublin printer's draughtsman, Frederick Bate, was the secretary.[23] Here, for instance, a concert and ball, or "Social Festival," was held on 13 February (admission 2 shillings). Tea-parties were held every Sunday. The Association also held quarterly meetings, lectures (which were free), classes in music, elocution, grammar, and singing, and possessed a circulating library. Sundays were also the occasion when candidates and members of the Association were examined by the "Senior Council" previous to their entry into the First and Second Classes. (Wallace, however, seems to have had no official affiliation with the movement, and thus apparently did not progress in this manner.) The Burton Street hall was at this time thus the centre of London Owenism; while meetings were also held at 94 John Street, where on 13 April, for instance, fifty members of the Community Friendly Society dined, and were joined by some sixty more for dancing, singing, and recitations.[24] In mid-1837 the London Owenites reorganized themselves, electing a new secretary, Robert Alger, and president, James Braby, and setting up a board of management for the "metropolitan branch." Momentum was clearly gathering, and by mid-August the newly-constituted Board reported 230 members and candidates in London.[25] In late October new premises were taken over at No. 69, Great Queen Street, Lincoln's Inn Fields; this was in July 1838 described as the "Social Institution" and "Parent Branch" in London.[26] Fifteen branches of the movement, with about 1,500 members, were reported at the end of 1837.[27] By late 1838 weekly lectures were being offered at no fewer than nine London locations.[28] By April 1840 a much-expanded "New Social Hall" capable of seating 1,000 people was opened at the John Street premises at a cost of £3,000 which appears to have become the new headquarters of Branch A1.[29] The John Street premises thus became the centre of London Owenite organization in 1840, supplanting the Burton Street hall and eventually other lecture and meeting sites after Wallace's departure.

What intellectual ambiance would Wallace have encountered at the John Street premises during early 1837, presumably from perhaps late January to June? This was in fact an exceptionally pregnant moment for social and political reformers, as both Chartism and Owenism grew steadily, and vied for working-class attention. London Owenism was as divided in this period as ever. There were Christian as well as secularist socialists, Chartists and those hostile to political involvement, and vigorous controversies respecting the recently-established New Poor Law, its Malthusian underpinnings, the vices of classical political economy, and most importantly, the necessity for a new system of society based on Owen's principles.[30] The "new views" of marriage were a favourite lecture topic, too, with Charles Southwell and Margaret Chappelsmith both addressing the subject at various later points. Wallace would certainly have encountered a conception of society in which competition was regarded as both the essential social mechanism and the final arbiter of human success, even individual worth, for this was precisely the world view which Owenism was most concerned to displace. In opposition to competition Owenism offered a principle of community of property because "without such a community, men must compete each one against all others for wealth, and because while they do so compete for individual's possessions, there must always exist, in every nation, a class of paupers or unsuccessful competitors."[31] This competitiveness was not regarded by the Owenites as instinctive as such; "We all inherit, constitutionally, a capacity for competing with each other, but we do not all inherit it in the same degree: the circumstances into which we are born, therefore, may by developing this capacity mould us all into competitors, but cannot endow us with *equal rigour and intentions*."[32] Debates about Malthus, marriage, and competition were thus constantly linked in Owenite circles. It has even been contended that Owen used Malthus's phrase "struggle for existence" (Slotten 2004, 145); whether he did or not, Owenites certainly did as early as 1839.[33] (But there is no evidence that Wallace himself actually read Malthus until 1844.[34])

One wonders, of course, what a lad of fourteen, even one precocious and curious, could have made of all this. Certainly he would not have been ignored, as a child, or really, by the standards of the time, a young adult. To the contrary, the Owenites viewed children as the rising generation, and were eager to cultivate their attention. As a leading London Owenite, Robert Alger, wrote, "The children are to be considered of equal, if not of superior importance to their parents or adult friends, because, in our anticipations of Community, we cannot conceal from ourselves the fact that the mistraining which the old society has induced, more or less, in all the existing adult generation, will be far more prejudicial to us than the rising generation, whose more plastic minds and habits may, by rational means, be so much more easily trained to an unerring conduct, and a cordiality of disposition."[35] At an influential age such exposure to a hospitable, but also intellectually daringly, and even socially modish, group could well have made the indelible impact on Wallace which some interpreters have presumed. A paternal, friendly rather than patronizing, attitude by the Owenites, who would not have been unsympathetic to another Welshman (Owen having been born in Newtown, Wales), would have made this ambiance naturally sympathetic to the young Wallace. For such comradeship was, effectively, the core of what the "new views" entailed: a sociable, as opposed to a competitive, way of coexisting with fellow human beings.

Socialism in the Later Development of Wallace's Thought

Following his "individualist" phase, Wallace made the transition in the 1870s, like so many others of his generation, towards accepting more collectivist solutions to social ills. In his case he maintained through the 1880s a commitment to land nationalization (see Chapter 15), without abandoning most of the other premises of liberal political economy, remaining thus "inclined to think that no further fundamental reforms were possible or necessary" (S729 1908, 326–27). Here, however, it was not Owen, but an American, Edward Bellamy, who effected Wallace's second conversion:[36]

> From boyhood—when I was an ardent admirer of Robert Owen—I have been interested in socialism, but reluctantly came to the conclusion that it was impracticable, & also to some extent repugnant to my ideas of individual liberty and home privacy. But Mr. Bellamy has completely altered my views on this matter. He seems to me to have shown that real, not merely delusive liberty, together with full scope for individualism and complete home privacy, is compatible with the most thorough industrial socialism.[37]

As he expressed it elsewhere:

> The book that thus changed my outlook on this question was Bellamy's "Looking Backward," a work that in a few years had gone through seventeen editions in America, but had only just been republished in England. On a first reading I was captivated by the wonderfully realistic style of the work, the extreme ingenuity of the conception, the absorbing interest of the story, and the logical power with which the possibility of such a state of society as that depicted was argued and its desirability enforced. Every sneer, every objection, every argument I had ever read against socialism was here met and shown to be absolutely trivial or altogether baseless, while the inevitable results of such a social state in giving to every human being the necessaries, the comforts, the harmless luxuries, and the highest refinements and social enjoyments of life were made equally clear. From this time I declared myself a socialist, and I made the first scientific application of my conviction in my article on "Human Selection" in the *Fortnightly Review* (September 1890).[38]

Wallace would thereafter often acknowledge this debt to Bellamy. He also gave publicity to Bellamy's scheme for the means of effecting a transition from capitalism to socialism, which included a mixture of the nationalization of chief public services, such as railways, and the municipalization of others, like electricity, gas, and water supply. Crucial to the transition was also to be the abolition of all inheritance of property by those not directly heirs of its owners, a principle Wallace had accepted many years earlier (S622 1905). He also accepted Bellamy's principle of "coupons," in which money alone was capable of "*representing productive work*" as the "only true standard of value and the best instrument of exchange"—which was a wholly Owenite ideal as well.[39] With this alteration in the currency mechanism, Wallace anticipated that those employed in government service would alone enjoy the advantage of having their wages exchangeable at government shops and for government services. Increasingly those possessing the old gold and silver currency would find themselves at a disadvantage; the old currency would become devalued, and its possessors would increasingly enter the government service (S622, 5). This is nothing like Owen's "Plan," though it bears some resemblance to proposals by writers like John Gray in the 1820s and 1830s, who had been inspired by Owen.

The starting point of Wallace's Bellamy phase was thus partly "to reply to the common objection against Socialism, that it would lead to a too rapid increase of population" (S497 1894, 315). This defined the specific perspective which Wallace would adopt towards socialism in this period. He was clearly aware that the principles of natural selection were widely perceived as having a human and social application, and that this had been broadly (indeed was increasingly) construed as supporting the virtues of individual economic competition as the principal means of weeding out the human "fit" from the "unfit." This conclusion Wallace would resoundingly dismiss, but without rejecting the case for a necessary improvement in human type, and this, too, in the relatively short term. The question was how to reconcile this with a desire for a more just and egalitarian society. To Wallace the "first scientific application of my conviction" entailed the proclamation that socialism alone could provide the means for regulating both the size and the quality of the human population. Thus he was able to refute, or at least to evade, the "gloomy" conclusion Darwin himself had drawn about the future of human evolution on the basis of the fact that "in our modern civilization natural selection had no play, and the fittest did not survive. Those who succeed in the race for wealth are by no means the best or the most intelligent, and it is notorious that our population is more largely renewed in each generation from the lower than from the middle and upper classes" (S727 1900, 1:509). Instead, Wallace contended, "when the course of social evolution shall have led to a more rational organization of society, the problem will receive its final solution by the action of physiological and social agencies, and in perfect harmony with the highest interests of humanity" (S727, 1:509–10). If there is an Owenite component here, it is ironically expressed. For while Owenism was widely and frequently accused of being a form of determinism, it was Wallace who rejected Darwin's scientific determinism, and an all-pervading, all-powerful system of natural selection, in favour of a voluntarist plea for social and political reform.

Wallace's "conversion" to socialism, even if partly driven by humanist or ethical sympathies, was thus in the first instance closely wedded to the need to solve the human evolutionary problem. It can only be understood, further, in the context of his unwillingness to accept Francis Galton's eugenicist solution to the problem of the supposed evolutionary degeneration of the human species. According to Galton and Weismann, Wallace surmised, intelligence could not be presumed to be cultivated or developed through life, and the improvements then transmitted. Hence "special skill derived from practice, when continued for several generations, is not inherited, and does not therefore tend to increase" (S727, 1:512). Darwin's cousin, Galton, whose *Hereditary Genius* (1869) began the eugenics movement, sought both to limit the marriages of the less "fit" by a scheme of marriage licences, and also to promote intermarriage amongst the more talented, even to the extent of proposing that they live separately in rural communities.[40] But Wallace, doubtless bearing in mind the crucial Darwinian problem of the fecundity of the poor, which it was assumed would inevitably induce the degeneration of the human species, thought this likely to be ineffective:

> Its tendency would undoubtedly be to increase the number and to raise the standard of our highest and best men, but it would at the same time leave the bulk of the population unaffected, and would but slightly diminish the rate at which the lower types tend to supplant or to take the place of the higher. What we want is, not a higher standard of perfection in the few, but a higher average, and this can best be produced by the elimination of the lowest of all and a free intermingling of the rest (S727, 1:513).

Wallace instead utterly rejected proposals to regulate marriage and restrict childbirth, such as those proposed by Hiram M. Stanley, similar to what Galton had mooted for some time, saying

> that nothing can possibly be more objectionable, even if we admit that they might be effectual in securing the object aimed at. But even this is more than doubtful; and it is quite certain that any such interference with personal freedom in matters so deeply affecting individual happiness will never be adopted by the majority of any nation, or if adopted would never be submitted to by the minority without a life-and-death struggle (S427 1890, 328–29).

Later, too, he remarked that "Segregation of the unfit ... is a mere excuse for establishing a medical tyranny. And we have enough of this kind of tyranny already ... the world does not want the eugenist to set it straight ... Eugenics is simply the meddlesome interference of an arrogant scientific priestcraft" (Marchant 1916a, 2:246–47). He took a similar view of the population issue generally, arguing that "under rational social conditions the healthy instincts of men and women will solve the population problem far better than any tinkering interference either by law or by any other means" (Marchant 1916a, 2:160–61).

Nor for Wallace were proposals to make marriage a free contract easily revokable, such as Grant Allen had offered, more acceptable. These were "in some respects the very reverse of the last, yet it is, if possible, even more objectionable. Instead of any interference with personal freedom he proposes the entire abolition of legal restrictions as to marriage, which is to be a free contract, to last only so long as either party desires" (S727 1900, 1:514). Wallace insisted upon highly restricted, rather than liberal, laws of divorce, and the essential maintenance of the nuclear, monogamous family as a pillar of civilization. Wallace, however, argued that advancement would in fact be impeded if divorce were liberalized, even more so if monogamy were abandoned: "*History* shows conclusively that where divorce has been easy, licentiousness, disorder, and often complete anarchy have prevailed. The history of civilization is the history of advance in monogamy, of the fidelity of one man to one woman, and one woman to one man" (S727, 1:518).

Wallace's distance from Owenism here could not have been greater, though he was not alone in this traditionalist stance at the time; David Maxwell's *Stepping-Stones to Socialism* (1891), for instance, contended that "There can be very little doubt that the natural and instinctive union of the sexes, the kind which promises the greatest individual happiness and the best results for the community, is permanent marriage" (Maxwell 1891, 18).

Allen's argument, however, was close to the Owenite position of the 1830s and 1840s, in which proposals for civil marriage ceremonies and ease of divorce were frequently caricatured as masking the aim of "community of wives" as the ultimate ideal of the socialist system, or less politely, "one common scene of vice and prostitution."[41] It was true that Owen, in assailing "Mr. Malthus and his ardent, but inexperienced disciples" in 1835, had attacked the "single-family" system, "the artificial union of the sexes as devised by the priesthood, requiring single-family arrangements, and generating single-family interests," by contrast to the communitarian system (Owen 1993a, 2:280–81). But the chief thrust of Owen's argument respecting marriage was in fact merely that it should be based upon affection rather than financial considerations. Trained to become the companions rather than slaves of men, women would possess equal rights (Owen 1841, 35), or as a leading Owenite put it, would "be brought upon an equality with man; she will be placed aloof from the sphere of want and poverty" (Morrison 1838, 10). There was, however, also a faintly biological character to these debates. Owen had written of the tendency of the existing marriage system for man to "continually degenerate in his physical, mental, and moral powers,"[42] as a consequence of the irrational nature of existing marital arrangements. He also promised that socialism would produce "a race of superior beings, physically, intellectually, and morally," even by the application of "the same general principles as those which men now pursue in obtaining superior vegetable and animal productions, by a scientific knowledge of the methods most proper to be applied to attain the object proposed."[43] Critics assumed this to mean that "the breed cannot be improved" nor a "superior race" of children raised unless socialism were introduced (Brindley 1840, 3). There was also a neo-Malthusian context to these debates. Owen had taught since the mid-1820s that Malthus's assertion that population invariably outstripped the provision of a means of subsistence was false, and that

> so far from its being true that the means of subsistence cannot be made to keep pace with the highest possible rate of increase in population, the very reverse of this proposition must hold good for at least many centuries to come; that by a system of union and cooperation, society will possess the power of creating wealth to an unlimited extent, and that it need no longer to be regarded as an object of contest or individual desire any more than water or air is at present (Owen 1993b, 67–68).

He would later insist, too, that his own plan could support four times the existing population, in much greater comfort than at present (Owen 1993c, 359–61). Discussion of Malthus was thus very common in Owenite and Chartist circles in the late 1830s. Indeed, the famous 1838 satire by "Marcus," *The Book of Murder!* (Claeys 2001, 1:383–437) echoing Swift's *Modest Proposal* of 1729, probably written by the Owenite printer George Mudie, certainly reflected the anxiety of Owenite authors to meet Malthusian criticisms of the impossibility of socialism. Thus marriage, socialism, and the improvement of the human species had clearly been linked in the Owenism to which Wallace was exposed in his youth.

Half a century later, however, Wallace would outline his position in somewhat different terms. Contending that the issue as raised by Stanley and Allen could not be solved so long the present social system was "not only extremely imperfect, but vicious and rotten at the core" (S727 1900, 1:516) Wallace then set forth his own argument, encapsulated in the phrase, "Social Advance will result in Improvement of Character." This presented socialism as the only plausible alternative to the eugenic restriction on marriage:

> It is my firm conviction, for reasons I shall state presently, that, when we have cleansed the Augean stable of our existing social organization, and have made such arrangements that *all* shall contribute their share of either physical or mental labour, and that all workers shall reap the *full* and equal reward of their work, the future of the race will be ensured by those laws of human development that have led to the slow but continuous advance in the higher qualities of human nature. When men and women are alike free to follow their best impulses; when idleness and vicious or useless luxury on the one hand, oppressive labour and starvation on the other, are alike unknown; when all receive the best and most thorough education that the state of civilization and knowledge at the time will admit; when the standard of public opinion is set by the wisest and the best, and that standard is systematically inculcated on the young; then we shall find that a system of selection will come spontaneously into action which will steadily tend to eliminate the lower and more degraded types of man, and thus continuously raise the average standard of the race. I therefore strongly protest against any attempt to deal with this great question by legal enactments in our present state of unfitness and ignorance, or by endeavouring to modify public opinion as to the beneficial character of monogamy and permanence in marriage (S727, 1:517).

This left two problems to be solved: "the increase of population, and the continuous improvement of the race by some form of selection which we have reason to believe is the only method available." Taking up Bellamy, Wallace insisted that such improved circumstances would lead to a relatively late marriage age, which would "besides be inculcated during the period of education [up to the age of twenty-one], and still further enforced by public opinion." Then, moreover, women would exercise a much more careful choice of mates than at present. The theme of sexual selection was to be central to his argument:

> The most careful and deliberate choice of partners for life will be inculcated as the highest social duty; while the young women will be so trained as to look with scorn and loathing on all men who in any way wilfully fail in their duty to society—on idlers and malingerers, on drunkards and liars, on the

> selfish, the cruel, or the vicious. They will be taught that the happiness of their whole lives will depend on the care and deliberation with which they choose their husbands, and they will be urged to accept no suitor till he has proved himself to be worthy of respect by the place he holds and the character he bears among his fellow-labourers in the public service (S727, 1:521).

This would increase the age of marriage, and reduce the risk that over-population would end the socialist utopia:

> in a state of society in which all will have their higher faculties fully cultivated and fully exercised throughout life, a slight general diminution of fertility would at once arise, and this diminution added to that caused by the later average period of marriage would at once bring the rate of increase of population within manageable limits.[44]

Finally, natural selection would increase the quality of human stock. Not only would fewer women marry once all had the means of achieving education and independence; those who did would choose their mates much more carefully than at present. Thus "a powerful selective agency would rest with the female sex":

> there can be no doubt how this selection would be exercised. The idle and the selfish would be almost universally rejected. The diseased or the weak in intellect would also usually remain unmarried; while those who exhibited any tendency to insanity or to hereditary disease, or who possessed any congenital deformity would in hardly any case find partners, because it would be considered an offence against society to be the means of perpetuating such diseases or imperfections (S727, 1:524).

Fertility would diminish even further as the expectation of a lengthier life became more widespread: "In a society in which women were all pecuniarily independent, were all fully occupied with public duties and intellectual or social enjoyments, and had nothing to gain by marriage as regards material well-being, we may be sure that the number of the unmarried from choice would largely increase" (S727, 1:523).

Hence a process not of natural, but artificial, selection would take place, in which women, no longer susceptible to the allurements of wealth and gaudy trappings of conspicuous consumerism, would exercise a crucial role in promoting character.

Wallace thus did not doubt that the socialist system was greatly superior to the eugenic. We see what a considerable distance there is here between Wallace's position and that of Francis Galton:

> This method of improvement, by elimination of the worst, has many advantages over that of securing the early marriages of the best. In the first place it is the direct instead of the indirect way, for it is more important and more beneficial to society to improve the average of its members by getting

> rid of the lowest types than by raising the highest a little higher. Exceptionally great and good men are always produced in sufficient numbers, and have always been so produced in every phase of civilization. We do not need more of these so much as we need less of the weak and the bad. This weeding-out system has been the method of natural selection, by which the animal and vegetable worlds have been improved and developed. The survival of the fittest is really the extinction of the unfit (S727, 1:525–26).

Thus, thought Wallace,

> In the society of the future this defect will be remedied, not by any diminution of our humanity, but by encouraging the activity of a still higher human characteristic—admiration of all that is beautiful and kindly and self-sacrificing, repugnance to all that is selfish, base, or cruel. When we allow ourselves to be guided by reason, justice, and public spirit in our dealings with our fellow-men, and determine to abolish poverty by recognizing the equal rights of all the citizens of our common land to an equal share of the wealth which all combine to produce—when we have thus solved the lesser problem of a rational social organization adapted to secure the equal well-being of all, then we may safely leave the far greater and deeper problem of the improvement of the race to the cultivated minds and pure instincts of the men, and especially of the Women of the Future (S727, 1:526).

When he restated this argument some years later, then, Wallace again explicitly made sexual selection contingent upon "a greatly improved social system [which] renders all our women economically and socially free to chose; while a rational and complete education will have taught them the importance of their choice both to themselves and to humanity" (S649 1912, 49).

He here contended that the "fittest" races triumphed over the less fit, with a "survival of the fittest among competing peoples necessarily leading to a continuous elevation of the human race as a whole, even though the higher portion of the higher races may remain stationary or may even deteriorate" (S727, 2:495). Within each race, moreover, "we cannot doubt that the prudent, the sober, the healthy, and the virtuous live longer lives than the reckless, the drunkards, the unhealthy, and the vicious" (S727, 2:495). Such processes, however, were interfered with, he thought, by artificial celibacy, and by "the system of inherited wealth, which often gives to the weak and vicious an undue advantage both in the certainty of subsistence without labour, and in the greater opportunity for early marriage and leaving a numerous offspring" (S727, 2:496). In "Human Progress, Past and Future" (S445 1892), Wallace thus proposed that since the effects of training, education, habits, and environment could not be proven to be heritable, human progress depended upon the natural elimination of the vicious by their inferior habits of life, and much more importantly, the greater education of women, such that

> When such social changes have been effected that no woman will be compelled, either by hunger, isolation, or social compulsion, to sell herself whether in or out of wedlock, and when all women alike shall feel the refining influence of a true humanizing education, of beautiful and elevating surroundings, and of a public opinion which shall be founded on the highest aspirations of their age and country, the result will be a form of human selection which will bring about a continuous advance in the average status of the race. Under such conditions, all who are deformed either in body or mind, though they may be able to lead happy and contented lives, will, as a rule, leave no children to inherit their deformity. Even now we find many women who never marry because they have never found the man of their ideal. When no woman will be compelled to marry for a bare living or for a comfortable home, those who remain unmarried from their own free choice will certainly increase, while many others, having no inducement to an early marriage, will wait till they meet with a partner who is really congenial to them ... In such a reformed society the vicious man, the man of degraded taste or of feeble intellect, will have little chance of finding a wife, and his bad qualities will die out with himself. The most perfect and beautiful in body and mind will, on the other hand, be most sought and therefore be most likely to marry early, the less highly endowed later, and the least gifted in any way the latest of all, and this will be the case with both sexes. From this varying age of marriage, as Mr. Galton has shown, there will result a more rapid increase of the former than of the latter, and this cause continuing at work for successive generations will at length bring the average man to be the equal of those who are now among the more advanced of the race (S727, 2:506–07).

To the eugenic contention that only the "destruction of the weak and helpless" would advance the cause of humanity, thus, Wallace argued that

> education *has* the greatest value for the improvement of mankind,—and that selection of the fittest may be ensured by more powerful and more effective agencies than the destruction of the weak and helpless. From a consideration of historical facts bearing upon the origin and development of human faculty I have shown reason for believing that it is only by a true and perfect system of education and the public opinion which such a system will create, that the special mode of selection on which the future of humanity depends can be brought into general action (S727, 2:508).

How original were Wallace's proposals concerning sexual selection? In "Human Progress: Past and Future" (S445 1892) Wallace emphasized the need for sexual selection amongst human beings in the future, terming this "by far the most important of the new ideas I have given to the world." But in *Looking Backward* (1888), Bellamy had explored this theme at length, envisioning precisely the role for women in human evolutionary strategy which Wallace would take up and embellish from 1890 onwards.[45] Bellamy laid stress on the labour-saving devices, such as communal kitchens and public laundries, which would reduce women's burdens, and enable them to be employed in their own "feminine" industrial army (albeit with shorter hours of labour and longer holidays).[46] Wallace, too, would envision the future in terms of economies of scale which included the provision of communal restaurants, laundry and cleaning services, in order to economize on household expenditure (S622 1905). Bellamy stressed further that the greater independence of women did not deter them from marriage (Bellamy 1927, 260–61). He envisioned instead that "The sexes now meet with the ease of perfect equals, suitors to each other for nothing but love." And he suggested that this would have a positive impact on evolution:

> ... the fact ... that there are nothing but love matches, means even more, perhaps, than you probably at first realize. It means that for the first time in human history the principle of sexual selection, with its tendency to preserve and transmit the better types of the race, and let the inferior types drop out, has unhindered operation. The necessities of poverty, the need of having a home, no longer tempt women to accept as the fathers of their children men whom they neither can love nor respect. Wealth and rank no longer divert attention from personal qualities. Gold no longer "gilds the straitened forehead of the fool." The gifts of person, mind, and disposition; beauty, wit, eloquence, kindness, generosity, geniality, courage, are sure of transmission to posterity. Every generation is sifted through a little finer mesh than the last. The attributes that human nature admires are preserved, those that repel it are left behind. There are, of course, a great many women who with love must mingle admiration, and seek to wed greatly, but these not the less obey the same law, for to wed greatly now is not to marry men of fortune or title, but those who have risen above their fellows by the solidity or brilliance of their services to humanity. These form nowadays the only aristocracy with which alliance is distinction. ... Perhaps more important than any of the causes I mentioned then as tending to race purification has been the effect of untrammeled sexual selection upon the quality of two or three successive generations (Bellamy 1927, 267–68).

Thus, concluded Bellamy, women now cultivated a "natural impulse to seek in marriage the best and noblest of the other sex," with the result that "Our women have risen to the full height of their responsibility as the wardens of the world to come, to whose keeping the keys of the future are confided. Their feeling of duty in this respect amounts to a sense of religious consecration. It is a cult in which they educate their daughters from childhood" (Bellamy 1927, 269). Later, in *Equality* (1897), which Wallace described as "the most complete and thoroughly reasoned exposition, both of the philosophy and the constructive methods of socialism" (S729 1908, 327–28), Bellamy responded to contemporary feminist criticisms by giving still greater freedom and autonomy for women. Here he reiterated the technological advances which would free women from drudgery, and the greater independence which the "national" system would extend them (Bellamy 1924, especially 130–43). He did not, however, reintroduce the theme of sexual selection, though he did contend that greater sexual equality had produced both a reduction in family size, commencing with the "cultured classes," stating that "it is necessary that the women through whose wills it must operate, if at all, should be absolutely free agents in the disposition of themselves, and the necessary condition of that free agency is economic independence" (Bellamy 1924, 412).

We see, then, why critics have asserted of Wallace's own account of sexual selection that "He borrowed it almost verbatim from Bellamy" (Fichman 2004, 268–72). But is anything novel about Wallace's use of Bellamy's principle? Wallace's marriage of natural selection to Bellamyite nationalist socialism, while unusual, was hardly unique (Claeys 2000). But Wallace did use an egalitarian socialist approach to sexual selection as a conscious alternative to eugenics, which Bellamy did not; and therefore as a critique of inegalitarian social Darwinism, and indeed of Darwin's own conclusions respecting human evolution. This may appear to be only a matter of emphasis. But Wallace was clearly far better versed in the evolutionary controversy than Bellamy, and if Wallace's idea was thus strikingly similar to Bellamy's, he developed it primarily as a means of solving the evolutionary problem, rather than as an auxiliary argument in support of a new social system. As we have seen, moreover, there remains the possibility of a residual Owenite influence in the linkage of the ideas of socialism, marriage, and natural selection, which had been suggested if not developed by Owen in 1834.

The Form of Government and Role of the State

There is a second area, moreover, in which the embrace of Bellamy also overlaps with Wallace's earlier Owenite sympathies. If Bellamy's nationalism offered Wallace a solution to the problem of human evolution, it also embroiled him in a heated controversy about the form socialism should assume, as well as the means by which it could be achieved. The most important aspect of this debate concerned the degree to which the future society should be centralized or decentralized, the role the state should play in its inception and development, and the degree of democratic control over any administrative mechanism. William Morris referred to Bellamy's scheme as "State Communism, worked by the vast extreme of national centralisation"; Bellamy in turn replied to Morris, reviewing *News from Nowhere*:

> Mr. Morris appears to belong to the school of anarchistic rather than to the state socialists. That is to say, he believes that the present system of private capitalism, once destroyed, voluntary co-operation, with little or no governmental administration, will be necessary to bring about the ideal social system. This is in strong contrast with the theory of nationalism which holds that no amount of moral excellence or good feeling on the part of a community will enable them to dispense with a great deal of system in

> order to coordinate their efforts as to obtain the best economic results. In the sense of a force to restrain or punish, governmental administration may no doubt be dispensed with in proportion as a better social system shall be introduced; but in no degree will any degree of moral improvement lessen the necessity of a strictly economic administration for the directing of the productive and distributive machinery.[47]

Yet Bellamy can also be seen as responding to the theme of over-centralization, for in *Equality* he described a tendency "countrywards" as having continued "until the cities having been emptied of their excess of people" and become park-like (Bellamy 1924, 294). This can certainly be understood as a move in Morris's direction. It was also one of the original proposals put forward by Owen (as well as Fourier and a number of other early socialists).

Wallace himself showed a similar desire to balance local with statist elements, which might be construed as evidence of a residual Owenite legacy. Though Wallace preferred *Looking Backward* to *News from Nowhere*, saying it also presented a "more enjoyable socialist regime than that sketched by Morris," this does not necessarily mean that he was unsympathetic to Morris's more communal and federative approach to socialism, with which he was certainly acquainted.[48] Bellamy had suggested that local government might retain "important and extensive functions in looking out for the public comfort and recreation, and the improvement and embellishment of the villages and cities," which would entitle it to a proportion of the aggregate sum of national labour available (Bellamy 1927, 209). Yet here we perceive some subtle variations on Bellamy's ideas in Wallace's views. Bellamy evidently did not, like Wallace, envision creating rural communities such as the Owenite colony in Ireland, Ralahine, which Wallace thought provided an excellent example of how land colonies could be instituted and organized. (Interestingly, he describes it as an organization which began despotically and ended up as self-governing [S655 1909, 9].) Wallace also praised Thomas Spence's view "that every parish should have possession of its own land, to be let out to the inhabitants, and that each parish should govern itself and be interfered with as little as possible by the central government, thus anticipating the views as to local self-government which we are now beginning to put into practice" (S498 1894, 171).

A second related issue was the role to be played by the state in introducing the new social system. Here, of course, Owenism had been resolutely opposed to a statist path, still more to any revolutionary strategy. Wallace, however, agreed with Frederick William Hayes and many other contemporary socialists that the new society would be best achieved by a parliamentary majority of socialists, an approach absolutely unacceptable to Owen and most of his followers (S501 1894, 53). Wallace thus also wished the government to commence the task of social reform: "what we insist upon now is, *that we declare war against every form of want, poverty, and industrial discontent, and that the Government must lead the way and set the pace*" (S734 1913, 31). Like both Owen and Bellamy (Bellamy 1927, 57), he was opposed to any revolutionary transformation, and also rejected the use of any compulsion in the introduction of socialism, writing that "If a considerable minority refuse to submit to the socialist regime it will be difficult, if not impossible, to compel them, while it would certainly be unjust and unsocial to do so" (S501 1894, 54).

A third heated and for our purposes still more complex issue was that of the specific design of the future society, and especially its mode of government. In *Looking Backward* only one corporation exists, the State, in which all are stockholders. The government organizes the system of production and distribution, and all belong, for a service of twenty-four years, to an "Industrial Army," headed by an elected president. In Bellamy's scheme much of the real supervision of society was performed by the leaders of "guilds," who were typically forty to forty-five years of age (Bellamy 1927, 190–91). The President was only to be elected by those unconnected with the industrial army, *e.g.*, those who had passed the age of forty-five, as well as members of the professions, though these were ineligible for election themselves. Judges were also to be appointed only from those aged forty-five onwards (Bellamy 1927, 204–05). This would have ensured a near-gerontocracy, the more so as life-expectancy was expected to extend considerably, though this essentially anti-democratic feature of Bellamy's system, sometimes called "patriarchal," and identified with a nostalgic conception of pre-industrial New England (Thomas 1983, 253), has been largely neglected in the literature.[49] Indeed some serious misconceptions have been put forward on this issue.[50] How far Bellamy's patriarchalism might have been indebted to Owen's paternalism is thus certainly a subject fit for conjecture; we know Bellamy referred to Owen in various book reviews and editorials in the *Springfield Union* in the 1870s, as well as Fanny Wright, Robert Dale Owen, Josiah Warren, and others associated with the Owenite movement (Bowman 1986, 46). Like Owen he was opposed to democratic elections as "wearisome vanity . . . a delusion and a snare."[51] He may well have been acquainted with what Owen termed his "paternal" system of government, which in its chief expression assigned government of socialist communities to those aged thirty to forty, with "external affairs" placed in the hands of those aged forty to sixty, as a means of avoiding the electoral process which Owen and later Bellamy found so distasteful. Such proposals, for government by a committee of colonists of a certain age, had been mooted by Owen since the mid-1820s, and had been described by William Sargant, among others (Sargent 1860, 235). So Wallace was certainly aware of them. But the "probable debt" of Owen on Bellamy has been hinted at, rather than substantiated.[52]

With respect to his ideas on government the influence of Owen on Wallace can only be described as "probable," as well, for Wallace's citation of the scheme for government by age makes reference only to Bellamy. In his 1897 discussion of another community, at Frederiksoord in Holland, Wallace proposed that

> as time went on, and a generation of workers grew up in the colony itself, a system of self-government might be established; and for this purpose I think Mr. Bellamy's method the only one likely to be a permanent success. It rests on the principle that, in an industrial community, those only are fit to be rulers who have for many years formed integral parts of it, who have passed through its various grades as workers or overseers, and who have thus acquired an intimate practical acquaintance with its needs, its capacities and its possibilities of improvement (S512 1897, 21).

Elsewhere his account of Bellamy includes the observation that:

> As in a well-regulated modern family the elders, those who have experience of the labours, the duties, and the responsibilities of life, determine the general mode of living and working, with the fullest consideration for the convenience and real well-being of the younger members, and with a recognition of their essential independence. As in a family, the same comforts and enjoyments are secured to all, and the very idea of making any difference in this respect to those who from mental or physical disability are unable to do so much as others, never occurs to any one, since it is opposed to the essential principles on which a true society of human brotherhood is held to rest (S727 1900, 1:519).

Yet there were other instances in which Wallace praised the use of a full adult suffrage to manage communal affairs, notably in his 1882 review praising of E. T. Craig's system of management at the Irish Owenite colony at Ralahine, which Wallace regarded as proving the worthiness of working people for self-government, though he also described it as analogous to Owen's supervision at New Lanark (S727 1900, 2:461, 471). And elsewhere Wallace proclaimed that "Our object should be to train up self-supporting, self-respecting, and self-governing men and women; and we should aim at this by developing the conceptions of solidarity and brotherhood" (S655 1909, 21). Generally, however, Wallace seems to have believed, with Owen and Bellamy, that popular sovereignty as expressed through universal suffrage was not conducive to an orderly polity or efficient productive system. "Democracy," in this sense, was usually used by Wallace in the sociological sense, as describing the majority or working class. Even in his *The Revolt of Democracy* (S734 1913), thus, we find no discussion of the suffrage or forms of constitution.

Wallace and Owenism: The Balance Sheet

Let us now draw together the various threads of discussion presented so far. We have seen that the standard view of the influence of Owenism on Wallace was that there was an early socialist phase in Wallace's thinking, followed by a subsequent period of individualism, the embracing of spiritualism, then succeeded by a later socialist phase. The issue of continuity respecting a range of social, political, and philosophical issues across this long period—1837 to 1913—is a complex and perplexing one. A generally plausible or likely imputation of influence to Owenism is, as we have seen, uncontroversial, but is at the same time so vague as to lack much explanatory value. Marchant comments on this issue that "he was so deeply impressed with the reasonableness and practical outcome of these theories that, though considerably modified as time went on, they formed the foundation for his own writings on Socialism and allied subjects in after years" (Marchant 1916a, 1:16). Elsewhere he says Owenism "greatly impressed" the young Wallace (S734 1913, xiii). There is no reason, as we have seen, to reject this view. But neither does it deepen our understanding of Wallace very much, unless we wish to be somewhat more conjectural. It is not too daring to speculate that Wallace did pick up an essential humanitarianism from the Owenites which gave him a markedly more tolerant attitude towards "savage" or "uncivilised" races, one which was to demarcate him very sharply from Darwin, who by the mid-1870s had come to accept the doctrine that the "uncivilised" races were in many senses inferior to the "civilised," and would inevitably perish in their engagement with the latter. Wallace, on the contrary, argued that "there is no good evidence of any considerable improvement in man's average intellectual and moral status during the whole period of human history, nor any difference at all in that status corresponding with differences in material civilisation between civilised and savage races today." Thus he insisted that "the supposed great mental inferiority of savages is equally unfounded" (S649 1912, 33, 43–44). Yet such sympathies could of course also have been derived from other sources, notably Wallace's own extensive observations of "savage" behaviour. But again, a predisposition to regard "savages" as essentially no different from Europeans could have been provided by an Owenite bias.[53]

What, then about the "strong" argument? Greta Jones's recent account of the Owenite influence on Wallace claims that this assumed three main forms: a general bias in favour of environmentalism, rather than heredity; a sympathy for ideas of home colonization, or communitarian settlement on the land; and a critical interpretation of Malthus which stressed that population growth need not outstrip the provision of a subsistence (Jones 2002, 73–96). Jones emphasizes the continuity between Wallace's earlier and later encounters with socialism, stressing that even during the land nationalization phase Wallace still upheld "his original utopian vision of the self-governing community based on Owenite principles" (Jones 2002, 75). (His agreement in 1892 to become vice-president of the Freeland movement to colonize the highlands of Kenya, which was led by Theodor Hertzka, might also be cited in this context.) Jones particularly insists that the central Owenite dogma, that an improvement of circumstances surrounding the individual would lead to an improvement in the individual's behaviour, must have influenced Wallace, thereby giving him a sense of the plasticity or malleability of human character which Darwin, amongst others, lacked. This is as we have seen a plausible if unprovable hypothesis. Jones does not, however, see the Owenites, including both Owen himself and his eldest son, Robert Dale Owen, as neo-Malthusians—which the latter certainly was[54]—or discuss their view of divorce, marriage, and the family, which was markedly different, as we have seen, from Wallace's. Nor does she treat Wallace's applause for Bellamy's paternalism in relation to Owenism, as we have done. Moreover, Jones ignores both the statist and urban biases in Wallace's later thought, which counterbalance any residual communitarian ideals he may have recalled from his youth. The "strong" case for Owenite influence on Wallace, then, needs to be balanced by a judicious appreciation of both the proximity and distance between Bellamy's ideas and those of the Owenites. If—but it is a big "if"—Wallace was exposed both to the ideas of the environmental determination of character, and the hint that sexual selection might play a part in this process, the Owenite component in his thought could be described as central and definitive, for he would then have been strongly predisposed towards Bellamy's restatement of these themes in the 1880s. But this remains an unproven, and possibly unprovable, hypothesis.

Finally, there is the question of spiritualism, in which Wallace's interest commenced in the summer of 1865. Owen, of course, much to the disgust of some of his more materialist and secularist leading disciples, had followed a similar track from the early 1850s until his death in 1858 (see S727 1900, 2:521 ff.). Wallace wrote that "the completely materialistic mind of my youth and early manhood" had been "slowly moulded into the socialistic, spiritualistic, and theistic mind I now exhibit." This implies a conscious rejection of Owenism as Wallace had understood it in his youth, in favour of Owen's own views of the 1850s, and those of Robert Dale Owen. It has been recognized that spiritualism provided for Wallace an assured goal for evolutionary progress, and that in 1865 he began to look for evidence of some form of "higher intelligence," which he confirmed to Darwin in an 1869 letter (Durant 1979, 47; Kottler 1974). This has been described as based on Owen's teachings (and Owen reportedly visited Wallace during one séance: Wilson 2000, 16). But it cannot be associated with Owenism as such. To the contrary, most Owenites were fairly alarmed at this development of Owen's ideas, preferring the secularist bias of Owen's early career.

In conclusion, then, the Owenite influence on Wallace can be described as assuming seven forms: a focus on the central desirability of educating the majority to a high level; an emphasis upon feminism; an environmentalist approach to character; a paternalist approach to management under socialism; a general applause for communitarianism, later intermixed with statism; secularism, later supplanted by spiritualism; and neo-Malthusianism. Wallace most notably departed from Owenism in his rejection of ease of divorce, and the degree of his acceptance and promotion of a more statist approach to socialism. He also gave much greater support to the individualist, or traditional, ideal of the family, than Owen. And his residual Owenite sympathies were, as we have seen, tempered by the influence of Bellamy in particular from the late 1880s onwards, though Owenism would have predisposed him to accepting several leading elements of Bellamy's scheme, notably sexual selection and government by age.

Notes

1. One of the builders' leaders, James Morrison, was a prominent Owenite.
2. Nevertheless though Wallace claimed that Spencer propelled him towards individualism, he also noted that Spencer's view of the land "made a permanent impression on me, and ultimately led to my becoming, almost against my will, President of the Land Nationalisation Society" (S729 1908, 319).
3. The chief general history remains J. F. C. Harrison (1969).
4. *New Moral World*, vol. 1, no. 1 (1 November 1834), 7; no. 2 (8 November 1834), 16.
5. *New Moral World*, vol. 1, no. 3 (15 November 1834), 24.
6. *New Moral World*, vol. 1, no. 6 (6 December 1834), 48.
7. *New Moral World*, vol. 1, no. 9 (27 December 1834), 72.
8. *New Moral World*, vol. 2, no. 68 (13 February 1836), 121.
9. *New Moral World*, vol. 1, no. 11 (10 January 1835), 85.
10. *New Moral World*, vol. 1, no. 51 (17 October 1835), 404.
11. On the character of Owenite branch life see Claeys (1988, 19–32).
12. *New Moral World*, vol. 2, no. 89 (9 July 1836), 296. The first report of the Society is reprinted in *New Moral World*, vol. 3, no. 132 (6 May 1837), 220–21.
13. On London Owenism in this period see J. F. C. Harrison (1969, 224–25). Harrison also terms this the "John Street Institute" at the time of Wallace's attendance.
14. *New Moral World*, vol. 3, no. 138 (17 June 1838), 274.
15. The first mention of a London "Hall of Science" appears to be in early 1838, but no address for it is given (*New Moral World*, vol. 4, no. 171 [3 February 1838], 116.) A slightly later mention gives the address of the "Hall of Science" as Commercial Place, City Road, Finsbury (*ibid.*, no. 174 [24 February 1838], 140, no. 14 [26 January 1839], 224.) In early 1839 it was reported that these premises had been purchased for nine years (*ibid.*, no. 14 [26 January 1839], 219). The John Street premises were still described as a "branch association" in this period (*ibid.*, no. 189 [9 June 1838], 261). In June further premises were opened at Theobald's Road, Red Lion Square, and it was announced that a "Temperance Free-and-Easy" had been opened at the "Community Coffee-rooms at 94 John Street" for the purpose of consolidating the energies of the socialists in the western part of the metropolis, supplying the rooms in Theobald's Road with efficient lectures, issuing tracts, and the furtherance of other business connected with spreading a knowledge of the Social system (*ibid.*, no. 192 [30 June 1838], 284). In July 1838 Branch A1 was reported as separate from Branch 16, which was called the "Hall of Science" (*ibid.*, no. 196 [28 July 1838], 315).
16. *New Moral World*, vol. 3, no. 115 (7 January 1837), 81.
17. On Sunday 5 February, for instance, the topic was a comparison between Socialism and the Jesuit system (*New Moral World*, vol. 3, no. 120 [11 February 1837], 123).
18. *New Moral World*, vol. 3, no. 116 (14 January 1837), 90; no. 122 (25 February 1837), 140, 142.
19. *New Moral World*, vol. 3, no. 142 (15 July 1837), 308.
20. It is reprinted in *New Moral World*, vol. 3, no. 142 (15 July 1837), 305.
21. *New Moral World*, vol. 3, no. 118 (28 January 1837), 107.
22. *New Moral World*, vol. 3, 131 (29 April 1837), 212.
23. Bate would later contribute some £12,000 he had inherited from two spinsters to the Queenwood community.

24. *New Moral World*, vol. 3, no. 130 (22 April 1837), 202.
25. *New Moral World*, vol. 3, no. 146 (12 August 1837), 337.
26. *New Moral World*, vol. 4, no. 160 (18 November 1837), 29; no. 193 (7 July 1838), 294; no. 203 (15 September 1838), 378.
27. *New Moral World*, vol. 4, no. 164 (16 December 1837), 57.
28. *New Moral World*, vol. 4, no. 5 (24 November 1838), 75.
29. *New Moral World*, vol. 7, no. 69 (15 February 1840), 1110; Supplement (23 May 1840), 1243; *New Moral World*, vol. 7, no. 78 (18 April 1840), 1253; no. 79 (25 April 1840), 1263.
30. For a general survey of some of these debates see Dean (1995).
31. *New Moral World*, vol. 3, no. 121 (18 February 1837), 132.
32. *Ibid.*
33. See, *e.g.*, "Proceedings of the Fourth Congress of the Association of All Classes of All Nations" (1840), in Claeys (2006, vol. 7).
34. S729 1908, 123. For the evidence see Moore (1997, 290–31).
35. *New Moral World*, vol. 3, no. 147 (19 August 1837), 346.
36. Wallace did cite other sources for his socialist ideas occasionally; he described Robert Blatchford's *Merrie England*, for instance, as containing: "more important facts, more acute reasoning, more conclusive argument, and more good writing than are to be found in any English work on the subject I am acquainted with" (S498 1894, 197).
37. S431 1891, quoted in Fichman (2004, 251–52). On Bellamy and his followers see in particular Bowman *et al.* (1962, 111–12), which notes Wallace's praise, and Lipow (1982).
38. S729 1908, 327–28. On the reception of Bellamy's writings see Bowman *et al.* (1962) and, more generally, MacNair (1957).
39. S622 1905, 5. On Owen's view of the exchange mechanism, see Claeys (1987b, 34–60).
40. Galton wrote a brief utopia devoted to the subject, much of which was destroyed after his death. See F. Galton (2001, 188–233).
41. Brindley *c.*1840, 3. In fact, most socialists assumed Owen chiefly sought ease of divorce; see, *e.g.*, Jones and Bowes (1840, 12).
42. Owen 1834, 7. Owen specifically noted that in former times, especially in Greece and Rome, individuals possessed "greater physical powers, more intellectual strength and vigour, and more exalted notions of public virtue, and of truth and sincerity, than are to be found in modern times," the moderns having "gradually degenerated, through all ranks and degrees, into mere pedlars and panders for money gains." He thus contrasted a regime in which "natural connexions being formed between those whose sympathies or qualities of mind and body are in harmony with each other" to one which made "wealth, family, titles or privileges of some kind" to be "the artificial uniting motive" (p. 8). This is very close to the type of contrast Wallace later envisioned.
43. Owen 1836, 18, 22–23. Owen elsewhere wrote that "Under these united deranging circumstances of the existing malformation of the character of the present generation, and of the consequent malformation of society, it is impossible that the parents of infants can be now in a condition, either of mind or body, to procreate children to have the superior physical, mental, and moral organization, which it is so desirable that all infants should possess, for their own happiness, and for the benefit of the state and of the world" (*ibid.*, part three, 1838, p. 11).
44. S727, 1:522. The point has generally been sustained, though property prices need to be factored into the equation.

45. It should be noted that Bellamy too has been accused of lacking originality in these proposals. It has been suggested that Bellamy was indebted to August Bebel's well-known *Women in the Past, Present, and Future*, first published in 1883, and translated into English in 1885. See Mrs John B. Shipley, *The True Author of "Looking Backward"* (New York, John B. Alden, 1890), as cited in Griffith (1986, 70). See also Shipley (1890, 3). Bebel certainly took up Darwinist arguments, as applied to marriage, in order to contend that human evolutionary strategy demanded a socialistic system. But while he commended complete marital freedom in the future, and linked this obliquely to neo-Malthusianism, this was not linked to any evolutionary strategy dependent on such choice. See Bebel (1885, 127–29, 229). However, we know that Bellamy had in any case begun to wed feminism and Darwinism in his *Springfield Union* articles in the mid-1870s. As early as 1873, specifically, he published an editorial entitled "Who Should Not Marry," where he doubted that "legal restrictions on the subjects are desirable or practicable," preferring a system of constraint by public opinion. It is reprinted in Widdicombe and Preiser (2002, 181–83).
46. Bellamy 1927, 257. It has also been suggested that he derived the idea from an early American utopia, Marie Howland's *Papa's Own Girl* (1874). See Morgan (1944, 221). See further Griffith (1986, 70). Howland suggested that a reduction in the number of children would increase female independence, as well as advocating women's rights in general, intermixed with some communitarian experimentation.
47. Quoted in Bowman *et al.* (1962, 87).
48. Both had contributed to Burnett *et al.* (1886), in which Morris offered a sketch of "The Labour Question from the Socialist Standpoint."
49. An important exception is Arthur Lipow (1982), who views the Nationalist movement generally as an antidemocratic "authoritarian middle class reaction against capitalism" (p. 8), and notes the absence of universal suffrage and democratic control in Bellamy's scheme (pp. 24–29), emphasizing that "Even by the most generous standard, there is no democracy to be found in it" (p. 29).
50. *e.g.*, Rooney 1985, 59: "The principle of democracy was rarely questioned; in fact it was the lack of consideration for the will of the majority that was cited as the chief problem."
51. Quoted in Bowman (1986, 61). Bellamy has been cited as referring to a number of Owen's works in an article entitled "Literary Notices," *Springfield Union*, 23 October 1875, 6 (*ibid.*, 134 n. 5). But this reference appears to be incorrect. (Thanks to Maggie Humberston of the Springfield Library for verification.)
52. Morgan 1944, 222, 367–68, 370. In one of the very few works to link Bellamy to Owen, the latter's influence is described as consisting "chiefly in his catholic support of virtually every means then extant of wiping out existing evils, and in his emphasis on the peculiar virtues inhering in co-operation" (Barnes and Becker 1961, 2:631), which has no bearing on the central issues addressed here. Some biographers have asserted that "Bellamy's socialist society bore a close resemblance to Robert Owen's New Lanark" (Slotten 2004, 436).
53. It may be noted that Robert Dale Owen also opposed the association of primitivism with virtue, arguing that "The half-civilized Indian ... may, even in his degradation, be considered, not happier or better indeed, but nearer permanent virtue and happiness, than when he roamed the woods, untempted and unseduced" (Owen 1840, 9).
54. A rather dated account is Himes (1928), 627–40.

14

Wallace, Women, and Eugenics

Diane B. Paul

• • • • • •

> **These concluding chapters stamp Mr. Galton as an original thinker, as well as a forcible and eloquent writer; and his book will rank as an important and valuable addition to the science of human nature.**
>
> Wallace's review of Francis Galton's *Hereditary Genius*, printed in *Nature* (S161 1870, 503).

> **The world does not want the eugenist to set it straight. Give the people good conditions, improve their environment, and all will tend towards the highest type. Eugenics is simply the meddlesome interference of an arrogant, scientific priestcraft.**
>
> Frederick Rockell's interview with Wallace, printed in the *Millgate Monthly* (S750 1912, 663).

In the *Millgate Monthly* interview, conducted the year before Wallace's death, the interviewer expressed surprise at the intensity of his subject's anti-eugenic feeling. Wallace explained that he was sensitive on the point, having recently been described in a scientific publication as an enthusiast for eugenics. Scornfully insisting that nothing could be further from the truth, he asserted: "Not a reference to any of my writings; not a word is quoted in justification of this scientific libel. Where can they put their finger on any statement of mine that as much as lends colour to such an assertion? Why, never by word or deed have I given the slightest countenance to eugenics. Segregation of the unfit, indeed! It is a mere excuse for establishing a medical tyranny. And we have enough of this kind of tyranny already" (S750, 663).

Yet Wallace's objections notwithstanding, a contemporary might be excused for reading him as an advocate of eugenics, albeit not of the kind that involved segregation or other forms of negative selection. As John Durant (1979, 51) commented almost thirty years ago, misunderstandings on the point are "perhaps forgivable" given ambiguities in Wallace's thought. In particular, Wallace wrote one of the few favorable contemporary reviews of Francis Galton's *Hereditary Genius* (1869), a book that argued the urgent need for eugenics (though Galton did not actually coin the term until 1883).

How do we square Wallace's positive assessment of Galton's work in 1870 (a judgment he never repudiated) with his fierce denunciation of eugenics in the 1912 interview and in other conversations and several publications?[1] Had Wallace's stance shifted over time, perhaps accompanied by an unconscious reconstruction of the past? Or could he reasonably be characterized either as endorsing or opposing eugenics depending on which features of his thought were emphasized and what the evaluator understood by "eugenics"? To at least make a start on answering this question, this essay explicates Wallace's attitudes towards efforts to control human breeding and attempts to situate these attitudes in the context of both his scientific views on the nature of heredity and selection and his broader socio-political commitments, especially his radical egalitarianism, his anti-statism, and his views on marriage and the capacities and condition of women. It is hoped that such an exploration will illuminate not only aspects of Wallace's thinking but also some underappreciated complexities in late nineteenth and early twentieth-century debates over the nature and meaning of innate human differences.

Galton's *Hereditary Genius* and Wallace's Response

In Charles Smith's list of the most important people in Wallace's intellectual life, Galton ranks eighth—below Darwin, Herbert Spencer, and Henry Walter Bates, but above many individuals assumed to have been far more consequential for him, such as Robert Owen (**http://www.wku.edu/~smithch/wallace/mostcite.htm**). On the face of it, the ranking seems curious, and it would be natural to wonder if it resulted from an aberration in the weighted referral system employed by Smith; or since the rankings are based on the number of times that Wallace refers to various individuals in the main text of his writings, perhaps reflected the existence of numerous *hostile* mentions of Galton's work. But Wallace's comments on Galton were always respectful, and when he raged against eugenics, as he did frequently in his later years, it was apparently not with Galton in view. Even in "Human Selection" (S427 1890), a major statement on human breeding written after his conversion to Edward Bellamy's version of socialism, Wallace treated Galton's views as worthy of thoughtful consideration. Moreover, on several important matters, including issues related to what Galton would later term the "nature-nurture" debate, the two men were implicit allies. At least some of their commonalities (as well as divergences) are evident in Wallace's review of *Hereditary Genius*. So let us now turn to the argument advanced in that book and to Wallace's response.

Galton's researches in human heredity had been inspired by Darwin's 1859 publication of *The Origin of Species*. "I am sure I assimilated [the *Origin*] with far more readiness than most people,—absorbing it almost at once, and my afterthoughts were permanently tinged by it. Some ideas I had about Human Heredity were set fermenting and I wrote *Hereditary Genius*" (Pearson 1924, 70; see also Pearson 1924, 82, 206, 357; Galton 1908, 287–88). Although the controversy-shy Darwin chose not to discuss human evolution in the *Origin*, Galton immediately found in the book a scientific explanation for humans' seeming inability to live up to their moral ideals. The insight was that man's imperfect nature, explained by theologians as a consequence of original sin, was actually a product of natural selection. Human beings were not fallen angels, but incompletely-evolved apes with inclinations that often clashed with their worthier judgments. In Galton's new view "the development of our nature, under Darwin's law of Natural Selection, has not yet overtaken the development of our religious civilisation. Man was barbarous but yesterday, and therefore it is not to be expected that the natural aptitudes of his race should already have become moulded into accordance with his very recent advance" (Galton 1865, 327).

Galton supposed that as a product of selection, human morality and intellect could be rapidly improved through breeding.[2] The need for progress was urgent, given not only the complexity of modern civilization but the apparent easing of

Figure 27 Portrait of Francis Galton.

Early photo taken from Karl Pearson's book *The Life, Letters, and Labours of Francis Galton* (1924). Out of copyright.

the process of natural selection. For in his view (and Darwin's), medical care and public and private charity now salvaged many of those who earlier would have succumbed to cold, starvation, or disease. Moreover, hereditary paupers, dullards, and criminals bred at an alarming rate—while the competent members of society married late and produced few offspring. Now that the weak were no longer being relentlessly culled from the stock, Galton worried that evolutionary progress could come to a halt. It seemed obvious to him that the solution was to breed from the best—although he was extremely vague as to how this aim might be accomplished, as in the following passage:

> The time may hereafter arrive, in far distant years, when the population of the earth shall be kept as strictly within the bounds of numbers and suitability of race, as the sheep on a well-ordered moor or plants in an orchard-house; in the meantime, let us do what we can to encourage the multiplication of the races best fitted to invent and conform to a high and generous civilisation, and not, out of a mistaken instinct of giving support to the weak, prevent the incoming of strong and hearty individuals (Galton 1869, 356–57).

But as to what it is that we can and should do, Galton was largely silent.[3]

Galton's primary task in *Hereditary Genius* (an expanded version of his two 1865 papers, "Hereditary Talent and Character") was to prove scientifically that human mental and moral qualities—and not just physical ones—were hereditary. Using data obtained from biographical reference works, Galton showed that high achievement runs in families; *i.e.*, that scientists, statesmen, military commanders, literary men, poets, judges, musicians, painters, and divines prominent enough to be listed were more likely than members of the population as a whole to have near male relatives who were also sufficiently eminent to be listed. Galton knew that skeptics would protest that the experiences and connections of the progeny of these high achievers would differ from those of persons chosen at random. But he dismissed the idea that social circumstances could explain their success, at least in science and other fields he considered meritocracies. Those with natural ability would succeed, no matter how adverse their environment, while those who lacked it would fail, however favorable their start in life or influential their social connections.

What was true of individuals applied equally to groups. *Hereditary Genius* included a chapter analyzing the comparative worth of different races. According to Galton's calculations, which were based on estimates of the proportion of eminent men in each race, black Africans on average ranked at least two and Australian aborigines three grades below whites in natural ability. But Galton did not consider these or other "savage" races a threat to Anglo-Saxons or Teutons since the stronger races would inexorably eliminate the inferior in a natural process that was already well underway (Stepan 1982). Of greater interest was the considerable variation found among white races, and especially within the

Anglo-Saxons. To Galton, it was obvious that the ancient Greeks, and especially the sub-race of Athenians, were the ablest people in history. Unfortunately, the most accomplished Athenian women often failed to marry and bear children while both emigration and immigration weakened the race (Galton 1869, 331). Thus Galton feared that even very superior races could deteriorate and ultimately disappear.

Contemporary reaction to the book was generally tepid and sometimes hostile. (When it was reissued in 1892, the response was much warmer [Gillham 2001, 171–72].) The exception was men of science. In her diary, Galton's wife Louisa's wrote: "Frank's book not well received, but liked by Darwin and men of note" (Pearson 1924, 88). Darwin was indeed enthusiastic, writing his cousin that: "I do not think I ever in all my life read anything more interesting and original" (Darwin and Seward 1903, 41). Darwin also wrote Wallace, saying that he agreed with every word in the latter's favorable review in the journal *Nature* (Marchant 1975 [1916], 206). So what did Wallace like—and dislike—in the book?

First, and perhaps most unexpectedly given his socialist leanings, Wallace thought that Galton had proved that reputation could serve as a measure of natural ability. Thus he wrote:

> that notwithstanding all the counteracting influences which may repress genius on one side, or give undue advantage to mediocrity on the other, the amount of ability requisite to make a man truly "eminent" will, in the great majority of cases, make itself felt, and obtain a just appreciation. But if this be the case, the question of whether "hereditary genius" exists is settled. For if it does not, then, the proportion of mediocre to eminent men being 4,000 to 1, we ought to find that only 1 in 4,000 of the relations of eminent men are themselves eminent. Every case of two brothers, or of father and son, being equally talented, becomes an extraordinary coincidence; and the mass of evidence adduced by Mr. Galton in the body of his work, proves that there are more than a hundred times as many relations of eminent men who are themselves eminent, than the average would require (S161 1870, 502).

Wallace also wrote approvingly of Galton's comments on the decline in innate quality since the Greeks of Pericles' time. Apropos the claim that ancient Athenians were at least two grades of ability higher than modern Britons, Wallace remarked: "Well may Mr. Galton maintain that it is most essential to the well-being of future generations that the average standard of ability of the present time should be raised."[4] And it is clear that Wallace concurred with Galton in blaming the Church for his compatriots' low intellectual and moral state on the grounds that its enforcement of celibacy selected against those men and women with the most gentle natures while its persecution of freethinkers selected against the bravest and the most truthful and intelligent. (Although Wallace does not mention it, he would surely also have approved of Galton's assertion [1869, 362] that, with respect to the goal of race improvement, the best form of society was one in which incomes were "not much [derived] from inheritance.")

The review expresses only one mild disagreement. Galton had taken issue with the Malthusian claim that marriage should be delayed until the husband could adequately support a family. In Galton's view, only the prudent would follow this advice, resulting in an increase in the (hereditarily) imprudent, who would both produce larger families and more generations in a century. But Wallace the naturalist pointed out that although the impulsive may marry earlier than the judicious, an increase in population is less dependent on the number of offspring that are born than the number that manage to survive to adulthood. In his view, the prudent man may marry late, but often weds a much younger woman. And he will in any case tend to leave more healthy offpsring than will "the ignorant and imprudent youth, who marries a girl as ignorant and imprudent as himself" (S161 1870, 502).

Thus already manifest in the review are Wallace's beliefs that mental and moral qualities can be inherited, that the level of mentality and morality differs among nations as well as individuals, and that the standard of his own society is not what it should or could be. As contemporary reviews of the book show, the first two claims at least were hardly self-evident. Thus the philosopher and economist John Stuart Mill was a particularly vehement critic of the view that either individual or group differences in mentality or morality are attributable to differences in heredity. Indeed, in his influential *Principles of Political Economy*, first published in 1848, Mill wrote: "Of all the vulgar modes of escaping from the consideration of the social and moral influences on the human mind, the most vulgar is that of attributing the diversities of conduct and character to inherent natural differences" (Mill 1965, 319).

Wallace is actually much closer to Galton than he is to Mill on the issue of innate differences (although he and Mill would shortly become allies on the issue of land reform).[5] To understand better why this should be so, we now turn to Wallace's thinking on the evolution of human character.

Wallace on Human Evolution

As with so much else in Wallace's life, the best place to start is with his encounter with Robert Owen. Wallace was first introduced to Owen's theories when, after leaving school at the age of thirteen, he went to work as a builder's apprentice in London.[6] As Greg Claeys notes elsewhere in this volume, "A connection with Robert Owen and Owenism runs through much of Wallace's long life, starting in early adolescence." As Claeys also indicates, one result of that encounter was to plant the seeds of religious skepticism. A second and related result was acceptance of the central Owenite principle that "the character of every individual is formed *for* and not by himself ..." (S729 1908, 46–47). But contra Claeys, this principle does not reflect an environmentalist perspective, but rather an anti-religious and determinist one. In Owen's view, human character was a product both of heredity, which accounts for humans' natural powers, dispositions, and tendencies, and of environment, which can either reinforce or deflect their proclivities.[7] Human character is thus a product of natural and social forces, including education, but decidedly *not* a product of individual will.

That is a significant point of contact with Galton. Wallace's exposure to Owenism would have primed him to be sympathetic to Galton's claim that imperfections in human character are not due to original sin and, more generally, to his determinism. In *Hereditary Genius*, Galton expressed his impatience with the view, especially evident "in talks written to teach children how to be good," that individuals succeed through their own diligence and moral effort. In Galton's view, this was nonsense, since success is a function of natural abilities. But the implication (which shocked many of his contemporaries) that there were no grounds for assigning personal responsibility—that individuals deserve neither blame for their vices nor credit for their virtues, since both are beyond their control—is the same irrespective of whether control was exerted by nature or nurture or both. Thus the Owenite influence would have been as likely to favorably dispose Wallace to Galton's arguments as it would to prejudice him against them.

Wallace first discussed the heredity of human intellect and morality in "The Origin of Human Races and the Antiquity of Man Deduced from the Theory of Natural Selection" (S93 1864). This was the first important paper applying Darwin's theory to humans, appearing a year before Galton's essay. The aim was to resolve the continuing dispute between monogenists and polygenists by proposing that human evolution had passed through an early stage of purely physical development, during which distinct races appeared, and a later stage, when selection acted mainly on mind. The monogenist and polygenist positions were thus made congruent; although the races had a common origin, divergence had occurred in the distant past before the evolution of humans' most distinctive traits, their intellect and character. Once selection began to act on the brain, humans were able to transcend their physical environment and the evolution of human physical form effectively ceased.

Wallace argued that sympathy, a tendency to cooperate, and foresight would provide an advantage in the struggle among groups; that is, groups in which those characteristics were prevalent would thrive, while their competitors would diminish in strength and numbers and eventually disappear. The higher—more intellectual and moral—races would supplant the lower in a process of selection that continues to the present and explains why Europeans have consistently prevailed whenever they have come into contact with "low and mentally undeveloped" native populations. The point is reiterated in the conclusion to *The Malay Archipelago*, first published in 1869, where Wallace writes that the true Polynesians are doomed as are the feisty Papuans. "If the tide of colonization should be turned to New Guinea," he wrote, "there can be little doubt of the early extinction of the Papuan race. A warlike and energetic people, who will not submit to national slavery or to domestic servitude, must disappear before the white man as surely as do the wolf and the tiger." (These comments on the inevitable extinction of many aboriginal populations are similar to those of Darwin, who while sometimes deploring both the methods and mores of the colonizers, assumed the inevitability of the stronger eradicating the weaker, and comforted himself with the thought that the ultimate results would be salutary [see Keynes 1988, 172, 408; Darwin 1969, 1:316].)

Thus Wallace downplayed the existence of a struggle for existence among individuals while stressing the importance of the struggle among groups—a move characterized as the displacement of internal with external social Darwinism by Durant (1979, 42), who also notes the paper's unmistakable debt to the writings of Herbert Spencer, especially his *Social Statics.* It was an argument that greatly impressed Darwin (Greene 1981, 103–04), and also William Rathbone Greg, a Scottish essayist and son of a prominent mill-owner. In "On the Failure of 'Natural Selection' in the Case of Man," an influential article published anonymously in *Fraser's Magazine* (1868), Greg quoted Wallace at length, pointing to his claim that in humans, selection had come to center on the mind rather than body. He also agreed with Wallace's claim that natural selection continues to operate in the struggle among tribes, nations, and races, noting that: "Everywhere, the savage races of mankind die out at the contact of the civilised ones." But Greg was primarily interested in his own society, where he felt that the beneficent process of selection had been halted. In civilized societies like England, medicine and indiscriminate charity allowed the least valuable members of society not only to survive but to propagate their kind. As a result, paupers and imbeciles were outbreeding the middle class.

Wallace's initial reaction to Greg's proto-eugenic article is unclear. But he did comment on a critique of Greg that appeared several months later in *The Quarterly Journal of Science.* According to the anonymous author of the critique, natural selection, including of moral qualities, *does* continue to operate both among and within societies. Thus "there is no excuse for speaking of a failure of Darwin's law or of 'supernatural' selection." The author continues:

> We must remember what Alfred Wallace has insisted upon most rightly—that in man, development does not affect so much the bodily as the mental characteristics; the brain in him has become much more sensitive to the operation of selection than the body, and hence is almost its sole subject. At the same time it is clear that the struggle between man and man is going on to a much larger extent than the writer in "Fraser" allowed. The rich fool dissipates his fortune and becomes poor; the large-brained artizan does frequently rise to wealth and position; and it is a well-known law that the poor do not succeed in rearing so large a contribution to the new generation as do the richer. Hence we have a perpetual survival of the fittest. In the most barbarous conditions of mankind, the struggle is almost entirely between individuals: in proportion as civilization has increased among men, it is easy to trace the transference of a great part of the struggle little by little from individuals to tribes, nations, leagues, guilds, corporations, societies, and

> other such combinations, and accompanying this transference has been undeniably the development of the moral qualities and of social virtues (Anon. 1869, 153).

On reading this commentary, Wallace wrote Darwin (20 January 1869) asking if he had seen "the excellent remarks on *Fraser's* article on Natural Selection failing as to Man?" and remarking that: "In one page it gets to the heart of the question, and I have written to the Editor to ask who the author is."

Several months earlier, in comments following a paper delivered at the British Association for the Advancement of Science meetings (S142a 1868), Wallace suggested for the first time that Darwinian natural selection could not fully account for human intellectual and moral evolution, a point he reiterated in the better known essay on Lyell's geology (S146 1869). Then in "The Limits of Natural Selection as Applied to Man" (S165 1870), he famously shocked Darwin, among others, by arguing more specifically that natural selection alone could not explain the development of certain intellectual powers, such as abstract reasoning, or higher moral sensibilities (or certain physical traits such as hairlessness), and that these qualities could only be explained by the operation of some "unknown higher law."

The reasons for Wallace's announcement that natural selection is not a sufficient explanation of human evolution have been extensively debated by scholars, and need not concern us here.[8] What is relevant is that his view that "more recondite" forces were also operating in the realm of mentality and morality did not imply abandonment of the views expressed in 1864. Indeed, as Charles Smith has noted, he chose to publish a revised version of the essay under the title "The Development of Human Races Under the Law of Natural Selection" in his *Contributions to the Theory of Natural Selection* (S716 1870). Had Wallace fundamentally changed his view, it is hard to understand not only why he would choose to reprint that essay but how he could favorably review Galton's 1869 book with its selectionist account of human intellect, talent, and character. As Smith (2003–, Introduction) notes, Wallace's views were broad enough "to accommodate both natural selection and spiritualism ..."

In any case, we can now understand the tone of the review as a reflection of views that Wallace and Galton shared in 1870: that both individuals and groups differ in their innate endowments, that contemporary Britons are less capable than ancient Greeks, that behavior is determined and imperfections in character are not explainable by the doctrine of original sin. Both are critical of a system of inheritance that interferes with meritocracy. And they would move closer together as Wallace came to side with Galton (and August Weismann) on the question of whether acquired characters are heritable and hence whether selective breeding is a *sine qua non* of human improvement.

Of course there were already differences, and as we will see, new ones would emerge and/or sharpen. While both Galton and Wallace were determinists,

Wallace's Owenite-inspired outlook allowed environmental reforms—and especially education—to be an indirect cause of hereditary improvement. They would part ways on spiritualism. Moreover, Wallace would come to situate women at the center of his scheme, whereas for Galton, women were always on the periphery. Wallace was much more libertarian than Galton, even though the latter was a Whig, and thus inclined to keep the functions of the state to a minimum. And Wallace's egalitarian convictions separated him sharply from the elitist Galton.

Those convictions were nowhere more evident than in Wallace's concluding comments in *The Malay Archipelago*, published in 1869, which discusses the lessons that can be learned from "savage" man. In contrast with Darwin, Wallace generally admired the aboriginals he encountered in his travels as a naturalist and ethnographer, and he often favorably contrasted their values and behavior with those of his compatriots. Thus he famously commented in a letter home: "The more I see of uncivilised people, the better I think of human nature on the whole, and the essential differences between so-called civilised and savage man seem to disappear" (S22 1855). He ends *The Malay Archipelago* with a description of the ideal social state to which he thinks the higher races have always been and are still tending. That condition is one of "individual freedom and self-government, rendered possible by the equal development and just balance of the intellectual, moral, and physical parts of our nature,—a state in which we shall each be so perfectly fitted for a social existence, by knowing what is right, and at the same time feeling an irresistible impulse to do what we know to be right, that all laws and all punishments shall be unnecessary" (S715 1891, 456).

Wallace then suggests that a social state close to attaining this ideal actually exists in some aboriginal communities, where everyone is law-abiding and virtually equal in wealth and knowledge. In such communities there are no masters and servants and the division of labour is muted, as is competition. As a result, there is little incentive to major crime, and petty crime is repressed, partly by public opinion, but mostly by "that natural sense of justice and of his neighbour's right, which seems to be, in some degree, inherent in every race of man." And Wallace goes on to suggest that while Europeans may have progressed far beyond savages in intellectual achievements, the mass of the population has "not at all advanced beyond the savage code of morals, and have in many cases sunk below it." He concludes:

> We should now clearly recognise the fact, that the wealth and knowledge and culture of the few do not constitute civilization, and do not of themselves advance us towards the "perfect social state." Our vast manufacturing system, our gigantic commerce, our crowded towns and cities, support and continually renew a mass of human misery and crime absolutely greater than has ever existed before. They create and maintain in life-long labour an ever-increasing army, whose lot is the more hard to bear, by contrast with the pleasures, the comforts, and the luxury which they see everywhere

around them, but which they can never hope to enjoy; and who, in this respect, are worse off than the savage in the midst of his tribe.

This is not a result to boast of, or to be satisfied with; and, until there is a more general recognition of this failure of our civilization—resulting mainly from our neglect to train and develop more thoroughly the sympathetic feelings and moral faculties of our nature, and to allow them a larger share of influence in our legislation, our commerce, and our whole social organization—we shall never, as regards the whole community, attain to any real or important superiority over the better class of savages (S715 1891, 457).

Thus Wallace's passionate egalitarianism is already evident and distinguishes his social views from those of Galton (or, for that matter, Darwin). The divergences would only deepen over time. Over time as well, eugenics was transformed from a utopian ideal with no very clear practical ramifications to a concrete program to control human breeding, including proposals to segregate or sterilize the hereditarily unfit.

Wallace on "Positive" and "Negative" Eugenics

Although Galton did not object to this approach, his own concrete proposals all involved "positive" eugenics: the encouragement of breeding by those with favorable traits. More specifically, Galton wished to encourage members of the hereditary elite to marry each other and at a young age. Among his proposals to accomplish this end was an 1890 scheme to give Cambridge University women judged especially superior in physique and intellect £50 if they married before age twenty-six and £25 on the birth of each child (McWilliams Tullberg 1998, 85). This kind of non-coercive eugenics seemed to Wallace inoffensive, if futile. Thus he commented in *Social Environment and Moral Progress* (S733 2007 [1913], 141–42): "Sir F. Galton's own proposals were limited to giving prizes or endowments for the marriage of persons of high character, both physical, mental, and moral, to be determined by some form of inquiry or examination. This may, perhaps, not do much harm, but it would certainly do very little good." It would be ineffective since, on Wallace's view, natural selection worked by purging the worst, rather than by improving the good. As he said in an interview published as "Woman and Natural Selection" (S736 1893): "This method of improvement by the gradual elimination of the worst is the most direct method, for it is of much greater importance to get rid of the lowest types of humanity than to raise the highest a little higher. We do not need so much to have more of the great and the good as we need to have less of the weak and the bad. The method by which the animal and vegetable worlds have been improved and developed has been through weeding out. The survival of the fittest is really the extinction of the unfit."

Wallace was usually at pains to distinguish Galton's version of eugenics from the proposals for "artificial selection by experts, who would certainly soon adopt methods very different from those of the founder" (S733 2007 [1913], 142). Coercive methods to prevent the less desirable types from breeding were anathema to Wallace, an ardent libertarian. But Wallace also rejected the doctrine of inheritance of acquired characters, and with it the view that correcting unhealthy conditions and habits could directly modify heredity. Thus some form of selection was required for race improvement. But again, the methods could not be coercive. Asked in an interview whether "in view of the iron law of heredity" it would not be desirable to prohibit criminals and the diseased and deformed from marrying, Wallace replied that the answer lay not in legislation, but in the woman of the future. When they became the selective agents in marriage, the unfit would be gradually eliminated from the race (S737 1894).

In *Social Environment and Moral Progress* Wallace wrote:

> I protest strenuously against any direct interference with the freedom of marriage, which, as I shall show, is not only totally unnecessary, but would be a much greater source of danger to morals and to the well-being of humanity than the mere temporary evils it seeks to cure. I trust that all my readers will oppose any *legislation* on this subject by a chance body of elected persons who are totally unfitted to deal with far less complex problems than this one, and as to which they are sure to bungle disastrously (S733 2007 [1913], 143–44).

But the ostensible problem that eugenics addressed—the need to improve the hereditary quality of the race—was very real to him. Wallace never wavered in his beliefs that mental and moral traits were inherited and that since heredity was not directly alterable by the environment, the path to improvement necessarily involved selective breeding. This perspective did not obviate environmental reform—on the contrary, it was an absolutely essential prerequisite for hereditary improvement. But the role of reform was indirect: It created the conditions under which selective breeding could positively and effectively modify the human race.

Wallace's own scheme was designed to avoid both the Scylla of coercion and the Charybdis of "free love" (the abolition of marriage). In the view of many political and social radicals, the key to race improvement lay in abolishing the institution of marriage and allowing women complete freedom in choosing their mates. Advocates of free love considered marriage dysgenic since the choice of a partner was so often based on financial or other considerations unrelated to heredity. In particular, women's need for economic security induced them to marry men who were physically, mentally, or morally deficient. In a socialist society, women would no longer need to marry for base reasons, and if they threw off the shackles of marriage, the process of sexual selection would be allowed full play. There would be no need for political authorities or scientific experts to decide who should and should not breed. Women would naturally choose to mate with the fittest men and their collective choices would elevate the race.

But as Martin Fichman (2004, 256–57) notes, Wallace was a conservative when it came to marriage and sexuality, and he feared that free love would undermine family life and long-term parental affection. In "Human Selection," the first essay in which he publicly declared his socialism, Wallace characterized arguments for free love on eugenic grounds as "detestable" (S427 1890). Indeed, he treated these arguments as scornfully as he did proposals to legislate segregation or sterilization of the unfit.

His solution was to de-couple sexual selection from free love. As is well known, Wallace had a eureka moment after reading Edward Bellamy's utopian novel *Looking Backward* (1960 [1888]), which imagines a classless society in which gold would no longer "gild the straitened forehead of the fool." Instead of marrying the wealthiest men, women would choose those who were the bravest, the kindest, and the most generous and talented, thus assuring the transmission of these traits to posterity. Although Wallace had dismissed the importance of sexual selection (or at least the mechanism of female choice) in respect to other animals, he followed Bellamy in arguing that if full equality of opportunity for women were established, its operation would spontaneously and continuously raise the standard of the human race.

Those men and women who were physically, mentally, and morally superior would marry earliest and in consequence, produce the most children. Of course this scheme assumes that the rejected individuals would not be able to gratify their sexual desires outside of marriage. Wallace acknowledges that for men, who have stronger passions than women, this assumption may seem problematic. It is a problem he rather implausibly resolves by assuming that, in a reformed society, men will have no means of gratifying their passions outside of marriage. In any case, the result of unleashing the process of sexual selection would "be a more rapid increase of the good than of the bad, and this state of things continuing to work for successive generations, will at length bring the average man up to the level of those who are now the more advanced of the race" (S736 1893, 3).[9]

Conclusion

Wallace was nothing if not an independent thinker. That is as true in respect to eugenics as other scientific and social matters. Wallace is often characterized as a fierce opponent of eugenics, but that is not quite right. His disagreement with Galton was based on different understandings of both how natural selection worked and what kind of improvement was needed. As he wrote in *Social Environment and Moral Progress* (S733 2007 [1913], 152), defending his Bellamy-inspired perspective on sexual selection, "this mode of improvement by elimination of the less desirable has many advantages over that of securing early marriages of the more admired; for what we most require is to improve the *average* of our population by rejecting its lower types rather than by raising the advanced types a little higher." Wallace and Galton were both concerned with the hereditary quality of the population, which they considered badly in need of improvement. Wallace rejected Galton's solution because he thought it ineffectual, not immoral.

Like Galton, he rejected the doctrine of the inheritance of acquired characters, and so unlike many on the political left, Wallace could thus not simply rely on social reform to do the job of race improvement. Although economic, political, and educational reforms were imperative, they could not by themselves modify human heredity. (Contesting the common view that a hard view of heredity had pessimistic implications, Wallace stressed that were Lamarckism true, bad habits and social conditions would have continuously degraded humanity [*e.g.* S737 1894].) And since Wallace agreed with Galton that mental and moral differences were largely attributable to differences in heredity, improvement of the human race necessarily involved some form of selective breeding.

But he was morally opposed to the two alternatives on offer, both of which involved interference with marriage, to Wallace an almost-sacred institution. Thus "free love," which appealed to so many political and social radicals, held no attraction for him, and legislation to prevent the unfit from breeding was if anything even more repugnant. Wallace's view of how natural selection worked combined with his libertarian-socialist commitments and views on women to yield a solution that was distinctly his own.

There is no obvious right answer to the question of whether that solution constitutes "eugenics," nor would the answer be of any significance. Eugenics is a notoriously protean concept, sometimes defined (as by Galton) expansively, and sometimes narrowly—depending both on prevailing attitudes and the aims of the writer or speaker. By some definitions, Wallace was an advocate; by any definition, he was also a critic. That was also true of a host of left-leaning biologists in the late nineteenth and first three decades of the twentieth centuries, such as J. B. S. Haldane, Julian Huxley, and H. J. Müller. But Wallace's unique blend of hereditarianism, egalitarianism, and anti-statism provides a particularly potent challenge to conventional categories. No simple label will do justice to his intriguingly complicated views.

Notes

1. For example, Marchant (1975 [1916], 467) reports that in a discussion of "the teachings of some Eugenists," Wallace said: "change the environments so that all may have an adequate opportunity of living a useful and happy life, and give woman a free choice in marriage; and when that has been going on for some generations you may be in a better position to apply whatever has been discovered about heredity and human breeding, and you may then know which are the better stocks."
2. However, with his later discovery of the principle of regression to the mean, Galton came to believe that he had exaggerated the potential speed of improvement (see Galton 1908, 318).

3. In "Hereditary Talent and Character" (1865) Galton had imagined a utopia in which the state instituted a system of competitive examinations designed to identify the country's most talented young men and women. (The exams for women took into account beauty, good temper, and "accomplished housewifery," as well as intelligence and character.) Eugenic marriages would be rewarded monetarily and with a lavish ceremony in Westminster Abbey. But this was clearly a fantasy.

   Galton was not opposed in principle to coercion; rather, he recognized that it was irrelevant to positive eugenics and not politically viable in respect to negative measures. His views are most clearly detailed in *Memories of My Life*. Aiming to defend eugenists from the charge that they promoted "compulsory unions, as in breeding animals," he insisted that eugenic marriages could only be promoted through social influence and recognition. But he also wrote: "I think that stern compulsion ought to be exerted to prevent the free propagation of the stock of those who are seriously afflicted by lunacy, feeble-mindedness, habitual criminality, and pauperism ... I cannot doubt that our democracy will ultimately refuse consent to that liberty of propagating children which is now allowed to the undesirable classes, but the populace has yet to be taught the true state of these things. A democracy cannot endure unless it be composed of able citizens; therefore it must in self-defence withstand the free introduction of degenerate stock" (Galton 1908, 311). In general, Galton was very circumspect on the issue of compulsion in his published writings.
4. He later employed this example as an argument against Lamarckian inheritance, noting that "all the accumulated effort of thousands of years has not made us greater men, intellectually, than the ancients, clearly proving that there has not been a continuously progressive development in the race" (S737 1894, 83).
5. Mill sought Wallace out to join the Land Tenure Reform Association. According to Mason Gaffney (1997, 612–13), Wallace saw land inheritance as a dysgenic factor giving an artificial advantage to unfit heirs, although the point is not made explicitly in Wallace's 1882 book *Land Nationalisation*. Mill stands tenth in Smith's statistical ranking.
6. James Moore (1997, 300–03) argues that it was the rural misery Wallace witnessed in Wales, where peasant grievances, especially against rent charges (which replaced the ancient right to pay tithes in kind) had turned violent, that cemented Wallace's budding socialist sympathies.
7. This was a standard "Lamarckian" view at the time (and Owen, like almost all nineteenth-century writers, assumed the inheritance of acquired characters). For example, in his famous 1874 study of the "Jukes" family, Richard Dugdale assumed that family members had inherited a proclivity for criminal behavior, but that their hereditary tendency to crime could be easily diverted, especially through education, to more productive ends (Dugdale 1877; see also Paul 1995, 43–44).
8. Smith (2003–) and Fichman (2001, 2004) believe that Wallace always viewed natural selection as a law subservient to more profound forces. A long-standing view, however, is that Wallace experienced an abrupt change of heart on the matter of the evolution of human mental and moral traits—linked to his embrace of spiritualism and/or disenchantment with the domestic political uses being made of his work—sometime in the mid-1860s; for examples of this perspective, see Kottler (1974) and Slotten (2004).
9. Wallace also predicts that the large surplus of women over men, which acts as another hindrance to the operation of sexual selection, will disappear in a more egalitarian

society. Unlike Darwin, Wallace believed that, in respect to mental abilities, women were the equal of men, which explains why provided educational opportunities, they often proved their superiors in performance. But by the same reasoning, he concluded that women lacked the "inherent faculty" to compose music; after all, women receive a better musical education than men but have produced no great composers (see S737 1894).

15

Out of "the Limbo of 'Unpractical Politics' ": The Origins and Essence of Wallace's Advocacy of Land Nationalization

David A. Stack

In the winter of 1839–40 Wallace and his brother William were surveying in Kington and Radnorshire. When their work at Llanbister was complete they travelled ten miles south to undertake a task that, according to his 1905 autobiography *My Life* (S729), was new to Wallace: "the making of a survey and plans for the enclosure of common lands." He was later to describe enclosure as "a legalised robbery of the poor for the aggrandisement of the rich," but in 1840 Wallace thought nothing of the "simple robbery" he was helping to perpetrate on the tenants, leaseholders, and scattered cottagers of Llandrindod Wells. The work was interesting, and he "took it for granted that there was *some* right and reason in it," and that the land would be rendered more productive.[1] When he returned to the district, over half a century later, he saw how wrong he had been. The land had been neither drained nor cultivated. The area of common land ostensibly reserved for the use of the poor had become a golf-links, whilst the local population, stripped of their right to keep animals on the moor and mountain, suffered from a "scanty and poor" supply of milk and were dependent on butter supplies from Cornwall and Australia. The only beneficiaries had been the landowners, who had increased the size and value of their estates. The whole proceeding, Wallace concluded, had been "unjust, unwise, and cruel" (S729 1905a, 1:150–58).

The story of how Wallace had grown from a naive surveyor aiding enclosure in 1840, to become the founding inspiration of the Land Nationalization Society (LNS), forty years later, was told in Chapter 34 of the second volume of his autobiography. The narrative structure of that chapter—which begins in 1853 with Wallace reading Herbert Spencer's *Social Statics* (1851) and culminates in his 1889 embrace of socialism—has left its imprint upon all subsequent accounts of Wallace and land nationalization. This is understandable and, to a degree, inevitable, but it is also unfortunate. As much as any other autobiography—"the least convincing of all personal records" (Kitson Clark 1967, 67)—*My Life* conforms to a set of contemporary conventions and literary tropes that distort our understanding. First, it privileges the individual over the sociological. Despite his grounding in Owenite necessitarianism, Wallace charted his developing thoughts on land nationalization as part of an individual odyssey, and largely neglected the broader political and philosophical context. Second, Wallace overemphasized the continuity in his intellectual development by treating it as a largely unproblematic, linear progression. This was typically Victorian, but highly tendentious, especially when Wallace implied that his socialism followed logically from land nationalization. Third, instead of taking his readers through the labyrinthine strands connecting his scientific and social preoccupations with land, Wallace chose to recount his interest in land nationalization in a discrete chapter, and in so doing fostered the impression that the subject could be compartmentalized from his wider thought. This inadvertently encouraged most, though by no means all, of Wallace's subsequent biographers to misrepresent land nationalization as a distraction from his science (*e.g.* George 1964, 219–21; Raby 2001, 230).

Land, in fact, was integral to all elements of Wallace's thought—social and scientific. His career was suffused with land-related questions, and all his most important work was concerned with the ecological interaction of men, animals, and their natural environment. As Mason Gaffney put it: "Wallace's insights were not just into man and nature, but man and nature in relation to *land*" (Gaffney 1997, 611). It was not just that Wallace the land surveyor and Wallace the biogeographer were indubitably the same person: as Martin Fichman has remarked of Wallace's *The Geographical Distribution of Animals* (S718 1876), the book is "permeated with the language and metaphor of surveying and boundaries" (Fichman 2004, 220). It was also that prior to taking up the cause of land nationalization, Wallace had mulled over the land question in a number of contiguous fields. Even in his writings on natural selection the hovering Malthusian spectre provides a constant reminder that Wallace is concerned with species life in relation to the land (Young 1969; Moore 1997). And he first made public his interest in land reform in the final paragraphs of "one of the most important natural history books of the nineteenth century" (Bastin 1989, vii): *The Malay Archipelago* (S715). The context is significant in confirming the connection between Wallace's scientific and his social thought. The timing, coming at the end of the decade in which Wallace struggled to apply the tools of Darwinism to social life, is equally so.

I

As an addendum to the book's final chapter delineating "The Races of Man in the Malay Archipelago," Wallace provided "a few observations on a subject of yet higher interest and deeper importance, which the contemplation of savage life has suggested." These amounted to an inversion of the Orientalist order, with the heretical suggestion that the West could learn from the East, and a castigation of the "social barbarism" of civilized society. In the main text Wallace did not explicitly contrast systems of land ownership. He contented himself with citing the relative lack of a division of labour or a competitive struggle for existence in "savage" societies, and endorsed their absence of vast inequalities. But lurking behind Wallace's remarks was the Malthusian image of overpopulation, evident in his horror at the poverty "which the dense population of civilised countries inevitably creates." His unease, moreover, at the breaking of the connection between man and the land was revealed in his concern that manufacturing, commerce, and "our crowded towns and cities, support and continually renew a mass of human misery and crime *absolutely* greater than has ever existed before" (S715 1891, 457). Thus it made perfect sense when, in a further final note, Wallace's critique culminated in a searing indictment of English land ownership.

Having once again noted the prevalence of pauperism, vice, and crime amidst the greatest wealth a nation had ever known, Wallace found one more example to justify his use of the term "social barbarism":

> We permit absolute possession of the soil of our country, with no legal rights of existence on the soil, to the vast majority who do not possess it. A great landholder may legally convert his whole property into a forest or a hunting-ground and expel every human being who has hitherto lived upon it. In a thickly populated country like England, where every acre has its owner and its occupier, this is a power of legally destroying his fellow-creatures; and that such a power should exist, and be exercised by individuals, in however small a degree, indicates that, as regards true social science, we are still in a state of barbarism (S715 1891, 458).

Two connected points immediately come to mind when reading the closing passages of *The Malay Archipelago*. First, one is struck by the incongruity of this—at first sight—gratuitous rant at the end of an otherwise ostensibly apolitical scientific treatise, cum-travelogue. Second, one is left in no doubt as to the depth of Wallace's debt to Herbert Spencer. The criteria by which Wallace judged the relative merits of "civilisation" and "barbarism" was that of the "ideally perfect state" which, he said, "our best thinkers" defined as "individual freedom and self-government, rendered possible by the equal development and just balance of the intellectual, moral, and physical parts of our nature" (S715 1891, 456). The "thinkers" he had in mind were George Combe, author of the *Constitution of Man* (1828) and champion of an almost neoclassical balance between man's intellectual, moral, and physical nature, and Spencer, who had himself been influenced by Combe (Stack 2008). Wallace added his own assertion that the social ideal was more nearly approached "among people in a low stage of civilisation," but appropriated Combe's concern with poverty and inequality in the midst of wealth, and adopted Spencer's disquiet with a system of land ownership that potentially excluded the bulk of the population from a resource necessary to their continued existence.

Spencer's objections had been set out in Chapter 9 of his *Social Statics*: "The Right to the Use of the Earth." Spencer opined that all men had been born into a world adapted to their gratification; and "the law of equal freedom" gave each the right to use the earth provided his actions did not prevent others from doing the same. This freedom—as intrinsic to man's existence as access to the air—was fatally compromised by a system of private land ownership, which left the landless dependent on the sufferance of the landowners. Having made this argument for access, and established the inherent injustice of private property in land, Spencer recommended that "Society" become its own landlord, through a merger of separate ownerships into "the joint-stock ownership of the public." This was land nationalization, not as a prelude to socialism—men would continue to compete for the tenancy of vacant farms and so on—but to permit mankind's "resumption" of its right to the soil. Spencer made his case with the optimism and searing certainty of youth, reinforced by a prose style that self-consciously eschewed "measured movement" in favour of clear statements of principle. He offered, however, no practical measure by which nationalization might be effected (Spencer 1851, 114–24).

In *My Life* Wallace claimed that he first read *Social Statics* upon his return from the Amazon in 1853, and that Chapter 9 "made a permanent impression" upon him. He was, he later recalled, puzzled as to a practical solution and returned to considering land reform "at intervals" thereafter (S729, 2:238). This is a plausible enough account: Wallace was undoubtedly aware of Spencer and his writings in the 1850s (see 29 October 1858 letter from Wallace to his agent Stevens, discussed in Chapter 4). But it nonetheless seems reasonable to speculate that the crucial reading of *Social Statics*, which led to Wallace's practical interest in land reform, took place—not upon his return from the Amazon in 1853—but upon his return from the Malay Archipelago in 1862. This alternative chronology would help explain the sixteen-year lacuna between Wallace's reading of *Social Statics* and his first public comments on land reform, and the otherwise curious decision to first publicly commit himself to land reform in the pages of *The Malay Archipelago*. This makes more sense if Wallace was actively pondering Spencer's critique at the same time as he began work on preparing the text of *The Malay Archipelago*. Wallace himself seemed to suggest as much in the 1880 article in which he announced his conversion to land nationalization, where he referred to having read *Social Statics* eighteen years earlier (S329 1880, 735). This does not mean that Wallace necessarily got his dates wrong in *My Life*.[2] In 1880 he referred to having read, rather than *first* reading, *Social Statics* in 1862; but it does suggest that it was the 1862 reading that put Wallace on the path to land nationalization.

The crucial point is that land reform assumed a centrality in Wallace's thought in the 1860s as part of a package of ideas that included his developing social Darwinism and his embrace of spiritualism. We know that Wallace was most strongly drawn towards Spencer in the early 1860s, at precisely the time he was seeking to apply the insights of Darwinism to social questions. No writer offered a better guide to the intertwining of the social and the scientific, and in a note to his 1864 essay on "The Origin of Human Races and the Antiquity of Man" Wallace explicitly credited Spencer's *Social Statics* with having inspired his thought (S93, clxx). In 1862 or 1863 he and Bates, both "immensely impressed" by Spencer's *First Principles*, had first called upon the philosopher, and by the mid-1860s, Wallace was a frequent guest at Spencer's Bayswater home (S729, 2:23–24). Michael Shermer accused Wallace of "a rare act of sycophancy" in naming his first son Herbert Spencer Wallace (Shermer 2002, 240). But it might be more perspicaciously understood as mark of adulatory esteem, in the English radical and Chartist tradition of naming one's children after admired political figures. As evidence of Wallace's enduring regard we might also note that in 1900 he described Spencer's *First Principles* (1862) as "one of the greatest intellectual achievements of the nineteenth century" (S589, 4). More pertinently, in 1873 he was still thanking Spencer for the "permanent effect" that "the illustrative chapters" of *Social Statics* had produced "on my ideas and beliefs as to all political and social matters" (Shermer 2002, 239). Wallace's views on the land were undoubtedly part of this more general development.

The connection between Wallace's social Darwinism and his support for land nationalization is easily identified: he was concerned with a land monopoly that stunted the full development of human potential and abhorred the dysgenic effects of unequal inheritance. The connection with his spiritualism, however, is less immediately obvious. Even the most dogged proponent of continuity in Wallace's thought, Martin Fichman, left this connection unexplored. He acknowledged that Wallace's views on land nationalization were "integral elements" in his "system of social evolutionism, in which biological and socio-political convictions reacted on one another," but said little about the link to spiritualism (Fichman 2004, 219–20). A similar silence pervaded his earlier article on Wallace's theism, in which land nationalization featured only as a potential distraction (Fichman 2001, 235, 243). Yet not only did Wallace succeed in promoting spiritualism and land reform at the same time, for example during his US tour of 1886–87, but philosophically the two were linked. The crux of Wallace's spiritualism, the teleological assumption that man and civilization are progressing to a set destination was, as we have just seen, the starting point for his strictures on land reform in *The Malay Archipelago*. And Wallace's whole approach to political reform, his conception of "social duty—of what constitutes justice in social life" was, as he explained, dependent on his view of "man's spiritual nature" (S545 1900, 521). While understandably reticent, for strategic reasons, about preaching spiritualism to land nationalizers, for fear of alienating potential sympathizers, Wallace had no compunction in championing land nationalization to spiritualists, as can be seen in his address to the 1898 International Congress (S545).

A further connection between Wallace's spiritualism and land nationalization might be seen in terms of his initial reluctance to openly advocate either cause. The tenor of Wallace's remarks in *The Malay Archipelago* suggests that he had already privately settled upon some form of land nationalization, as the only adequate solution to the land question. Yet he refrained from any public advocacy until 1880, and in 1870 seemed prepared to bide his time in a free trade reform body, the Land Tenure Reform Association (LTRA), of which more below. At this stage all we need note is that his tentative approach to land nationalization fits with Fichman's depiction of Wallace's similarly tentative approach to spiritualism in the mid-1860s, and the pattern by which in the 1880s he gradually "became more confident in expressing publicly views that he chose to deemphasise, for professional and strategic purposes, in the 1860s and early 1870s" (Fichman 2001, 245 n.). Wallace's own explanation, however, was that it was only in 1880 that he had struck upon the solution to Spencer's unanswered question of *how* to nationalize the land.

II

The central theme in Wallace's writings on land nationalization was defined in the title of his very first article on the subject: "How to Nationalize the Land" (S329 1880). The question as to *why* land might be nationalized was, of course, dealt with, but very much as a second order problem. Indeed at times Wallace writes as if the *why* question is no question at all; that the case for nationalization is settled and all that remains is to find a means to enact it. For Wallace the theoretical case established by Spencer was overwhelming. The only difficulty lay in taking land nationalization out of "the limbo of 'unpractical politics'" by hitting upon a "practical mode" to carry it into effect. This was precisely what Wallace thought he had done. His unique contribution, as he saw it, lay in resolving the long-standing conundrum of how to pass from the prevailing system of private property to a system of state ownership, in a manner which would be affordable, minimize any injustice to landowners, and avoid the potential pitfalls of state jobbery and mismanagement. He considered himself an innovator on the question of *how* land was to be nationalized. The "key," as Wallace described it, first in his article published in the *Contemporary Review* of November 1880, and then at greater length in his 1882 book, *Land Nationalisation: Its Necessity and Aims* (S722), lay in distinguishing two values in the land: inherent value and the value added by improvements.

Wallace invoked his training as a land surveyor to argue that accurate estimates of the two values were possible. Inherent value, he said, depended upon both natural conditions, including geological formation, natural drainage, climate, aspect, surface, and subsoil, and social factors, such as density of population, the vicinity of towns, ports, railways, and roads. Such value was not the product of any individual, and thus should not accrue to any individual. Value added as the result of improvements, by the labour or outlay of the owners and occupiers, by contrast, could be maintained, improved, or destroyed by the actions or inactions of the occupiers of the land. By distinguishing these values, and assigning a distinct payment for each, said Wallace, it became possible to nationalize the land without creating an unwieldy system of state management. He envisaged, rather, a system of occupying ownership, as the land passed to state ownership, but not state management. The land would become state property, and all occupiers would pay a *ground* or *quit-rent* on its inherent value. The rest of the value, however, would be the property of the occupier, who would be able to buy, sell, subdivide, and bequeath their *tenant-right* like any other piece of property. The only significant restriction being that no individual would be able to become a landlord (S329 1880, S722 1882).

The procedure by which land was to pass from individuals to the state differed between "How to Nationalize the Land" and *Land Nationalisation.* In the earlier article Wallace had recommended a tentative, twofold programme. First, in cases of intestacy, the laws of inheritance were to be altered. Where there were no immediate heirs and, therefore, no (in Bentham's phrase) "just expectation" of inheritance, land was to pass to the state, rather than to distant blood relations (S329, 269). Second, Wallace proposed a measure that would lead to land coming into state ownership at a gradually accelerating rate over a long period. At a set date, land would remain private property only for three more transfers of ownership, after which point it would become the property of the state. In "How to Nationalize the Land," Wallace made a virtue of a process, which could take well over a century to complete, arguing that "the very gradual acquisition of the land" would provide a good opportunity to test the state machinery and operation of the scheme (S329, 274). By the time *Land Nationalisation* was published, however, Wallace was far less cautious. He now proposed, as LNS policy, the passing of a general Act providing for the nationalization of all the land, to come into operation five or ten years hence.

Under this more accelerated scheme the question of compensation for "existing owners and their expectant heirs" loomed much larger. Compensation, in the form of a one-off payment was ruled unnecessary. The land, Wallace asserted, had always been the property of the state, and had never legitimately existed as private property. It had only been appropriated by individuals on the sufferance of the community, which periodically asserted its rights when, for example, compulsorily taking land for railway constructions. Nonetheless, he recognized that the state had no right to diminish "the *income* which any living person does or may derive from it." Thus he proposed payments in the form of annuities, for the *quit-rent* now payable to the state. It was "a matter of detail" as to how long these payments should be made, but there was to be no question of perpetually keeping a class of "pensioned idlers, living upon the labour of others." Thus Wallace suggested that annuities should only be paid for three generations, or only to those born before the decease of the present owner (S722, 197–98). Of more interest to Wallace was how existing tenants would become occupying owners.

Following nationalization, the present occupiers would remain in place upon the continued payment of a *quit-rent* to the state. To become a state tenant, however, they would have to purchase their *tenant-right*, direct from their existing landlord, either by private arrangement or via a "land court" (similar in function to those being proposed under the Irish Land Bill). If a tenant were unable to afford such a purchase then he would be able to negotiate a terminable rent, repayable over a period of fourteen to forty years. In addition, Wallace envisaged allowing every man, "*once in his life*," to select a plot for his personal occupation. His choice would be limited in various ways: he could only select agricultural or waste land; the land had to be bordered by a public road; be of a quantity of between one and five acres; be close to his present dwelling; and could not be upon either very small holdings or an estate from which 10 per cent of the land had already been taken. Whether these conditions were met or not would be determined by a local court, but for Wallace the key point was that nationalization would allow all men access to a plot and thus obviate the dangers of the monopoly of the land that he had identified in *The Malay Archipelago* (S722, 202–06, 215–20).

III

In preparing to write *Land Nationalisation* Wallace had immersed himself in voluminous reports on agriculture, Irish famines, the plight of the Highland Crofters, and a host of other land related publications. As he leafed through the pages he must have noticed that many of these were well-thumbed documents. Interest in the land question had been building, at an accelerating rate, from the mid-1870s, and Wallace was far from alone in seeking a solution. Official recognition of the gravity of the problem had come in the form of a Royal Commission on Agriculture, chaired by the Duke of Richmond, which was sifting through its evidence at precisely the same time as Wallace. More immediately, Wallace had been prompted to write his book by the attention his article "How to Nationalize the Land" had received from A. C. Swinton, Dr G. B. Clark, Roland Estcourt, and others, and their success in persuading him—"much against (Wallace's) wishes"—to become the founding president of the LNS in March 1881 (S729, 2:240). *Land Nationalisation* was written as a statement of the Society's aims and appeared amidst a plethora of publications addressing the question of land reform. A full understanding of Wallace's writings on the land, therefore, is only possible with at least an outline appreciation of the three factors that framed late nineteenth-century Britain's fixation with land: railways, Ireland, and agricultural depression.

The latter had arrived in the late 1870s as the cheap imports that had followed the opening of the Midwest prairies combined with an outbreak of epidemic diseases among cattle and sheep, and a series of unusually wet summers, to bring a quarter century of relative economic prosperity to a juddering halt. The dark storm clouds of Irish unrest, of course, had a longer provenance, but by the 1870s the "Irish question" had largely transmogrified itself into a "land question." The crisis in English agriculture, and more especially the poverty and distress of the agricultural labourer, was to be the sustaining motif of Wallace's writings on land nationalization. But it was Ireland, and the successive attempts at Irish land reform, that provided the greater intellectual fillip. It was a proposal from the Irish Land League to convert tenants into peasant proprietors that prompted Wallace's article in the *Contemporary Review.* Even more significantly, Ireland was rapidly changing the political terms of trade in which the land question was being discussed. As the situation across the water deteriorated so Gladstone and others were emboldened to make policy proposals in which the sanctity of private property was clearly violated and landowner security was increasingly regarded as contingent. Indeed, the notion of *tenant-right* and a general valuation of the land, central elements in Wallace's programme for nationalization, were elements in Gladstone's proposals for Irish land reform (S483 1893; Douglas 1979; Ward 1976).

Not that Wallace himself was overly concerned with Ireland: the plight of the Irish featured in his writings only as a convenient club with which to beat English landlordism. But Ireland was important in providing practical ideas and, along with the grain rotting in England's fields in 1879, giving impetus to a rapid recalibration of land reform politics. Proposals that had once seemed radical, such as Joseph Chamberlain's 1873 free trade programme demanding the abolition of primogeniture, repeal of the laws of entail, revising the laws of enclosure, and acknowledging a tenant's title to any improvements, had within six years been rendered moderate, and seemingly inadequate. The apparently unseemly haste with which Wallace sped from tentatively proposing a gradual programme of nationalization for Ireland in November 1880, to president of the LNS and advocate of a much bolder programme a mere four months later, was only a more extreme trajectory of a similar movement found among many of his contemporaries.

In practical terms the foundations of the sanctity of private property in the land had been subject to fifty years of subsidence, triggered by the compulsory purchase schemes that accompanied the spread of the railways (Kostal 1994). By the 1870s the "special" status of landed property was a commonplace among liberal commentators. Thus when Wallace came to consider "How to Nationalize the Land" in 1880 he was able to head the piece with a quote from James Froude, which distinguished property in "moveable things" from property in land, and argued that because men must live on the land in order to live, land itself could never be private property but must always be "really the property of the nation that occupies it." Wallace was also fairly conventional in the arguments he made in favour of nationalization. He was justified in stressing the originality of his answer to the question of *how* land might be nationalized, but when it came to considering *why,* his answers resounded with themes familiar from a long tradition of English radicalism stretching back at least as far as the late eighteenth century. Wallace himself barely acknowledged this heritage, however, or the even older theological argument that the land was God's gift to all men and, therefore, the individual property of none.

In *My Life* he genuflected to the late eighteenth- and early nineteenth-century land nationalizers, William Ogilvie and Thomas Spence, but stopped short of allowing them any influence on his own thought, and said nothing of the broader radical tradition of land reform (S729, 2:240–41). In one sense this was fair. It is unlikely that he had read Ogilvie and Spence prior to embracing land nationalization, and his entrée had come, as he always said, from reading Spencer. But Wallace denied too much when he claimed that it was "induction" alone that had led him to identify the contrast between "landlordism and tenancy with a pauperised and degraded population" on the one hand, and "occupying ownership with a thriving and contented one," on the other (S722, 183–84). This dichotomy, which sat at the heart of Wallace's land reform politics, was a familiar theme in the arguments of preceding generations of radical land reformers. An article in the *Westminster Review* of 1870 had neatly summarized the long-standing radical conclusion "that, wherever the land is of easy access and widely distributed among the inhabitants of the country, the soil is well cultivated and the people industrious, prosperous and contented. On the other hand, wherever the land is in the hands of a few proprietors, cultivation is checked, and the mass of the people are idle, indigent and improvident" (Anon. 1870). Thus Wallace was hardly a pathfinder in dedicating *Land Nationalisation* to demonstrating that to the extent men enjoyed access to the land so would the country be "free from poverty and the people prosperous and contented" (S722, 17). Rather, in answering the question *why* land should be nationalized, Wallace regurgitated arguments about "monopoly," access, independence, material abundance, and the "moral elevation" of free, rural workers, which were recurrent among radicals (Chase 2003).

At the heart of his critique was a demand to end the monopoly of the land and to permit access to all. It was, Wallace wrote, "the birthright of every British subject to have the use and enjoyment of a portion of his native land, with no unnecessary restrictions on that enjoyment other than that implied by the equal right of others" (S364 n.d., 4). It was true that he differed from most radicals in arguing that nationalization was the necessary precondition for this access, but he shared their view that access was itself the precondition of true freedom, as only access to the land could render men independent. As Malcolm Chase has shown, the same desire for independence lay behind the Chartist Land Plan of the 1840s (Chase 1996). Wallace also shared the Chartist insight into the interdependence of land reform and political democracy. It was a constant refrain of both that landlordism—a system in which the few exercised power, through threats of eviction, over the many—was incompatible with the independence necessary for true democracy. Quite simply, Wallace argued, there could "be no real freedom under landlord-rule":

> So long as the agricultural labourer, the village mechanic, and the village-shopkeeper are the tenants of the landowner, the parson, or the farmer, religious freedom or political independence is impossible. And when those employed in factories or workshops are obliged to live, as they so often are, in houses which are the property of their employers, that employer can force his will upon them by the double threat of loss of employment and loss of a

> home. Under such conditions a man possesses neither freedom, nor safety, nor the possibility of happiness, except so far as his landlord and employer thinks proper. A secure HOME is the very first essential alike of political freedom, of personal security, and of social well-being (S512 1895, 4).

Wallace's arguments for nationalization also shared in the economic and moral assumptions of earlier nineteenth-century land reformers. In particular, he echoed earlier radical critiques of luxury and inequality—previously voiced by William Godwin and Charles Hall, among others—and combined them with the common anti-Malthusian contention that the potential material abundance of the land was held in check by artificial social arrangements (Chase 1985). Thus the introductory sections of *Land Nationalisation* dwelt upon the "disturbing agency" of "excessive wealth (accumulated) in the form of landed or funded property," which permitted "a large and ever-increasing class of non-producers" to levy "a perpetual and heavy tax" on "productive workers" by diverting them "from the production of use and beauty" (S722, 13–14). Nationalization would render the soil more productive by giving "every labourer freedom to enjoy and cultivate a portion of his native soil." This would reduce inequality, and secure "independence," and produce "comparative affluence" (S722, 17). It was, thought Wallace, axiomatic that when the cultivator of the soil was its virtual owner, and all products of his labour remained his own, then food production would be maximized (S722, 19). This, in itself, might be deemed sufficient justification for land nationalization, but according to Wallace any increase in production was merely "incidental" (S722, 229).

The ultimate justification for land reform lay not in its economic efficiency, but in "the improvement it would effect in the condition of labourers and producers of all kinds, an improvement which would be social and moral as well as merely physical, and would raise the status and add to the well-being of the whole community" (S722, 229). Just as the *independent* labourer was superior to his *dependent* cousin, so a nation of proudly independent men would be infinitely superior to its present degraded and dependent incarnation. This, of course, harked back to a centuries-old notion of regarding England's independent yeomanry as the moral backbone of the nation and suggested nostalgia for an idyllic, rustic golden age. Certainly Wallace was unapologetic in his hope that land nationalization would, "to some extent, re-establish that village life the destruction of which by landlord rule, Mr. Thomas Hardy, and other writers so much deplore, and also bring back a sample of those independent yeomen cultivating their own farms (for these would be practically their own) which historians, politicians, and philanthropists agree in considering to have been a strength to the country" (S371 n.d., 3). Wallace, as we shall see below, was far from being a straightforward rural nostalgic, but what he shared with earlier radicals was a deep unease with urban living and the environmental destruction wrought by industrialism, and a hankering after a more "natural" rural existence (S722, 9). The great desideratum for Wallace was not rural life *per se*, but an independence that could only be secured by occupying ownership.

This was why he dedicated so much time to explaining how individuals would be able to claim small plots, so that even those who were not farmers had access to the land. And this too was why he so fiercely opposed subletting, mortgages, and even the provision of allotments: each compromised independence. Subletting would recreate, albeit on a small scale, the "evils" of landlordism. As would mortgages, which gave the lender a landlord's power over the mortgagee. The difficulty here was that under his scheme of nationalization, occupiers might not have the money to purchase their tenant-right. To circumvent this, Wallace countenanced the use of fixed term loans from local authorities or loan societies (S722, 202), and the relatively benign view he took of the fledgling building societies accords with Chase's argument that these organizations were well regarded among radical land reformers (S329, 276, 277, 281; Chase 1991). Far less benign in Wallace's opinion was the provision of the 1894 Local Government Act that allowed local authorities to take wastelands and rent portions of them on an annual tenancy. The best that this could produce, protested Wallace, was a system of allotments that further negated the freedom of the labourer. Citing John Stuart Mill, Wallace argued that allotments tended to reduce incomes, by encouraging employers to reduce even subsistence wages, safe in the knowledge that their labourers would not starve. More than this they tied the labourer, who soon became dependent upon his allotment, to the land, and thus further weakened his bargaining position (S495 1894, 4–5). At root, Wallace, in a long line of radical land reformers, sought independence for the labourer. He differed only in the rather significant detail that unlike most—although by no means all (see Plummer 1971)—of his predecessors he thought nationalization was the necessary precondition.[3]

IV

Wallace's debt to the English radical tradition has generally been underplayed. His connections, and supposed debts, to John Stuart Mill and Henry George, by contrast, have been overstated. Both had walk-on parts in the story of Wallace and land nationalization, but neither had a major impact on his thought. George, after all, was the embodiment, and most influential spokesman, of the late nineteenth century's fascination with land reform; all other land reformers, Wallace included, were swept along in his slipstream. Mill's toying with land nationalization in his later career, meanwhile, has long interested historians and intrigued his contemporaries too, including Wallace. It was Mill, moreover, who, after reading Wallace's searing denunciation of "social barbarism" in *The Malay Archipelago*, wrote from his second home in Avignon to enlist Wallace in the LTRA's fight against the evils of land monopoly (S729, 2:235). Wallace and George, meanwhile, were, from the moment of the latter's arrival in England in 1884, even closer. Wallace always referred to the Georgeite Land Restoration Society as allies, and when in New York in the autumn of 1886 he spoke in support of George's bid to become mayor of that city (S403 n.d.). These connections, however, are not enough to justify the erroneous assumptions that Wallace "learned" from Mill or played "second fiddle" to George (Gaffney 1997).

In the case of Mill we would need to substantiate a philosophical link, because the personal connections were ephemeral and the organizational links nugatory. A few letters were exchanged relating to land policy, but any influence seems to have been exercised by Wallace, who attempted to toughen up LTRA policy (S729, 2:235–38). When the two men finally met Wallace was clearly unimpressed (S729, 2:236–37). It could be contended that Wallace learnt about land reform agitation in the LTRA, but we have very few facts with which to substantiate such a view, partly because the Association appears to have been almost stillborn. In *My Life* Wallace minimized its importance and suggested that Mill's death in 1873 "put an end" to it (S729, 2:238). A similarly early demise is ascribed to the Association in most general histories of the land question (Douglas 1979, 18–19). This may be inaccurate. David Martin's study of Mill and land reform suggests that the Association staggered on well into the 1870s, while James Marchant claimed that Wallace only "retired" from the LTRA following the formation of the LNS (Martin 1981, 41; Marchant 1916a, 2:143). But whatever the precise details it is clear that the Association was, at best, moribund in the 1870s and provided little in the way of any positive influence on Wallace. Indeed one might argue that its only role was to provide an impediment to Wallace's embrace of land nationalization. If it did then this was more than serendipitous for the free traders. Mill and his allies had devised the LTRA as part of an explicit strategy "intended to draw support away" from land nationalization and to keep the issue of land reform under a middle-class, radical leadership and out of the hands of the more extreme Land and Labour League. With this in mind Mill self-consciously set out to recruit—or more pejoratively entrap—working-class radicals, and perhaps the same motive prompted his initial letter to Wallace (Claeys 1987a, 141).

Certainly the LTRA represented a brand of liberal politics that Wallace would later denounce as inadequate and inefficacious. Despite the controversy aroused by Mill's policy of state appropriation of the increment from any future increases in land value, the LTRA remained an essentially mid-nineteenth-century liberal organization. It was pledged to the removal of primogeniture, entail, and other impediments to a freer trade in land, shared its offices with the Cobdenite National Education League, and "deliberately sought to continue the work of Cobden." To drive this last point home invitations to the LTRA's inaugural conference featured a quote from Cobden's last major speech, in which he had said that if he were a young man he would form a league for free trade in land (Martin 1981, 39). In addition, it might be noted that while the LTRA was a self-consciously middle-class organization, Wallace's LNS pitched itself, not always successfully, at a working-class audience. Its leadership, most obviously in the form of Wallace himself, and Mill's stepdaughter Helen Taylor, were of a similar social composition to that of the LTRA, but its arguments—as Wallace made clear in *Land Nationalisation*—were dedicated "To the Working Men of England." It was to reach this audience that Wallace had written a book that was "clear and forcible, moderate in bulk, and issued at a low price" (S722, viii), and the Society's famous yellow vans toured the country in "the great work of convincing the highest and best-organised among the manual workers" (S729, 2:240).

The possibility that Mill might have been a significant influence upon Wallace's advocacy of land nationalization stems from the fact that Mill's writings on land reform were "so eclectic" that "both individualistic peasant proprietors and co-operative projects" could draw succour from his work (Martin 1981, 42). Wallace himself was alive to their ambivalence. In *My Life* he blamed Mill, and other individualist thinkers, for delaying his own embrace of nationalization (S729, 2:238–39), but in *Land Nationalisation*, and with less conviction in *My Life* as well, Wallace presented Mill as a thinker who hovered on the precipice of land nationalization, and who would have embraced it if only he could have seen his way to a "practical and just mode of abolishing landlordism," which avoided the problems of state management (S722, 209–10, 185). If Mill could, that is, have known of Wallace's scheme. The more positive side of Mill's duality—for Wallace at least—was the distinction he drew between private property, which theoretically was absolute, and property in land, which must always remain contingent. The fact that the distinction was most clearly expressed in Mill's posthumous and unfinished *Chapters on Socialism* (1879) (which was edited, it might be noted, by Helen Taylor) encouraged the belief that had he lived Mill would have committed himself to land nationalization (Mill 1989; Ottow 1993). Wallace, after all, proceeded from the necessarily contingent status of land ownership—which, because it was essential to the life of all could only be held on the sufferance of society—to reach precisely this conclusion.

In no sense, however, could Wallace be said to have "learned" this distinction from any one source. Mill was hardly unique in making a special case of landed property. Theoretically, a distinction between property generated by labour, and property in land, had a long radical and socialist heritage, stretching back through Paine, Godwin, Hodgskin, and Owen (Chase 1985; Bronstein 1999). Indeed the distinction was present in Mill's own writings from at least his 1848 *Principles of Political Economy*, and Mill himself had "learned" it from the Saint-Simonians, who distinguished unearned income and inheritance in land from income from labour, and Coleridge, who argued that land ownership entailed a power over other human lives which precluded it ever being absolute (Claeys 1987a, 139–40). Even if we were able to prove that Wallace had derived his understanding of this distinction direct from Mill, we would still be faced by a gap between Mill's apparently reluctant acceptance of the logic of land nationalization and Wallace's wholehearted advocacy. In crude terms this boiled down to Mill making utility the fundamental test for any policy of land reform, and Wallace working from the premise of an *a priori* concept of "justice."

Except, of course, such a dichotomy fails to do justice to the nuance (and occasional confusion) of either man's argument: both regularly blurred their criteria.[4] Mill generally led with a utilitarian argument, but was prone to identify utility with justice. In his *Principles of Political Economy* (1848), for example, he wrote that: "When private property in land is not expedient, it is not just" (Gray 1979). Wallace, in contrast, generally made a moral case for land nationalization, rooted in his conception of justice, and thus argued that private property in land could never be just. But he was not immune, as we have seen, from supplementing this point with a utilitarian list of the benefits he expected to follow from occupying ownership. Nonetheless, and without wishing to overstate the significance of the divergence for practical politics, there was an important distinction between Mill and Wallace's view of the land. As Jonathan Medearis has recently shown, Mill's approach to land nationalization (and socialism) was premised upon an abandonment of first principles, including the Lockean labour-based justification of property, in favour of the second order, "consequentialist" sanction of utility (Medearis 2005). Wallace, by contrast, always gave precedence to an argument from moral principle. In this he self-consciously echoed Spencer's assertion that the land question was one of "pure equity": "Either men *have* a right to make the soil private property, or they *have not*. There is no medium." Spencer, therefore, was not only the most likely source for the distinction that Wallace drew between landed property and other forms of private property; Spencer was also, according to Wallace, a writer "as far ahead of John Stuart Mill as John Stuart Mill (was) of the rest of the world" (Marchant 1916a, 2:150).

In terms of social background and political instinct Wallace was closer to Henry George. Both were lower-middle-class autodidacts, with a shared interest in evolution (Hill 1997; Laurent 2005). And when Wallace first read *Progress and Poverty* he recognized a kindred soul. Swept along by the powerful prose, Wallace gushingly described it "as the most startling and original book" of the past twenty years, and immediately recommended it to Darwin and Spencer. What most impressed Wallace was the power of George's indictment of poverty amidst unprecedented wealth, the pre-eminence he gave to land reform as the solution, and his demonstration of the potential tax ameliorating powers of land revenues. Each of these points had been anticipated, albeit more briefly, in Wallace's own writings. George's leitmotif, the prevalence of poverty amidst wealth had formed the backdrop for Wallace's indictment of land ownership in the concluding passages of *The Malay Archipelago*, while in "How to Nationalize the Land" Wallace had first broached the potential tax benefits of land reform (S329 1880, 276). These points of precedence are important. Wallace did not encounter *Progress and Poverty* until 1881, and there is no reason to doubt his assertion that "the greater part of the manuscript" of *Land Nationalisation* had already been written (S722 1882, 9 n.). Reading *Progress and Poverty* undoubtedly invigorated Wallace's sense of social injustice—as it did for thousands of others—and its effect can be seen in the language of *Land Nationalisation*, but chronologically it came too late to serve as a significant intellectual influence. It was not "repressed jealousy" that led Wallace to "cast George as simply a theorist who confirmed Wallace's inductive argument" (Gaffney 1997, 614), but a relatively accurate statement of their relationship.

Wallace did not play "second fiddle to George." Even in his initial rush of enthusiasm for *Progress and Poverty* he never lost sight of the distinctiveness of his own position, or what he saw as George's shortcomings. Indeed, by the time he wrote *My Life* in 1905 Wallace's regard for George was fairly muted, and he pointedly praised Robert Dick's comparatively obscure *On the Evils, Impolicy, and Anomaly of Individuals Being Landlords and Nations Tenants* (1856), for "anticipating the main thesis" of *Progress and Poverty* (S729, 2:255, 241). In *Land Nationalisation* his tone had been warmer but even at this stage Wallace was at pains to emphasize that his and George's books possessed "a totally distinct line of argument and proof." And when Henry Fawcett, in 1883, made the mistake of treating Wallace and George's arguments as substantially the same, Wallace reacted angrily, claiming that their respective positions were "absolutely distinct and unlike" (S365 1883, 4). George, after all, envisaged a future of peasant proprietorship and free trade, accompanied by a single tax on ground rent (Gaffney and Harrison 1994). Wallace, by contrast, had publicly entered the land debate precisely to challenge the efficacy of peasant proprietorship in Ireland, and built his arguments for occupying ownership and land nationalization around a critique of the inadequacy of free trade solutions.[5] This was the distinction Wallace had in mind when, in objecting to Fawcett's identification of his and George's policies, he had described the foundation of the LNS as "the formation of a distinctively English school of land reformers" (S365, 4).

V

For Wallace the politics of the LNS were qualitatively different from those of all contemporary liberal land reformers, including George. It was not the detail of distribution that Wallace questioned, but the very principle of private land ownership. In common with Mill, Spencer, and a host of preceding radicals, he drew "a broad distinction between the products of men's labour which are and should be private property, and land, the gift of nature to man and the first condition of his existence, which ever remain the possession of society at large." But he went beyond most of them in arguing that however freely traded or equally distributed, land could never "equitably become private property"; only production by human labour could confer that status (S365, 5–6). It was on the basis of this Lockean labour theory of property that Wallace distinguished a legitimate private ownership, and therefore free trade, in the buildings, drainage, and other adornments of the land, from an illegitimate and unjust trade in the land itself. Wallace, that is, viewed the land as a resource above and beyond the dictates of the free market. This was neither a classically liberal nor a necessarily socialist argument. Wallace subscribed to an individualist theory of property, but argued that the land—and other natural resources—must be exempted from the ever-encroaching process of capitalist commodification.

Even prior to his first public advocacy of land nationalization—and long before his embrace of socialism—Wallace developed an ecological and conservationist critique of the internationalization of production and the commodification of natural resources. Wallace's attempt to amend the LTRA programme, to include a demand for the state to be able to take possession, for the purpose of preservation, of all natural objects and artificial constructions of historical or artistic interest, along with his failed bid to be appointed superintendent of Epping Forest, can be seen in the broader purview of his increasingly forthright espousal of the need for a sphere of life protected from the pressures of commercial society. Sometimes, as with his arguments for land nationalization, Wallace made his case in terms of a first principle. In other instances, however, Wallace came closer to a utilitarian test in arguing that free trade, and the operations of the market, should be restrained on grounds of expediency.

Wallace's first systematic critique of free trade liberalism came in a letter he sent to the *Daily News* in 1873, entitled "Free-trade Principles and the Coal Question" (S231). In the 1860s and 1870s "the coal question"—or, "how to indefinitely sustain an economy built upon a finite natural resource"—was a recurrent worry for economists, including Mill and William Jevons. Wallace's contribution fed upon a shared Malthusian paranoia with the apparent foolhardiness of exporting finite, mineral deposits. Wallace was not alone in finding a partial solution in export duties. What distinguished his contribution, and prefigured his later approach to the land question, was his use of the example of coal to enunciate a more general critique of private ownership and free trade. He began by identifying the fuel, along with water and land, as a resource essential to human existence. In such cases, he continued, private ownership, and with it control of supply by the few, created an inherent injustice to the many, which made free trade unsuitable. While it was "an axiom with all liberal thinkers" that free exchange between nations was universally good, Wallace argued that there were "certain commodities," including coal, "which we have no right to exchange away without restriction" (S231, 138).

At this point Wallace's conventional "political" argument against monopoly slipped seamlessly into an environmentalist and ecological case for conservation. The so-called "rights" of the owners to trade freely had to be held in check to prevent them wreaking an injustice that would extend across generations. In environmental terms the unchecked production and free exchange of coal would disfigure the country, reduce animal and vegetable life, and produce unsightly slag-heaps and cinder-tips. Ecologically it represented a positive crime: the selfish exhaustion of an irreplaceable resource, "held by us in trust for the community, and for succeeding generations," for this generation's wealth and luxury without a thought for the future. Where a resource took more than a generation to replenish and renew it was a "duty," Wallace argued, to check "further exhaustion" (S231, 144). Economically such restraint also made sense. An unchecked free exchange of coal, by encouraging overproduction and the export of a precious resource would permanently increase the costs of a chief necessity of life, create a workforce dependent upon ultimately unsustainable levels of production, and reduce the nation to economic dependency (S231, 138–39).

The crux of Wallace's case, which was most fully expressed in his 1879 article "Reciprocity and Free Trade," was that "no commercial principle, however good in itself, can be of universal application in an imperfect human society" (S306, 169). Free trade, that is, was not "a moral truth" "to be sought after for its sake," but a "maxim of expediency"; the "mere commercial advantages" that flowed from its practice had to be weighed against its potential moral and ecological costs—for man and nature (S306, 168, 170). In the case of man, Wallace displayed a Combeian concern with stability and balance, and a republican—or civic humanist—suspicion of specialization (see Pocock 1975). Thus while happy to concede the principle of comparative advantage as "commercially sound," Wallace felt that an international division of labour "must always be subordinated to considerations of social, moral, and intellectual advantage" within each nation (S306, 173). "Man," he declared, echoing Combe, "has an intellectual, a moral, and an aesthetic nature," and these various faculties were unlikely to be exercised and gratified in a workforce stultified by excessive specialization, however cheap their imports of cotton, silk, and claret (S306, 174, 179, 171). "Free trade under the guidance of capitalism," Wallace continued, would not only create an imbalance in the development of individuals. Specialization, by encouraging excessive production of one or two goods, or resources, would also destroy "the beauty and enjoyability" of nature (S306, 172).

Such indictments of the expediency of free trade were not, in themselves, essential to Wallace's positive case for land nationalization. As we have stressed, Wallace treated the question of land ownership as a first principle, not as a matter to be judged on its utility. Where his critique of the inefficacy of the market mechanism became relevant, however, was in framing negative arguments against the free trade solutions of his liberal contemporaries, including Mill and George. This was most obvious in Wallace's[6] *Land Lessons from America* (S403 n.d.), a pamphlet which lauded George, but the argument of which could be simply summarized as: *free trade does not work*. America, Wallace noted, was free of the "special disadvantages" that so exercised British liberals. There was no law of primogeniture or entail; the transfer of land was cheap and simple; and a complete register was kept of all land sales and mortgages. In addition, in the US the "complete taxation of ground rents," that the more advanced reformers called for, was already in place. Most of all America possessed "an almost inexhaustible extent of land," often of "marvellous fertility." Yet, despite its advantages "all the evils of landlordism" which scarred Europe were present. Speculation was "everywhere excessive"; in the cities and suburbs, "where working men live" houses were crowded together, often without gardens, and rents were high. In rural districts, meanwhile, the independent American farmer was in decline, usurped by a landlord-tenant system, which brought in its wake the wholesale eviction of tenants on an Irish scale (S403, 7–12).

The US thus offered conclusive proof that free trade reforms would "utterly fail" (S403, 12) because they did not go to the root of the problem (S385 n.d., 3). The liberal desire for small, independent farmers could not be realized until the "monstrous wrong" of treating land as a commodity to be bought and sold for a profit was ended. Even if it were possible to establish a body of peasant proprietors, Wallace explained, markets were dynamic, not stable, and possessed an inherent tendency towards monopoly and concentration of ownership (S385, 6). Some farmers would inevitably have better land than others, which would lead to "unequal competition," and enable the owners of the better land to drive owners of the worse out of business. "The system, therefore, contains within itself the elements of decay and failure" (S385, 7). Far from ensuring equity, free trade "would simply enable those capitalists who desire land to obtain it more easily" (S365 1883, 20). The only way to *sustain* a fairer distribution of the land, Wallace concluded, was to cease treating it as a commodity, by placing it beyond the free market. Only when land was *let*, not *sold*, would it be possible to enable all labourers and mechanics to have an acre of land to live on and an acre or two to cultivate, with the opportunity of getting a small farm of 10–40 acres at some point in the future.

VI

For those attuned to regard individualism and collectivism as antithetic, there seems to be an uncomfortable contradiction at the core of Wallace's advocacy of land nationalization. On the one hand, his arguments were rooted in an individualistic, English radical tradition, which upheld a Lockean theory of property, and celebrated the independence of small-scale farming. On the other, land nationalization entailed a collectivization of ownership, and the creation of an economic sphere above and beyond the market, which seemed to signal a first step towards socialism. Moreover, as we have just seen, before 1880 Wallace had developed a critique of the limits of liberalism and free trade. After 1889 he described himself as a socialist. The temptation to link these two facts together, with Wallace's nine intervening years as a land nationalizer providing the coupling, is both understandable and commonplace. As J. A. Hobson noted in 1897, "both theoretic students of society and the man in the street regard Land Nationalisation as a first step in the direction of Socialism." But we need to be wary of being beguiled by the superficially socialistic implications of the term "nationalization." The "organised Socialists," as Hobson went on to explain, were often suspicious. Some indulged in a brief flirtation with the policy, before discarding it, but in all but a few cases land nationalization remained peripheral to socialist thought. It was not to socialists that the policy particularly appealed, but to "a certain little knot of men of the lower-middle or upper-working class, men of grit and character, largely self-educated, keen citizens." Men, in other words, like Wallace, who saw nationalization of the land as "a plain moral sanction," and "a 'natural right,' essential to individual freedom" (Hobson 1897, 841–42).

In general terms, as I have argued elsewhere (Stack 2003), Wallace's socialism was consistent with his Darwinist science and spiritualist interests. Specifically in relation to land reform, however, there was no inevitable or necessary connection with socialism—in either direction. Socialism had not led Wallace to land nationalisation. We have been able to trace the roots of Wallace's arguments without reference to his 1830s immersion in Owenite socialism. In this our argument is consistent with the earlier chapter in this volume, in which Greg Claeys notes that Wallace's later transition from an "individualist" to a more "collectivist" perspective on social ills was not predicated upon his earlier Owenism. But what of the obverse claim that land nationalization led to socialism? Wallace certainly implied this chain of causation in *My Life*. For example, he commented favourably upon the socialists' adoption of a policy of "free access to land, with a view to its future nationalisation" as "a first step" towards socialism (S729 1905a, 2:255). But note the subtle elision and the sequence: it is socialists who are adopting land nationalization as a first step to socialism; not land nationalizers being led step-by-step into socialism. In his more considered remarks, such as his 1895 Presidential Address to the LNS, Wallace contented himself with the neutral observation that there was "no antagonism between Land Nationalisation and Socialism." His reticence, on this occasion, may have reflected the sensibilities of his predominantly non-socialist audience, who saw no need to venture beyond their "special reform." But his argument—that there was "nothing whatever in our principles that points to individual as opposed to collective occupation of land"—implicitly conceded the indisputable corollary that there was nothing in the Society's principles that necessarily pointed to the collective, as opposed to the individual, occupation of the land either (S512 1895, 19).

After 1889 Wallace's practical plans for the land developed a distinctly socialist hue. Alongside his earlier vision of individual occupying-ownership, he increasingly expounded schemes for small-scale self-sustaining collectives. In *Suggestions for Solving the Problem of the Unemployed, etc., etc.* (S512 1895), for example, he detailed how such communities could be organized and extolled the benefits likely to follow. But philosophically his case for land nationalization was unchanged. To get to the bottom of this apparent contradiction we need to return to the beginnings of Wallace's embrace of land nationalization: Herbert Spencer. The role of the "synthetic philosopher" in Wallace's thought was complex. According to Wallace, Spencer made him a land nationalizer but, for many years, prevented him becoming a socialist. Yet when Wallace embraced socialism, he justified his conversion on self-consciously Spencerian principles of "justice" and "equality of opportunity" (S545 1900). This was possible because individualism and collectivism lay side-by-side, as uneasy bedfellows in Spencer's thought. It was only as the nineteenth century drew to a close that Spencer self-consciously sought to eliminate all incipient collectivist implications from his writings, and created his enduring image as an arch-individualist. At the same time, however, many of his readers, including Annie Besant, Beatrice Webb, and Jack London were engaged in a collectivist exegesis of Spencer's writings (Stack 2003). Wallace was part of this latter movement. The real connection between Wallace's land nationalization and his socialism was that he presented both in Spencerian terms. His socialism was not a rejection of Spencer but, as he saw it, a case of going "to the root of the matter" and following Spencer's principles "to their logical conclusion" (S729 1905a, 2:253, 272).

VII

Wallace's plan to nationalize the land was presented in precisely these terms; as a practical mechanism through which Spencer's analysis of the land question in *Social Statics* could have been carried into effect. Given this, we might have expected the men's personal friendship to blossom into political cooperation. Yet when, as one of his first acts as president, Wallace invited Spencer to join the LNS in April 1881, Spencer politely declined: "As you may suppose, I fully sympathise with the general aims of your proposed Land Nationalisation Society; but for sundry reasons I hesitate to commit myself, at the present stage of the question, to a programme so definite as that which you send me" (Marchant 1916a, 2:154). Spencer returned to his criticism of Wallace's peremptoriness in July, describing the LNS as "at present premature," and repeating that he was "disinclined to commit to any scheme of immediate action." This lack of enthusiasm, and a respect for Spencer's desire to avoid association with a topic that would provide his enemies with "more handles" against him, explains Wallace's reticence, thereafter, in citing Spencer as an influence (Marchant 1916a, 2:154–55). Whereas Wallace had loudly trumpeted the connection in his *Contemporary Review* piece, Spencer was barely acknowledged in *Land Nationalisation.* A quote from *Social Statics*, along with similar sentiments from Froude, Mill, George, Newman, and Gladstone appeared on the title page, but in the text Spencer was only mentioned in passing and did not merit an index entry.

It was not until his 1892 Presidential Address to the LNS, that Wallace again began citing Spencer's authority. On this occasion, in addition to claiming that it was from *Social Statics* that he had "first derived the conception of the radical injustice of private property in land," he mischievously added that to Spencer was "primarily due the formation of the Land Nationalisation League" (S450 1892, 15). This was calculated to rile Spencer, who for the previous three years had been desperately attempting to disassociate his name from land reform. For while Wallace had been circumspect in citations of Spencer, other land reformers were less restrained. Henry George, in particular, had "extensively and approvingly quoted" from *Social Statics* in both *Progress and Poverty* and his lesser-known work *The Irish Land Question* (1883). This was embarrassing enough for the increasingly conservative Spencer, but matters came to a head in 1889 when *The Times* reported an LNS meeting at which Spencer's name was invoked. Spencer's letter of rebuttal to the newspaper provoked a lengthy correspondence in which Spencer found himself ranged with Auberon Herbert and the Liberty and Property Defence League against Thomas Huxley and a series of disillusioned disciples (Levy 1890). In a desperate attempt to finally sever any connection between his name and land reform Spencer added an appendix on "The Land-Question" to his 1891 book *Justice*, the fourth volume in his *Principles of Ethics*. In the appendix his opposition to land nationalization was as stark as his endorsement of it had been forty years before. All proposals for land reform, Spencer argued, should be rejected as likely to lead to a state "less desirable ... than the present." First, because the levels of compensation payable would be too great; second, because the "violence" of repossession without compensation would be greater than the "violence and fraud" by which the land was taken; and third, because of "the badness" of public administration compared to private management (Spencer 1891, 266–70, 91).

Wallace regarded Spencer's conclusion, "that individual ownership, subject to State-suzerainty, should be maintained," as an apostasy (Spencer 1891, 270). While the possibility had remained that Spencer might be recruited to the LNS, or at the very least remain neutral, Wallace had been careful not to offend a friend—or provoke a potential enemy. Once Spencer had declared himself an opponent, however, Wallace was determined to show that the philosopher had "not refuted his own work, and that it is his later and not his earlier writings that are illogical, and are even inconsistent with the main principles of his own philosophy" (S450 1892, 16). This was to be the constant refrain in all of Wallace's subsequent comments on Spencer. It was a criticism that fed upon a real and obvious change in the tone and content of Spencer's work. The early Spencer had been a vibrant, iconoclastic radical; the later incarnation was a more ponderous character, his prose more prosaic, and his philosophy increasingly hedged in qualifications, nuance, and doubt. Whether this represented the simple betrayal that Wallace, and George, identified, however, is moot (George 1893). Certainly Spencer understood his own "drift to conservatism" not as an apostasy, but as a process of maturing away from "absolute" to "relative" ethics. Even his analysis of the land question in *Social Statics*, it could be argued contained some "rudimentary loopholes," in terms of an acknowledgement of the difficulties of remedying the injustice, "which he later enlarged and, in his own estimation at least, slipped through" (Wiltshire 1978, 121).

From our perspective, the really interesting feature of this is that Wallace and Spencer were able to develop the arguments of *Social Statics* in divergent directions. This seems to confirm the insight of Michael Taylor that the tension between Spencer's inveterate individualism and the collectivist tendencies in his conception of justice, were increasingly played out in the 1880s and 1890s. And, as Taylor noted, nowhere was the tension more obvious than in the case of private property in land, where "justice" seemed to suggest a necessarily collectivist solution (Taylor 1992, 247). Wallace and Spencer, that is, were developing the different potentialities in a shared body of thought. That Wallace himself had some inkling of this can be seen in his dual characterization of Spencer—and to a lesser extent Mill—in *My Life*. Both are charged with delaying Wallace's embrace of land nationalization, with their dire warnings about "the inevitable jobbery and favouritism" that accompanied state administration. Both are also charged with delaying his embrace of socialism, with their "loudly proclaimed dogma that without the constant spur of competition men would inevitably become idle and fall back into universal poverty." Yet at the same time, Wallace credited Spencer—and again to a lesser extent Mill—with a positive role in leading him to towards both land nationalization and socialism. In the case of Spencer, moreover, Wallace justified both land nationalization and socialism by reference to a Spencerian concept of "justice"—although the details of each argument were distinct.

In relation to land reform, Wallace and Spencer's conception of "justice" was an individualist assertion of man's right of access to the means of life. The "justice" that provided Wallace's rationale for socialism, however, was more complex: "The use by every one of his faculties for the common good, and the voluntary organisation of labour for the equal benefit of all" (S545 1900, 521–26). This confirms two elements in our argument. First, that there was no necessary connection between land nationalization and socialism. Second, Wallace's case for land nationalization, even after he became a socialist, rested upon radical, individualist assumptions. This is worth reiterating because it highlights the irony that, in relation to land reform, Wallace was aligned with the radical providentialist Spencer of *Social Statics*, rather than the scientific evolutionist author of the *First Principles*. Thus when he claimed his inspiration for land nationalization was from "the first eminent Englishman of science" Wallace was being disingenuous. Certainly Wallace, like Spencer, believed that ethical truth was "as exact and as peremptory as physical truth," and this insight was essential to his advocacy of socialism (Wiltshire 1978, 123–25). But it had little directly to do with his case for land nationalization. As we have stressed, Wallace made his argument on two bases. First, a natural rights argument for access, which he justified by appeals to Spencer's *Social Statics*, but which had a far longer heritage. Second, a utilitarian claim that land nationalization would benefit society as a whole. The two arguments did not hold equal weight in Wallace's writings: the argument for access was a first principle; the utilitarian claim was of a second order. Indeed, as his 1892 strictures on Spencer demonstrate, if anything he became more wedded to his natural law first principle not less. He was dealing with a question of justice rather than economics, and took offence at Spencer's slippage from a similar position of principle to one of utilitarian expediency.

VIII

If we follow Wallace in regarding access to the land as an inviolable first principle then any question as to the enduring relevance of his ideas is easily answered. Justice and natural rights are eternal, and as relevant today as in Wallace's time. More problematic, however, might be making a positive case for Wallace's utilitarian claims. In Wallace's mind land nationalization assumed the status of a universal panacea. His list of the advantages that would follow its enactment was almost endless: taxation would be abolished; both wages and profits would rise; labourers would become "a body of industrious, honest and sober men"; villages would revive; the production of food would increase; towns would become "less congested"; rents would fall; and so too would crime. In short, addressing "the enormous magnitude of the evils" produced by private ownership of the land would inexorably raise the "general well-being of the whole community" (S722, 229). The basis of this would be a simultaneous restoration of "the land to the people" and "the people to the land." As they were drawn back "from the towns to the country" (S495 1894, 6), Wallace envisaged men living a "natural" existence in which food needs were met locally and regional economic autarky was matched by a seasonal pattern of labour in which the whole population was available at harvest (S512 1895). All of which rather suggests that the supposed benefits of land nationalization depended upon a backward looking rural nostalgia with little contemporary relevance.

There is an element of truth in this criticism. A pastoral setting was the most obvious and appealing antidote to the erosion of independence. Industrial production, by its nature, entailed an ever-diminishing sphere of self-reliance in work—as too did large-scale agricultural production—which could be reversed by reinvigorating small-scale farming. Indeed, even after his conversion to socialism Wallace was always predisposed to agrarian living, which offered a relatively simple solution to his civic humanist horror of specialization. His labour colonies "for solving the problem of the unemployed," moreover, were to be arranged in villages in which work patterns were determined by seasonal and climatic patterns. Instead of agricultural produce being left to rot, at harvest time "the whole population would be available to supply whatever assistance the head-farmer required." Adults were not to have one occupation, but a "pleasant and healthy variety" of which at least one would be sedentary and one active and laborious. This more "natural" sense of balance, compared with the rigid specialization of urban industrialism, was also to apply to children. They would spend no more than 3–4 hours a day on school work, and pass the rest of their time helping with the "simpler agricultural processes" (S512 1895, 12).

Wallace, however, was no simple-minded rural nostalgic. All but eight years of his life had been spent living in the country and however much he eulogized over the need for men to be connected to the land, he never lost sight of the inconveniences, hardships, and injustices often entailed in rural life (Marchant

1913, xxxi). Thus he judged William Morris's *News from Nowhere* "a charming poetical dream, but as a picture of society almost absurd, since nobody seems to work except at odd times when they feel the inclination" (S729 1905a, 2:267). It is true that Wallace's radical protest against centralizing tendencies in politics and the economy sometimes led him—as it had earlier nineteenth-century radicals—into an incorrigible anti-urbanism. This was evident in *The Malay Archipelago*, and remained with Wallace until his death. But this attitude had as much to do with the increasingly pervasive contemporary critique of modernism as it did with any lingering, archaic agrarianism. Indeed, far from making "nostalgia into public policy" (Hunt 2004), the demands of land reformers such as Wallace, and even George, reflected the "continuing urbanisation of politics" by reformulating the land question into one about the economic appropriation of rent and power, rather than of ownership *per se* (Chase 2003).

For Wallace the land question was always an urban as well as a rural issue. From his very first intervention he had raised the issue of private dwelling houses in towns (S329 1880, 281–82), while in *Land Nationalisation* he had argued that the "evils" of landlordism, which by consolidating farms and destroying cottages had driven labourers into the towns, had been "more severely felt in towns and cities" than in the countryside (S722, 210, 213). Moreover, Wallace was able to look beyond "old corruption" and "the evils of landlordism" to locate economic injustice in "the capitalistic system" (S512, 11–12). And even as he railed against "the over-crowding of towns and the depopulation of rural districts," he was able to recognize the social advantages of town and city living. Thus he specified 5,000 as the ideal size for his colonies as this would be large enough to supply "most of the relaxations and enjoyments of the town, such as music, theatricals, clubs, reading rooms, and every form of healthy social intercourse" (S512, 14).

Finally, rather than simply wanting to return men to the land, Wallace wanted to foster a thoroughly "modern" ecological connection between man, nature, and the land. It is in this vision that the contemporary relevance of Wallace's land politics lies. His autarkic communities and economic nationalism seem dated, but rephrased as "locally-sourced produce," "fair trade," and "anti-globalization," they soon acquire a contemporary ring. Wallace's toying with colonialism is less easily excused (S412 n.d.). His arguments for access also lack an immediate resonance, but establishing a "right to roam" is still a live political issue, and "key worker" housing schemes illustrate the enduring need to curtail the private monopoly of the land. As for contemporary movements for conservation and preservation, Wallace can justly be regarded as a pioneer. He would, no doubt, delight in the UN programme of "World Heritage Sites," as much as he would be appalled at the prospect of developers desecrating the Green Belt around London. Drawing direct parallels like this can be a dubious, not to say facile, exercise for the historian. It is especially dangerous with a thinker like Wallace who was consistently open to new ideas and keen to develop and update his own thought and understanding. What we can say with certainty is that if Wallace were alive today he would not repeat the same thoughtless disregard he showed the residents of Llandrindod Wells in 1840. He would be on the side of the poor and the dispossessed as they struggled to be housed, to conserve, or to simply live off the land.

Notes

1. Peter Raby (2001, 218) was mistaken to suggest that Wallace was "disturbed" by his work.
2. Wallace also referred to reading *Social Statics* in 1853 in S450 (1892).
3. It is possible, although there is no evidence, that Wallace might have encountered the land nationalization programme of the Chartist James Bronterre O'Brien; either directly, as a young man, from reading O'Brien's Chartist journalism, or later from the London "O'Brienites" who were active in the 1860s. See Plummer (1971).
4. On the philosophical divergence see Becker (1977) and Waldron (1988).
5. George's single-tax policy was not, as Avner Offer mistakenly characterized it, the "Siamese twin" of land nationalization (Offer 1981, 184–85).
6. Wallace's disappointment in the American dream was familiar trope on the British Left. See Bronstein (1999, 3).

16

Alfred Russel Wallace and Anti-Vaccinationism in the Late Victorian Cultural Context, 1870–1907

Martin Fichman

Introduction

> The successive Vaccination Acts were passed by means of allegations which were wholly untrue and promises which have all been unfulfilled. They stand alone in modern legislation as a gross interference with personal liberty and the sanctity of the home; while as an attempt to cheat outraged nature and to avoid a zymotic disease without getting rid of the foul conditions that produce or propagate it, the practice of vaccination is utterly opposed to the whole teaching of sanitary science, and is one of those terrible blunders which, in their far-reaching evil consequences, are worse than the greatest of crimes (S726 1898b, 315).

Alfred Russel Wallace was one of the most innovative and controversial thinkers of the late Victorian era. In the early 1880s, with characteristic vigour, he launched himself into the centre of the bitter debate over the unpopular and wide-ranging English compulsory vaccination laws. In 1883 Wallace wrote his first public letter denouncing this practice, to the International Anti-vaccination Congress held in Berne. This public denunciation of England's increasingly strict compulsory Vaccination Acts (1853, 1867, 1871) was succeeded by a series of booklets, pamphlets, book chapters, and articles detailing years of intensive research into the question of vaccination's effectiveness and Wallace's original statistical work on the issue (S368 1883, 160; S374 1885; S726 1970 [1898]; S551 1898; RCOV 1890). For a man admired by Charles Darwin, Sir Charles Lyell, Joseph Dalton Hooker, and Charles Sanders Peirce as one of the keenest minds of the Victorian age, Wallace's public conversion to the anti-vaccination camp was a coup d'état for the various English anti-vaccination leagues: his was an authoritative voice in their campaigns.

Yet Wallace's deep commitment to the anti-vaccination movement, and its scientific, ideological, economic, and socio-political dimensions, has remained largely unexplored.[1] His cogent and influential defence of anti-vaccinationism has been relegated to the historical sidelines, as has the pervasive and powerful anti-vaccination movement itself. Recent work, however, by Jennifer Keelan (2004) and Nadja Durbach (2005)—along with a growing body of articles by various authors—has begun to alter dramatically the contours of the magnitude and significance of the vaccination debates (1870–1914). These debates helped redefine the boundaries of medical expertise and state control over its citizens' bodies (Farley *et al.* 1987; Arnold 1993; Hardy 2000; Keelan 2004; Bhattacharya *et al.* 2005; Colgrove 2005; Durbach 2005). Similarly, some historians in recent years have begun to free themselves from the caricature of Wallace as a brilliant scientist who unfortunately "lapsed" into non-scientific or questionable crusades (Scarpelli 1992; Smith 1991, 202–16; Vetter 1999; Moore 1997; Berry 2002; Camerini 2002; Jones 2002; Fichman 2004; Slotten 2004).[2]

It was precisely Wallace's scientific training combined with his social and cultural activism that sharpened his alertness to the flaws in the medical arguments supporting vaccination and the compulsory Vaccination Acts. Assessing Wallace's engagement in anti-vaccination activities is an important step toward understanding the significance of Victorian debates over vaccination. Pro-vaccinationists argued that vaccination was effective, smallpox was ubiquitous, and that the risk of catching smallpox and dying from the disease greatly outweighed the rare complications from vaccination itself. The few deaths from vaccination were necessary to protect the interests of the public. Anti-vaccinationists produced an astute risk calculus of their own: smallpox was neither ubiquitous nor infection inevitable, vaccination did not provide sufficient protection, was as risky as smallpox itself, and there were safer and more reasonable alternatives to a state-enforced compulsory medical intervention (Fichman and Keelan 2007).

Anti-vaccination science had credibility in the public realm and to be an anti-vaccinationist was not necessarily to be anti-science. Of course, the terms "scientist" and "anti-vaccinationist" are themselves somewhat anachronistic for the Victorian period. Science and its specialized disciplines were in the process of becoming professions, but this development was still in its early stages in the Victorian period (Barton 2003, 73–119, esp. 73–74, 100, 108–18).[3] Wallace's conception of the scope of scientific knowledge and inquiry included elements that some of his contemporaries were seeking to dissociate from the nascent and still fluid definition of professional science—such as social, political, economic, religious, and ethical consequences of scientific advances. Wallace differed from a number of his celebrated contemporaries who regarded these latter components as crucial additions to science *in* culture, but not as integral parts of the late nineteenth- early twentieth- century corpus of scientific knowledge itself. Wallace rejected these distinctions and made a key decision, early in his career, to incorporate science into a broader ethical and socio-political framework (Fichman 2004, 4–7).

The logic for compulsory vaccination crossed the permeable boundaries between scientific reasoning and political reasoning (Keelan 2004; Durbach 2005). Since vaccination policy targeted populations, not individuals, it in turn embedded the implementation and enforcement of compulsory vaccination into the social and political machinery of the state. Thus for Wallace any attempt to separate the scientific reasoning for vaccination from the socio-political reasoning would have been meaningless in any practical sense. As the like-minded anti-vaccinationist Walter Hadwen asserted:

> I say that the very moment you take a medical prescription and you incorporate it in an Act of Parliament, and you enforce it against the wills and consciences of intelligent people by fines, distraints and imprisonments, it passes beyond the confines of a purely medical question—and becomes essentially a social and political one (Hadwen 1896).

Wallace's investigations into vaccination reflected his holistic approach to the natural and social world, or what has been described as Wallace's evolutionary cosmology. His evolutionary cosmology—which had its origins in his theoretical and field studies in biology—developed from the 1850s onward into a comprehensive world view. By embedding natural selection within the framework of a theistic evolutionary teleology, Wallace viewed seemingly disparate domains, such as human evolution, spiritualism, land reform, and medical ideas about man's natural habitat, as interconnected (Fichman 2004, 6–7, 204–05). All these, according to Wallace, fell within the proper purview of the scientists. Thus, he was opposed to any simplistic or uni-causal treatment of complex phenomena—including the control of a living evolving disease with a single intervention like vaccination.

In the last decade, historians have increasingly paid attention to the interplay between societal context and individual thought and action, dramatically enriching the historiography of science and medicine.[4] By identifying and examining relations among his most fundamental convictions, Wallace's evolutionary cosmology gains a compelling clarity. Many of the paradoxes and unorthodoxies of which he was habitually charged are seen to fall into his broader and integrated pattern of belief and behaviour. A major benefit of Wallace's evolutionary cosmology was his ability to deploy scientific expertise on behalf of causes that he regarded as indispensable to the definition of an equitable and ethical life/society. A major risk was having certain aspects of his research marginalized by some influential voices in the scientific community—most notably the so-called scientific naturalists (S729 1969 [1905], 2:280; Soderqvist 1996, 49–53, 60–65, 70–74; Raby 2001, 218–22).[5]

While Wallace was committed to science and its methodology as one of humanity's grandest achievements, he also recognized that the unbridled embrace of scientific and technological developments in the name of material "progress" was misguided and potentially destructive. Science, and its increasingly potent industrial applications, had to be tamed. What Wallace termed, with pointed irony, the "wonderful century" was marked by both successes and dangerous failures. Wallace's own definition of individual and societal progress was at odds with some of the most fundamental precepts of Victorian capitalism and imperialism (S726 1970 [1898], vii–ix and 232–34). In *The Wonderful Century: Its Successes and Failures* (S726 1898b), Wallace provides an idiosyncratic but penetrating ranking of the nineteenth century's great breakthroughs as well as its notable defects. Among the glories of the nineteenth century Wallace describes the theoretical and practical accomplishments in physics, chemistry, astronomy, geology, and biology—the latter including both the theory of natural selection and the germ theory of disease and a nascent conception of the body's system of immune defences (S726 1970, 143–49). Alongside these symbols of progress, Wallace lists what he regards as the nineteenth century's most egregious failures: the neglect of phrenology, the opposition to hypnotism and psychical research, militarism ("the curse of civilization"), the plunder of the earth, and mandatory vaccination programmes. The latter constitutes the single longest chapter in the book (110 pages): "Vaccination a Delusion: Its Penal Enforcement a Crime" (S726 1970, 213–323).

Wallace's anti-vaccinationism is notable for two major reasons. First, Wallace developed a convincing critique of some of the most frequently deployed theoretical and, particularly, statistical arguments of the pro-vaccination movement (Fichman and Keelan 2007). Second, Wallace embeds his scientific critique of pro-vaccination statistical methodology within the broader framework of Victorian culture. Wallace opposed those in the emerging medical establishment who promoted closer ties to increasingly state-sanctioned monopolistic and interventionist politics and practices. As he declared on the last day of his testimony before the 1890 Royal Commission on Vaccination, "Liberty is in my mind a far greater and more important thing than science" (RCOV 1890, Question 9654, p. 127).

Vaccination Science and Statistics

Most "men of science" held to some version of the view that smallpox vaccination (if performed competently with good lymph) provided immunity against smallpox. In adopting anti-vaccinationism Wallace was, however, not the only high profile Victorian scientific figure to do so. Charles Creighton (1847–1927), one of the most eminent Victorian epidemiologists, and pathologist Edgar March Crookshank (1858–1928) both brought together several troubling lines of evidence that suggested that the early nineteenth-century experiments on vaccination and immunity were inconclusive in the light of the natural history of smallpox, late nineteenth-century pathology, new bacteriological taxonomy, and contemporary standards of evidence (Creighton 1888, 1889; Crookshank 1889). They were critical of the assertion that Edward Jenner and his contemporaries had successfully performed rigorous "controlled" experiments proving an infection with vaccine lymph protected against smallpox and, in turn, they provided an alternative interpretive framework for the existing epidemiology and vaccine statistics (RCOV 1891, Question 11794, p. 91).

It is not surprising that anti-vaccinationists had a receptive audience for their claims that vaccination was an expensive distraction from more proven approaches to disease control such as rigorous sanitation and quarantine of infectious cases, improvement of personal health through better nutrition, exercise, and a cleaner environment. The statistics produced to support these holistic medical ideas about smallpox were widely available in the form of brochures, pamphlets, and scientific articles. Wallace's contributions to a more appropriate use of statistics in the contentious dispute over vaccination science were innovative, influential in shaping the course of the late Victorian debates, and have had a lasting impact (Fichman and Keelan 2007).[6] Statistics themselves, however, could not resolve the vaccination controversy; cultural arguments (contextualizing the statistics) such as those Wallace provided were also highly significant. Anti-vaccination logic which framed anti-vaccination statistics had an intuitive appeal for those like Wallace who were interested in broad social and political reform. Anti-vaccinationism attracted followers whose interests clustered around several key reform movements, notably those of a philosophical and socio-political character. Five of the most recurring allegiances of many, though not all, anti-vaccinationists were to: (1) social/socialist reformism (including the complex relationships of the politics of anti-vaccinationists to the so-called battle between "old" and "new" Liberals in the last decades of the Victorian era), (2) spiritualism, (3) Swedenborgianism, (4) vegetarianism, and (5) anti-vivisection (Durbach 2005, 41–47). In addition to theoretical and philosophical kinship, these movements also served to provide the anti-vaccinationists with potent models for institutional organization (leagues, organized debates, mass meetings, pressure group tactics). By the 1860s and, increasingly, in the last three decades of the nineteenth century, the anti-vaccination movement operated quite effectively at both national and local levels, with membership that was both significant in numbers and distributed broadly throughout England (Durbach 2005, 38). While it is not possible to identify any of these five affinities as causal agents in turning Wallace toward the anti-vaccinationist cause, it is clear that these allegiances provide an important framework for his statistical research. This cluster of affinities made him open to the arguments that vaccine injuries were under-reported, that vaccination science was supported by a particular interventionist medical tradition that had a long history of making patients worse rather than better, that universal vaccination was a simplistic approach to a complex problem of infectious disease, and that compulsory vaccination placed an unjust burden on the poor and working class (as did protectionist trade and private land ownership). Collectively, these affinities made him suspicious of high-handed government medical experts and any polemically-motivated narrow "scientific consensus," such as support of universal vaccination or the rejection of spiritualism (Lightman 1997; Fichman 2004, Chapter 4).

Wallace, Social Reformism, and the Anti-Vaccination Movement

In the early 1870s, a global pandemic of smallpox swept across Europe and the UK. The epidemic aggravated existing tensions over compulsory vaccination and seems to have further polarized popular opinion about vaccination along class lines. When enforcing vaccination, local magistrates targeted the working class because it was both politically and economically awkward to enforce the law among the gentry (their social superiors).[7] This reinforced the notion that resistance was largely a working-class phenomenon and played on traditional cultural stereotypes that portrayed the working-class as the locus of disease. After the 1853 Act was passed making vaccination compulsory, the legislation was expanded first in 1867 to allow officials to repeatedly fine recalcitrant parents; in 1871, following the great epidemic, every Board of Guardian across the country was required to hire vaccination officers to enforce compulsion. Fines for refusing to vaccinate a single child often surpassed £30, an astronomical sum for a labourer. In the case of working-class resistors, fines were paid by auctioning off the convict's possessions or sending the father to jail—events that evoked sympathy for the anti-vaccinationists (Durbach 2005).

Historians seeking to elucidate the extent and impact of anti-vaccination activism have rediscovered an important and generally overlooked Victorian movement. For example, resistance to vaccination became rapidly absorbed into late nineteenth-century British working-class consciousness and culture and became a touchstone issue for labour activists. Many working-class people came to interpret compulsory vaccination as a violation of their bodies and a form of political tyranny (Durbach 2000, 45). The struggle for workers' ability to literally have control over their own bodies also neatly intersected with reform movements of the late nineteenth century that emphasized the need to nurture self-discipline, temperance, and moral reform among the lower classes. Jennifer Keelan (2004) has convincingly demonstrated that the science and politics of anti-vaccinationism were also significant forces in late Victorian North American, particularly Canadian, cultural history of medicine.

Anti-vaccinationism can be linked to a broad spectrum of progressive reform movements which, taken together, create a fascinating profile of the typical nineteenth-century anti-vaccinationist. Prominent anti-vaccination activists were often simultaneously involved in universal suffrage, early animal rights activism (anti-vivisectionists), and holistic food reforms such as vegetarianism and the whole food movement (whole grain breads versus white flour). They were also supporters of restricting the consumption (among the working class) of expensive luxury items associated with modern urban life: tobacco, coffee, and tea (Keelan 2004, 155, 168, 170–71; Durbach 2005, 41–46, 122–23). Although Wallace never gave up meat entirely, he deemed vegetarianism to be the best diet for humans ultimately (S745 1909, 282).

With respect to anti-vivisection, Wallace believed that there was a great chasm between the capacity of pain/suffering of humans and that of other animals.

Accordingly, his particular opposition to vivisection was motivated by his belief that

> The bad effect on the operator and on the students and spectators remains; the undoubted fact that the practice tends to produce a callousness and a passion for experiment, which leads to unauthorized experiments in hospitals on unprotected patients, remains; the horrible callousness of binding sufferers in the operating trough, after the experiment, by careless attendants, brutalized by custom, remains; the argument of the uselessness of a large portion of the experiments, repeated again and again on scores and hundreds of animals, to confirm or refute the work of other vivisectors, remains; and, finally, the iniquity of its use to demonstrate already-established facts to physiological students at hundreds of colleges and schools all over the world, remains (S732 1910, 381).

It is worth noting that Darwin was both a pro-vaccinationist and not opposed to vivisection. At least one source for this particular set of differences between the two men was their differing evolutionary epistemologies, most notably their divergent views of the limits of natural selection as the main mechanism of evolutionary change (S716 1870, 332–71; Scarpelli 1992, 114, 127; Fichman 2004, 123–26, 200–02).

Anti-Vaccinationism and Limits to State Intervention

Some laissez-faire economists and free-trade proponents such as the English economist Alfred Milnes objected to government-supervised compulsory vaccination because it undermined the promotion of a strong independent citizenry who had an absolute right to choose medical treatment and to determine, in good conscience, the best means to achieve good health for themselves and their children (Milnes 1897). As one Canadian anti-vaccinationist grumbled, "A paternal state was an infernal state" (Ross 1893, 229, quoted in Keelan 2004). However, many anti-vaccinationists held conventional views about the role of government as a necessary agent in certain large public works projects to create a generally healthy living environment. They saw state intervention as the most effective way to ensure a clean water supply, functional sewers, and the restriction of overcrowding in urban housing. Many anti-vaccinationists subscribed to widely held medical theories that contagions were opportunistic and thrived best in the bodies of the young, malnourished, and unclean and spread quickly through densely packed tenement housing. In contrast to Milnes and his followers, these anti-vaccinationists restricted their condemnation of state paternalism to those areas in which government sought to curtail traditional political liberties. (This tension between individual and state continues, to be sure, to the present.) Wallace's own evolving position with respect to both Liberalism and socialism provides one striking case study of the crucial importance of political ideology and activity within the anti-vaccination movement.

Wallace's commitment to obliterating socio-economic and class inequities in Britain dates from his youth (Moore 1997; Jones 2002, 73–96). This commitment became more overt in the period 1870–1900, when Wallace became a leader in the fight for land nationalization and an outspoken advocate for socialism. These socio-political convictions led Wallace to oppose mandatory vaccination schemes vigorously and to regard their penal enforcement as nothing short of a "crime" committed by the state against its citizens, notably its poorer members. Wallace had become an active and influential social reformer in a turbulent period in British history (Fichman 2004, 211–13). The bitter controversy over Irish landlordism that intensified during 1879–80 provoked Wallace to assume a more assertive role in the agitation for land reform. He was now prepared to go public with his plan for more radical and thoroughgoing, and hence lasting and systemic, socio-economic change (Gaffney 1997, 612–13). All land, Wallace proposed, would revert to the state, while the improvements or increased value given to the land—such as buildings, drains, plantations—would remain the saleable property of the present owner (now "state-tenant"). The management of the land would devolve not to the state but to the actual tenant proprietors. The publication of these views in an article in the *Contemporary Review* (November 1880) immediately attracted the attention of those who desired land reform but opposed increased state intervention in land management (S329 1880). The Land Nationalisation Society (LNS), with a programme based on Wallace's principles, was formed in 1881 with Wallace as its president (S729 1905a, 2:239–40). At this stage of his political career, Wallace was still a Liberal, albeit one situated at the extreme left of the Liberal spectrum. He was not yet uncomfortable with Liberal domination of the land reform movement (Offer 1981). *Land Nationalisation: Its Necessity and Its Aims* (S722) was published the following year, 1882. But although he remained committed to the goals of the LNS, Wallace's march toward socialism in the 1880s led him to move beyond even radical Liberal land reform strategies. Wallace adopted an increasingly socialist tone in efforts at this stage to put forth policy suggestions mitigating the widespread pauperism, vice, disease, and crime of large portions of the English labouring classes "which strike foreigners with the greatest astonishment." The fact that many landholders were also magistrates, Wallace argued, further enhanced their power to coerce tenants into conformity with the landholders' own political and religious opinions (S722 1906, 100, 129–35, 176–79). Wallace's views on land reform paralleled his attitude toward vaccination policies: both are evidence of his final move from Liberalism to socialism. A fundamental component of Wallace's conception of socialism was the sanctity with which he clothed the concepts of individualism and personal "home privacy." Wallace specifically underlined the phrase "liberty is as dear as equality or fraternity" in his annotated copy of *Looking Backward*, Edward Bellamy's 1888 socialist utopia novel (Fichman 2004, 250–52)[8] —the work he credited with being the most decisive influence on his espousal of socialism in the last two decades of his life.

Wallace wanted to integrate his deeply-held convictions regarding the sanctity of the individual—particularly with respect to matters of home, family, and health—with his realisation that the state did have a crucial, if circumscribed, role to play in the increasingly complex industrial society of late nineteenth-century Britain. This challenge for Wallace was reflected in the complex attitudes of the anti-vaccination movement(s) to the broader political debates over the nature of "liberalism" and the trajectory of the Liberal Party in the 1880s and 1890s and beyond (Durbach 2005, 69). Anti-vaccinationists opposed the increasingly compulsory and, in their perspective, discriminatory, administrative implementation of vaccination legislation that targeted the poorer segments of the population (with fines and imprisonment), and this spurred some of the most vocally aggressive episodes of the anti-vaccination campaign (Durbach 2005, 36 ff.). Thus, when Alfred Milnes argued that the compulsory vaccination acts were "never demanded by the people" and the London Society for the Abolition of Compulsory Vaccination asserted "the people are with us," populist discourse had become a potent political tool to blunt "the class tensions inherent in the [anti-vaccination] movement. [Such discourse] found its clearest expression in the language of citizenship that emphasized the rights of [all] freeborn Englishmen" (Durbach 2005, 70–71). The opponents of compulsory vaccination could side with Milnes on this point; they were particularly skilful at deploying the concept of the "rights of citizens"—not only to agitate for political and legal equality, but also against infringement on the "sanctity of the home" (Gibbs 1856, 3). Wallace's increasingly socialist critique of the so-called "New Liberals" paralleled in striking manner the critical views of many anti-vaccinationists on the complex transformation of the Liberal agenda in the critical decades of the 1880s and 1890s (Fichman 2004, 328–30; Durbach 2005, 87–89).

Spiritualism, Swedenborgianism, and Anti-Vaccination

To round out the nexus of convictions held by many individual anti-vaccinationists, mention must be made of their spiritualist and Swedenborgian beliefs. During the Victorian period—as, indeed, during most eras, including our contemporary one—medical and religious beliefs and practices were often intertwined. If nineteenth-century anti-vaccinationism is frequently characterized as an alternative or dissenting medical movement, then it is scarcely surprising that religious nonconformity frequently motivated its members. "Many who resisted vaccination did so for deeply held religious, as well as for political or medical reasons ... Nonconformist anti-vaccinators were deeply opposed to the alliance between the state and the established Church of England and likened vaccination to a sacrament zealously enforced on the people. Compulsory vaccination seemed little different from compulsory baptism," these anti-vaccinators argued (Durbach 2005, 44–45). Swedenborgianism and, particularly, spiritualism were potent avenues of dissent in the late Victorian period. Once again, Wallace shared these convictions with many anti-vaccinationists.

Wallace's spiritualist convictions have been one of the most intensively studied aspects of his life, particularly after the late 1860s (Barrow 1986, 186–88; Owen 1990, 131–32).[9] It is pertinent to note, therefore, that Wallace credited the prominent anti-vaccinationist William Tebb, also a spiritualist, as providing (in the late 1870s/early 1880s) a major stimulus to the writing and publication of several of Wallace's tracts defending anti-vaccination (S729 1905a, 2:351–52). A number of Swedenborgians such as William White and James Garth Wilkinson were drawn to the anti-vaccination movement, in part, because the procedure and after-effects of vaccination seemed particularly odious—and avoidable—intrusions into one's body. Within the framework of Swedenborgian theories of the relationship between physical and spiritual health, damaging the body could also endanger the soul. White, a Swedenborgian bookseller, is significant since he, with Tebb (a wealthy merchant) and the pharmaceutical chemist William Young, founded the socially diverse London Society for the Abolition of Compulsory Vaccination (LSACV) in February 1880. The LSACV played a major role in the attempt to bring together middle-class and working-class anti-vaccinationist proponents (Durbach 2005, 39–41, 45–46).[10]

Wallace likely first learned of Emmanuel Swedenborg (1688–1772) from Robert Dale Owen's (the son of Robert Owen) *Footfalls on the Boundaries of Another World* (1861). Dale Owen's book, and subsequent writings, elaborated a tamer version of the doctrines of the eighteenth-century Swedish theologian, scientist, philosopher, and mystic. Dale Owen integrated essentials of Swedenborgianism with spiritualism and political reformism, especially Owenite socialism. Wallace, thus, had an early exposure to Swedenborg that subsequently influenced certain of his own views on the mind–body connection. Wallace, like his contemporaries the American pragmatists William James and Charles Sanders Peirce, drew on elements of Swedenborgianism in fashioning their own philosophies of nature. Wallace's, James's, and Peirce's views, clearly, were more consonant with the findings of nineteenth-century science, notably evolutionary theory (Fichman 2004, 160–61).

For Wallace, a key component of Swedenborg's philosophical system was the "spirituous fluid." Swedenborg's influential two-volume treatise *The Animal Kingdom, Considered Anatomically, Physically and Philosophically* (English trans. 1843–44) is predicated on the assumption that the fundamental "substance of the animal kingdom is the spirituous fluid."[11] In his chapter on "The Human Soul," Swedenborg declares that the basis of both biology and philosophy is the maxim that the spirituous fluid is, also, man's mental life. He asserts that from "the anatomy of the animal body we clearly perceive, that a certain pure fluid glances through the subtlest fibres ... and nourishes [and] actuates ... everything therein" (Swedenborg 1843–44, 2:35, 211, 216, 233). Thus, if the human soul resides in the body, its physical embodiment is the spirituous fluid. Accordingly, anatomy and moral philosophy become integral elements of a more comprehensive world view. Swedenborg further maintained that "if this fluid be regarded as the purest of the organs of its body, and the most exquisitely adapted for the reception of life" then it has a theological as well as biological function. Wallace applauded this notion of the spirituous fluid as the nodal point for integrating the diverse phenomena of nature, including the workings of human physiology (Fichman 2004, 112–17).[12]

Wallace's interest in the metaphysical mind–body connection clearly complemented his understanding of the complex nature of both health and disease. Wallace's view of the cause of smallpox was conventional: smallpox was a contagious disease that spread by overcrowding and an unclean and unsalubrious environment. While anyone was susceptible to the disease, it disproportionately struck the poor, the young, the sick, and the aged. However, the complex relationships between environment and the reproduction of the smallpox contagion in its host, apparent to Wallace, made any simple medical intervention to eradicate smallpox seem naive. The disputed theoretical claims surrounding vaccine-induced immunity made Wallace suspect that the phenomenon was highly unlikely. Any vaccine effect was a product of the wishful thinking of the medical profession (Keelan 2004). There were far more obvious explanations for susceptibility to the disease, and conversely, for barriers or protections from smallpox. The vast social and cultural changes that brought about the Victorian sanitary movement, and improved wages and nutrition for the poor and working class, made it highly improbable that the dramatic decline in smallpox from the mid-eighteenth century to the late nineteenth century had been principally brought about by a medical intervention, vaccination.[13]

While Swedenborgism itself did not lead inevitably to anti-vaccinationism, any more than did anti-vivisectionism or socialism, the constellation of ideologies described above permits for a provisional profile of the Victorian anti-vaccinationist; we can, moreover, locate scepticism about the procedure in the fertile ground of these socio-political and cultural reform movements. These movements shaped how individuals like Wallace approached the contentious issue of the causal relationship between vaccination and epidemic diseases like smallpox. The anti-vaccinationists' world view made certain solutions to the smallpox problem appear more logical than others. This caused them to ask fundamentally different questions from their pro-vaccinationist rivals, and in turn led to distinct empirical and statistical analyses. Indeed, while each side presented "controlled" case studies to support their assertions, without an unambiguous test to measure or demonstrate vaccination's effectiveness, the anti-vaccinationists continued to mount credible statistical critiques of vaccination science. It is clear that anti-vaccinationism at the close of the nineteenth century demonstrates the problematics associated with casting it as anti-science (Keelan 2004). Both pro- and anti-vaccinationists participated in negotiations that were crucial to the early development of this particular branch of medicine by aiding and defining what data were or were not appropriate. By their differing voices and strategies, both sides demonstrate how closely intertwined were culture and Victorian vaccination science (Fichman and Keelan 2007, 604).

Conclusion

Six years before Wallace's death, the anti-vaccinationists won a conscience clause that effectively dismantled the compulsory vaccination laws. Statistical arguments against vaccination—including Karl Pearson's notable critique of typhoid vaccine in 1904—continued to be effective against efforts to expand the vaccination program in England during the first two decades of the twentieth century (Hardy 2000). It was only with the advances in immunology and serology between World Wars I and II—in particular, the use of anti-toxin to combat diphtheria—that the central disputes that plagued Victorian vaccine science, statistics, and epidemiology *seemed* to be resolved. The development of sulfa drugs and antibiotics played a similar role in making the ideology of universal vaccination (or a magic bullet) to combat infectious disease close to a cultural given for subsequent generations of both scientists and the public in the second half of the twentieth century.

These critical advances in twentieth-century immune therapies and the rise of bacteriological medicine and serology, however, have tended to minimize, or obscure the nature of, the problems involved with claiming a causal relationship between the implementation of vaccination and the decline of smallpox in the nineteenth century. In general, medical historians, until only very recently, have downplayed or dismissed the powerful arguments made by figures like Wallace, Creighton, and Crookshank: the impressive success of twentieth-century vaccination programmes cast a long shadow over its more shaky beginnings. This essay has explored the sources and strength of the sociocultural roots of Victorian anti-vaccinationism, focusing specifically on their influence on Wallace. Keelan has demonstrated the potency of the scientific, particularly statistical, critiques put forward by the Victorian anti-vaccinationists (Keelan 2004, 2005). It is now clear that science and culture served the architects of both sides of the vaccination debates.

For Wallace, statistics was a formidable tool in dealing with the massive, and often conflicting, data that confronted pro- and anti-vaccinationists alike. His critique was twofold. First, he successfully demonstrated that many of the statistical assumptions upon which the pro-vaccinationists rested their theoretical and empirical claims—as well as the adjoining policy recommendations—were deeply flawed. For example, the fact that most of the pro-vaccinationist data used highly selected subpopulations (*e.g.*, smallpox hospitals and state prisons) meant to Wallace that the impact of vaccination was kept artificially independent of other relevant variables such as class, age, and general ill health. Second, Wallace approached the raw data of vaccination with refined statistical "categories" and models that proved more appropriate than some of those used by the pro-vaccinationists (Keelan, in Fichman and Keelan 2007, 596–89, 603; see Fig. 28). He was a particularly prominent voice on the anti-vaccinationist side precisely because he brought both brilliant scientific credentials and a cultural authority to these debates.

Wallace's critique of vaccination science and *its* particular socio-cultural framework has been reborn in the complex modern biopolitics of universal vaccination.

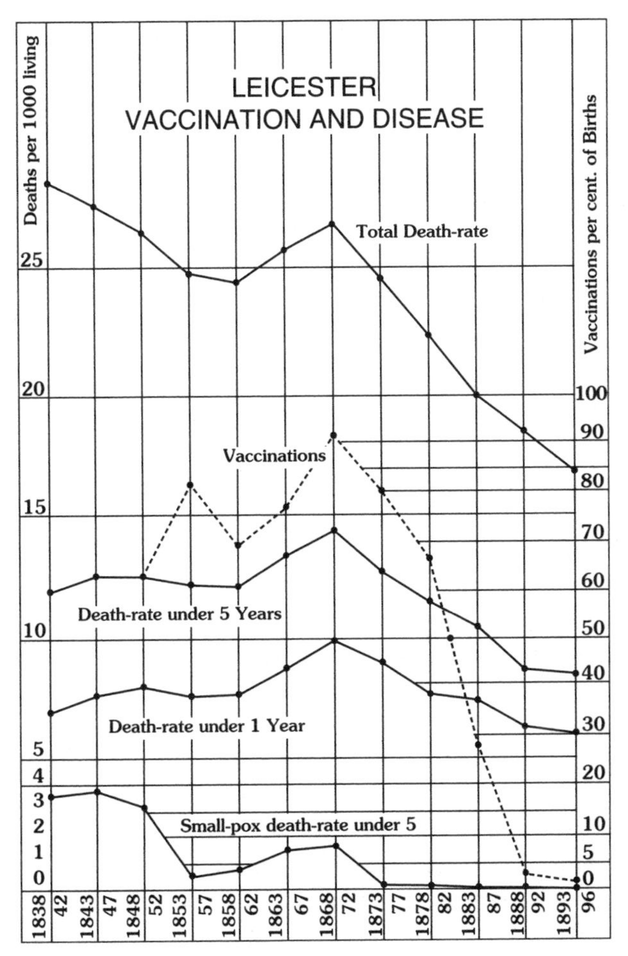

Figure 28 Graph (redrawn) showing relationship between vaccination and incidence of smallpox in Leicester, UK, from Wallace's book, *The Wonderful Century.*

Present day anti-vaccinationists emphasize, among other factors, vaccine toxicity and side-effects, the troublesome link between vaccines and pursuit of financial profits by the massive pharmaceutical industry, and the age shift in infectious disease occurrence (*i.e.*, that vaccine induced immunity tends to wear off, as opposed to natural infection which tends to induce lifelong immunity—thus resulting in a larger population of non-immune adults who, as research shows, have severe effects from diseases that in children are usually mild; Link 2005, esp. Chapter 4). Vaccination critic Barbara Loe Fisher's writings demonstrate a profound suspicion of the objectivity of medical research and the medical-commercial complex. At its core is a critique—similar to Wallace's—of the aims and goals of universal vaccination as *the* method to combat contagious disease (Coulter and Loe Fisher 1991). Further, popular sentiment against vaccination tied to arguments structurally similar to Wallace's and his contemporaries have resurfaced since the 1980s. Resistance has generated public debate over the safety and necessity of routine childhood vaccination and has had a significant impact on public health policy. It is still technically difficult to provide a clear demonstration of the effectiveness of vaccination using statistics derived from the poorly controlled setting of human populations. Moreover, the optimistic predictions of the end of infectious disease prevalent in the mid-twentieth century have, in the last twenty-five years, been dramatically deflated. Three developments in particular have chastened the pro-vaccinationist mantra (Link 2005, 124–26): the resurgence (especially in the developing world) of epidemic diseases once deemed eradicated by vaccination, novel or emerging infectious agents such as Ebola or HIV, and highly adaptable germs (antibiotic resistant tuberculosis and influenza). The underlying rationale of universal vaccination (generating specific immunity to a specific disease contagion) seems simplistic and its safety increasingly suspect in many sectors. As we become more familiar with the protean nature of many pathogens and their co-evolution with the environment, Wallace's evolutionary cosmology appears increasingly germane. Diseases with simple external causes and cures are understood to be the exception to the rule and multifactorial paradigms are now dominant in the culture and science of medicine (Fichman and Keelan 2007, 605). Doctors, biologists, and public health authorities, among others, invoke—as did Wallace—metaphors drawn from environmentalism and molecular biology where the boundary between humans and their environment is more permeable. Wallace's anti-vaccinationism—as many of his other positions analysed in the other essays in this volume—seems as insightful at the start of the twenty-first century as it did in the Victorian era. His approach to the complex issues surrounding vaccination, moreover, is reflective of Wallace's broader commitment to an evolutionary and ecological perspective on the "world of life."

Notes

1. A notable exception is Scarpelli (1992). The subject is also treated, albeit only briefly, in Slotten's recent lengthy biography (2004, 422–36).
2. Wallace's opposition to mandatory vaccination programmes was shared by a considerable portion of the population in both Europe and North America. Wallace's name and formidable power of argumentation became an important tool for the anti-vaccination movement. His writings generated considerable interest among both the proponents and detractors of anti-vaccination. Wallace's sophisticated statistics-based critique of the medical efficacy of, and dubious public health safeguards relating to, vaccination are briefly discussed in Smith (1991, 202–16), Scarpelli (1992), Keelan (2004), and most fully by Jennifer Keelan in Fichman and Keelan (2007).
3. The term "man of science" was far more widely employed than the terms "scientist" or "professional scientist" throughout the Victorian era. If one refers to members of the emerging scientific community as "men of science," a more inclusive/less exclusive

signification becomes appropriate. This, indeed, is what the "men of science"—and their publics—understood to be the case.

4. A representative sample of the growing contexualist literature on Victorian science history includes: Jardine *et al.* (1996); Lightman (1997); Desmond (2001); Golinski (1998); Secord (2003); Barton (2003); Lightman (2004); Fichman (2004); Endersby (2008).
5. It must be emphasized, however, that the monolithic term *scientific naturalism* has been deconstructed by recent scholars. T. H. Huxley and John Tyndall, to cite two of the most polemic and well-known spokesmen for Victorian scientific naturalism, have been shown to have far more complex views on the nature of science (especially with respect to the interaction between science and religion/ethics) than traditionally has been portrayed. See, *e.g.*, Barton (1987).
6. Jennifer Keelan provides a detailed and precise analysis of Wallace's statistical contributions in Fichman and Keelan (2007, 596–604).
7. Various working-class resisters described a very uneven application of the penal code and argued that it was largely the working class that was targeted while the magistrates allowed upper-class citizens to not conform to the laws if they chose. See the testimony contained in the sixth report of the RCOV (1897).
8. Wallace's annotated copy of *Looking Backward* is in the Alfred Russel Wallace Library, special collections, at Edinburgh University Library; the annotation is on pages 137–38 of Bellamy's book.
9. On the link between spiritualism and anti-vaccinationism generally see Barrow (1986) and Owen (1990).
10. The LSACV dissolved in 1896 to form the National Anti-Vaccination League (NAVL), the same year in which the Royal Commission on Vaccination was releasing its final report (RCOV 1897). "The NAVL was an alliance of local leagues across the United Kingdom that sought to combine their funds and efforts to present a united front at a key moment in the anti-vaccination campaign" (Durbach 2005, 40).
11. Garth Wilkinson, anti-vaccinationist, was the editor and translator of this English edition. See Swedenborg (1843–44).
12. His Swedenborgian concept of the spirituous fluid that links all bodies as part of the unity of life lent itself, also, to feminist approaches to humanitarian and animal-rights issues. Significantly, toward the end of the nineteenth century a number of middle-class female social reformers began to absorb anti-vaccinationism into their feminist platform (Durbach 2005, 45–46). Wallace, though holding on to certain patriarchal tenets of Victorian culture, was more supportive of feminism and the cause of women's rights than were many of his male contemporaries (Fichman 2004, 276–79).
13. Anne Hardy (1983) argued that the decline of smallpox was caused by a complex series of bureaucratic, sanitary, and administrative public health technologies, though she clearly acknowledges a role for vaccination.

17

The Universe and Alfred Russel Wallace*

Steven J. Dick

Alfred Russel Wallace is best known for his work as a naturalist and evolutionist, and for his general interest in life on Earth, taking "life" in the broadest sense of the word to include both biology and culture. So it comes as a surprise to many to learn that he had an early fascination with astronomy, remained "deeply interested" in astronomical discoveries throughout his life, and late in life wrote two books on one of the most sensational and important aspects of the subject—life on other worlds. This raises stimulating questions for the historian of science: Were Wallace's astronomical views incidental or fundamental to his life and thought? If the former, why did he write those books? And if the latter, just how did his astronomical ideas fit into his own world view? And did the influence in developing his world view go from astronomy to biology or from biology to astronomy?

In part the answers to these questions are to be found in Wallace's ideas on purpose in the universe, ideas that relate to what we would today call anthropic reasoning and intelligent design. These ideas would culminate with his book *Man's Place in the Universe: A Study of the Results of Scientific Research in Relation to the Unity or Plurality of Worlds* (S728). Published in 1903, that book sheds unique light on Wallace's anthropocentric world view, and illuminates his entire career, including his belief that the human mind must be set off from animals in the evolutionary process. In accordance with the world view expressed there, in 1907 Wallace also wrote a more narrowly focused book on the habitability of Mars. Taken together, these works tell us about Wallace's views of the universe and its relation to humans, placing his better-known work in a much broader context. And the considerable reaction to those works reveals how others viewed Wallace and his world view, both in its terrestrial and extraterrestrial aspects.

* Parts of the section entitled "Man's Place in the Universe" are reprinted with permission of Cambridge University Press from Chapter 2 of Steven J. Dick, *The Biological Universe: The Twentieth Century Extraterrestrial Life Debate and the Limits of Science* (New York: Cambridge University Press, 1996).

Wallace and Astronomy

In Chapter 8 of his autobiography (S729 1908, 101–02), Wallace spoke of an early interest in practical astronomy. He had learned the use of the sextant for surveying, aided by a book on nautical astronomy borrowed from his brother. When he was about eighteen, he had constructed a small telescope, with which he observed the Moon, Jupiter's satellites and star clusters. These activities, he noted, "served to increase my interest in astronomy, and to induce me to study with some care the various methods of construction of the more important astronomical instruments; and it also led me throughout my life to be deeply interested in the grand onward march of astronomical discovery." But, Wallace continued in the next sentence, "what occupied me chiefly and became more and more the solace and delight of my lonely rambles among the moors and mountains, was my first introduction to the variety, the beauty, and the mystery of Nature as manifested in the vegetable kingdom."

Thus did Wallace concentrate on his great life's work in biology and natural history, work that at times was physically rigorous and required the vigor and carefree attitude of youth. Contemplating the bigger picture of the universe as an octogenarian did not require fieldwork, simply a view of the heavens overhead, a voracious appetite for reading about the latest developments in astronomy, and correspondence with some of the leading astronomers of the day.

Wallace's numerous recent biographers barely mention his interest in astronomy, yet hint at the importance it might have held for his overall world view. In a single paragraph Raby (2001, 271–72) notes that "All Wallace's subjects, however disparate they might seem, or driven by circumstances, were closely connected, first teasing out the relationship between man and the rest of animate life, and then moving on to consider the physical conditions that gave rise to life in time and space; or seeking systems to improve the ways individuals and races shared the earth's resources more fairly, so that moral and intellectual progress, and happiness, could follow." It was logical, Raby continues, for Wallace to address the bigger issues of astronomy that he had broached in *The Wonderful Century: Its Successes and Its Failures* (S726 1898b), and this he did in *Man's Place in the Universe* (S728 1903), published ten years before his death in 1913. The thesis of the latter book, Raby tells us, is "in essence, a philosophical or theological, assertion of intelligent cause and design over chance ... Wallace would not accept that life was accidental, because he refused any explanation whose corollary was that man would die out by the continued operation of the same laws that had allowed man to evolve in the first place." As a kind of afterthought, or specific instance of this belief, Raby notes, Wallace's much slimmer volume *Is Mars Habitable*? (S730 1907) argued against Percival Lowell's claim that intelligence exists on Mars.

Slotten (2004, 457–61) is only a bit more expansive in his biography of Wallace, devoting a half dozen of its 500 pages to Wallace and astronomy. After reviewing Wallace's arguments in *Man's Place in the Universe*, Slotten summarizes its thesis as follows: "that human beings, the culmination of conscious organic life, had arisen on our planet alone in the whole vast material universe. If human beings were the unique and supreme product of this vast universe, then some controlling Mind or Intelligence had conceived us for this very purpose. The immensity of the stellar universe, the long and slow and complex progress of nature, and the vast aeons of time that had passed before our development served as the raw materials and the spacious workshop for a Mind that 'produced' the planet that eventually resulted in humankind."

Despite the rejection of Wallace's arguments by some astronomers, Slotten suggests that the state of astronomy was unsettled enough at the time to make Wallace's position plausible, thereby giving solace to Wallace's refusal to believe that "humanity, with its faculties, aspirations, and powers for good and evil, was a simple by-product of random forces—that human beings were merely animals of no importance to the universe and requiring no great preparation for their advent." Like Raby, Slotten sees Wallace's Mars treatise as an expansion of these arguments. Underlying Wallace's argument against life on Mars, he says, "was an idiosyncratic view of the origin of the universe—one might call it teleological evolution, the belief that a controlling Mind or Intelligence manipulated natural laws for distinct ends. It was this belief, more than anything else, that had made so many of his scientific contemporaries uncomfortable. Essential to his theory was his anthropocentrism—a belief in humanity's unique position in the universe. Nothing, barring the discovery of intelligent life elsewhere in the universe, could change his mind on that issue" (Slotten 2004, 474–75).

Yet another recent biography (Shermer 2002) is similarly brief on Wallace and astronomy, but its author emphasizes how the basis for *Man's Place in the Universe* stretched back to Wallace's early days, and stretched forward to one of his last books, *The World of Life: A Manifestation of Creative Power, Directive Mind and Ultimate Purpose* (S732 1910). Wallace's *raison d'être*, in Shermer's view, was "a belief in a purposeful cosmos that under the direction of a higher intelligence inexorably led to the appearance of humans who were capable of perfectibility and would, in time, achieve immortality of spirit. It was a consilient worldview that tied together his many and diverse interests and commitments, ideologies and philosophies, and was ultimately grounded in a unique form of Wallacean scientism" (Shermer 2002, 230–31). Like Raby and Slotten, Shermer connects *Is Mars Habitable?* with his earlier work: "Because Wallace had already committed himself several years earlier to the position that humans are unique in the cosmos, he could not let such apparent contradictory evidence be presented without a challenge" (Shermer 2002, 295).

Considering the fundamental nature of the questions these biographers raise in connection with Wallace's astronomical writings, it is surprising those works have not been subjected to more detailed analysis. Though these writings came late, we shall argue that the ideas expressed in his two astronomy books are integral to his thought, and that the influence was from biology to astronomy rather than the reverse. Despite the fact that *Man's Place in the Universe* was written shortly after Wallace had moved houses, was desperate for money and verging on bankruptcy (Raby 2001; Slotten 2004)—and that a book on the popular subject of life on other worlds would help him financially—much more than money was involved. As Slotten noted, though the fragile state of Wallace's finances served as the proximate inspiration for his writings on astronomy, this was true of much of his work. Wallace himself noted in his autobiography "I feel that without the spur of necessity, I should not have done much of the work I have done" (Slotten 2004, 457–58). For Wallace the heavens and the Earth were united by an anthropocentric world view, teleological evolution, and a special role for humans in the great chain of being.

Man's Place in the Universe

In tackling the astronomical problem of life on other worlds Wallace was tapping into a tradition well known in the world of the naturalist. "The question of questions for mankind—the problem which underlies all others, and is more deeply interesting than any other," T. H. Huxley wrote with regard to Darwin's theory of evolution, "is the ascertainment of the place which Man occupies in nature and of his relations to the universe of things" (Huxley 1971 [1863], 71). Wallace certainly knew of this passage in Huxley's famous work *Man's Place in Nature*. And he certainly knew also of the larger plurality of worlds tradition that placed this question in the context of the larger universe. This controversy (Crowe 1986; Dick 1982) has a long tradition, and had raged especially in Britain since the publication of William Whewell's book on the subject at mid-century (Whewell 2001). That work went through five editions by 1859, with another printing in 1867. Concerned about the implications for Christian doctrine—especially redemption and incarnation—Whewell had argued that the Earth was the only world with life. Quickly disposing of one of the chief philosophical arguments of the pluralists, the teleological argument that the vast spaces of the universe must have some purpose, Whewell argued that confining intelligence to the "atom of space" that was the Earth was no worse than confining humanity to the "atom of time" that geology revealed intelligence had existed on Earth. On the more empirical side he argued that no proof existed of other solar systems, that the stars might not be exactly similar to our Sun, and that in any case many of them were binary stars whose putative planets would therefore not have conditions conducive for life. In our own solar system, he argued, only Mars approached the conditions of the Earth, and it was just as likely as not that Mars was still in a condition of "preintelligence." Finally Whewell cautioned against the unbridled use of the analogy argument in science. Crowe (1986) has discussed Whewell's pluralism in great detail, and Heffernan (1978) has compared the arguments of Whewell and Wallace. Although the nineteenth-century philosopher and the nineteenth-century naturalist differed in some of their arguments, particularly those having to do with purpose in the universe, they agreed in their conclusion that humans were likely the only intelligent life in the universe.

Both Whewell's and Wallace's antipluralist positions were very much in the minority in the context of nineteenth-century science and natural theology. Whewell's treatise generated a tremendous amount of debate, but did little to weaken support for a plurality of worlds among scientists or the religious. Crowe (1986) documents twenty books and some fifty-four articles and reviews written in response to Whewell; of these about two-thirds still favored pluralism despite Whewell's arguments. Treatises such as Sir David Brewster's *More Worlds Than One: The Creed of the Philosopher and the Hope of the Christian* (1854) continued to be driven by an attachment to teleology and natural theology. It went through ten printings by 1871, and a third English edition in 1874 was reprinted in 1876 and 1895. By that time at least seven editions of Richard Proctor's *Other Worlds Than Ours* (1870) had appeared. Although reconciliation with the doctrines of incarnation and redemption was never achieved, for most the benefits to natural theology overwhelmed Whewell's objections.

All of this, and its connections to the problem of anthropocentrism, did not escape Wallace's notice. As an evolutionist it was only natural that Wallace, like Huxley, should be interested in the issue of humanity's place in nature, and the plurality of worlds tradition allowed him to explore the subject on the broader canvas. Wallace had become aware of recent astronomical advances while writing four chapters for *The Wonderful Century* (S726 1898b), and it is in this work that we find the source of his anthropocentric view of the universe. Here Wallace related his amazement at discovering the view of John Herschel, Simon Newcomb, and Sir Norman Lockyer that our Sun was situated near the center of the Milky Way system. Other research had shown that this system was finite, implying that our Sun and its accompanying planets were situated in the center of the entire universe. The startling fact of this privileged position, together with the indication based on planetary environmental conditions that the Earth is the only inhabited planet in the solar system, led Wallace to wonder whether the Earth is the only inhabited planet in the whole universe. In addition, Wallace added,

> For many years I had paid special attention to the problem of the measurement of geological time, and also that of the mild climates and generally uniform conditions that had prevailed throughout all geological epochs; and on considering the number of concurrent causes and the delicate balance of conditions required to maintain such uniformity, I became still more convinced that the evidence was exceedingly strong against the probability or possibility of any other planet being inhabited (S728 1903b, v–vi; see also S741 1903).

Wallace was particularly influenced by the British astronomer and historian Agnes Clerke's books *The System of the Stars* (1890) and *History of Astronomy* (1885), the American astronomer Simon Newcomb's *The Stars* (1902), Sir John Herschel's *Outlines of Astronomy* (1869 edition), and by the work of Lord Kelvin.

As early as March 1901 Wallace had corresponded with Clerke about the possibility of a book on the subject of other worlds. But first would come an article published simultaneously in Britain and the United States in February and March 1903 (S602), six months before the book appeared in the autumn of that year. Beginning with the Copernican system and continuing with the vast Newtonian universe and the revelations of larger telescopes, Wallace argued that the importance of the Earth and its inhabitants had diminished. He argued that our Sun is located in the center of a cluster of suns, itself at the center of a finite stellar universe, and that only this central position in the stellar universe is suitable for life.

His striking conclusion was:

> that our position in the material universe is special and probably unique, and that it is such as to lend support to the view, held by many great thinkers and writers to-day, that the supreme end and purpose of this vast universe was the production and development of the living soul in the perishable body of man (S602, 474).

By the end of August 1903 Wallace's book was ready for the press. The manuscript of the book, preserved in the British Library in London, tellingly shows that the original title "Universe for Man" was deleted and replaced with "Man's Place in the Universe." The book's "connected argument" for the Earth as the unique home of life in the universe began only after five chapters on the background of contemporary astronomy, chapters that Clerke later praised as a brilliant summary of the latest results in astronomy (Clerke 1904). From that point on, Wallace's main argument for the uniqueness of the Earth is based on three indispensable arguments: (1) life can exist only around our Sun or the cluster of suns surrounding it; (2) no life exists on planets around other suns in the solar cluster; and (3) no life exists in our solar system beyond the Earth.

It is no exaggeration to say that the first argument was the dominant and catalyzing argument for the whole book. In supporting it, Wallace claimed that he was simply espousing the view of the most eminent astronomers of his day, a claim exaggerated but not completely wide of the mark. When Wallace wrote in 1903, all stars, and indeed all observable phenomena in the universe, were widely believed to be part of a single system, physically associated through the gravitational force, perhaps several thousand light years in diameter (compared to the 100,000 light years now estimated), with the Sun in a nearly central position. The "island universe" theory, which postulated many such systems, had been in gradual decline since the 1860s and had completely fallen from favor by the late 1880s (Smith 1982; Berenzden Hart and Seeley 1976).

It is therefore not surprising that Wallace viewed the universe as a single system of stars with our solar system at the approximate center. From the equality of star counts on both sides of the plane formed by the Milky Way—"the fundamental phenomenon upon which the argument set forth in this volume primarily rests"—Wallace argued that we reside in the central part of this plane and that the stellar

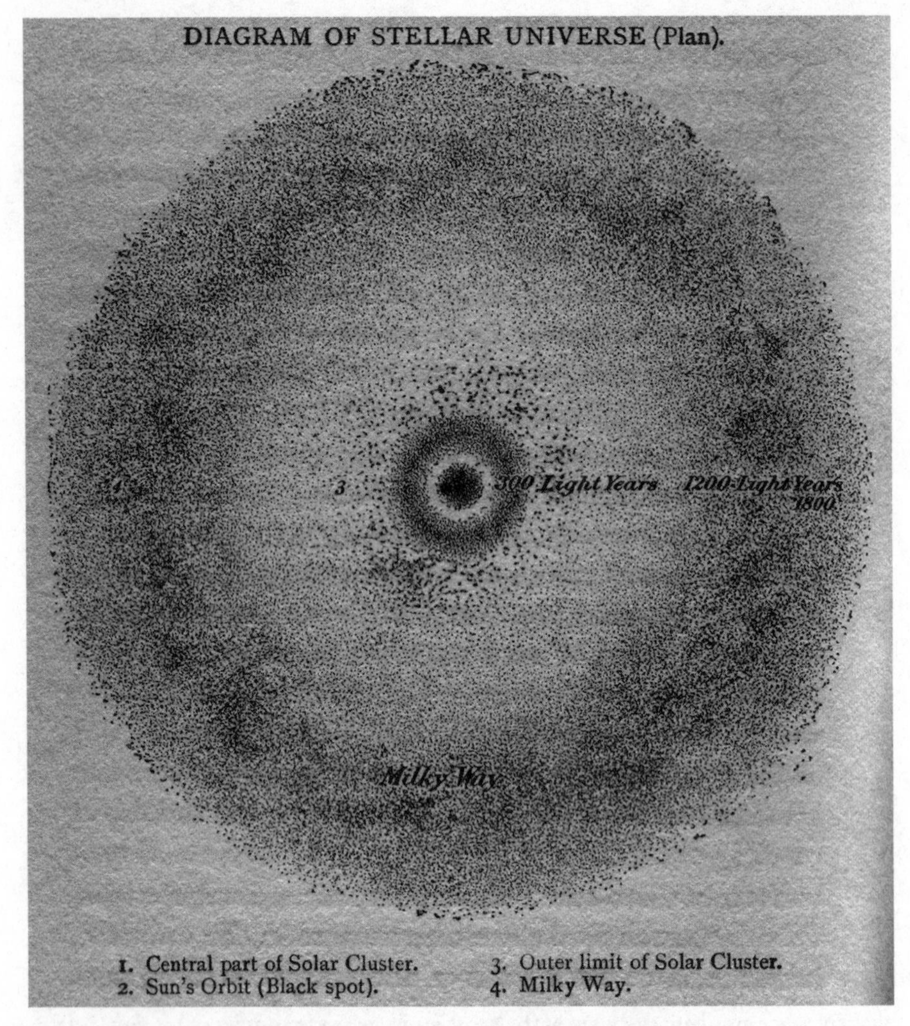

Figure 29 Diagram of the Stellar Universe from Wallace's book *Man's Place in the Universe.*
Out of copyright.

system is spherical. Embedded in this general spherical structure of stars was a "solar cluster" consisting of a group of several hundred to several thousand stars surrounding our sun, and which seemed to form a condensed group of stars separate from the rest of the Milky Way. Wallace set the dimensions of the entire system at 3,600 light years, placed the Sun at the outer margins of the solar cluster, and put it in orbit around the center of gravity of the cluster (Fig. 29).

The centrality of the Sun was "the very heart of the subject" of Wallace's inquiry for more than philosophical reasons, for on that position rested much of his argument against other worlds. The importance of centrality for Wallace was that only in such a position could a star such as the Sun generate a uniform heat supply over the long period needed for the original development of life on Earth, which he put at several million years. The origin of the Sun's power was a long-standing problem whose solution lay two decades in the future with an understanding of atomic processes in the Sun. Wallace's own theory must be judged to be as plausible as any other at the time. Drawing on Lord Kelvin's meteoric hypothesis, he argued that the solar cluster would, through gravitation, draw inflowing matter toward it from the outer region of the star system. This inrushing matter would furnish the energy for the formation and maintenance of luminous suns. Because this process was gravitationally most efficient near the center of the system, stars in other regions would presumably have much shorter lives, not sufficient for the development of life. This view of the universe thus reduced to perhaps several hundred the number of candidate stars that might harbor life-bearing planets.

In arguing against planetary systems around stars of the solar cluster, Wallace played a game of successive elimination. Some stars would be too small or too large for the appropriate long-term heat required for the development of life. Others were just in the process of forming. Others, recent research had shown, were gravitationally bound double stars inimical to the development of life. Even if planets existed around a few stars, they might not be the proper distance from the Sun, or have the proper mass. Looking at all these required conditions for life, Wallace concluded, "I submit that the probability is now all the other way" (S728 1903b, 278–85).

This brought the problem down to our own solar system, and here Wallace launched into a detailed discussion of the complexity of life on Earth, which he saw as a prerequisite for understanding the possibility of life beyond Earth. Here Wallace brought to bear some of the problems on which he had spent a lifetime of contemplation, thus offering a unique view. As Clerke commented, "No great biologist has ever before seriously considered the possibilities of cosmic life, and they can only be fully discussed in the light of expert biological science" (Clerke 1904). This was a difficult task; in many ways biology did not even become a science until the nineteenth century, in part through events surrounding Darwin and Wallace themselves, and the study of physiology was still at the organic rather than the microscopic level (Allen 1975).

Nevertheless, Wallace plunged ahead with one of the greatest questions of all: What is life? Rejecting philosophical definitions given by Aristotle, Spencer, and others as vague and abstract, and dissatisfied as well with definitions based on general properties, Wallace quickly focused on protoplasm, which Huxley had called "the physical basis of life." He marveled that it is composed basically of only the four elements hydrogen, carbon, nitrogen, and oxygen, and that the variety of life is produced by the amazing ability of carbon to form compounds. The chemical reactions in the protoplasm, Wallace argued, determined the physical conditions necessary for the development and maintenance of life on Earth: a regular heat supply, resulting in a limited range of temperatures; a sufficient amount of solar light and heat; water in great abundance and universally distributed; an atmosphere of sufficient density; and alterations of day and night to keep the temperature in balance.

Life, Wallace believed, must have a temperature range from 32 to 104 degrees Fahrenheit, while solar light is necessary for its essential role in decomposing carbon dioxide into carbon and oxygen in plants. Water not only constitutes a large proportion of the living body of both planets and animals, it is also essential in producing a limited range of temperatures. And the atmosphere must be of sufficient density to store heat and to supply oxygen, carbon dioxide, and water vapor for life; he estimated that a reduction of the density by one-fourth would probably render the Earth uninhabitable. Thus, any change in this great variety of conditions would be inimical to life. In order to achieve these conditions on a planetary scale, the essential factors were the distance from the Sun, the obliquity of the ecliptic (causing the seasons), the persistence of a mild climate through geological time, and the distribution of water.

The question, then, is whether this complex combination of conditions is found beyond the Earth. Wallace leaves no doubt of his opinion: considering all the causes "it seems in the highest degree improbable that they can *all* again be found combined either in the solar system or even in the stellar universe" (S728 1903b, 310). Pointing out that there was no reason to believe life could thrive except under conditions similar to those on Earth, Wallace concluded "We may therefore feel it to be an almost certain conclusion that—the elements being the same, the laws which act upon and combine and modify those elements being the same—organized living beings wherever they may exist in this universe must be fundamentally, and in essential nature, the same also." And as a corollary "Within the universe we know, there is not the slightest reason to suppose organic life to be possible, except under the same general conditions and laws which prevail here" (S728 1903b, 182–89).

Turning to the planets of our solar system, Wallace next argued that planetary habitability depends primarily on the mass of the planet, for this determines whether or not it can retain the molecules that compose an atmosphere. Thus Mercury and Mars are too small to retain water vapor. Because of their low density the larger planets "can have very little solid matter" on which life might develop. This left only Venus, but Venus was believed to always keep the same face toward the Sun, so one side is too cold, the other too hot. (This is now known not to be true; Venus is uniformly hot at a temperature of about 900 degrees Fahrenheit due to the greenhouse effect of its atmosphere—thus ruling out life in any case.) Moreover, the lifetime of the Sun renders it impossible that these planets now unsuitable for life might have been suitable in the past or could be in the future:

> We are, therefore, again brought to the conclusion that there has been, and is, no time to spare; that the *whole* of the available past life-period of the sun has been utilised for life-development on the earth, and that the future will be not much more than may be needed for the completion of the grand drama of human history, and the development of the full possibilities of the mental and moral nature of man (S728 1903b, 275).

Although this ended Wallace's "connected argument," he also put forth very briefly a philosophical argument from purpose. If the Earth is the only inhabited planet in the universe, Wallace asserted, this may be seen either as a coincidence or as a very important conclusion indicating that the universe was brought into existence for the ultimate purpose of the development of man on Earth. While Wallace believed the majority of scientific men would call it coincidence, he left no doubt that he was of the opposite opinion, and that this opinion was contrary neither to science nor to religion. Life on every planet

> would introduce monotony into a universe whose grand character and teaching is endless diversity. It would imply that to produce the living soul in the marvellous and glorious body of man ... was an easy matter which could be brought about anywhere, in any world. It would imply that man is an animal and nothing more, is of no importance in the universe, needed no great preparations for his advent, only perhaps, a second-rate demon, and a third or fourth-rate earth (S728 1903b, 317).

This conclusion was diametrically opposed to natural theology, which had consistently argued for two centuries that other worlds would declare the glory of the Creator. By contrast, Wallace believed the immensity of space and time "seem only the appropriate and harmonious surroundings, the necessary supply of material, the sufficiently spacious workshop for the production of that planet which was to produce, first, the organic world, and then, Man" (S728 1903b, 317–18). Although isolated and briefly stated, this argument from purpose perhaps played a greater role as a driving force than Wallace would admit. As Heffernan (1978, 96) concluded, for both Whewell and Wallace "their private interest in plurality sprang from extra-scientific convictions as to the purpose of the universe." The same is true not only of their interest, but also of their conclusions.

A final argument, "An Additional Argument Dependent on the Theory of Evolution," was added to the 1904 fourth edition of Wallace's book. Especially interesting because Wallace was so closely involved with the evolution arguments of his day, it is independent of the three connected scientific arguments and may be seen as another aspect leading to the same conclusion. Wallace argued that since humanity is the result of a long chain of modifications in organic life, since these modifications occur only under certain circumstances, and since the chances of the same conditions and modifications occurring elsewhere in the universe were very small, the chances of beings in human form existing on other planets was very small. Moreover, since no other animal on Earth, despite the great variety and diversity of forms, approaches the intelligent or moral nature of humanity, Wallace concluded that intelligence in any other form was also highly improbable:

> If the physical or cosmical improbabilities as set forth in the body of this volume are somewhere about a million to one, then the evolutionary improbabilities now urged cannot be considered to be less than perhaps a hundred millions to one; and the total chances against the evolution of man,

> or an equivalent moral and intellectual being, in any other planet, through the known laws of evolution, will be represented by a hundred millions of millions to one (S728 1904, Appendix, 334–35).

By its deletion of the idea of purpose from his central argument and by its inclusion of biological aspects, Wallace's book marked a signal advance in the debate about other worlds. It asked some of the same fundamental questions about life which are still asked today in the field of astrobiology (Dick and Strick 2004). Its approach to eliminating other life sites based on conditions believed necessary for life, and concluding that Earth was unique or rare, was one adopted almost a century later by reputable scientists (Ward and Brownlee 2000), and his view that life might exist, but not intelligence, was similar to their conclusion.

Yet, the book's thesis about the location of humanity at the center of the universe was recognized as a mistake even before the book was published. Reviewing the article on which the book was based, the British astronomer E. Walter Maunder concluded that "Every one ... of Dr. Wallace's demonstrations falls to the ground" (Maunder 1903). H. H. Turner, Savilian Professor of Astronomy at Oxford, charged that Wallace "seems to me to have unconsciously got his facts distorted, and to indicate practically nothing wherewith to link them to his conclusions" (Turner 1903). In the United States Harvard astronomer and Lowellian W. H. Pickering rejected Wallace's argument, characterizing it as "not science, and ... not very satisfactory" (Pickering 1903). And in France, pluralist *par excellence* Camille Flammarion surprised no one with his conclusion that "An examination of Mr. Wallace's plea in favor of his geocentric and anthropocentric theory has not convinced me; on the contrary, it seems to me to give a more solid basis than ever to the opposite opinion" (Flammarion 1903). Wallace's book received more than forty reviews, most of them, with the exception of a few such as that of Agnes Clerke, highly skeptical.

Though *Man's Place in the Universe* went through seven editions by 1908 and another in 1914, and was translated into German in 1903 and French in 1907, it had little influence beyond the second decade of the twentieth century. The reason is not far to seek. Within fifteen years of Wallace's death in 1913, most of his central assumptions had been rendered obsolete by an emerging new cosmology. In 1918 the American astronomer Harlow Shapley reported, based on his study of the distribution of globular clusters of stars, that our solar system was located in a very eccentric position in the Galaxy, at its periphery rather than its center. By 1924 Edwin P. Hubble had demonstrated to the satisfaction of most astronomers that many other galaxies exist outside our own, galaxies that he showed a few years later are fleeing from one another in an "expanding universe" (Smith 1982). And beginning in the 1920s Arthur S. Eddington and others devised energy-producing mechanisms for stars that swept away any need for an infalling matter theory such as Wallace proposed. We now know that we live in a universe billions of light years in extent, characterized by an interrelation among parts and the whole that astronomers characterize by the term "cosmic evolution." Though Wallace recognized the evolution of the stars based on the contemporary work of astronomers (S728 1903b, 128–34), neither he nor they could have known the extent of full-blown cosmic evolution, ranging from the Big Bang to the present and covering some 13.7 billion years of time (Chaisson 2001).

In the end, the failure of *Man's Place in the Universe* is marked by the dominance of the anthropocentric world view over all other arguments. Despite his best intentions and protestations to the contrary, in the end Wallace's theory was simply too *ad hoc*, less driven by probabilities than by deep-seated anthropocentrism and teleology. Convinced by 1898 of the nearly central position of the Sun, Wallace first sought and found the significance of this fact in the uniqueness of life, and then adduced arguments in favor of the view that advanced life was found beyond the Earth neither in our solar system nor in others. Although his theory was based on what he believed to be the fact of the Sun's centrality, the pioneer of evolution nevertheless fell victim to the effect of a world view with insufficient proof. In his concept of the plurality of worlds, no less than in his ideas about the evolution of humanity, he found it necessary to set humanity apart. Although adopting a scientific approach, Wallace's book serves as a lesson on the limits of science when world views dominate empirical evidence. It is a lesson we still need to remember in the twenty-first century.

Life on Mars

As Wallace's biographers hint, his treatise on Mars was a more specific instance of his broader claims for an anthropocentric universe, which meant a universe without intelligent life beyond Earth. Indeed, Wallace himself makes the connection in the opening words of *Is Mars Habitable?* (S730 1907): "This small volume was commenced as a review article on Professor Percival Lowell's book, *Mars and Its Canals*, with the object of showing that the large amount of new and interesting facts contained in this work did not invalidate the conclusion I had reached in 1902, and stated in my book on *Man's Place in the Universe*, that Mars was not habitable" (S730, v). These words, and the book's subtitle, "A Critical Examination of Professor Lowell's Book 'Mars and Its Canals,' With an Alternative Explanation," place the volume squarely in the plurality of worlds tradition and more specifically in the context of the claims of its most spectacular adherent, Percival Lowell. The treatise is of interest not only for these reasons, but also because it illustrates Wallace's insistence on physical evidence for a theory, and for its criticism of Lowell's predisposition toward an idea and his use of the concept of purpose, both of which led him astray. Ironically, despite Wallace's laudable insistence on physical evidence, it was the latter two ideas, in the form of anthropocentrism and the purpose of the universe, that (we now know) also led him astray in *Man's Place in the Universe.*

Percival Lowell the man and his ideas have by now been well researched. His most recent (and best) biographer, David Strauss, has emphasized how Lowell was one of the principal American disciples of Herbert Spencer, Wallace's British contemporary, principally known for his evolutionary theories applied to culture (Strauss 2001, 6, 97–113). Wallace was also a Spencerian in many ways, and even named his first son Herbert Spencer Wallace (Shermer 2002, 240–42). Strauss shows in detail how Lowell's work was "shaped by his lifelong commitment to the realization of Herbert Spencer's cosmic evolutionary project. Lowell embraced Spencer's concept of a cosmos governed by the law of evolution to explain the development of both the natural and the social world. He applied Spencer's evolutionary scheme during his travels in East Asia and again as he launched his search for extraterrestrial life" (Strauss 2001, 97).

That search for life beyond Earth centered on Mars and its supposed canals. The controversy has its roots in the observations of the Italian astronomer Giovanni Schiaparelli, made during the favorable opposition of Mars in 1877 (the same year its two moons were discovered), that the planet was crisscrossed by a network of straight lines. The Schiaparellian Mars that Lowell inherited was a planet with two polar caps composed of snow and ice, seas and continents arranged very differently from those on Earth, and an atmosphere rich in water vapor. The vaporous atmosphere he believed was supported not only by changes in the polar caps requiring a transportation mechanism for water vapor, but also by spectroscopic observations. It was also a planet of change, for the melting polar caps seemed to produce a temporary sea around the northern cap. Schiaparelli believed this water was distributed over great distances by "a network of canals, perhaps constituting the principle mechanism (if not the only one) by which water (and with it organic life) may be diffused over the arid surface of the planet." How such a system of lines could originate had led some to see them as the work of Martians. Schiaparelli was inclined to consider them as produced by the evolution of the planet, similar to features such as the English Channel on Earth. But he left the door open to the artificial hypothesis: "I am very careful not to combat this supposition, which includes nothing impossible" (Dick 1996, 69–70).

Lowell walked through this open door without hesitation. In January 1894 Lowell made the decision to finance an expedition to Arizona, where he founded the Lowell Observatory. Its sole purpose was to study the Mars problem, and in particular the origin of the canals with the possibility of demonstrating their artificial nature. August of that year was another particularly favorable opposition of Mars, and Lowell made the best of it. Already in 1895 Lowell published his first book on the subject, entitled simply *Mars*, where he argued that Mars was not only habitable but inhabited, by intelligent Martians who had built the canals. This was followed by *Mars and Its Canals* (1906)—which precipitated Wallace's book—and then by *Mars as the Abode of Life* (1908) and numerous other publications in which Lowell attempted to prove the artificial nature of the canals. The details of that controversy have been given elsewhere (Hoyt 1976; Crowe 1986; Dick 1996), but let us now return to Wallace and his place in the debate.

Wallace's reaction to Lowell's work is characterized by its insistence on good physical evidence and sound arguments based on that evidence. We need not go through those arguments in detail, but Wallace's summary at the end of his volume will suffice to show the nature of his arguments:

> (1) All physicists are agreed that, owing to the distance of Mars from the sun, it would have a mean temperature of about −35° F. (= 456° F. abs.) even if it had an atmosphere as dense as ours.
>
> (2) But the very low temperatures on the earth under the equator, at a height where the barometer stands at about three times as high as on Mars, proves, that from scantiness of atmosphere alone Mars cannot possibly have a temperature as high as the freezing point of water; and this proof is supported by Langley's determination of the low *maximum* temperature of the full moon.
>
> The combination of these two results must bring down the temperature of Mars to a degree wholly incompatible with the existence of animal life.
>
> (3) The quite independent proof that water-vapour cannot exist on Mars, and that therefore, the first essential of organic life—water—is non-existent.
>
> The conclusion from these three independent proofs, which enforce each other in the multiple ratio of their respective weights, is therefore irresistible—that animal life, especially in its higher forms, cannot exist on the planet.
>
> Mars, therefore, is not only uninhabited by intelligent beings such as Mr. Lowell postulates, but is absolutely UNINHABITABLE.

Wallace's insistence that evidence be primary is ironic because Lowell prided himself on his empirical observations of Mars. As detailed by both Dick (1996) and Crowe (1986), the problem in the canals-of-Mars controversy is not with the lack of observations but with their difficulty of interpretation. As Lowell's recent biographer has pointed out, Lowell's observations were theory driven, and thus "reflected in important ways deep-seated intellectual preferences on his part. To be sure, Lowell undertook painstaking empirical investigations of the canals of Mars over twenty-two years ... And he was certainly following accepted scientific practice in using his data to confirm a theory. What distinguished Lowell from some of his counterparts, however, was his plan to use the data to support Spencer's all-embracing evolutionary scheme. In attempting to link data and theory, Lowell frequently engaged in thinking of an imaginative sort. Thus, he confirmed Spencer's theory to his own satisfaction, but he acquired a reputation among his peers for selectivity in collecting and interpreting data" (Strauss 2001, 97).

The canals-of-Mars debate embodies one of the great problems in history and philosophy of science: the relation between theory and observation. Wallace was no stranger to this relationship, nor was Darwin. They also had to adduce a mass of facts in constructing their theory of evolution by natural selection. Their theory, however, was at a different level than Lowell's, and therefore of a different nature, and the differences highlight the ambiguities of the word "theory." Lowell's theory of artificial canals on Mars has been definitively disproven by better observations—spacecraft to Mars have shown no features resembling canals, much less artificial constructions (the "face on Mars" notwithstanding). The theory of evolution by natural selection, while based on a mass of empirical facts, is much more subtle in not representing a single planetary object at a given time, but numerous biological outcomes developed over a great length of time.

There is another problem that Wallace detects in Lowell's argument, the problem of teleology. Speaking of Lowell's supposed canals Wallace notes:

> Again, he urges the "purpose" displayed in these "canals." Their being *all* so straight, *all* describing great circles of the "sphere," all being so evidently arranged (as he thinks) either to carry water to some "oasis" 2000 miles away, or to reach some arid region far over the equator in the opposite hemisphere! But he never considers the difficulties this implies. Everywhere these canals run for thousands of miles across waterless deserts, forming a system and indicating a purpose, the wonderful perfection of which he is never tired of dwelling upon (but which I myself can nowhere perceive) (S730, 103).

In this criticism Wallace was correct. But it is indeed ironic in light of Wallace's own tendencies toward a teleological world view on a much grander scale, which we have seen displayed in *Man's Place in the Universe*, and to which we now return.

Wallace and Purpose in the Universe

Although Wallace's anthropocentric geography (or uranography) of the universe was quickly outdated, his more general idea of purpose in the universe—represented in his statement that "the supreme end and purpose of this vast universe was the production and development of the living soul in the perishable body of man"—rather surprisingly made a comeback in a different guise in scientific circles during the last quarter of the twentieth century. Moreover, in the last chapter of *Man's Place in the Universe* Wallace wrote as follows:

> Lastly, I submit that the whole of the evidence I have here brought together leads to the conclusion that our earth is almost certainly the only inhabited planet in our solar system; and, further, that there is no inconceivability—no improbability even—in the conception that, in order to produce a world that should be precisely adapted in every detail for the orderly development of organic life culminating in man, such a vast and complex universe as that which we know exists around us, may have been absolutely required (S728 1903b, 306).

These ideas relate to what today is called anthropic reasoning and its counterpart, intelligent design. In order to see just how they relate we need to understand the modern idea of the anthropic principle. The basis for the principle is the increasing realization, founded on modern observational data, that the universe appears to be fine-tuned for life. In other words, if the values of some of the physical constants, the strengths of the fundamental forces, and the masses and charges of subatomic particles were even slightly different, stars and planets could not form, and could not give rise to the conditions necessary for life. This leaves two possibilities: either an intelligent entity (which may or may not be the "God" of terrestrial history) designed the universe to be this way (intelligent design), or there is some natural explanation. The latter is often framed in terms of what is called the multiverse (Davies 2007; Rees 1997, 2000, 2001), namely that our universe is just one of an ensemble of universes, some of which are suitable for life, and others of which are not. We happen to be in one that is so finely tuned; in the words of Davies (2007) we have hit the "cosmic jackpot," and so we should not be surprised. Each of these two scenarios has its weaknesses: the first is a supernatural explanation (if one invokes a supernatural entity), while the second is natural, but thus far just as empirically unverified as intelligent design.

These themes are controversial for both scientific and religious reasons. As Wilson (1993, 11) wrote, "What makes the anthropic principle approach provocative is that it hints at the existence of deep connections between the universe and humanity, makes striking claims about mankind's place in the universe, and is reminiscent of classical arguments for a divine cosmic designer. The principle involves additional philosophical and religious themes as well: the significance and value of human life, the goal-directedness of nature, and the degree to which the universe is anthropocentric."

The concepts of an anthropic principle, anthropocentrism, and design in Nature each have their own history, but at the same time are related to each other. One can be anthropocentric (human-centered) without invoking design or the anthropic principle. It is difficult, however, to invoke the anthropic principle and not be anthropocentric, unless one plausibly redefines it as a biocentric principle (see below). And the anthropic principle implies design to some people, especially the theologically inclined (Harris 1991, 1992; Ross 1993), who use the fine-tuned universe to argue for design by a creator God. Others, however, look for a natural, rather than a supernatural explanation for such a universe.

The design argument is very old indeed, as Barrow and Tipler (1986) detail in their massive volume on the anthropic cosmological principle. It was often used in the natural theology arguments of the seventeenth through nineteenth centuries, and is commonly invoked even today to explain those things not fully understood in astronomy or biology, as in the idea of "irreducible complexity" (Behe 1996). The more specific idea that the world is fine-tuned for life is not so old. It was anticipated by Wallace's American contemporary, the Harvard biochemist Lawrence J. Henderson, who marveled at the fitness of the terrestrial environment for life, especially because of the properties of water (Fry 1996). In the final paragraph to his *The Fitness of the Environment*, Henderson (1970, 312), even extended his argument to the universe at large:

> The properties of matter and the course of cosmic evolution are now seen to be intimately related to the structure of the living being and to its activities; they become, therefore far more important in biology than previously suspected. For the whole evolutionary process, both cosmic and organic, is one, and the biologist may now rightly regard the universe in its very essence as biocentric.

The "Biocentric Principle" would have been a better term for the fitness of the universe for life, since it does not imply that terrestrial life forms, including humans, could be the only form of life. The universe was fine-tuned for life, including beings of higher consciousness, not only for humans, even if the only intelligent life in the universe we yet know is human. But when these questions were once again pondered six decades later the term "anthropic principle" was coined, almost by accident. This occurred in a path-breaking article on the subject by astronomer Brandon Carter (1974), who first formulated what he called the weak anthropic principle: "what we can expect to observe must be restricted by the conditions necessary for our presence as observers." Thus the gravitational constant is constrained by the fact that we exist, otherwise the conditions for life could never have arisen. Carter went on to expound a "strong anthropic principle," namely, "that the universe (and hence the fundamental parameters on which it depends) must be such as to admit the creation of observers within it at some stage," a much more problematic claim. Although such reasoning turned the deductive method on its head, Carter argued that it might be considered a kind of explanation for why our universe is the way it is.

These two versions of the anthropic principle embody two interpretations of the old idea of "final causes": is the universe just biofriendly, or is life written into the laws of the universe so that the existence of life and intelligence is predetermined? Moreover, they also embody the age-old question of chance and necessity in the workings of Nature. The question is still a very active one in astrobiology, cosmology, origins of life studies, and science in general, especially since the French biologist and Nobelist Jacques Monod stated in 1971 that "The universe was not pregnant with life, nor the biosphere with man. Our number came up in a Monte Carlo game" (Monod 1971). To which another Nobelist replied almost twenty-five years later:

> The Earth is part, together with trillions of other Earth-like bodies, of a cosmic cloud of "vital dust" that exists because the universe is what it is. Avoiding any mention of design, we may, in a purely factual sense, state that the universe is constructed in such a way that this multitude of life-bearing planets was bound to arise ... The universe is not the inert cosmos of the physicists, with a little life added for good measure. The universe is life, with the necessary infrastructure around; it consists foremost of trillions of biospheres generated and sustained by the rest of the universe (De Duve 1995).

There is no doubt that Wallace denied chance in connection with the origin and fate of humans. And with all his discussion about the conditions necessary for life,

Wallace would have embraced the weak anthropic principle, though he had no idea of the quantitative underpinnings in the physical constants. But he never came close to stating the strong version, that the universe could not exist without life. In fact, because Wallace believed he had proved the Earth to be the only abode for life, the term "anthropic" is much more aptly applied to his beliefs than to today's formulation, which is really a biocentric principle. A universe fit for life does not necessarily imply that life is the purpose of the universe, as Wallace stated. But Barrow and Tipler (1986) in a later version of the anthropic principle made exactly this teleological argument: that habitability is the goal of the universe. In this version, they argued, life could never die out, because then the goal of the universe would cease to exist.

Because Wallace opted for a "final causes" explanation for the purpose he saw in the universe, it might seem he would also have accepted the idea now known as intelligent design (Numbers 2006, 373 ff.), the modern version of the design argument. That such is not the case seems clear from his "Notes Added to the Second Edition of *Contributions to the Theory of Natural Selection*" (S716 1871). Here Wallace explicitly distinguishes between a supernatural God or Deity and a superior or controlling intelligence, especially when it comes to the origins of humans:

> Now, in referring to the origin of man, and its possible determining causes, I have used the words "some other power"—"some intelligent power"—"a superior intelligence"—"a controlling intelligence," and only in reference to the origin of universal forces and laws have I spoken of the will or power of "one Supreme Intelligence." These are the only expressions I have used in alluding to the power which I believe has acted in the case of man, and they were purposely chosen to show, that I reject the hypothesis of "first causes" for any and every *special* effect in the universe, except in the same sense that the action of man or of any other intelligent being is a first cause. In using such terms I wished to show plainly, that I contemplated the possibility that the development of the essentially human portions of man's structure and intellect may have been determined by the directing influence of some higher intelligent beings, acting through natural and universal laws (S716 1871, 372–72A).

In this opinion, Wallace anticipated the view of one of the giants of twentieth-century astronomy, Fred Hoyle, who made a similar argument in his book *The Intelligent Universe: A New View of Creation and Evolution* (Hoyle 1983). Here Hoyle, a self-proclaimed atheist, concluded that the simplest explanation for the biofriendly universe was that a natural (not supernatural) superintellect had engineered the universe to make it fit for carbon-based life and intelligence. This view has recently been elaborated in two books by complexity theorist James Gardner, whose subtitles reveal their thesis: *Biocosm—The New Scientific Theory of Evolution: Intelligent Life is the Architect of the Universe* (Gardner 2003), and *The Intelligent Universe: AI, ET, and the Emerging Mind of the Cosmos* (Gardner 2007). It is thus clear that the concept of "God" as a supernatural entity and the concept of "superior intelligence" can be separated. Alternately, they could also be united by the unconventional idea of a "natural God" (Dick 2000), though Wallace himself never took this step.

Although written more than three decades earlier, Wallace's 1871 "Note" needs to be kept in mind in the context of his later astronomical volumes. In an interview one month after the publication of *Man's Place in the Universe*, Wallace was asked if he agreed with the view of Sir Oliver Lodge that "the attempt to explain the universe by chance has absolutely failed. It must have had a designer," which the interviewer described as an Infinite Being. "Certainly," Wallace replied, "My whole argument tends in that direction, though my object in writing 'Man's Place in the Universe' was purely scientific, not religious" (S741 1903, 177). Again, Wallace's appeal to science rather than religion indicates this "designer" may be seen as a non-supernatural superior intelligence.

Nevertheless, in the same interview, as elsewhere in his books on natural selection, Wallace clearly set himself apart from Darwin, who believed evolution applied to all aspects of humans, including the physical, mental, and moral. Wallace even had other exceptions to Darwinian thinking: "I do not think it is possible to form any idea beyond this, that when man's body was prepared to receive it, there occurred an inbreathing of spirit—call it what you will. I believe this influx took place at three stages in evolution—the change (1) from the inorganic to the organic, (2) from the plant to the animal, (3) from the animal to the soul of man. Evolution seems to me to fail to account for these tremendous transitions" (S741 1903, 177). These ideas were part and parcel of his anthropocentric and teleological world view. This is clear even from the title of Wallace's 1910 volume *The World of Life: A Manifestation of Creative Power, Directive Mind and Ultimate Purpose* (S732).

Intelligent design must also be starkly distinguished from the idea of late twentieth-century creationism (today called "scientific creationism"), which implies that the world was created in less than ten thousand years (Numbers 2006, 8–9). With his vast knowledge of biology and geology—and evolution—Wallace knew better than that, putting the time scale for the history of life at about one hundred million years. Nonetheless, it seems today ironic, in a world where evolutionists are often set against the intelligent design movement, that one of the founders of evolution by natural selection should embrace final causes, though neither intelligent design in its modern meaning, nor creationism.

Conclusion: Wallace and the Connections between Biology and Cosmology

Lest there be any doubt of the connections between Wallace's own work on biology and astronomy, we may compare two passages of his writings. The first comes from his book *Darwinism* (S724), which Shermer calls Wallace's "definitive statement" on the subject, published in 1889 when he was sixty-six. Wallace rails against those who think humans "are but the products of blind eternal forces of the universe," who came into being by chance and would once again disappear when the Sun lost its heat energy:

> As contrasted with this hopeless and soul-deadening belief, we, who accept the existence of a spiritual world, can look upon the universe as a grand consistent whole adapted in all its parts to the development of spiritual beings capable of indefinite life and perfectibility. To us, the whole purpose, the only *raison d'être* of the world—with all its complexities of physical structure, with its grand geological progress, the slow evolution of the vegetable and animal kingdoms, and the ultimate appearance of man—was the development of the human spirit in association with the human body (S724, 476–77; see also Shermer 2002, 230–31, and see also p. 33).

The second passage we have already cited from the article "Man's Place in the Universe," written fourteen years later and providing the foundation for the book by the same name. There Wallace argued that modern astronomy now showed "that our position in the material universe is special and probably unique, and that it is such as to lend support to the view, held by many great thinkers and writers today, that the supreme end and purpose of this vast universe was the production and development of the living soul in the perishable body of man" (S602 1903, 474).

These passages clearly show that by the end of his life, Wallace's biology and astronomy were part and parcel of the same anthropocentric and teleological world view—one that incorporated final causes, but not first causes in the sense of a supernatural deity. Despite a lifelong interest in astronomy, because his astronomical writings came so late in life, it seems likely that his biological world view affected his cosmology rather than the reverse. But in either case the publication of Wallace's two books on astronomy reveals a consistent and consilient world view, one that must have been a great satisfaction to him in his final years. Within a few years after his death developments in astronomy would shatter his anthropocentric world view of man's place in the universe. But his teleological world view remains very much alive in current discussions of the anthropic principle and intelligent design, both—undoubtedly to the delight of the now disembodied Wallace—still considered out of the mainstream of modern science.

In the end the significance of Wallace's astronomical writings is not in their failure to grasp the geography of the universe. In that he was a man of his times, grappling with the limited observations available at the turn of the twentieth century. The significance is rather in his joint treatment of astronomy and biology, in anticipating not only some of the arguments and content of the current discipline of astrobiology, but also the profound questions of the cosmological connections between the two, bearing on whether the universe is by its very nature biocentric. The concept of a biological universe (Dick 1996) encompasses both the question of the existence of life in the universe, and why the universe should be so biofriendly. Today, even as astrobiologists search for life, others ponder the mysteries of why life can exist at all. British Astronomer Sir Martin Rees (1997, 2000, 2001), physicist and astrobiologist Paul Davies (1982, 2007) and many others are now exploring the ramifications of a possibly deep and profound relationship between biology and cosmology.

In this connection twenty years ago the physicist Freeman Dyson called for the need to build bridges between biology and cosmology: "Only a few heretics... dare to express the view that the structure of the universe may not be unambiguously reducible to a problem of physics," he wrote in his iconoclastic book *Infinite in All Directions*. "Only a few romantics like me continue to hope that one day the links between biology and cosmology may be restored." He suggested further that

> The prospects are bright for a future-oriented science, joining together in a disciplined fashion the resources of biology and cosmology. When this new science has grown mature enough to differentiate itself clearly from the surrounding farrago of myth and fiction, it might call itself "cosmic ecology," the science of life in interaction with the cosmos as a whole. Cosmic ecology would look to the future rather than to the past for its subject matter, and would admit life and intelligence on an equal footing with general relativity as factors influencing the evolution of the universe (Dyson 1988, 50–51).

The heretical Wallace might have agreed with this sentiment had it been limited to humans, which he believed to be the sole intelligence in the universe. As astronomers search the universe for life, we cannot be so confident of such an anthropocentric and teleological world view today. We do not know if the universe is full of intelligence, or if humans are its ultimate creation—and therefore whether the connections between biology and cosmology, if they exist, will reveal an anthropic or a biocentric universe. But if and when such a science of cosmic ecology does develop, Wallace—along with his many other roles—may be seen as a significant early precursor, even if his belief in final causes may not lead to final answers.

18

Wallace's Unfinished Business*

Charles H. Smith

In February 1858 a then little-known bird and insect collector named Alfred Russel Wallace (1823–1913) was struck with a startling revelation while fending off an attack of malaria in the Moluccas. As soon as the fit passed he prepared an essay on the idea—natural selection—and sent it off to a man he figured would be interested in the concept: Charles Darwin. The rest of the story is well enough known not to bear repeating; in the end it was Darwin whose name became most associated with the principle, with Wallace relegated to "other man" status, and his ideas to the dustbin of history.

Whether all this was fair or even represented some kind of conspiracy against Wallace has been debated for many years (*e.g.*, Brackman 1980; Brooks 1984; Berry 2002), but most observers seem to feel that, all told, things worked their way out pretty well. Certainly, natural selection was revealed to the world at the earliest possible juncture, and even Wallace benefited to the extent that he was immediately welcomed into the highest echelons of scientific discourse, along the way becoming one of the most famous men of his time.

But in truth the premature reading of Wallace's brainchild may also have had some negative effects on the longer term development of evolutionary theory. The more one reads and digests the full body of Wallace's work, the more one realizes that Wallacean natural selection is quite a distinct animal from Darwinian natural selection, and that the two men's views on evolution overall were more different yet. And, whereas every word of Darwin's writings has been run through the philosophical grist mill and thoroughly digested, much of what Wallace wrote has yet to receive its rightful full appraisal.

Although a fair amount has been published on Wallace and his ideas over the years, a sharp increase in interest has been evident of late (see Raby 2001; Shermer 2002; Fichman 2004; Slotten 2004). While there are likely many reasons for this, surely one of the most important has been the re-examination of hundreds of what

* Reprinted by permission from Volume 10, Number 2 of *Complexity* (Wiley Periodicals, Inc.), copyright 2004.

might be termed "lost" writings of his. Many of these have revealed clues as to what he actually had in mind both in the years preceding the "Ternate essay" on natural selection, and those following it. I have discussed this subject in considerable detail elsewhere;[1] here, this new interpretation of Wallace's intellectual evolution is summarized with the ultimate object of illuminating a possible new direction in evolutionary and biogeographic studies it suggests.

Although the considerable impact on Wallace of Charles Lyell's *Principles of Geology* and Robert Chambers's *Vestiges of the Natural History of Creation* in 1844 or 1845 has been noted by just about everyone who has written on him, it is less well known that at that point he had already been entertaining evolutionary views of a nonbiological nature for several years. In early 1837, when just fourteen years old, he fell in with a group of Owenite socialist utopians, and was profoundly influenced by their views on how to bring about progressive social reform (see Chapter 13). Wallace was especially taken with their ideas on social justice, and in turn with the relation between belief and just cause and, ultimately, the intrinsic advantages of absorbing and applying varied forms of knowledge. Apart from Wallace's own recollections on these matters in his autobiography *My Life* (S729 1905) we know of these influences because three of his earliest writings, from the period 1841–43 (S1 1905, S1a 1845, S623 1905) have survived. Two of these even extend the "varied knowledge" notion to a prescription for success for the evolution of whole societies.

The centrality in Wallace's thinking of his views on belief, in particular, and how this is related to social and natural change, cannot be overemphasized. The following passage, from an 1861 letter sent to his brother-in-law Thomas Sims while Wallace was still in the Malay Archipelago, is lengthy, but tells the whole story:

> ... You intimate that the happiness to be enjoyed in a future state will depend upon, and be a reward for, our belief in certain doctrines which you believe to constitute the essence of true religion. You must think, therefore, that belief is *voluntary* and also that it is *meritorious.* But I think that a little consideration will show you that belief is quite independent of our will, and our common expressions show it. We say, "I wish I could believe him innocent, but the evidence is too clear"; or, "Whatever people may say, I can never believe he can do such a mean action." Now, suppose in any similar case the evidence on both sides leads you to a certain belief or disbelief, and then a reward is offered you for changing your opinion. Can you really change your opinion and belief, for the hope of reward or the fear of punishment? Will you not say, "As the matter stands I can't change my belief. You must give me proofs that I am wrong or show that the evidence I have heard is false, and then I may change my belief"? It may be that you do get more and do change your belief. But this change is not voluntary on your part. It depends upon the force of evidence upon your individual mind, and the evidence remaining the same and your mental faculties remaining unimpaired—you cannot believe otherwise any more than you can fly.

> Belief, then is not voluntary. How, then, can it be meritorious? When a jury try a case, all hear the same evidence, but nine say "Guilty" and three "Not guilty," according to the honest belief of each. Are either of these more worthy of reward on that account than the others? Certainly you will say No! But suppose beforehand they all know or suspect that those who say "Not guilty" will be punished and the rest rewarded: what is likely to be the result? Why, perhaps six will say "Guilty" honestly believing it, and glad they can with a clear conscience escape punishment; three will say "Not guilty" boldly, and rather bear the punishment than be false or dishonest; the other three, fearful of being convinced against their will, will carefully stop their ears while the witnesses for the defence are being examined, and delude themselves with the idea they give an honest verdict because they have heard only one side of the evidence. If any out of the dozen deserve punishment, you will surely agree with me it is these. Belief or disbelief is therefore not meritorious, and when founded on an unfair balance of evidence is blameable.
>
> Now to apply the principles to my own case. In my early youth I heard, as ninety-nine-hundredths of the world do, only the evidence on one side, and became impressed with a veneration for religion which has left some traces even to this day. I have since heard and read much on both sides, and pondered much upon the matter in all its bearings. I spent, as you know, a year and a half in a clergyman's family and heard almost every Tuesday the very best, most earnest and most impressive preacher it has ever been my fortune to meet with, but it produced no effect whatever on my mind. I have since wandered among men of many races and many religions. I have studied man and nature in all its aspects, and I have sought after truth. In my solitude I have pondered much on the incomprehensible subjects of space, eternity, life and death. I think I have fairly heard and fairly weighed the evidence on both sides, and I remain an *utter disbeliever* in almost all that you consider the most sacred truths. I will pass over as utterly contemptible the oft-repeated accusation that sceptics shut out evidence because they will not be governed by the morality of Christianity. You I know will not believe that in my case, and *I* know its falsehood as a general rule. I only ask, Do you think I can change the self-formed convictions of twenty-five years, and could you think such a change would have anything in it to merit *reward* from *justice*? I am thankful I can see much to admire in all religions. To the mass of mankind religion of some kind is a necessity. But whether there be a God and whatever be His nature; whether we have an immortal soul or not, or whatever may be our state after death, I can have no fear of having to suffer for the study of nature and the search for truth, or believe that those will be better off in a future state who have lived in the belief of doctrines inculcated from childhood, and which are to them rather a matter of blind faith than intelligent conviction. (Marchant 1975 [1916], 65–67).

One can only conclude from this entirely transparent argument that Wallace felt a belief in false things—and possibly even unreal things—was unproductive; that is to say, "personally nonadaptive." And yet false beliefs both existed, and could be overcome: the pattern of human history seemed to prove as much. What, in turn, did he suppose the pattern of biological change might prove?

On reading Chambers about 1845, Wallace very quickly figured out how to demonstrate that evolution did in fact take place: through the study of the traces of the speciation process left in the fossil record and in current distribution patterns. He was not so quick, however, to recognize how individual adaptations fit into the overall picture. The problem, possibly beginning as early as this 1845 period, was his initial position on utility as it related to adaptation. At this time it would appear that, contrary to Wallace's well-known post-1858 position, he believed many adaptations served no necessary utilitarian purpose. There are remarks to that effect in his 1853 book *Narrative of Travels on the Amazon and Rio Negro* (S714 1889, 58–59), and this position is even more plainly stated in the little known work "On the Habits of the Orang-utan of Borneo," published in 1856:

> Do you mean to assert, then, some of my readers will indignantly ask, that this animal, or any animal, is provided with organs which are of no use to it? Yes, we reply, we do mean to assert that many animals are provided with organs and appendages which serve no material or physical purpose. The extraordinary excrescences of many insects, the fantastic and many-coloured plumes which adorn certain birds, the excessively developed horns in some of the antelopes, the colours and infinitely modified forms of many flower-petals, are all cases, for an explanation of which we must look to some general principle far more recondite than a simple relation to the necessities of the individual. We conceive it to be a most erroneous, a most contracted view of the organic world, to believe that every part of an animal or of a plant exists solely for some material and physical use to the individual,–to believe that all the beauty, all the infinite combinations and changes of form and structure should have the sole purpose and end of enabling each animal to support its existence,—to believe, in fact, that we know the one sole end and purpose of every modification that exists in organic beings, and to refuse to recognize the possibility of there being any other. Naturalists are too apt to *imagine*, when they cannot *discover*, a use for everything in nature . . . (S26, 30)

Wallace probably arrived at this anti-utilitarian position on the basis of two main considerations. First, and as suggested earlier, he had undoubtedly observed that many human beliefs and behaviors existed that were anything but progressively utilitarian. Yet these nevertheless existed, had come into being somehow, and even once operating did not always prevent society from moving forward. In like fashion, one could imagine a biological process in which adaptations emerged not as the feature innovations of evolutionary advance, but instead in some manner making them a byproduct of, or perhaps even just "correlated" with, it. Second, and following Chambers's idea that it made better sense to envision an evolutionary process operating on the basis of natural law than unknowable forces, Wallace was rejecting the notion that each individual adaptation served a prior purpose in the overall scheme of things—that is, arose as a first cause. In the passage from "On the Habits ..." quoted above, his concern in this regard is obvious in the three concluding sentences.

Strange as it may sound, it is thus not likely that Wallace's significant concerns during the pre-1858 period included identifying adaptive structures that were ... adaptive. This did not stop him from believing that there was an evolutionary "progression," however, and he was also making it his business to identify its final cause. Despite the existence of what appeared to be nonadaptive behaviors and structures, there had to be "some general principle far more recondite" (as he describes it in the quote given above) that was driving evolutionary change—some force or set of forces, perhaps climatological or geophysical in nature, that subtly overrode the clutter of detail apparent at the adaptational level, inexorably acting to propel change at a slow, grandiose scale. Perhaps if while in the field he examined enough particulars of form and function, he might be able to figure out what this final cause was. Note, however, that at no time before 1858 did he imagine that he would ever be able to understand how or why all adaptational structures *individually* came into being: looking back at this matter in his 1905 autobiography *My Life*, he wrote: "My paper written at Sarawak rendered it certain in my mind that the change had taken place by natural succession and descent—one species becoming changed either slowly or rapidly into another. But the exact process of the change and the causes which led to it were absolutely unknown and appeared almost unconceivable" (S729 1905a, 1:360).

The "Sarawak" paper he speaks of, "On the Law Which Has Regulated the Introduction of New Species" (S20), was published in 1855. It signaled, as he says above, his final recognition of evolutionary descent as a biological reality. But the "Every species has come into existence coincident both in space and time with a pre-existing closely allied species" model it famously embraced only described the *results* of the process, not its *causes*. Indeed, he was actually no closer to an understanding of either the final or immediate causes of evolution than he had been ten years earlier (as evidenced by the orang-utan paper, written and published about a year later). All of this changed in early 1858, and the famous bout with malaria during which he thought out the principle of natural selection.

Consider, now, how the survival of the fittest concept most likely would have struck Wallace at this point. Contrary to the way it has commonly been portrayed, it was not at all the "logical conclusion" of Wallace's earlier attempts at dealing with evolution. Indeed, one might reasonably argue that it was their absolute antithesis, and, accordingly, "On the Tendency of Varieties to Depart Indefinitely from the Original Type" (S43 1858) not only does not refer, even obliquely, to the Sarawak paper or law, it doesn't refer to any of the several following writings Wallace published that applied that model (S37 1857, S38 1858, S40 1858, S41 1858). For Wallace, the central revelation of early 1858 was his ability now to envision a single, generalizable process through which any individual adaptation could continually be selected for or against. He would have to give up the idea that some adaptations had no utility (unless they were somehow integrally connected with ones that did), however. This was not such a problem biologically (he actually had no way to prove whether any given adaptation was utilitarian, anyway), but he felt unable to budge on the matter of human thoughts and beliefs, which yet seemed to afford too many instances both of "nonprogressive belief," and higher attributes such as mathematical abilities that had come into existence before useful applications could be found for them. In consequence, the paper he sent to Darwin makes no mention of humankind: and not because he didn't wish to point to the situation with humans as being an exception, but for the very reason that he *did*.

Previous to 1858 it had thus been Wallace's peculiar perspective that evolution was in a general sense progressively adaptive, but that some of the individual adaptive structures produced through it were not. Now, through natural selection, he could believe that *all* strictly biological adaptations were in an ecological sense adaptive, but not necessarily evolutionarily adaptive: *i.e.*, a species' adaptive suite might serve to support it in the environment of one era, but then fail it in the next, leading to extinction. Was this a helpful elaboration? Wallace apparently thought so, despite the fact that it neither shed any light on why some human attributes yet seemed to be nonadaptive, nor helped him to understand what the final cause of evolution was—or for that matter, whether it was still even necessary to think in terms of final causes. He would test the waters on this new idea by circulating the draft of an essay on what seemed to be the one element of the question that was tightly defendable, the "special case" of natural selection as it applied to non-human species.

But before Wallace knew it, the paper, which included thoughts rather closely resembling some of those held by the man he had sent it to for possible forwarding to Charles Lyell, was read publicly and set to print. He was informed only after the fact. He was now viewed by—everyone—as "Darwinian," despite the fact that his ideas actually extended to well beyond what that tag represented.

How Wallace would extricate himself from this situation is a subject I have taken up elsewhere (Smith 2003, 2003–; see also Chapter 21 in this collection); for the present let us shift the discussion away from history and toward today's science. We can begin by suggesting that the attention that has been lavished on debating whether Darwin might have committed intellectual theft from Wallace should be refocused on a matter of substantially greater import: whether nearly 150 years of largely ignoring Wallace's world view has been in our best interest.

In claiming that Wallacism has a right to be considered on its own terms, and as more than just a historical satellite to Darwinism, we may look in the first instance for elements of Wallace's framework that might have significant relevance to today's efforts to model large-scale evolutionary processes. One such element harkens back to Wallace's law-like model of natural selection, which distinguishes between ecological and evolutionary outcomes in a manner contrasting in certain important respects with Darwin's solution to the problem.

One of the most intriguing passages in Wallace's Ternate essay likens the action of natural selection to a governor on a steam engine:

> The action of this principle is exactly like that of the centrifugal governor of the steam engine, which checks and corrects any irregularities almost before they become evident; and in like manner no unbalanced deficiency in the animal kingdom can ever reach any conspicuous magnitude, because it would make itself felt at the very first step, by rendering existence difficult and extinction almost sure soon to follow (S43 1858, 62).

In his work *Steps to an Ecology of Mind*, anthropologist Gregory Bateson made some interesting comments on this passage:

> The steam engine with a governor is simply a circular train of causal events, with somewhere a link in that chain such that the more of something, the less of the next thing in the circuit ... If causal chains with that general characteristic are provided with energy, the result will be ... a self-corrective system. Wallace, in fact, proposed the first cybernetic model ... Basically these systems are always *conservative* ... in such systems changes occur to conserve the truth of some descriptive statement, some component of the *status quo*. Wallace saw the matter correctly, and natural selection acts primarily to keep the species unvarying ... (Bateson 1972, 435).

Later, in the collection *Mind and Nature: A Necessary Unity*, Bateson added the following observations:

> If it had been Wallace instead of Darwin [who started the trend], we would have had a very different theory of evolution today. The whole cybernetic movement might have occurred one hundred years earlier as a result of Wallace's comparison between the steam engine with a governor and the process of natural selection ... (Bateson 1979, 43).

Bateson's point is a most remarkable one, but he and the others who have studied cybernetic relations in connection with evolution have never looked in any detail into how the 1858 Ternate model actually fit into Wallace's overall cosmology at that point. Without doing so, we can proceed no further in this direction: cybernetic theory notwithstanding, it is clear that no model of the greater evolutionary program can invoke a causal explanation resting *entirely* on negative feedback processes, as it is ultimately the breaking away from such recursive constraints that by definition leads to novel development. Actually, Bateson might have done more with his observation even at that point had he wished, as the evolutionary relationship between negative and positive feedback couplings had already been explored earlier in an important and influential systems paper by Magoroh Maruyama entitled "The Second Cybernetics: Deviation-Amplifying Mutual Causal Processes" (Maruyama 1963). In this work Maruyama describes how information imported from the environment represents feedbacks of two kinds: deviation-countering processes (negative feedbacks) which tend to enforce equilibrium conditions, and deviation-amplifying processes (positive feedbacks), which cause systems to change, either in a direction of greater or lesser order.

While this position is helpful to understanding how a living system might simultaneously be equilibrium conserving and equilibrium superseding, it does not specify the conditions under which directions "of greater or lesser order" might be obtained; *i.e.*, what is it in the longer term evolutionary sense that tips the scales in favor of greater order?

In writings published in the 1980s (Smith 1984, 1986, 1989) I argued that Wallacean natural selection was better suited to thinking in such systems terms than Darwin's model. Bateson had already pointed out (as indicated in the passages produced above) that natural selection might be considered a *conservative* process; that is, that it does no more than produce the net result of a return toward equilibrium for a system pushed toward disorder. All that was left to do was to identify the components of the complementary deviation-amplifying function. I posited that the overall thrust of organic evolution might be conceptually and practically studied by: (1) agreeing with Bateson and Maruyama, and regarding adaptive structures as operationalizing a process of negative feedback in which energy sources at the surface of the earth are temporarily diverted and captured, then applied to do chemical and physical work, then finally returned in degraded form to the physical environment envelope (and ultimately into space), maximizing system entropy, (2) treating the adaptive structures themselves as in the main a *potential* for effecting system change, and (3) most importantly, regarding that potential as enacted through the entry into new ecological associations through organismal/population behavior, movement, and dispersal (*i.e.*, as the positive feedback/deviation-amplifying part of the process capable of leading to net negentropy accumulation). Ultimately, evolution-serving deviation amplification is achieved by the tendency of individuals and populations to disperse through and interact with their environment *nonrandomly, in preferred spatial directions*: specifically, in those directions in which the relevant life support resources are being made available—occurring, and turning over—at more optimum rates.

The idea that adaptive structures are in the first instance negative feedback-relaying nodes is hardly a revolutionary one, as this function is necessitated by their role as mediators in the biogeochemical cycling of matter and energy, and the operation of the Third Law. In turning to his "governor" understanding initially Wallace was of course not thinking in such elaborate terms; instead, for him the important notion was that adaptations emerged on a "whatever" basis: that is, the process involved selection—at random—leading to *whatever* structures that might ultimately serve a population's persistence. As both Wallace and Darwin believed, natural selection could not produce *more* than what was needed to persist; instead it merely continued to reduce inefficiency of system operation by eliminating its weak links. But again, this in itself is not evolution. I submit that one has to understand the information that is part and parcel of organized adaptive structure at any given time as a *potential* only: that is, a potential that supports entries into new kinds of information-sharing networks at the ecological/environmental level.

There are a number of stumbling-blocks to evolution and evolutionary ecology that this kind of thinking directly overcomes; for the present, only one particularly obvious application can be noted briefly. This involves Wallace's supposed hyperselectionism (or closely related panselectionism). Writers such as the late Stephen Jay Gould have criticized Wallace for arguing that natural selection represents the only variation-accumulating mechanism, and for talking down the importance of mutations and Mendelian inheritance to biological change (Gould 1980). Two points need to be made in this regard. First, while it is true that Wallace did believe that all adaptive structures passed through the filter of natural selection and were maintained in that fashion, he also noted on several occasions that the "laws" of origin of the variations upon which the survival of the fittest operated were quite unknown. He was thus more interested in defending the primacy of natural selection as an evolutionary "shaping" agent than he was in debating how variations came about to begin with.

More importantly, moreover, it can be seen that through the model discussed here, there is nothing logically circular about the way Wallace treated adaptations to begin with. Regardless of whether adaptive structures may or may not be idiosyncratic in their purpose and function as related to organismal success, they serve an *evolutionary* function not in their deviation-countering (entropy maximizing) role, but instead in their potential to propel a deviation-amplifying process through environmental engagement. The latter represents a conceptually different evolutionary outcome—spatial interaction at the ecological/population level—than the adaptive structures themselves, and thus provides a venue for hypothesis testing that does not fall prey to circular reasoning.

In the 1980s such views fell on deaf ears; this was a period in which more interest was being shown in the irreversible thermodynamics modeling of E. O. Wiley and D. R. Brooks (Wiley and Brooks 1982; Brooks and Wiley 1986) among biologists, and in the cladistic methodologies being perfected by systematists and vicariance biogeographers (Hennig 1966; Nelson and Rosen 1981; Nelson and Platnick 1981). Both of these perspectives closely follow the generally Darwinian view that evolution is not much more than a matter of phyletic diversification—"tree-thinking" (O'Hara 1988, 1992), in the parlance of the period. This is not to suggest that either school of thought depends directly on classic Darwinian views on speciation and the like, but to acknowledge that each does tend to focus on organism—, adaptation-centered, rather than ecologically-centered, outcomes.

The need for a tempered revision of Darwinian "tree-thinking" is likely to become increasingly evident as the challenges of biodiversity conservation become ever greater. We cannot truly expect to become shepherds of the earth's biotic resources before we secure a firmer understanding of those supra-population forces that shape the evolutionary-ecological interface, and the mere documentation of phyletic diversification, including its further detailing into genomic inventories, is not enough to get the job done. Clearly, we must look to evolutionary models that are more environmental in their emphasis, or as Greer-Wooten described the matter back in 1972: "in analyzing the dynamics of systems, the researcher should place more emphasis on flows (of energy, materials, or information) between components of the system, and the system and its environment, than on changed attributes of the elements" (Greer-Wooten 1972, 17–18). Wallace himself understood this all the way back in the 1850s, ultimately reaching beyond his simple phyletic determinism model of 1855 to produce a more integrated one invoking environment-mediated stochasm: natural selection.

In now returning to Wallace's vision a final time here, the following observations seem relevant. It will be recalled that earlier I implied Wallace's adoption of a final-causes view of the organization of the natural world was not limited to his later career (as exemplified in his books *Man's Place in the Universe* [S728] in 1903 and *The World of Life* [S732] in 1910), but was integral to his pre-1858 positions as well. His search for a final cause relevant to human societal functions led him to adopt spiritualism and socialism (and, actually, for good reason[2]), but he never did give up on the idea that more "removed" forces might be channeling the direction of purely physical and biological nature as well. Hints of this leaning turn up in a variety of contexts: for example, in his frequently stated view that known laws of nature seem always to be subservient to more "recondite" (his term) factors, in his familiar argument (adopted by many to this day) that only Earth can possibly have observed the many physical/astronomical constraints that have led to the evolution of advanced life-forms, in his belief that natural selection often involves the change of less advanced creatures according to the needs of more advanced ones, and in his continued support of the overriding causal influence of Sclaterian faunal realm development.

Although Wallace's thinking never included esoteric notions of positive-negative feedback couplings or cybernetic relations, it seems to me that his juxtaposition of a "governor model" of organism-environment state-space (*i.e.*, natural selection) onto an assumed final causes-based evolution process is still both logical, and exploitable. Indeed, somewhat abstract models of this kind are currently being offered up by proponents both of the anthropic principle, and the Gaia hypothesis (*e.g.*, Barrow and Tipler 1986; Lovelock 1988; Lenton 1998). More revealing ecogeographic models are possible as well, I think, if we proceed generally as follows.

It should be apparent from the variety of positions taken by adherents of the anthropic and Gaia hypotheses that, philosophically speaking, the "final causes" concept has produced the gamut of teleological mindsets. We need not adopt the more extreme of these to suggest how a system as described here could find its way to higher levels of order, however. Suppose, for example, that the environment as it physically extends away from any given individual organism inherently presents statistically greater survival probabilities in some directions than in others. On this basis, individuals—and more importantly, populations—might tend to extend more easily in some spatial directions than in others, in so doing entering into new associations supported by new adaptations forced into existence by such extensions.

Let us further suppose that these survival opportunities are governed in the most general sense by the degree of optimality of turnover and rate of availability of certain fundamental resources, for example water. If we can make this argument, we might also be able to argue that the degree of specification of selection required to fit into the more ideal environments will be less than that required to fit into less ideal ones: that is, that because there is too much or too little of something vital at certain times and places, a good deal more selection must go into establishing adaptations that will continue to support morphostasis in those places. This latter kind of selection will ultimately lead to the kinds of specialized organisms that will be evolutionarily at risk should the environment change markedly at some future point.

What has just been described can be interpreted as a mild form of final causation: it suggests that *all* populations will *tend* to disperse in preferred directions, and in so doing nonrandomly perpetuate genetic flexibility. This is evolution—environmentally mediated (or even directed) evolution, to be sure, but not environmentally *determined* evolution: again, as in Wallace's thinking, that which is selected for to meet the challenge constitutes *whatever* can be genetically sorted out, in large part by trial and error, to support persistence. In earlier writings (Smith 1984, 1986, 1989) I imagined an environmental "potential information field" over which populations dispersed and evolved in this way. The magnitude of the potential was identified through an index combining idealness of a location's annual soil moisture budget with its degree of deviation from mean planetary temperature conditions (to produce a resource turnover rate surrogate conceptually linked to Van't Hoff's law). I theorized that, as a statistical whole, the shapes and orientations of geographic ranges of populations should reflect such a spatially-varying driving mechanism, and in fact was able to elicit empirical support to back this hypothesis.

In recent years other investigators (*e.g.*, Kerr and Packer 1997; O'Brien 1998; Kerr and Currie 1999, Kerr 2001; Hawkins *et al.* 2003a; Hawkins *et al.* 2003b; Hawkins and Porter 2003a; Hawkins and Porter 2003b) have been attempting to understand spatial variation in diversity patterns through approaches that share some of these objectives, but these efforts have so far lacked the dynamic modeling perspective that allows them to do more than correlate certain diversity characteristics with particular ambient environmental conditions. These are not, therefore, evolutionary models as they now stand, but it would not take much reorientation of purpose to turn them into such. Efforts of this kind might give us a much more interactive view of the meaning of biodiversity, and at the same time allow us to overcome the logical dilemma that R. C. Lewontin remarked upon in an essay published in 1984: "The process is adaptation and the end result is the state of being adapted ... The problem is how species can be at all times both adapting and adapted" (Lewontin 1984, 237-38). In fact, all we need do is follow Wallace's lead and understand that there *is* no "process of adaptation": only the *result* of stochastically accumulated adaptive structures that recapitulate past and present ecological associations, and that generate actions eventually playing out in space and time as responses to final causes inherent in the environmental delivery system. Through the deliberate elucidation of such causes we might finally elevate ourselves from the incomplete and under-nourishing evolutionary philosophy that a strict form of Darwinian "tree-thinking" produces—and at long last make an effort to attend to "Wallace's Unfinished Business."

Notes

1. See Smith (1991, 1992/1999, 2000–, 2003–, 2004a). Historian Martin Fichman has also adopted this interpretation in two recent works (Fichman 2001, 2004).
2. Once thought to be mere eccentricities, Wallace's excursions into spiritualism, socialism, and other radical ventures are increasingly being recognized as fitting logically into his overall philosophy of nature. See Fichman (2004).

19

Wallace in Wonderland*

James Moore

• • • • • •

... so many out-of-the-way things had happened lately, that Alice had begun to think that very few things indeed were really impossible ...

[Said Alice to the Cheshire Cat], "I wish you wouldn't keep appearing and vanishing so suddenly: you make one quite giddy."

"All right," said the Cat; and this time it vanished quite slowly, beginning with the end of the tail, and ending with the grin, which remained some time after the rest of it had gone.

"Well! I've often seen a cat without a grin," thought Alice; "but a grin without a cat! It's the most curious thing I ever saw in all my life!"

Lewis Carroll, *Alice's Adventures in Wonderland* (1865)

Today the world stands in awe of "Science" with a capital "S." It towers above us, mighty and austere, a colossus of "unnatural" knowledge, our secular providence (Wolpert 1992; Midgley 1992). This is a chimera. Scientists may send us to Mars, scientific research may root out Aids, but Colossus Science dazzles and distracts, blinding us to debates about natural knowledge-claims—about what may *count* as science—that took place over time. Two centuries ago there were only small-"s" sciences, local knowledge*s* claiming to be scientific. These raw materials were then

* Reprinted by permission from Volume 11 of the *Annals of the History and Philosophy of Biology* (Universitätsverlag Göttingen), copyright 2007.

This essay derives from lectures given at Bennington College, Bunkyo Gakuin University, Case Western Reserve University, the Open University, Oregon State University, the history of science section of the British Association for the Advancement of Science, the 2005 Shrewsbury Darwin Festival, and the 2003 conference Transformism, Evolutionism and Creationism, sponsored by the French Ministry of Foreign Affairs (Fond D'Alembert) and the Wellcome Trust Centre for the History of Medicine at University College London. The substance of the lectures was adapted from material prepared for the Open University foundation course A103, An Introduction to the Humanities.

shaped, pounded, and developed into the monolithic Science the world now reveres. As a biographer, I am as intrigued by the fabricating as the end-product: my job is to show how the Colossus was *made*. For Science is as much a natural product of history as any religious system and was formed by similar forces. This is now a commonplace among historians, and I want to illustrate it from the critical years when the British scientific mainstream acquired its anti-spiritual slant. I focus on the individual who, perhaps more than anyone, helped clinch this outcome, the brilliant, self-taught naturalist best remembered as Darwin's co-discoverer of natural selection, Alfred Russel Wallace (1823–1913).[1]

Biography makes science memorable. Who can forget the charming Ladybird "Lives of the Great Scientists" or the sumptuous TV biopics about Robert Oppenheimer and the Bomb, Rosalind Franklin and DNA, and the voyage of Charles Darwin? All these portraits, great and small, strengthen one of history's commonest preconceptions: that of the scientist as hero. Complex he or she may be; wracked by conscience, driven by ambition, tragically slighted or overlooked: nevertheless, like a lone mountaineer, the scientist-hero inspires and lifts our thoughts, leading us to new vistas of progress.

Heroes need human challenges as well as natural ones, men to master as well as mountains. Such is the drama of discovery endlessly scripted in our time. The scientist must have an adversary, an evil institution such as "the Church" or a corrupt individual, preferably a politician. This becomes his stumbling-block, frustrating free enquiry, impeding progress. Or a rival scientist may fit the bill: a vainglorious upstart, a grandiloquent imposter, a muddle-headed friend. Heroism consists in surmounting all such obstacles and, shunning self-pride and pretence, pursuing nature wherever "she" may lead (Golinski 1998, 192–94).

High in the first division of scientific heroes is the "Newton" of natural history, Charles Darwin (Ruse 1979, 31). It was he who slowly and methodically, with infinite patience and perseverance, singlehandedly solved that "mystery of mysteries," how living species originate. Or so hero-worshippers say. And to them Darwin had fiendish foes—bishops mostly, the odd politician, countless minor bigots—who fought in vain to stem the tide of truth. Less is heard of Darwin's professional rivals. Science with a capital "S" is supposed to be united; in heroic history, only renegades break ranks. Yet Victorian men of science did break ranks, and one of them is routinely cast as Darwin's opposite number, Wallace.[2]

Not that he is all ogre. As often as Wallace is made a whipping-boy or laughing-stock he is the genial seer or saint. Winsome and likeable, he sometimes plays constructive parts—the resourceful Tonto to Darwin's Lone Ranger, the helpful Watson to Sherlock Holmes. Even so, Wallace remains the anti-hero. In all roles he is the foil, a lesser light reflecting a greater glory. "In Darwin's shadow," according to a recent biographer (Shermer 2002), he is still "Darwin's moon" (Williams-Ellis 1966).

Wallace, no lunatic, is made a mere satellite because of his part in one of the most poignant ironies in the history of science. In February 1858 (the story goes) he had the brilliant misfortune to hit on a scientific "principle" twenty years too late. Unknown to him, Darwin, working privately, had got to it—natural selection—first, and when he heard of Wallace's work he scooped the kudos by rushing into print with *On the Origin of Species.* Wallace, his originality eclipsed, remained the perfect gent but stubbornly went his own way. Within a decade he and Darwin had parted company on a range of issues. Most remarkably, Wallace came out as a believer in disembodied spirits and gave them a role in evolution. In this he played the crank to Darwin's correctness and, it was said "lost caste terribly" (J. D. Hooker, in Colp 1992, 11). Only as an afterthought was he asked in 1882 to bear Darwin's coffin in Westminster Abbey (Moore 1982).

Ever since then, commentators on Wallace and Darwin have tended to take a moralizing line. They assess whose science was better or worse, superior or inferior, right or wrong. The question is always how much praise Wallace or Darwin deserves, how much credit the one should get relative to the other. This illustrates the peril of judging past science by present standards (Hardin 1960, 45; Ghiselin 1969, 150–51; Hardy 1984, 6–61, 64; White and Gribbin 1995, 233). The sense of science's *making*, of how the boundaries of science came to be drawn and of *how* Wallace and Darwin ended up on opposite sides over the spirits is lost by assuming in advance what "science" should mean.

Today's history of science tries not to take sides. It is anti-heroic, not because it favours anti-heroes but because it seeks to level the playing pitch and let everyone join in. The historian is not a referee, imposing our scientific rules on the past, but rather like a sports commentator, following the game of science as it was played, explaining strategies, describing the drama, the brilliant saves, the own-goals. The rules sometimes change, the goalposts move. Uproar ensues. Rival fans invade the pitch. Players are sent off and substitutions made. Order is restored. The game goes on, and its current state-of-play is known. But this knowledge should not skew the commentary. The match could have gone differently. By now another side might have been winning. Or in the future another will be—who knows?

So I propose to deal evenhandedly with Wallace and Darwin. By following the game of science as it was played in Victorian Britain, we not only learn about Wallace's efforts to make spiritual knowledge scientific; we also glimpse the processes by which what we now call "Science" was made.

New Scientists, New Science

Wallace does not wear labels lightly. Born in Wales to an impoverished lawyer, schooled to the age of thirteen, then apprenticed as a land surveyor, he was always awkward, independent, his mind shaped by a plebeian culture in which do-it-yourself science knew no bounds.

His opinions were typical of his age and class. Having renounced his parents' Anglicanism, he filled the void with utopian socialism and remained a radical freethinker for life. The latest would-be sciences attracted him. He learned to mesmerize from an itinerant performer and he gave demonstrations before small audiences. He took up phrenology and had his character "delineated" by getting his head shaved and "read." He hung out at Mechanics' Institutes, taught himself botany and geology, and in 1845, at the age of twenty-two, converted to the latest scientific heresy, evolution. Three years later he quit surveying and sailed for Brazil to search for a "theory of the origin of species." He drew a blank but persevered, sailing again in 1854 to the Dutch East Indies, where he hit on what Darwin already called natural selection (Moore 1997).

In 1862 Wallace returned to London like Rip Van Winkle waking to a new world. Famous now as Darwin's co-discoverer, but eight years out of touch, he made the rounds of scientific society, catching up on all the latest. Everywhere he heard about young men manoeuvring and new names rising. Victorian science was being transformed. The young guard called themselves "men of science" (only later "scientists"), and they owed their status more to merit than rank or wealth. Overworked and underpaid, they had won their spurs the hard way; engineers and naval doctors, surveyors and civil servants who had fought for funds and clawed their way to power. Some were taking top scientific jobs in the capital, pushing out the old genteel fat cats, their *bêtes noires*. The Oxbridge clergy with their cushy chairs, the City gents and dusty dilettantes all had divided loyalties; their science was yoked with God or mammon. The new men saw themselves as singleminded professionals, beholden to no one; a rising elite uniquely qualified to lead an emerging "scientific culture" (Yeo 1993, 32).

Their science matched their social ambition. It too was comprehensive, taking in life, the universe—everything. The older naturalists' world was split into material and spiritual parts and, the new men insisted, was destined for history's dustbin. Nothing now was sacrosanct, nothing taboo. Spiritual specialists like the Anglican clergy were worse than useless. Far from adding to knowledge, they blocked it. Asked how the universe was formed or living species originated, they answered "God." Asked about the human mind and morals, they dragged in the immortal soul. For such men all questions of origins and human nature lay shrouded in miracle and mystery, even despite the enormous gifts of science to material progress—steam traction, public sanitation, the electric telegraph, and more.

Science produced the goods—this was the new men's knock-down argument. All the great life- and labour-saving advances of the century had come from knowledge of natural law. Further progress would be made only as men like themselves discovered law and order everywhere, from nebular condensation to the evolution of humans. The origins and ends of things; life, mind, and morals—all would be shown to result from uniform material processes. Miracles and mysteries were finished; the spiritual specialists had to go. Britain's coming culture would be not just scientific but wholly secular.

Radical freethinkers had mooted this for decades. But here were respectable chaps, Fellows of the Royal Society, sounding off in public, baiting bishops, and even touting that old heresy, evolution. When Wallace arrived back in London it was the hottest scientific topic in town. Darwin had published his *Origin of Species* in 1859, arguing clearly and convincingly, with immaculate credentials, how living things had come into existence by a purely natural process. While older naturalists loathed the book, the new men loved it, and some were using Darwin's name to settle scores.

Darwin's book deliberately skirted that most sensitive subject, human origins. Not Thomas Huxley (1825–95), Darwin's self-appointed "bulldog," a naval doctor made FRS and now, still in his thirties, a new professor at the elite Royal College of Surgeons. He had his capacious jaws clamped on the darling of the Oxbridge divines, Richard Owen, head of the natural history collections at the British Museum. Owen at sixty was a brilliant anatomist and fossil expert—he invented the dinosaur-concept—and he had a fatal attraction for power. He was courtly, condescending, a Tory Anglican autocrat who hated evolution. His latest boast was that the brains of apes and humans are anatomically distinct; the latter could never have evolved from the former because the human brain had been "especially adapted to become the seat and instrument of a rational and responsible soul" (quoted in Desmond 1989, 288). Huxley let out a snarl and set upon him. The tussle spilled into the press. He savaged Owen for shoddy methods and betraying science. The structural differences between humans and the gorilla were in fact "not so great as those which separate the Gorilla from the lower apes," Huxley growled. Humans had evolved in body and brain, and the only theory with "any scientific" claim to explain it was the one "propounded by Mr. Darwin" (Huxley 1894, 144, 147).

The Origin of Mind

Wallace was no stranger to this controversy. An old radical himself, though barely forty years old, he had lived with apes in the Far East, comparing them with the native Dyaks, and hit on a "principle," which Darwin called natural selection, while pondering the origin of human races. He felt honoured to have prodded the great Darwin into print; overawed by the *Origin of Species*, he even admitted, "I really feel thankful that it has not been left to me to give the theory to the public" (Marchant 1916a, 1:73). Yet the men's names were inseparably linked, Wallace's modesty notwithstanding. The public counted him a member of the scientific avant garde and, like Huxley, a Darwinian defender of an ape ancestry for humans.

Not that the company Wallace kept was always respectable by Huxley's standards. He sought out radical allies in the new, men-only Anthropological Society of London. Its rooms were graced by a savage's skeleton and the meetings brought to order with a Negro's-head mace. Here all subjects were debated with virile directness, and in an unseemly reaction to the prudery of the age, the gents dwelt obsessively on such bare essentials as female "circumcision," phallic symbolism, and the anatomy of the "Hottentot Venus" (Stocking 1987, 247–54; Richards 1989; Qureshi 2004). The main theme, though, was race, and the tone fiercely racist. Wallace did not share the extremists' views, but the "Cannibal Club" (as members dubbed it) was the perfect place to stick his neck out. In a paper read before a meeting in March 1864 he became the first naturalist in Britain publicly to apply the theory of natural selection to the evolution of "man."

Rushing in where Darwin feared to tread, he tackled the fraught question of the origin of the human races. Were they separate species, as the racists claimed, with the Caucasian the highest and the Negro next to apes? Or did all races descend from a single ancestor and share a common humanity (as Darwin believed)? Wallace's answer was a clever compromise. He agreed with the extremists that the different *bodily* features of the human races—skin colour, hair texture, and so on—were developed from a homogeneous subhuman population in prehistoric times. These features had evolved by natural selection (or he would say, "survival of the fittest") as adaptations to different environments, just like the skins and furs of animals. But once the races acquired human *mental* qualities, their bodily evolution ceased. When humans began to control their environment, building shelter, making weapons, raising food, and above all aiding one another, all further advance was due to the power of mind. Natural selection now affected, not brawn, but brain. The fittest to survive were no longer physically the strongest, but mentally the brightest and most moral. Their "wonderful" faculties enabled them to escape the struggle for existence and enter a "social state" in which life was preserved and enhanced.

Where did the these mental faculties come from? If "that subtle force we term *mind*" (as Wallace called it) had brought about a "grand revolution … in nature," surely its appearance at a given "moment" was an even grander event, demanding explanation. Yet here he is strangely silent. His paper *nowhere* states how mind originated. Wallace is less concerned about the remote past than the not-so-distant future, when mind will be perfected. The fittest individuals, the brightest and most moral of every race will, he declares, transform the earth into "as bright a paradise as ever haunted the dreams of seer or poet." Utopia will simply evolve (S93 1864, clxvii–clxxx).

However, six years later, in 1870, Wallace's tune had changed. By now he realized that natural selection was not working as he had hoped. The "fittest" (by his standards) weren't surviving; "the mediocre" in "morality and intelligence" were swamping them. That year he reprinted his racial origins paper with a new conclusion, and here for the first time, publicly and dramatically, in a scientific text, he staked his utopian faith not just on the "glorious qualities" of mind that distinguish humans from animals, but also—amazingly—on "other and higher existences than ourselves, from whom these qualities may have been derived, and towards whom we may be ever tending" (S725 1891, 185). Not "which" but "whom"—Wallace now sought the origin of mind among *supernatural beings*.

Darwin's co-discoverer was the cuckoo in the new scientists' nest. Between 1864 and 1870 he turned traitor to their programme of universal explanation by natural law. Indeed, his "glorious qualities" and "higher existences" were reminiscent of the immortal souls touted by the parsons and Professor Owen. Just when their spiritual world was being banished, Wallace seemed to let it in through the back door. Why? Why did a respected naturalist suddenly break ranks and risk ridicule by embracing the supernatural? Was he stupid or just gullible? Maybe he was a crank after all. Instead of judging Wallace by today's scientific standards or those of his Victorian critics, we can try to account for his spiritual science in a constructive historical way.

Conversion

Wallace's interest in spirits had been piqued years earlier. While still in the tropics he heard about the parlour craze sweeping Britain and America, the weird rappings and rocking tables, the "miracles" and ghostly messages. Deploring superstition like a good freethinker, he resolved to investigate when he got home.

At first he was sceptical, and understandably. The spirit fad, or spiritualism, was the last wave in a tide of rural enthusiasms that engulfed upstate New York in the early nineteenth century. The wave spilled over like the rest and in the 1850s rolled into Britain behind a gypsy train of hucksters and hustlers calling themselves "mediums" (Cross 1950). These adepts claimed to have contacts beyond the grave; for a fee they would prove it with a séance. Parlours were darkened and hands held round a table. The spirits were invoked. Then the bumps would start, the table tilt, bells ring, breezes blow, candles burn, and objects would float in air. A message might be tapped out as if by telegraph or appear written on a slate. The showmanship was often spectacular, baffling unbelievers, who dismissed it all as fraud.

Yet many worthy persons were converted. Spiritualism satisfied the curious and soothed the bereaved (though some thought them self-deluded); equally it inspired political radicals, keen to underwrite their hopes. As a practical, empirical science, spiritualism served them well, like the old heresies phrenology and mesmerism. Its appeal, too, was direct and democratic—all were potential mediums, anyone could join a séance—plus it guaranteed an upward social evolution. Spirit was seen as a progressive force, immune to earthly failure; the supernatural world would transform the natural, bringing in the millennium (Barrow 1986).

Spiritualism also held a special place for women. From the society matrons who launched the movement to the working girls who joined it in the 1870s, females of every age and class thrived in the charmed circle of the séance. As a domestic circle it was of course one in which Victorian women were thought to function best, yet the spirits seemed specially drawn to them and would perform avidly in their presence. Shrewd ladies turned this to advantage, becoming star "public mediums." Impresarios such as Mrs Mary Marshall and Mrs Agnes Guppy put on theatrical displays of power. Besides bringing messages from the dead, they were known to cause objects—even themselves—to levitate, materialize, and disappear. Under spirit control such women apparently held sway over the material world in a manner only dreamt of by male scientists and politicians.

For this they were lionized, while radicals hailed the work of sister spiritualists as "invaluable ... for the furtherance of meaningful social reform" (Owen 1990, 28).

Make no mistake: the spiritual world of spiritualism was remote from that of the old men of science and the clergy. Their spiritual world was sacred and lay outside proper science. It could not be tampered with, for God and the soul were not experimental subjects. Also in this world, as in the material, a male elite officiated and policed the common boundary. There was no power-sharing with women. In spiritualism, by contrast, anyone could participate; the spirits could be beckoned and cajoled, and their antics manipulated. Females were spiritual specialists.

All this novelty, all this heresy, gave spiritualism a special allure, and Wallace was drawn to it helplessly after his paper on racial origins. This was a turning-point in his life. Single, shy, and over forty, he was also tall—six feet plus—gangly and gauche. Never mind, he resolved to marry, and was soon smitten with a woman of a superior class, still in her twenties. At first she resisted his attentions, but he persevered, and in mid-1864 her father agreed to an engagement and a wedding date was set.[3] Wallace himself takes up the story in a confessional letter to Darwin, dated 20 January 1865:

> For the last six months I have been doing absolutely nothing, & fear I shall not be inclined for work for some time to come. The reason is that I have suffered one of those severe disappointments few men have to endure. I was engaged to be married at Xmas, & had every reason to look forward to happiness, when at the last moment, when everything was arranged, & even the invitations sent out by the lady's father, all was suddenly broken off! No cause has been given me except mysterious statements of the *impossibility* of our being happy, although her *affection for me remains unchanged.* Of course I can only impute it to some delusion on her part as to the state of her health. You may imagine how this has upset me when I tell you that I never in my life before had met with a woman I could love, & in this case I firmly believe I was most truly loved in return.[4]

Wallace never again saw or heard of the woman or her family. Nor, he reflected, did he ever experience "such intensely painful emotion" (S729 1905a, 1:410).

Darwin's blunt advice—"banish painful thoughts" through "hard work"—was useless (Marchant 1916a, 1:160). Life had ground to a halt; serious work was impossible. There was nothing for it but to up stakes and start over, so just before Easter 1865, Alfred left lodgings in his sister's house and moved across London to live with his mother near the Regent's Park Zoo.

Three months later, as *Alice's Adventures in Wonderland* was published, Wallace tumbled headlong into the wonderland of séances.

At first he sat with friends and picked up the usual tapping and vibrations. Then in the autumn he visited the matronly Mrs Marshall, who astonished him by making a table levitate and revealing details about his long-dead brother Herbert.

Maybe an old mesmerist had such powers: Alfred practised at home for months without success. His sister Fanny came to the rescue; she said that her new lodger, young Miss Agnes Nicholl, could produce "curious phenomena." So in 1866 Alfred, Fanny, and friends began regular sittings with Miss Nicholl—later the redoubtable Mrs Guppy. They met on Friday nights, and now for the first time Alfred witnessed "miracles" in his own parlour. Stunned by the show, he immediately published detailed reports, giving the sitters' names and addresses lest sceptics doubt his word.

Here is Alfred on a cold December night. He sits in a stuffy shuttered room with a doctor, a lawyer, and other west London worthies, waiting expectantly. An hour has passed, the gas is turned down to a blue point and hands are joined in a circle.

> ... in a few moments several of the party said faintly that something was appearing on the table. The medium saw a hand, others what seemed flowers. These became more distinct, and some one put his hand on the table and said, "There *are* flowers here." Obtaining a light, we were all thunderstruck to see the table half covered with fern leaves, all fresh, cold, and damp, as if they had that moment been brought out of the night air. They were ordinary winter flowers which are cultivated in hot houses, for table decoration, the stems apparently cut off as if for a bouquet. They consisted of fifteen chrysanthemums, six variegated anemones, four tulips, five orange-berried solanums, six ferns of two sorts, one *Auricula sinensis* with nine flowers, thirty-seven stalks in all.

"Curiouser and curiouser." Wallace might have been botanizing in Brazil. He saw the "miraculous" with unblinking scientific eyes, and Miss Nicholl soon showed him that she could defy physics as well as biology by raising herself, chair and all, "instantaneously and noiselessly," to sit on a parlour table. She did so, he claimed, "some half dozen times, in different houses in London," before "at least twenty persons, of the highest respectability." This to Wallace was conclusive proof of spiritualism. He now knew not just "the reality of the facts," but also their implication: they had to be "the manifestation of some strange and preterhuman power" (S132 1867).

Why then did he convert to spiritualism? To say he was a fool would make a nonsense of his attainments in natural history; to say he was religiously motivated would deny his initial scepticism and rugged freethought. Spirits no more persuaded him of their existence than natural selection did of its reality. The conversion was neither a superhuman event, nor irrational, but merely a logical development of Wallace's long-term radical interests and convictions (Fichman 2001). He had been a phrenologist, a mesmerist, and an evolutionist for over twenty years. He believed that big brains in big skulls had enormous powers—unique powers, common to all humans without respect of race or class. Mesmeric phenomena showed him that mind was a "subtle force," evolution that this force arose in a body descended from apes, and natural selection that the human body ceased evolving when it acquired a mind. The unanswered question in his 1864 paper was, How?

Wallace made the mind a *spiritual* entity added to the body, not evolved, only after witnessing Mrs Marshall and Miss Nicholl produce phenomena that he himself could ascribe to nothing but some "preterhuman" power. In 1866 this was, for him, a scientific solution to a scientific problem. He sought out these women, however, only after a severe emotional crisis, which adds a further twist to the tale. Living with his mother, swayed by his sister, he remained vulnerable to forces—*social* forces—outside the male scientific establishment. No sooner had his fiancée misled him than he entrusted himself to female mediums, staking his reputation on their integrity. And he continued to do so after his mother's death in 1868, as when Miss Nicholl—now Mrs Guppy—revealed that her spirit would appear on a photographic plate. Sure enough, on 14 March 1874 Guppy produced the goods—Wallace describes the action (S717 1881, 190 n, for another photograph produced at this session, see Fig. 30):

> At the third sitting, after placing myself, and after the prepared plate was in the camera, I asked that the figure would come close to me. The third plate exhibited a female figure standing *close* in front of me, so that the drapery

Figure 30 A "spirit photo" taken on 14 March 1874 at Hudson's, with Mrs Guppy present as a medium.

> covers the lower part of my body ... [T]he additional figure started out the moment the developing fluid was poured on, while my portrait did not become visible till, perhaps, twenty seconds later ... [T]he moment I got the proofs, the first glance showed me that the third plate contained an unmistakeable portrait of my mother,—like her both in features and expression; not such a likeness as a portrait taken during life, but a somewhat pensive, idealised likeness—*yet still, to me, an unmistakeable likeness.*

In such events, perhaps, lies a deeper reason for Wallace's conversion to spiritualism, but fathoming it would require a full-scale biography.

Parting with Darwin

Wallace had seen the light. "The facts beat me," he insisted; "if I have now changed my opinion, it is simply by the force of evidence." Miss Nicholl's levitation had been the turning point, and afterwards he threw down the gauntlet: "Let those who believe it to be a trick, devote themselves to practise it, and when they are able to succeed in repeating the experiment, *under exactly the same conditions*, I will allow that some far more conclusive proof of the reality of these manifestations is required" (S132 1867).

So the onus of replication was now on sceptics. Wallace made sure by publishing a small pamphlet with a big title, *The Scientific Aspect of the Supernatural: Indicating the Desirableness of an Experimental Enquiry by Men of Science into the Alleged Powers of Clairvoyants and Mediums.* In late 1866 he rushed copies to the men of science he respected most, including Darwin's pit-bull terrier Thomas Huxley. "I have been writing a little on a *new branch* of Anthropology," Wallace explained disarmingly. "I fear you will be much shocked, but I can't help it; and before finally deciding that we are all mad I hope you will come and see some very curious phenomena which we can show you, *among friends only* ... We wish for the fullest investigation, and shall be only too grateful to you or anyone else who will show us how and where we are deceived" (Marchant 1916a, 2:187).

Huxley had had his fill of séances years before and he hated the supernatural. Spiritualism "may be all true, for anything I know to the contrary," he dissembled, "but really I cannot get up any interest in the subject." He was not shocked, nor did he think Wallace mad. The poor man had merely been duped by dotty ladies into wasting precious time. "I never cared for gossip," Huxley scowled, "... and disembodied gossip, such as these worthy ghosts supply their friends with, is not more interesting to me than any other." For *his* part, he had "half-a-dozen investigations of infinitely greater interest" to conduct, which left *him* no "spare time" for spiritualism. "I give it up for the same reason I abstain from chess—it's too amusing to be fair work, and too hard work to be amusing" (Marchant 1916a, 2:187–88).

Wallace was not amused, and he tried to raise the tone. He cared for gossip as little as Huxley did, "but what I do feel an intense interest in is the exhibition of *force* where force has been declared *impossible*, and of *intelligence* from a source the very mention of which has been deemed an *absurdity*." He invoked the name of Faraday, the foremost experimental physicist, who had declared that anyone who could prove the existence of "a power not yet recognised by science" would receive "applause and gratitude." "I believe I can . . . show such a force," Wallace insisted, adding that he now hoped physicists would "admit its importance" and look into it (Marchant 1916a, 2:188).

Here then were two accomplished naturalists of the same generation, both "Darwinians," disagreeing about the significance of spirit phenomena. Which of them was being "scientific"?

They *both* were, each according to his lights. Theirs was a dispute about what should *count* as science. To Huxley, being "scientific" meant investigating natural phenomena on naturalistic assumptions. Spiritualism violated these assumptions, so its phenomena could have no place in *his* science. To Wallace, being "scientific" meant investigating all alleged phenomena, even those deemed impossible or absurd. Spiritualism to him was proved by its phenomena, so they became integral to *his* science. This science included the super- or preternatural; Huxley's denied its existence. Wallace saw himself as working on "a *new branch* of Anthropology"; Huxley remained loyal to Darwin's. *Neither* anthropology, however, was seen as "scientific" by older naturalists and the clergy, who not only kept spiritual things out of science but rejected evolution also. The question for everyone in the 1860s was: Whose science shall win?

Ostracism

By the time Wallace's pamphlet reached Darwin, they had struck up a lively correspondence. Socially and intellectually Wallace was the junior partner, but his transparency, encyclopaedic knowledge, and persuasive powers impressed Darwin enormously—so much that in 1868 he confessed to Wallace, "I grieve to differ from you, and it actually terrifies me, and makes me constantly distrust myself" (Marchant 1916a, 1:227).

Conflicts were now emerging, mostly technical ones about the application of their theory. Wallace defended natural selection brilliantly, arguing only about how much it could explain. Darwin had admired his paper on racial evolution, but after receiving the spiritualist pamphlet, he worried more and more. In 1869, after learning that Wallace was to review the new edition of his old mentor Charles Lyell's *Principles of Geology*, he despaired. Lyell had crushed him by refusing to "go the whole orang" on human origins. Now Wallace threatened worse—to backslide. "I hope," Darwin shuddered, "you have not murdered too completely your own and my child" (Marchant 1916a, 1:241).

Infanticide it was. The review was brutal. Its conclusion seemed to Darwin so unscientific that he thought it might have been "added by someone else" (Marchant 1916a, 1:243). Wallace now argued that neither the mind *nor* all bodily features of humans could have evolved by natural selection. Primitive people possessed mental capacities far in *excess* of their survival requirements; they had physical features that were apparently *useless* except in a civilized state. Big brains, exquisite hands, naked skin, speech organs—such things must have evolved *prospectively*, long before they were needed, which showed intelligent foresight. Natural selection, being blind, could not have been the cause, so Wallace detected a supernatural "Power" guiding evolution.

Darwin, feeling betrayed, stabbed exclamation marks and scrawled indignant notes in the margins. "No!!!!" "I think the same argument could be applied to any animal—what use 5 toes to dogs foot"? Or what use a fine hand—"ties knots"—"opening fruit."[5] Suddenly he realized how "grievously" they differed; he was "very sorry for it" and told Wallace so (Marchant 1916a, 1:243). It was a watershed in their relationship. Never again would he fully trust his colleague's scientific judgement.

Wallace's review said nothing about spiritualism; his pamphlet only called for an "experimental enquiry" into its phenomena, but Darwin still read blank credulity between the lines. It was all too much, even if Huxley could see an advantage: proving spiritualism true would cut the suicide rate. "Better live a crossing sweeper," he laughed, "than die and be made to talk twaddle by a 'medium' hired at a guinea a *séance*" (in Anon. 1871, 230).

More sober scientists, or those with less to lose, took up Wallace's experimental challenge (Noakes 1999, 2002, 2004; Gay 1996). William Crookes, the analytical chemist who discovered the element thallium, constructed special apparatus and in 1870 began séances with the respected American medium Daniel Home. Darwin's true-blue cousin Francis Galton, a total sceptic, attended one and was "utterly confounded." In "full gas-light," with "perfect apparent openness," Home produced the most "extraordinary" phenomena—even the playing of an accordion suspended by one hand. This was no "vulgar legerdemain"; Crookes had taken "thoroughly scientific" precautions (Pearson 1924, 64). Galton begged Darwin to come and see for himself, but he refused, pleading ill health.

Darwin only once sat in a séance, in 1874, at his brother's house in London. The novelist George Eliot was present as well as Galton and other relatives. The performance was about to start when Darwin suddenly broke the spell, made excuses, and went upstairs to lie down. When he returned, he found the table stacked with chairs, which reportedly had been lifted over everyone's heads, with sparks flying, and wind rushing, and strange rapping. "The Lord have mercy on us all, if we have to believe in such rubbish," he groaned (F. Darwin 1887, 3:187). His wife Emma, who had seen the show, explained, "He *won't* believe it, he dislikes the thought of it so very much." Smiling sweetly, she branded him a "regular bigot" (Wedgwood 1980, 305). Only after Huxley had attended another séance and declared the medium a cheat did Darwin relax. It would now take "an enormous weight of evidence" to convince him that there was anything in spiritualism but "mere trickery" (Pearson 1924, 67).

The issues boiled up again at the 1876 meeting of the British Association for the Advancement of Science. By now Huxley's new model scientists dominated the annual roadshow, so there was uproar when Wallace, presiding over the Anthropology section, allowed a paper on thought transference to be read. The author was another physicist, W. F. Barrett, who had worked under Faraday. The row splashed into *The Times* and got linked with the case of Henry Slade, a American medium being sued by Huxley's protégé, the fiery Darwinian zoologist Ray Lankester. Lankester had caught Slade cheating in a séance; Wallace had sat with him and seen only miracles. The case ended up in London's Bow Street magistrate's court with Wallace as the star defence witness. Behind the scenes Darwin bankrolled the prosecution, which he considered a "public benefit" (Milner 1990, 29). Slade was convicted and fled the country, his career as a con-artist in ruins. Wallace wiped the egg from his face and walked out of the British Association.

This was a defining moment. Spiritual phenomena were now ruled scientifically out of bounds. At the British Association's 1878 meeting, Huxley took Wallace's place as chair of the Anthropology section and uttered a stern warning: "If any one should travel outside the lines of scientific evidence, and endeavour either to support or oppose conclusions which are based upon distinctly scientific grounds, by considerations which are not in any way based upon scientific logic or scientific truth ... I, occupying the chair of the Section, should, most undoubtedly, feel myself called upon to call him to order, and to tell him that he was introducing topics with which we had no concern whatever" (Huxley 1879, 576). Wallace, who was absent, had got the message already. His disgracing in the Slade trial had been deliberate; his spiritual science was beyond the pale. He now met with a ginger group of intellectuals who, in 1882, joined Barrett, Crookes, and fellow physical scientists to form the Society for Psychical Research (Gauld 1968; Haynes 1982). Its force was spent within a few decades.

* * *

For all his fame as Darwin's co-discoverer, Wallace was a radical round peg among the neat square holes of the rising scientific professions. He never fitted in, never specialized, never unlearned. His science sprawled untidily, from phrenology and mesmerism to spiritualism and socialism, confirming Darwin's fear that he might "turn renegade to natural history" (Marchant 1916a, 1:318). Even allies became uneasy. Oliver Lodge, a noted physicist and spiritualist, thought him "a good observer ... with a great deal of self-confidence in the midst of much simplicity and modesty," but equally "a crude, simple soul, easily influenced, open to every novelty and argument" (Hill 1932, 34).

Amiable and honest, stubborn and naive, this was the wonderful Wallace, author of ten major scientific tomes, who, when proposed for a Fellowship of the Royal Society in 1893, at the age of seventy, could not see why he should be so honoured: "I really have done so little of what is usually considered scientific work," he demurred politely (quoted in Durant 1979, 33).

Well, he had and he hadn't. Much of his "scientific work"—for which he was indeed elected FRS—belonged to the new Darwinian sciences of the late Victorian age. Much, too, belonged to the do-it-yourself, democratic sciences of the past. These sciences no longer counted in the new professionals' eyes, but for Wallace and many like him they remained vital. Understanding why this was so helps us to see how "Science" has changed, and it may give us hope. For in a day when Colossus "Science"—with a capital "S"—is revered, when scientists still draw boundaries, excluding competitors, defying critics; in a day when Intelligent Design is said to put the supernatural back in science and some evolutionists dare to see non-selective forces at work in evolution, Wallace's case reminds—and warns—us that sciences once made can be made again.

Said Alice to herself: "Dear, dear! How very queer everything is today ... Who in the world am I? Ah, that's the great puzzle!"

Notes

1. No attempt is made here to take into account the recent spate of Wallace books, the most substantial of which are biographical studies by Raby (2001), Shermer (2002), Fichman (2004), and Slotten (2004), and anthologies by Smith (1991), Berry (2002), and Camerini (2002). For a critical view, see Endersby (2003).
2. In his lately much celebrated textbook, *The Darwinian Revolution* (1979, 280), Michael Ruse explains: "Like nearly everyone else, I find myself relegating Wallace to the notes. This is unfair, since Wallace really did discover natural selection as an evolutionary mechanism, but not totally unfair. Wallace's creative work came twenty years afer Darwin's, he did not write out a full theory, and he did not form a party of supporters."
3. From manuscript notes for Wallace's autobiography, now in the Natural History Museum, London, Raby (2001, 170–71, 180–81) has identified the fiancée as Marion Leslie, elder daughter of Wallace's chess-playing friend Lewis Leslie, a London auctioneer and widower who lived in Kensington with offices in Mayfair.
4. Dar 106/7 (ser. 2), 20–21, Darwin Archive, Cambridge University Library.
5. Darwin's marginalia on his copy of Wallace's review are in the Darwin Reprint Collection, Cambridge University Library.

20

Wallace's Dilemmas: The Laws of Nature and the Human Spirit*

Ted Benton

Introduction

After many years of neglect, Alfred Russel Wallace has recently been the subject of a flood of scholarly studies. Earlier efforts often focused on one or another aspect of his exceptionally diverse intellectual and practical engagements, with a particular emphasis on the timing of his independent discovery of the key mechanism of organic evolution. More recent scholarship—most notably the work of Charles H. Smith and Martin Fichman—has recognized the intimate interconnections between his philosophical, political, spiritual, and scientific thinking. This holistic approach gives due weight to the phrase often used by Wallace when describing himself—"philosophic naturalist"—and will be adopted here.

However, there remain disputed areas of interpretation. Perhaps the most significant of these surrounds the emergence of a public disagreement between Wallace and Darwin on the topic of the origins and nature of the human species. While both agreed that humans had emerged from some primate ancestor in the remote past, Wallace became convinced that some "superior intelligence" had played a part in the development of the "higher" moral and mental faculties that raised humans far above other animal species. By contrast, Darwin remained committed to a thoroughly materialistic understanding of human evolution and distinctive character. This divergence is among the philosophical roots of the long, and now contested, historical split between evolutionary biology and what is sometimes called the "Standard Social Science Model" (Barkow *et al.* 1992; Pinker 1997): the controversial claim that the social sciences and humanities still work

* I am greatly indebted to George Beccaloni for illuminating discussions and for indispensable bibliographical help. Also an indispensable resource has been Charles Smith's *Alfred Russel Wallace Page* (Smith 2000–). I have, in addition, benefited greatly from his severe but fair editing of an initially overlength chapter. Thanks are also due to the staff of the Natural History Museum (London) library for giving me access to their Wallace archive.

with a pre-Darwinian view of human nature and (often) implausibly utopian visions of human potential (see Benton 1999b for a rejoinder).

One common interpretation of Wallace's departure from Darwinian orthodoxy on this topic has been to see it as an outcome of his growing "non-scientific" preoccupations. He had, by the late 1860s, become an ardent follower and advocate of the spiritualist movement. This coincided with a renewal of radical moral and political intuitions he had developed many years before through his experiences as a land surveyor, encounters with the socialist ideas of Robert Owen, and fondness for the writings of Thomas Paine. From the mid-1860s, on this line of argument, Wallace's spiritual and moral-political beliefs propelled him into a super-naturalistic view of human nature and prospects, and so to a radical "change of mind" away from his earlier scientific orientation.

Against this, Charles Smith has, in a series of scholarly publications (notably Smith 1992/1999, 2003–, 2004), sought to show that Wallace's philosophical, scientific, and political views formed a coherent framework of ideas that had already been established in its broad outlines as early as 1843–45. It would be anachronistic, he shows, to insist on a clear and consensual demarcation between scientific and non-scientific thought during this period. Following this line of thought, Smith provides a close analysis of Wallace's writings to show that there never was a major "change of mind," that Wallace's conversion to spiritualism was always predictable from what we know of his earlier framework of belief, and that the convergence with Darwinian evolutionism was always only partial. Since Wallace had never committed himself to the view that natural selection provided a complete explanation for everything human, his public departure from that position toward the end of the 1860s should not be seen as such a great transformation of his views. Smith's analysis is strongly supported by Martin Fichman, who has offered the following succinct summary: "Wallace's basic approach to the study of man and nature, however, was set in his mind well before he finally hit on natural selection. He maintained a consistent but evolving overall worldview in his writings over a span of seventy years. Neither natural selection nor spiritualism was a departure from this central vision. Momentous as the discovery of natural selection had been, Wallace was skeptical as to its competence to explain all of human evolution. He envisioned some additional explanatory model to resolve fully the question of human origins, their higher faculties, and their future evolution" (Fichman 2004, 194–95).

In what follows, I offer an interpretation of shifts in Wallace's "synthesis" through the highly significant period extending from the early 1850s to 1870. My reading adopts the view of Smith and Fichman that Wallace's scientific concerns were never fully separable from the moral, political, and metaphysical aspects of his (shifting) world-outlook. However, on the specific question of the alleged "change of mind" my reading suggests an interpretation somewhat at odds with both the prevailing views. His voluminous correspondence, reminiscences, critical interventions in public controversy, private notebooks as well as scientific works together bear witness to Wallace's continuing struggle to form a coherent synthesis reconciling his diverse intellectual and practical engagements. However, to recognize this is not to concede that his (always provisional) syntheses ever quite managed to achieve full internal consistency. There were several fault lines in Wallace's thought that issued in conceptual tensions and consequent repeated recastings of his outlook. Most significant among these were problems associated with the emergence of life itself, the link between higher animals and consciousness, the origin and nature of the "higher" human faculties, and the relation between the "civilized" and "savage" within the human species.

To anticipate, I will suggest that, indeed, throughout this 1850s to 1870 period, Wallace is best understood as a philosophic naturalist—one who was attempting to forge a coherent unity of thought on the full range of his intellectual, moral and metaphysical concerns. His efforts, I believe, exhibit three distinct phases, during each of which Wallace arrives at a provisional, but ultimately unstable synthesis. The three phases are: (1) a first phase extending from the early 1850s through to the "breakthrough" concept of natural selection (or as Wallace later preferred to call it, the "survival of the fittest") in early 1858, (2) a second phase beginning with the latter, and continuing on through his introduction to spiritualism, and (3) a final phase foreshadowed by a series of transitional writings from approximately 1866 through to the late 1860s, and a full synthesis appearing with Wallace's review of two works by Lyell (S146 1869) and his essay "The Limits of Natural Selection as Applied to Man" (S716 1870, 332–71). My tentative suggestion is that the "change of mind" interpretation gets its legitimacy from its focus on the shift from phase two to phase three. The alternative view gets its support from an emphasis on continuities between phases one and three, glossing over or reinterpreting some key texts of phase two. These phases are not in every case clearly demarcated, and there are some pieces of writing that appear to be transitional in character.

Wallace's Synthesis: Phase One (early 1850s to approximately 1857)

By the time of his departure on his second great foreign adventure in 1854, Wallace had long since been a "transmutationist" (from his critical but still sympathetic reading of Lamarck, and, more immediately, Robert Chambers's *Vestiges of the Natural History of Creation*). It seems likely, too, that he was already speculating on the origins of humans from ape-like ancestors. The possibility of encountering the orang-utan, on that hypothesis one of our closest living relatives, may well have influenced his choice of destination. Wallace's researches were also by now shaped by a relatively flexible but still definite set of broad principles and intuitions about the world and its workings. These were complemented by a view of both the moral/aesthetic value of the attempt to understand these workings of the world, and broad methodological and philosophical views on the nature, sources, and limitations of human knowledge. In common with the wider Victorian culture, he was also a believer in "progress," but gave to this idea a variety of different meanings in different contexts and at different points in his intellectual career.

Wallace's religious scepticism during this period is described in his autobiography (S729 1905a, 1:88–89). His position stopped short of a fully atheistic outlook, however, and there are fragmentary evidences of his not having ruled out the possibility of some sort of "afterlife." In particular, his notebooks and several publications from the early 1850s onwards indicate a continuing willingness to refer to the possibility of a "Supreme Being" or "Creator." As to Wallace's political outlook, his sense of outrage at injustice was seasoned by his experiences as a land surveyor in the late 1830s and early 1840s. These not only led him in the direction of natural history, but also introduced him to rural poverty and the injustice through the Enclosure Acts and "tythe commutations" he was required to enforce. Undoubtedly these experiences were a main inspiration for his later passionate commitment to the cause of land nationalization.

Potentially more of a test for Wallace's early egalitarianism might have been his encounters with people of other races and cultures—first on his Amazonian adventures, and then on his travels in the Far East. In fact, Wallace's characteristic open-mindedness, generosity of spirit, and critical orientation to his own civilization, left him full of awe and admiration after his first encounters with "absolute uncontaminated savages." Interestingly (and probably a legacy of his having read key works of the European Enlightenment), Wallace uses the categories of "civilized" and "savage" in his thinking about these encounters with humans who remain in "the state of nature," and there is certainly something of a Rousseauian celebration of the "noble savage" in his (partial) inversions of the more typical nineteenth-century value connotations of the contrast. As we shall see, the triple intersection of Wallace's direct experience of indigenous people with his political philosophy and evolutionary biology interests generates recurrent tensions in his later thought. In particular, the category of "progress" that spans these domains and seeks to integrate them with the normative and metaphysical dimensions of Wallace's outlook is subject to repeated recastings and transformations.

As to Wallace's "intuitions" about the nature of the world and its workings, one important early influence was the writings of Charles Lyell on what came to be known as "uniformitarian" geology: at its most basic this theory conveys a sense of the enormous diversity of things, events and processes of the world as forming an underlying law-governed unity. By the mid-1850s, on the evidence of his Species Notebook, the encounter with Lyell's geological work has convinced him of a more philosophically developed form of uniformitarianism that implies a related set of criteria by which to assess the scientific validity of explanatory hypotheses (*e.g.* Species Notebook, 45–53). There are three basic principles. First, that the forces shaping the earth in the distant past are broadly the same as those at work today; that is, there is historical constancy to the laws of nature. Second, that the changes that have occurred and still continue have taken place gradually, by small steps, over immensely long periods of time. Third, that the laws and forces that have operated and continue to operate in shaping inorganic nature are closely related to, if not identical with, those governing the succession of organic forms as disclosed by the fossil record. Wallace endorsed all these postulates, and, for him, they implied that successful scientific explanation unifies the widest possible diversity of classes of fact under the smallest number of interrelated universal laws.

This Newtonian conception of the unity and law-governed character of both organic and inorganic nature strongly disposed Wallace against the prevailing view of each species as a "special creation." His notebooks and publications of this period use four main arguments against this doctrine:

1. That it is "unphilosophical" (*i.e.*, inconsistent with the logic of the unity of nature) to suppose that the geological/environmental changes and species extinctions that the fossil record reveals should be due to secondary causes while species introductions should be treated as the work of a creative "Supreme Being." Exposing the inner contradiction in Lyell's creationism, Wallace argues: "It would be an extraordinary thing if while the modification of the surface took by natural causes now in operation & the extinction of species was the natural result of the same causes, yet the reproduction & introduction of new species required special acts of creation, or some process which does not present itself in the ordinary course of nature" (Species Notebook, 50–51).

2. That it fails to make sense of the details of distribution and affinities of species of animals and plants; notably, that structurally closely related groups are found in geographical proximity, though differing in their specific adaptations to local conditions, while very distant but ecologically similar places are inhabited by quite different groups of organisms (*e.g.* Species Notebook, 45–53).

3. That design by a creator, relying on functional relationships between organisms and their conditions of life, reduces to tautology: "Had the bats large eyes, it would be brought forward as an arrangement in exact accordance with their necessities, as purely nocturnal animals. But they have very minute eyes & apparently imperfect sight & they do very well without them" (Species Notebook, 12). Such pseudo-explanations are a mere cover for our ignorance: "We are like children looking at a complicated machine of the reasons of whose construction they are ignorant, and like them we constantly impute as cause what is really effect in our vain attempts to explain what we will not confess that we cannot understand" (Species Notebook, 32).

4. That such a view implies a low estimation of the powers of the Creator. To attribute the provision of simple functional relationships in organisms to a Supreme Being would be to impute to Him "a degree of intelligence only equal to that of the stupidest human beings. What should we think, if as a proof of the superior wisdom of some philosopher, it was pointed out that in building a house he had made a door to it, or in contriving a box had furnished it with a lid"

(Species Notebook, 31). He adds "Could the lowest savage have a more degrading idea of his God?"

Taken together these arguments sustain Wallace's disposition to the hypothesis of gradual modification, or "transmutation," of organic forms, such that new species emerged by descent from earlier, closely allied ones. The strongest non-theological argument against this view was the widely shared opinion that variation within a species was always confined within definite limits. This was Lyell's position at the time, and Wallace's notes record several strategies for dealing with it. One is to consider what extreme transformations have been brought about in domesticated animals and plants on a timescale far shorter than that of life on earth. Wallace clearly intends no analogy at this stage between transmutation in nature and changes brought about by domestication: the point is merely illustrative. But Lyell, while recognizing transmutation by domestication, insists only that it produces distinct varieties, not new species. For Wallace, on the other hand, the differences produced are so large that we have every reason to consider them new species. Finally, the notion that variation can never exceed the limits of species is mere prejudice—where there is evidence of it, Lyell merely defines the new form as a variety. Although Wallace later recognizes a strong distinction between species and sub-specific varieties, a relativistic view of the difference is important to his defence of the transformationist perspective at this stage.

However, in his seminal paper of 1855, "On the Law Which has Regulated the Introduction of New Species" (S20 1855), Wallace stops short of explicit commitment to the concept of transformation. He concedes that the mechanism by which a new species arises is quite unknown, but expresses the hope that his law is a step towards its discovery: "To discover how the extinct species have from time to time been replaced by new ones down to the very latest geological period, is the most difficult, and at the same time the most interesting problem in the natural history of the earth" (S20, 190; S716, 14).

How far Wallace was at this time from the achievement of this explanation is indicated by his strong opposition to explanations of organic traits invoking functional utility. This opposition was central to his case against special creation and the argument from design (Smith 1991, 1992/1999). His inclination in the Species Notebook from 1855–56 is instead to view structural features as prior to adaptive traits (Species Notebook, 53). This resistance to functional explanation is still clearer in the 1856 article "On the Habits of the Orang-Utan of Borneo" (S26). Wallace quickly dismisses attempts to explain the huge canine teeth of the orang in terms of its need to defend itself against predators. He goes on to list examples, such as the brilliant coloration of the plumage of some birds and the colours and forms of the petals of some flowers that, he claims, cannot be explained in terms of their utility to the bearer. Instead, we must look for some "general principle far more recondite."

Wallace goes on to suppose there must be some "general design" in the system of nature that determines the details. By a careful study of these "we may learn much that is at present hidden from us," and come to a "complete appreciation of all the variety, the beauty, and the harmony of the organic world," so long as we don't make the mistake of assuming that every "modification" exists solely for some use (S26, 30–31). These comments seem at odds with Wallace's criticisms of both Lyell and Knight in the notebook written only a few months earlier. There, Wallace chastised the proponents of design for the presumption of having "been behind the scenes at the creation & to have been well acquainted with the motives of the creator" (Species Notebook, 32). However, the reintroduction of the idea of design at this point remains consistent with the substantive case Wallace has made against design as evidence for the special creation of each species. In imputing to a creative agency design at the level of the "system of nature" as a whole Wallace remains true to his uniformitarian expectation that natural processes are to be explained as conforming to, as he puts it here, "general principles." The more "recondite" principles that will explain evidently non-utilitarian features in living beings are not to be arrived at by tautological pseudo-explanations or by hubristic claims to know the mind of the creator, but instead by painstaking empirical observations of the "details" of nature.

Wallace's acknowledgement of "design," moreover, is no retreat from his philosophical commitment to seek explanations for natural world features in terms of the operation of general laws or principles. There may be some form of ultimate purpose to the making of the world as a whole, but this will not license the use of teleological explanation at the level of *particular* events and processes. Further, the suggestion seems to be that while hypothetical thought combined with detailed observation and experiment may enable us to grasp something of the overall pattern or "design" of the world, this will not amount to a knowledge of the mind or intentions of whatever creative force it is that represents its ultimate author.

Next, there is the matter of Wallace's commitment to the idea of "progress." For him, as for many other Victorians, belief in progress was more than just a moral stance: he viewed it as some kind of endemic force or tendency inherent in the world itself. As such, progress entails change, of course, but change with both a direction and meaning, or value. We have already seen that Wallace's political views were progressive; he held to strong egalitarian principles and to a view of life as actualization of the potential inherent in the human faculties of goodness, love of beauty, and intellectual commitment.

During the mid-1850s the concept of "progress" is centrally at work in his nascent evolutionary hypothesis. In fact, the ideas of transmutation of species and of "progressive development" are frequently run together as if they were inseparable aspects of evolution (although in the 1855 paper he does make the analytical distinction: the law, he says, is "only one of gradual change" but that it is by "no means difficult to show that a real progression in the scale of organization is perfectly consistent with all the appearances, and even with apparent retrogression, should such occur" (S20, 191; S716 1870, 15). But what does "progress" amount to in the historical sequence of organic forms, and how is the fossil record to be considered consistent with it? The Species Notebook of 1855–56 contains detailed responses to palaeontological objections to "progressive development" presented by Lyell, the upshot of which is that Wallace takes "progressive development" at two distinct levels—in the increasing specialization and organizational complexity detectable in single lineages, and in the sequence from "lower" to "higher" in the evolutionary emergence of major taxonomic groupings of living beings as a whole. His great difficulty in sustaining the latter thesis is the relative lack of fossil evidence of transitional forms. Here, as elsewhere, Wallace takes refuge in the incompleteness of the fossil record as so far exposed and studied. However, an alternative source of evidence is the "rudimentary organs" found in some animals—the "minute limbs hidden beneath the skin in many of the snake-like lizards, the anal hooks of the boa constrictor, the complete series of jointed finger-bones in the paddle of the Manatus and whale" (S20, 195–96). Wallace clearly views these as transitional forms, as anticipations of future developments, suggesting, here, teleological explanation in the sense of determination of a present change by approach to an anticipated future state. In his subsequent republication of the essay he acknowledges his error in this (S716, 24).

Neither Wallace's published writings nor his notebooks from this period directly confront the matter of human origins. However, his uniformitarian philosophy and meta-theoretical commitment to "progress" might be taken to imply a naturalistic view of humans as subject to the same laws as their fellow animal species. Much of Wallace's writing from his *Narrative of Travels* in 1853 (S714 1889) onwards is consistent with this expectation. His accounts of the races of the Amazon in that book are written in much the same "taxonomic" mode as his descriptions of the other forest species—though, in addition to detailed accounts of stature, colour, physiognomy he adds equally dispassionate descriptions of character, dress, beliefs, and material culture. Much the same is true of his treatments of the racial and cultural groups he encountered in the "Malay Archipelago," though there, and in his eventual organization of the material for publication, the taxonomic anthropology is deployed as part of his biogeographical argument. But this, too, suggests a willingness to treat human groups alongside other species.

That Wallace was prepared to argue for an evolutionary relationship between humans and the orang is strongly suggested by a passage from his 1856 essay on the habits of the orang-utan:

> ... with what anxious expectation must we look forward to the time when the progress of civilization in these hitherto wild countries may lay open the monuments of a former world, and enable us to ascertain approximately the period when the present species of Orangs first made their appearance, and perhaps prove the former existence of allied species still more gigantic in their dimensions, and more or less human in their form and structure! (S26, 31).

Wallace's interest at this time in the question of human origins is also suggested by a 22 December 1857 reply from Darwin to one of his letters: "You ask whether

I shall discuss 'man.' I think I avoid the whole subject, as so surrounded with prejudices; though I fully admit that it is the highest and most interesting problem for the naturalist" (Darwin 1887b, 2:109). Some comments in the Species Notebook also suggest a softening of the boundaries between human and animal. In his discussion of "instinct" he doubts that humans have any, and suggests that other species—notably birds in their migratory and nesting habits—act with foreknowledge, and are capable of learning from their fellows (Species Notebook, 166) Again, in the same notebook he quotes Lyell's opposition to the possibility of a gradual transition from irrational to rational, such a "leap" being different in kind from the "passage from the more simple to the more perfect forms of animal organisation & instinct" (Species Notebook, 149). Wallace's comment is: "Here the absolute distinctness of reason & instinct is assumed; the argument depends on the terms rational and irrational which imply no gradation." Taken together with Wallace's comments on the mental life of birds and insects, this strongly suggests an inclination to include humans within the scope of his transformationist hypothesis.

But there are three unresolved areas of tension in Wallace's thinking during this period. The first is his acknowledged failure to that point to hit on a scientifically acceptable mechanism of organic change. For Darwin, strongly affected by the argument from design, this quest was circumscribed by the need to equate change with adaptation. In Wallace's case, the situation was very different: whatever was at work in bringing about the transmutation of species would have to provide an explanation of the presence of many apparently *non-utilitarian* characters, including beauty of form and colour.

A second contradiction lies in Wallace's responses to the orang. There is no doubting his great anticipation and then excitement at his first encounter. For Wallace 19 March 1855 was "a white letter day": his first sighting of the "Orang utan or 'Mias' of the Dyaks in its native forests" (Species Notebook, 5). However, his overriding concern is to recruit native help in tracking and killing as many of the "monsters" as possible, so their carcasses can be measured, skinned, and prepared as biological specimens (Fig. 31). The hunts were ruthless and apparently devoid of any moral reflection on Wallace's part; nevertheless, even in describing his Mias hunts Wallace cannot avoid attributing emotions and intentions to his quarry. A "huge black head" looks down "surprised at the disturbance," during one chase, and, on another occasion, a Mias throws down branches in its fury. Still more significant, are Wallace's tellings of the story of his attempt to raise an infant Mias, having shot dead its mother (S23 1856; S715 1869a, 1:53–57; S729 1905a, 1:343–45). His own strong paternal feelings for the creature are unmistakable: "From this short account you will see that my baby is no common baby, and I can safely say, what so many have said before with much less truth, 'there never was such a baby as my baby,' and I'm sure nobody ever had such a dear little duck of a darling of a little brown hairy baby before" (S729 1905a, 1:344–45. See Benton 1997 for a more detailed analysis of this episode).

Figure 31 Orang-utan attacked by Dyaks, from Wallace's book *The Malay Archipelago*.
Out of copyright.

Martin Fichman, however, has argued that Wallace "still seems to have regarded the gulf between these Great Apes ... and humans as conceptually unbridgeable." He continues: "Did these observations of the orangutan during the 1850s plant one of the seeds that would prompt Wallace to declare in the 1860s that human evolution was guided by factors other than natural selection and contribute to his refusal to leave the divine out of the history of *Homo sapiens*' development?" (Fichman 2004, 39). In view of the points made just above, Fichman's reading seems hard to justify. A more defensible interpretation might be that the orang was, for Wallace, an anomalous being, one that exposed unresolved tensions in his philosophical, moral, and scientific framework of thought.

Further tensions are revealed in Wallace's responses to the indigenous cultures he encountered on his travels. Alongside his descriptive anthropology, he has no hesitation in assigning "higher" or "lower" status to the "savage" and "semi-civilized" peoples he encounters on his travels. Often the criteria he uses are cultural ones, suggesting he felt that indigenous peoples are not inherently inferior to civilized Europeans. This, too, is implied in his favourable view of Dutch colonizers who made an effort to educate and "civilize" their native subjects. Still, this is tempered by a paternalistic judgement that at least the better versions of colonial rule are good for the natives.

Many years later, in his autobiography, Wallace recalled the "unexpected sensation of surprise and delight" at his "first meeting and living with man in a state of nature—with absolute uncontaminated savages!" (S729 1905a, 1:288). In the final, anthropological, chapter of his *Narrative of Travels on the Amazon and Rio Negro* he adds a lament about the indigenous "Indians":

> ... they seem capable of being formed, by education and good government, into a peaceable and civilized community. This change, however, will, perhaps, never take place: they are exposed to the refuse of Brazilian society, and will probably, before many years, be reduced to the condition of the other half-civilized Indians of the country, who seem to have lost the good qualities of savage life, and gained only the vices of civilization (S714 1889, 361).

Though Wallace never responded with such wonder to the indigenous peoples of the Malay Archipelago, he was still able to write home early on in his stay in Borneo: "The more I see of uncivilized people, the better I think of human nature on the whole, and the essential differences between civilized and savage man seem to disappear" (S729 1905a, 1:342–43).

Wallace's critical relationship to the materialism and inequalities of his own "civilized" culture are, perhaps, at work in his admiration for "man in a state of nature." Like some of the more radical figures among the philosophers of the Enlightenment, Wallace is not so much in love with the achievements of "progress" that he fails to see what is lost on entry into "civil society." Still, this admiration for the life of the noble savage, while it coheres well with his egalitarianism, seems hard to square with his belief in "progress." As with progress in the history of organic life, Wallace's account of progress in human history cannot be a simple narrative: from original savage barbarism to civilized European modernity: "uncontaminated savages" exhibit high moral virtues, while civilized Europe has its deep vices as well as, at its best, the capacity to rule others with benevolence.

But this ambiguity also affects the options available to Wallace in thinking about human origins. If there is no essential difference between civilized and savage peoples, and the latter are capable of being educated into the civilized state, then what is the content of the epithet "lower" as applied to other races? They cannot, it seems, be relegated to the status of relicts of our own evolutionary past, as evolutionary stepping-stones between ourselves and our pre-human ancestors (as implied by the use of the terms "higher" and "lower" in relation to non-human taxa). While this suggests a high degree of unity among the different races of humanity as a whole, it opens up a clear gap between humans and other species. But this, in turn, sits uncomfortably with Wallace's ambiguous feelings and thoughts about the orang, his speculations about consciousness in other species, and also his commitment to uniformitarianism.

Wallace's Synthesis: Phase Two (1858 to 1867/68)

Wallace's crucial "breakthrough" in resolving the fundamental question of the mechanism of organic change came early in 1858 and was quickly written up and posted off to Darwin. Although it is known that both Wallace and Darwin at this time were greatly interested in the question of human origins (see Darwin 1887b, 1:109) Wallace's seminal paper "On the Tendency of Varieties to Depart Indefinitely From the Original Type" (S43 1858) makes no mention of humans. The argument of the paper is strongly dependent on Malthus's "law of population" and a consequent "struggle for existence," which has the effect of accumulating favourable variations, generation by generation. Given a deterioration in environmental conditions, the less favoured varieties will become extinct, while the better adapted will survive and flourish. Unlike Darwin, Wallace does not arrive at this hypothetical mechanism by way of an analogy with domestication ("artificial selection"). Instead he contrasts the situation of domesticated breeds with the intensity of the struggle for existence confronting animals (and plants) in the wild, necessitating full use of all their "energies and faculties," so that even slight advantages will affect their bearers' chances of surviving and providing for their offspring.

The story of the impact of Wallace's paper on Darwin and his associates, the joint reading of their papers at the Linnean Society in 1858 and the stimulus this gave to Darwin's completion of his great work in 1859 is very well known (though still controversial in certain respects). Suffice it to say that Wallace declared himself delighted that his paper had been so well received, and in future years frequently expressed his great admiration for Darwin—specifically, for his *Origin of Species*. Interestingly, Wallace would credit Darwin with both a new science *and* a new philosophy. For Wallace natural selection was not merely one mechanism among others bringing about organic change, but, rather, a universal principle, uniting many diverse classes of fact, and thus comparable with the great Newtonian synthesis. It may also be that Wallace was so strongly attached to this concept (more strongly, perhaps, than Darwin himself) in part because of his own claim to its independent discovery.

Whatever the cause, Wallace's main writings in the next decade are devoted to applying the idea of natural selection to a great range of hitherto paradoxical or simply unexplained phenomena, always citing in respectful ways Darwin's supreme achievement. The most fundamental shift in Wallace's outlook related to his adoption of natural selection is his subsequent systematic rejection of the persistence of non-utilitarian characters. In his monumental 1864 study of the

Malayan Papilionidae as exemplifying the theory of natural selection, for example, Wallace appeals directly to Darwin's concept of sexual selection for an explanation of the "large canine tusks in the males of fruit-eating Apes" (S716 1870, 156), directly contradicting his earlier insistence on their non-utilitarian character in his 1856 paper (S26).

A bit earlier, in 1863, Wallace had first confronted the subject of human diversity. His paper "On the Varieties of Man in the Malay Archipelago" (S82 1865) does not directly deal with human origins, although the treatment of the human races is thoroughly naturalistic and taxonomic, and of a piece with his treatment of the diversity and geographical distribution of the non-human inhabitants of the Archipelago. Whilst there is no explicit reference to natural selection, Wallace does take advantage of current beliefs about the antiquity of humans to suppose that the current diversity of racial groups can be explained by the "slow but certain effects of the varying physical conditions," so that "we need no new power to introduce rapid changes of physical form and mental disposition" (S82 1865, 213).

But by March 1864 Wallace was prepared to face head-on the question of human origins. Interestingly, his paper "The Origin of Human Races and the Antiquity of Man Deduced From the Theory of 'Natural Selection' " (S93 1864) was delivered to the Anthropological Society of London, a polygenist and racist breakaway from the Ethnological Society (see Raby 2001, 176; Fichman 2004, 154). As the title of the paper unmistakably indicates, the argument is for a thorough-going application of the theory of natural selection to the human case. Ostensibly the paper is an attempt to reconcile the rival mono- and polygenist views of human origins and persisting racial differences, but this turns out to be inseparable from the question of whether natural selection applies to human evolution itself, or whether there is "anything in human nature that takes him out of the category of those organic existences, over whose successive mutations it has had such powerful sway" (S93, clxi).

Wallace's answer to this question is a brilliantly original piece of dialectic: human distinctiveness is, indeed, the result of the action of natural selection alone, but that distinctiveness, once produced, progressively erodes the power of the very force that brought it about. Humans are, indeed, "taken out of" the category of organic beings subject to the sway of natural selection, but what has taken them there is natural selection itself! Wallace notes two developmental trends in human evolution. First, the development of "social and sympathetic dispositions," that check the power of natural selection to eliminate the weak and sick, but at the same time, through the development of social cooperation and the division of labour, confer advantages in the struggle for existence on the communities that can adopt them. Second, the application of intelligence—to develop new weapons, clothes, or shelter—introduces a far more rapid and reliable means of adaptation to environmental conditions.

So, as humans develop both their intelligence and their "social habit," natural selection increasingly operates to augment their mental and moral attributes, rather than their bodily form. This explains the great gulf that now exists between human moral and mental attributes and those of even their closest primate kin, whilst so much is common at the level of physical form. It also resolves the question of the origin of racial difference. In so far as there are differences of physical form between the races, these must have been established very early on in human evolution, before natural selection shifted its operation from bodily to mental attributes.

Wallace's application of natural selection to humans in the context of a dispute about racial difference takes him in the direction of an account of evolutionary progress through the continued operation of the "struggle for existence" between the human races. The consequence is the gradual extermination of lower, "savage" races in the face of the mental and physical superiority of the European race. Once the "lower and more degraded" races have been displaced, the earth will again be inhabited by a "single homogeneous race." The paper concludes with Wallace's imagining of a future utopia based on the superior moral and intellectual character of the victorious race, combined with a reversal of the power of natural selection: though formerly products of natural selection, humans are now destined to replace it as the key force, selecting and so shaping the future development of (terrestrial) nature: "We can anticipate the time when the earth will produce only cultivated plants and domestic animals; when man's selection shall have supplanted 'natural selection'" (S93 1864, clxviii).

The argument is a brilliant one, enabling Wallace to remain true to his commitment to the power of natural selection, to his uniformitarianism, and even to his belief in evolution as progressive development, whilst giving due recognition to what is distinctive about humans and to their "true grandeur and dignity"—their status as a "being apart." There is even a place, now, as the prime outcome of evolutionary development, for the free, equal, and cooperative vision of the future society to be realized as an "earthly paradise."

However, all this is achieved at deep cost to other aspects of Wallace's earlier (*i.e.*, "phase one") belief-system. Most obviously, what has to be sacrificed is his generosity of spirit and admiration for "absolute uncontaminated savages": his recognition of the high moral character of some of their social arrangements and cultural achievements, and their form of knowledge of their surrounding natural environment—knowledge he frequently depended on for his own researches. The unequivocal commitment to European racial superiority, especially, goes against the whole tendency of his radical, critical orientation to the materialism and injustice of his own society.

The reader must wonder, too, at how the naturalist who trembled with excitement at the sight of a swallowtail butterfly, and whose whole career to date had been devoted to the passionate search for birds of paradise, beetles, and butterflies, could admit a future in which only cultivated and domesticated species survived. Beyond this there is a disturbing irony connected to his argument: if the future society is to host a race of humans of the most benevolent dispositions and highest moral virtues, it is hard to see how such beings could have emerged from the sort of racial "struggle for existence" that Wallace describes in the essay. His response to discussion of his paper suggests he entertained no moral qualms on this score:

> Now, it appears to me that the mere fact of one race supplanting another proves their superiority. It is not a question of intellect only, nor of bodily strength only. We cannot tell what causes may produce it. A hundred peculiarities, that we can hardly appreciate, may cause the one race to melt away, as it were, before the other. But still there is the plain fact that two races came into contact, and that one drives out the other. This is a proof that the one race is better fitted to live upon the world than the other (S93 1864, clxxxiii).

About fifteen months after delivering this paper Wallace became immersed in an intensive study of spiritualism (Smith 2003– , Chapter 5). This issued in July of 1866 as a serialized essay, "The Scientific Aspect of the Supernatural" (S118) which was subsequently (late 1866) circulated in pamphlet form. Wallace presents his aim in this work as solely to invite serious attention to the topic on the part of scientists, but it seems clear that he is already convinced of the reality of the phenomena of spiritualism, and their inexplicability in terms of existing materialist science. He is, too, clearly strongly attracted to the metaphysical belief-system advocated by key spokespersons for the spiritualist movement. However, his argument is that there is nothing in these experiences or the leading interpretations of them that either contradicts known laws of nature or lies beyond the reach of scientific investigation.

Much of the pamphlet is given over to what he takes to be well-authenticated examples of spirit-communication, extraordinary coincidences, table-turning, autonomous slate-writing, and the like, and to the testimony of well-respected and honourable converts. But there is also some theoretical discussion. Wallace distinguishes two sorts of "miracles," as apparent exceptions to the laws of nature. The less interesting of these are the results of hitherto unknown laws of nature, but the more interesting ones, the ones gathered together in the spiritualist literature, have a quite different explanation. These, too, are not really exceptions to the laws of nature, but are, rather, to be explained on the hypothesis of the existence of disembodied and imperceptible intelligent beings who are capable of acting on matter. Wallace thinks that the hypothesis of such existences is "not inconceivable," but would be seen by scientific orthodoxy as improbable and requiring sound evidential support. This, Wallace thinks, has already been provided, by a variety of facts that cannot be explained by any rival hypothesis. Wallace argues that the "spirit" is the essential part of all sensitive beings—it is spirit that feels, thinks, and perceives—while the body constitutes the "machinery and instruments" for the interaction between spirit and the external world. He recognizes that "the spirit is in general inseparable from the living body," but takes the view that at death it leaves the body, "retaining its former modes of thought, its former tastes, feelings and affections." After death these spiritual personalities continue the journey they started during their embodied phase, but now governed by a "progression of the fittest" that replaces the "survival of the fittest." Under certain conditions (not specified) these spiritual beings are capable of intervening and communicating with those of us still in our earthly "clothing."

Although Wallace appears to have become fully committed to spiritualism less than a year after writing this essay, his output on a variety of evolutionary topics are initially largely unaffected. His commitment to Darwinian natural selection and "utilitarian" explanation of the traits of living organisms remains as trenchant as ever, as does his defence of Darwinism against those who would continue to see beauty and other apparently functionless traits as the result of direct interventions by some creative power. In 1867 he is in print defending Darwin's *Origin* against the criticisms of the Duke of Argyll. His account of his and Darwin's theory gathers together the general facts of nature from which the theory of natural selection is deduced. He considers it probable that these "primary facts or laws" are themselves "but the results of the very nature of life, and of the essential properties of organised and unorganised matter." He defers to Spencer on this, but goes on to state the question at issue: Can the variety, beauty etc. of nature be explained on the basis of these facts and laws, or do we have to believe in the "incessant interference" of the Creator (S140 1867, 143; S716 1870, 267)? Wallace goes on to insist on the self-regulating character of the universe, and the "inherent power" of life-forms to adjust to each other and to surrounding nature. He then repeats his much earlier objections to the "special creation" view that it is demeaning to the Creator to imagine He could not have established the laws of nature with foresight as to their outcome. Whilst not disputing the idea of a Creator, Wallace remains consistent with his earlier agnosticism in his objections to Argyll's presumption to present the mind of the Creator in human terms. A case in point is Argyll's explanation of non-utilitarian beauty in nature as evidence of the Creator's love of beauty for its own sake. Wallace again calls on Darwin's concept of sexual selection to demonstrate the utilitarian explanation of beauty, and argues *ad hominem* that if the Creator was a lover of beauty, why was there so much ugliness in the world (S140, 482; S716 1870, 284–85)? As late as 1868 in his paper on the "Theory of Birds' Nests" (S139 1868; S716 1870, 231–63) Wallace is clear in his rejection of teleological and non-utilitarian explanations of colour as something "given to an animal not to be useful to itself, but solely to gratify man or even superior beings—to add to the beauty and ideal harmony of nature" (S139 1868, 88; S716 1870, 262). If this were the case, then these phenomena would be an exception to the ordinary course of nature and not the outcome of general laws. We would thus have to give up trying to explain them since *ex hypothesi* they are "dependent on a Will whose motives must ever be unknown to us."

It seems clear, then, that up to the late 1860s Wallace remains strongly committed to the explanatory power of a Darwinistic form of natural selection, on that basis rejecting both any notion of *specific* interventions in nature on the part of "superior" beings, and any attempt to claim knowledge of the mind of the Creator.

Wallace's Synthesis: Phase Three (from 1869 on)

Wallace released the first of what Fichman (2004, 157) refers to as his "bombshells" in 1869, with his review of two new editions of books by Charles Lyell (S146 1869). This was followed a year later by the essay "The Limits of Natural Selection as Applied to Man," included as the final chapter of his *Contributions to the Theory of Natural Selection* (S716). These texts are the first significant indicators of a difference of view on a scientific matter between Wallace and Darwin. They also, arguably (but controversially), mark a distinct shift on Wallace's part from previously published views, notably on the process of transition from ape-like ancestor to modern humans.

The Lyell review is almost wholly devoted to a celebration of his uniformitarian geology and an admiring welcome to Lyell's belated conversion to the theory of natural selection—which Wallace himself continues to endorse. The "bombshell" is contained in the last few pages, in which Wallace explains his reasons for having come to the conclusion that natural selection, while undoubtedly central to the evolution of man "the animal," is insufficient to account for certain of our higher mental and moral faculties, as well as certain (associated) physical characteristics. But still more potentially explosive is his indication, at the end of the text, that the complementary power to the action of natural selection is some form of purposive guidance connected to an "Overruling Intelligence." Both lines of argument are developed in the more extended essay that concludes *Contributions.*

The most fully worked-out argument in both essays is the claim that "savage" races and, so far as could be told, early humans, have or had a mental capacity only slightly less than that of the average of civilized Europeans, but far greater than that of the nearest non-human relatives, the man-like apes. By contrast, there is a great gulf between the actual intellectual, aesthetic, and other cultural *achievements* of the civilized races and those of the savages and early humans. It follows that our ancestors were endowed with higher mental faculties that were not used, and whose value has only been discovered much later as a result of the progress of civilization. Since natural selection can only fix in a population characteristics that are of use, and, indeed, confer an advantage in the struggle for existence, it cannot account for the surplus of unused mental potential present in our ancestral forms (and contemporary "savages").

This argument rests on two assumptions (other than its account of the necessary effects of natural selection). One is that mental capacity can be measured by cranial capacity, implying that the extent and kind of mental activity of which a being is capable is determined by the size and structure of the brain. Wallace is explicit in his support for this materialist view of the determination of mental capacity, which he regards as scientifically consensual. The second assumption is that early humans (and "savages" now) have modes of life in which the demands for survival are barely greater than those of the animals with which they compete, so that a parsimonious natural selection might have endowed them with barely more mental capacity than that of the apes. What goes along with this is an astonishingly crude caricature of savage life: "What is there in the life of the savage, but the satisfying of the cravings of appetite in the simple and easiest way? What thoughts, ideas, or actions are there, that raise him many grades above the elephant or the ape?" (S165 1870, 342).

Another argument, deployed in both texts, concerns the relatively small amount of hair covering human bodies. Specifically, it is difficult to see how a hairy coat could ever have conferred such a disadvantage to early humans for it to have been lost—and never recovered as humans migrated to colder climates. Wallace provides some anthropological evidence that many contemporary savages feel the lack of hair on the back, especially, and try to cover it by one means or another. In the review article Wallace includes the nakedness of human skin within a wider account of erect posture and beauty of physical form, none of which could have conferred any positive advantage in the struggle for existence.

Further, the marvellous delicacy and flexibility of the human hand looks to far more sophisticated uses than are implemented in the crude material cultures of savages and our Stone Age ancestors. Even in apes we can see the development of the hand as an organ far more developed than is required. And the immensely complex mental and physical organs involved in speech—and the beauty of voice involved in music—could never have been of use to the lowest savages, so they, likewise, could not have been evolved by natural selection acting alone.

Finally, the capacity for abstract reasoning, contemplation of infinity, "wide sympathy" with the whole of nature, sense of the sublime, and other attributes of civilized man are generally (but not universally) absent from "savage" cultures, despite the latter having the *potential* to develop these powers, given appropriate civilizing influences. Our moral sense, too, though to some extent favouring communities that possess it, could not have acquired the power of conscience, and even of sanctity, above merely utilitarian considerations.

Although there are some internal tensions and *non sequiturs* in this accumulation of arguments, they did—and some still do—pose a significant set of challenges for any purely materialistic account of the evolution of distinctively human attributes. However, the greater shock contained in these texts is the new metaphysical theory that Wallace postulates to supersede the insufficiencies of natural selection. The central device is the supposition that human evolution has been modified by what Wallace variously describes as an "Overruling Intelligence," "higher" or "controlling" intelligences. There are two steps in Wallace's argument here. First, he argues that the various non-utilitarian potentials bestowed on early humans subsequently came into their own as necessary preparations for the full development of humanity: "... (W)hat we can hardly avoid considering as the ultimate aim and outcome of all organized existence—intellectual, ever-advancing, spiritual man." So, for example, our nakedness played a part in stimulating our intellectual development in the provision of clothing, and our moral development in the introduction of personal modesty. The second step is the claim that these various aspects of our nature, though inexplicable in terms of natural selection, do become intelligible on the hypothesis of the intervention in human evolution of a superior intelligence that "has guided the development of man in a definite direction and for a special purpose" (S165 1870, 359). The final aid to our imagining this is an ingenious reverse use of the Darwinian analogy with domestication and cultivation: the higher intelligence or intelligences have guided our evolution for purposes of which we have been unconscious, in a way analogous to the purposive guidance given to organic change in the course of selective breeding of other species carried out by humans. In accord with his studies on spiritualism he projects the possible existence of a whole hierarchy of spiritual intelligences between humans and the Ultimate Being, filling in the gap, so to speak, between God and embodied humanity.

As if this were not already enough of an extension of thought, Wallace goes on to speculate on the nature of both matter and consciousness, concluding that matter is a form of force, and that force is ultimately "not improbably" "will-force"—so that force is a product of mind. By a series of such logical leaps Wallace finally confirms the Cartesian thesis of the unprovability of matter (as usually conceived) but demonstrability of "self-conscious, ideal existence" (S165, 369). In one breathless series of thoughts, we are taken from a set of interesting and debatable claims as to the limits of natural selection in the human case, to a radical objective idealist theory of the nature of matter and force, by way of a hypothetical purposive "superior intelligence" and guide.

Wallace maintains from this point on that these arguments are consistent with scientific agendas, do not involve any retreat from his commitment to the notion of a law-governed universe, allow him to continue to support "Darwinism" as an explanation for non-human organic change, and are even consistent with uniformitarianism, or "continuity" as he now calls it.

Wallace's Evolution: How Much Change?

Wallace's two most respected correspondents, Darwin and Huxley, were famously unimpressed by this apparent turn in Wallace's thought and, as Wallace expected, not at all convinced that his new synthesis fell within the acceptable bounds of science. But there are further questions on the internal consistency of Wallace's new synthesis as well as the extent of its departure from his earlier thinking.

First, there are questions of epistemology. Wallace continues to pay lip service to his earlier conception of scientific explanation as evidence-based, and thus limited in scope to the empirically testable. However, the leap from identifiable limits to the explanatory power of natural selection to the hypothetical purposive intervention of disembodied intelligences, their hierarchical organization and purposes in shaping human evolution, with the further leap to an objective idealist metaphysic, is full of logical *non sequiturs*, related to bodies of evidence only in a highly tendentious way, and speculative in the extreme. Moreover, the claim to consistency with uniformitarian principles is problematic. In both of his earlier "phases" Wallace had rejected "special intervention" explanations of particular natural phenomena. Yet here is a significant departure—a purposive intervention on the part of some creative intelligence into the "ordinary course of nature," and particular to the evolution of one species. On matters of ontology and metaphysics, the most striking shift is in the endorsement of the existence of disembodied intelligences. These were already declared as at least conceivable as early as the 1866 essay (S118) on spiritualism, where, though their hypothetical relation to evolution is made explicit, they are assigned no explanatory role in organic evolution. By 1870, however, spiritual beings have been assigned a crucial role in the formation of the human species. This final acknowledgement that disembodied individuals and purposive intelligences are *real* represents a major rupture with both his own earlier thinking, and prevalent materialist assumptions in scientific thought. Even in the core argument for the limitation of natural selection in the human case, Wallace relies on a thoroughgoing mind/body materialism. Unless we suppose that cranial capacity is a key indicator of mental powers, his argument falls.

There is also (much disputed) textual evidence that Wallace himself recognized that his encounter with spiritualism was something for which there was no place in his prior "fabric of thought." He acknowledges in an 1869 letter to Darwin that he would himself, a few years previously, have looked upon his current views on man as "equally wild and uncalled for." This subject will be taken up in the last chapter of this collection, where Charles Smith will argue for a different interpretation of Wallace's words.

The fact of Wallace's going public on the limitations of the explanatory power of the idea of natural selection was itself something of a shock, especially coming from so staunch an ally of Darwin. For both Smith and Fichman, however, the main issue is not whether the shift in Wallace's position circa 1869 represented *any* kind of change—it certainly did—but whether that change represented a *reversal* of position. Both writers note strong continuities between Wallace's whole framework of thought prior to 1858 and his position from the end of the 1860s. This is undoubtedly true, especially as regard Wallace's progressive moral and political views, his overall sense of evolution as a inherently "progressive," and his insistence on the significance of non-utilitarian features of the organic world. However, I would argue that even these continuities are not complete. Through both phases two and three Wallace remains solidly opposed to the presence of non-utilitarian characters in non-human species. The insistence in his 1856 paper (S26) on the habits of the orang-utan that animals are, indeed, endowed with characteristics that are of no use to them is sharply reversed after 1858 and remained in place. Only in the human case, and then only temporarily, did he allow of this possibility. Arguably, too, Wallace's "phase three" framework breached his earlier commitment to uniformitarianism, in its treatment of human evolution as a special case, and its introduction of a "guiding force" in human evolution that acts in a qualitatively different way from the other mechanisms at work in the "ordinary course of nature." Finally, Wallace's long-standing scepticism that the motives and intentions of a supposed "Supreme Being" can be knowable contrasts with his repeated assertions from 1869 onwards of the progressive evolutionary purposes toward which humanity is guided.

But if there are discontinuities between Wallace's earlier and later (in my terms, phase one and phase three) syntheses, there are still deeper apparent shifts involved in the transition from the mid- to late 1860s—from phase two to three. It is these that have provided the most support for the strong "change of mind" interpretation. Apart from the apparent retreat from uniformitarianism and his earlier conception of the nature and limits of scientific explanation, Wallace's apparent retreat on the role of natural selection in human evolution and recruitment of non-material purposive agencies to fill the gap does look rather like a serious "change of mind." Smith and Fichman's view of the continuity of Wallace's thought relies quite heavily on continuities between what I have called phases one and three. Whilst they are right to note these continuities, I've suggested that even these are not complete.

In Fichman's analysis, especially, the continuity thesis is maintained by rather glossing over the period from 1858 to 1864 (my "phase two"), when Wallace's published writings are at their most unequivocally Darwinian (and also strongly influenced by Spencer). So, for example, Fichman draws on the anti-utilitarian elements in Wallace's 1856 paper on the orang-utan as evidence that Wallace believed there were forces other than strictly material ones at work in the evolution of animals as well as humans. He refers to this essay as written when Wallace's "mind was gearing up for the articulation of natural selection," so that natural selection would become one major key to understanding the harmony of the natural world, with spiritualism as the other (Fichman 2004, 158–59). However, what this fails to take account of is that, far from Wallace's mind "gearing up" for natural selection in 1856, the discovery of natural selection in 1858 and adoption of Darwin's programme from 1859 produced an explicit and frequently reiterated *rejection* of the anti-utilitarian argument of the 1856 paper. It also fails to address the significance of Wallace's mid-1860s writing on human races and origins, none of which give any suggestion there are non-material forces at work in the evolution of humans or animals. The rightly most discussed of these works is the 1864 paper on the origin of the human races (S93). Oddly, Fichman reads this paper as already committing Wallace to the role of spiritual agencies in human evolution: "The continued action of natural selection and spiritual agencies destined *Homo sapiens* to an ever higher level of existence" (Fichman 2004, 156). Fichman, like Raby (2001, 177 ff.), seems to draw on Wallace's 1870 revisions in interpreting the view expressed by Wallace in the 1864 version. There is, of course, no specific mention of spiritual agencies in the paper as written in 1864, and the whole thrust of Wallace's argument as indicated by the title is to demonstrate the explanatory power of natural selection as applied to the human case.

Smith argues that Wallace never thought that natural selection could explain the higher human faculties and virtues, but lacking an alternative explanation preferred to remain silent until spiritualism provided him with the explanation he needed (*e.g.* Smith 1992/1999, 30). However, it is hard to read the 1864 paper as anything other than a highly original way of deploying the idea of natural selection to do just that—the key devices being the role of social processes, including racial conflict, as selective forces, and the shift in the target of selective forces from bodily to mental capacity and organization. Wallace is absolutely clear in his responses to discussion that by the latter he "always include[s] the brain and skull—the organ of the mind—the cranium and the face" (S93 1864, clxxxi). The idea of natural selection alone is thus used to explain the "true grandeur and dignity of man," as a being now unique in nature as exempt from the very pervasive power that produced "his" distinctive attributes. Smith's view is that from 1858 through to the "Origin" paper of 1864, and then beyond, Wallace continued to struggle with the question of the persistence of "survival-unrelated" characters, and that it was spiritualism that finally provided him with an answer. Smith's position is strengthened by the mere fact that there are no writings by Wallace after 1858 and before the 1864 "Origin" paper that bear on this matter, but in itself this is hardly conclusive evidence. Despite Wallace's relaxation of this view in the late 1860s with respect to human evolution only, his commitment to utility for non-human species becomes, if anything, even stronger with his growing divergence from Darwin on the significance of sexual selection.

Wallace's alterations to his 1864 paper for inclusion in the *Contributions* collection of 1870 (S716) feature a much-discussed piece of evidence concerning the vexed question of his "change of mind." There are several significant changes, and a comment from Wallace in his Preface. In the latter Wallace says he had considered extending the piece, but had in the end decided to remove some "ill-considered passages." Relevant to our theme here there are three changes worthy of comment. One is the change of title—now "The Development of Human Races Under the Law of Natural Selection"—a considerable weakening of the claim contained in the original. Another is a small but significant change of wording from "his mental development had correspondingly advanced" to " . . . his mental development had, from some unknown cause, greatly advanced" (S716 1870, 321). Finally, Wallace omits the final promise of a future earthly paradise from the original version and replaces it with an acknowledgement of the slowness of progress, and a denunciation of the low moral and intellectual status of the civilized nations, rendering them incapable of making the best use of technical advance. In such societies natural selection will favour the lowest morally and intellectually. This is turned into yet another evidence of the action of some other guiding force in human history, since there is, despite everything, still a steady advance in morality and intellect.

Taken together these changes do seem to indicate that Wallace, while remaining proud enough of the original essay to wish to include it, nevertheless could not do so without indicating that his current view differed from that held in 1864. Smith's analysis is that in 1864 Wallace had no purchase on the causes of the origin and connection to evolution in general of humans' "above nature" qualities, and was experimenting, temporarily, with a materialist position. By 1870 he did have an answer to this question in terms of "final causes" and the above-mentioned changes are the result. An alternative interpretation is, however, feasible. It is that the 1864 paper did address the question of human's superior qualities, presenting an explanation of them in terms of the gradual emergence of a new target—brain and mind—for the action of natural selection. In the original paper Wallace would have felt no need to explain the advance of human mental development other than by way of random variation (which he always took to be universal in organic beings, and capable of being taken in any direction and accumulated by selective pressures).

This leaves the intriguing question of the deletion of his earthly utopia in favour of a more measured and critical characterization of his own civilization. My guess here is that Wallace must have become uncomfortable with racist applications of the survival of the fittest, and unequivocal view of the racial superiority of Europeans to which the 1864 essay had committed him. The newer "Limits" synthesis, despite the crude reductionism of his caricature of "savage" life, at least allows him to view humanity as a whole progressing toward a brighter future, under the guidance of the superior, but benign spiritual influences suggested by his spiritualist beliefs.

The tensions within Wallace's "fabric of thought" produced by his most rigorous attempt to apply the theory of natural selection and "struggle for existence" to the human subject may thus have played some part in his later retreat from a fully materialistic account of human evolution and future prospects. Similar moral revulsion in the face of reductive and racist forms of social Darwinism in the latter part of the nineteenth century also led some of the key "founding figures" in the modern social sciences to conceptualize human cultures and societies as "*sui generis*" causal orders, to be studied in their own right and without reference to our primate origins. The seminal works of Boas, in relation to cultural anthropology, and Weber and Durkheim, foundational thinkers for sociology, are foremost here (*e.g.* Weber 1949; Durkheim 1964; Stocking 1987). This legacy continues to haunt these disciplines, limiting their ability to address increasingly pressing socio-ecological issues and problems of the adaptive limitations of human subjects in the face of contemporary forms of social life that have intensified and globalized the defects already denounced by Wallace (Catton and Dunlap 1978; Benton 1991). A central question for these disciplines now is: can a fully naturalistic understanding of humans be combined with the sort of progressive, egalitarian, and cooperative vision of the future that Wallace envisioned, and upon which human survival now depends (Benton 1993, 1999a; Soper 1995; Rose 1997; Dickens 2004)?

21

Wallace, Spiritualism, and Beyond: "Change," or "No Change"?

Charles H. Smith

In Chapter 18 the reader received a short introduction to elements of what I have termed the "no change of mind" model of Alfred Russel Wallace's evolving thought on the place of humankind in the evolutionary process. It was there pointed out, obliquely, that most previous writers had unjustifiably assumed: (1) as of 1858 Wallace thought human evolution was a function of the same basic causal influences as had effected plant and animal change, and (2) that even after that date his position on the necessary utility of adaptations had remained similar to his earliest thoughts on the matter. Regarding the latter point, it seems quite clear, on the basis of his own writings before 1858,[1] that the reasoning spelled out in the "Ternate" essay "On the Tendency of Varieties to Depart Indefinitely From the Original Type" (S43 1858) instead represents a sharp break from his earlier position. One can reasonably argue, in fact, that the main intellectual breakthrough expressed in "On the Tendency ..." is Wallace's unanticipated linking of a necessary utility argument to Malthusian thinking.

My impressions of the early evolution of Wallace's thoughts on utility appear in some detail elsewhere (Smith 1991, 1992/1999, 2003–), and inasmuch as Wallace's own words during the pre-1858 period state the case clearly enough, we will not dwell on this theme much further here. Instead, our attention will focus largely, after a short treatment of a few more relevant aspects of the Ternate essay, on events from the years 1862 through 1869, including Wallace's adoption of spiritualistic beliefs. It is my intent to show how a close study of Wallace's personal and professional involvements during that period provides considerable support for the "no change of mind" interpretation.

"On the Tendency ..."

The notion that Wallace "changed his mind" about (actually, *reversed* himself on) natural selection's all-sufficiency in explaining the evolution of humankind's "higher" attributes (morality, mathematical, and artistic abilities, etc.) arises in large part from two common misappraisals of the 1858 Ternate essay. The first is the uncritical assumption that the work represented a simple progression of thought from his 1855 "Sarawak law" essay (S20). Nothing could be further from the truth. In fact, deceived by his long-standing conclusions on the non-necessary utility of adaptive structures, Wallace had given up, at least for the time being, on finding any kind of functional link between adaptation and the evolutionary process,[2] and instead was concentrating his attention on spatio-temporal aspects of the natural record of speciation. Probably his earlier experience as a surveyor and mapmaker contributed to the latter emphasis; eventually he was able to visualize a process resulting in a geological and geographical distribution of forms that strongly invited an evolutionary interpretation. Wallace's conclusions impressed some workers,[3] but only to the extent that they accounted for *results* emerging from some yet unknown vehicle of change. Indeed, "On the Law ..." contains not even the slightest allusion to possible actuating mechanisms—that is, to anything akin to natural selection. This "results-driven" approach is continued in his works published between 1855 and 1858 (notably, S25 1856, S26 1856, S38 1857, S40 1858, and S41 1858) that apply the model to actual biogeographical situations.

I have suggested in the works mentioned above, along with my other essay here, originally published elsewhere (Smith 2004b), that Wallace's inattention to the dynamics of the adaptive process for so many years may in part have stemmed from his rather strict position on the nature of belief: specifically, how one's beliefs cannot be overturned by less than a productive confrontation with new, countermanding, information. (See in this connection the long and classic quotation from the letter to Thomas Sims, his brother-in-law, reproduced in Chapter 18.) In parallel fashion, he may have thought that adaptive structures merely emerged, idiosyncratically and unpredictably,[4] and that only once in place were they then secondarily shaped by remotely constituted forces—probably large-scale, long-term environmental ones such as climate and geological change. This Bauplanesque approach (actually somewhat Buffonian in character) at least obviated any need to view the link between adaptive structure and ecological function as first causes-mediated—probably a more pressing concern for him at the time.

The idea that there were always more "recondite" (one of his favorite terms) forces lying behind natural organization thus developed early on in Wallace's mindset, and he never really got off this train. Even the breakthrough on necessary utility resulting in the Ternate essay had no perceptible effect on his thinking in this regard; in particular, he probably still felt that human actions and beliefs came about and were maintained for reasons that were not always strictly utilitarian. It has long been thought, and is no doubt true, that as of 1858 Wallace had been giving much thought to the forces that might be contributing to human evolution, but on this correlative basis alone it has been assumed in most quarters that the Ternate essay described a process fully incorporating humans into the equation.

This brings us to the second misappraisal. In reality, there is no evidence in the essay itself—or anywhere else, for that matter—that this was the case. First, and straightforwardly enough, humankind is not mentioned in it. Second, Wallace himself never claimed later he ever intended such an interpretation—and in all likelihood, wouldn't he have done so, at some point, had he? Lastly, the whole essay is couched in a "special case" kind of argument involving domesticated forms ("It will be observed that this argument rests entirely on the assumption, that *varieties* occurring in a state of nature are in all respects analogous to or even identical with those of domestic animals, and are governed by the same laws as regards their permanence or further variation": S43, 54)—one in which it is not clear, perhaps deliberately, which outcome is meant to represent the exception and which the rule. Recall that Wallace had invoked Malthusian logic to help him contextualize the dynamics of change, and that, in his later words, "it then occurred to me that these checks must also act upon animals" (S726 1898b, 140). Had he perhaps been thinking about the kinds of forces that might transcend Malthusian kinds of control *first*, only secondarily recognizing how its strict application to animals invited "natural selection"? If so, he may also have been considering other analogues to domestication—including the later rejected picture of man as "God's domestic animal" (S716 1871, 372)—while feeling his way toward the survival of the fittest concept.

I believe it can now be considered as given that "Tendency" at the least represented a major break in Wallace's approach to the subject of necessary utility—but beyond that it may also have been, through his avoidance of discussion on the matter, his first step toward developing an argument that humankind rises above biologically material controls. This would explain, as mentioned below, Wallace's strange apology, made five times in print over a four-decade period, that he hadn't been allowed to view his proofs before the paper was rushed into print: the origins of humankind's higher attributes remained an issue for him after 1858, and he found it a bit grating (or exasperating) that everyone had put words in his mouth.

The implication is that Wallace must have found himself in a bit of a quandary as he headed home to England from Singapore in the spring of 1862. Yes, in outline the hypothesis he had written home about in 1858 did in many ways closely resemble the one Darwin had come up with, but he may well have felt he had been outmaneuvered and now could not speak his mind fully on the subject. Any of several reasons might explain why he didn't, at least just then, go ahead and do so. First, he had not yet reflected on the full implications of his collections; thus it was possible he had initially overlooked something that only later might provide direction. Second, the people he was dealing with were clearly, in more than one way, The Establishment, and could have made things very difficult for him had they felt he was crossing them. He might also have felt, probably quite rightly, that creating a fuss so soon after the initial victories could have put the whole evolution by natural selection theory in jeopardy—something he clearly would not have wanted. Further, the appearance of a reversal so soon after the initial presentation of the idea would have damaged his credibility, perhaps even making him look foolish. Lastly, there was the matter of Darwin's priority on the subject, established (in Wallace's eyes, anyway) by an attention of twenty years' duration. On the basis of any or all of these reasons, it is hardly surprising that at that point he expressed no dissenting views.[5]

Nevertheless, this really only explains why he posed no *contrary* views. As detailed in Smith (2003–), the degree of Wallace's *complete* silence for so many years on the theory to which he had helped give birth is remarkable in itself. It is a fact that he no more than names or alludes to the natural selection concept in his writings after February 1858 until he springs into action in the critical essay "Remarks on the Rev. S. Haughton's Paper on the Bee's Cell, and on the Origin of Species" (S83), as late as October of 1863. No past writer has ever paid any attention to this matter, though it would appear that this is one instance where a lack of action is very telling (Benton's disagreement in Chapter 20 notwithstanding[6]). Simply, it seems he could not decide which way to turn next. At first choosing to remain in the field for another four years, he bided his time until he could return to civilization, absorb the full meaning of his collections, and give fuller attention to the "higher attributes" issue at that point.

Directly on returning to England in early 1862 Wallace set sail again, this time launching himself into the uncertain waters of the London intellectual stream. This is most plainly evidenced by his regular appearances—either to sit or comment, or present papers—at the meetings of no fewer than seven different professional societies.[7] For a couple of years most of his time was devoted to reviewing his collections of birds and insects, but he undoubtedly was also contemplating his special problem. An important new development in this connection was a sudden increase in his interest in the writings of Herbert Spencer. He read Spencer's new work *First Principles*, and an older one, *Social Statics*, and even went to visit Spencer with his old friend Bates, as memorably recounted in *My Life* (S729 1905a, 2:23–24), for inspiration regarding the search for the "origin of life." Spencer shied away from comment. Still, the combined effect of Darwin and Spencer was enough to turn him temporarily down a materialist path, the most obvious fruits of which were his papers "The Origin of Human Races and the Antiquity of Man Deduced From the Theory of 'Natural Selection' " (S93) and "On the Phenomena of Variation and Geographical Distribution as Illustrated by the Papilionidae of the Malayan Region" (S96), delivered within weeks of one another in March 1864 to meetings of the Anthropological and Linnean Societies, respectively. But even the general success of the first paper (including the expressed approvals of both Spencer and Darwin), with its glowing coda on the future evolution of humankind, was ultimately not enough to convince him of its full validity. The search went on for "more recondite" forces.

Happily, Wallace's subsequent progression of thought on these matters is transparently evident in several of his lesser known publications from this period, and his concurrent pattern of professional attentions. We can now turn to this subject in some detail.

Wallace the Questioner

In September 1864, after a rather quiet six-month period broken only by a publication on the systematics of Eastern parrots (S102), Wallace presented a paper entitled "On the Progress of Civilization in Northern Celebes" (S104) at the annual meetings of the British Association for the Advancement of Science. It is in this paper that he begins to connect human to societal evolution—specifically, to those elements of the process that cannot be conceived simply in physical body terms. In Smith (2003–, Chapter 1) I discuss Wallace's long-standing support of the notion that a "many-directioned experience" is vital to one's progressive development. (Interestingly, the *very first line* of his first published work, written about 1841 for a town history [S1a 1845], is a quotation from Bacon: "Knowledge is power"!) This is certainly the central theme of his "The Advantages of Varied Knowledge" (S1 1905), a lecture written in 1843, and such ideas are extended obliquely to how whole civilizations advance in "The South-Wales Farmer" (S623 1905), written the same year. In the "Progress" paper he exposes his usually guarded Eurocentric side by arguing that in order to advance, wholly uncivilized peoples might benefit from a mild, if well-meaning, attitude of despotism:

> ... there is in many respects an identity of relation between master and pupil, or parent and child, on the one hand, and an uncivilised race and its civilised rulers on the other. We know, or think we know, that the education and industry, and the common usages of civilised man, are superior to those of savage life; and, as he becomes acquainted with them, the savage himself admits this. He admires the superior acquirements of the civilised man, and it is with pride that he will adopt such usages as do not interfere too much with his sloth, his passions, or his prejudices. But as the wilful child or the idle schoolboy, who was never taught obedience and never made to do anything which of his own free will he was not inclined to do, would in most cases obtain neither education nor manners; so it is much more unlikely that the savage, with all the confirmed habits of manhood, and the traditional prejudices of race, should ever do more than copy a few of the least beneficial customs of civilisation, without some stronger stimulus than mere example.[8]

Much of Wallace's discussion in this essay can be linked to his belief that the uncivilized inhabitants of the area had been positively affected by the introduction of coffee plantation culture by the Dutch. Nevertheless, he is also contemplating the kinds of forces that might *in general* help raise people's consciousness levels—in particular, those that might sponsor a form of "informed belief" (*i.e.*, systems of knowledge based on valid assumptions) useful to societal evolution.[9]

Wallace's next exploration of an "informed belief"-relatable theme came in an essay-like letter printed in *The Reader* issue of 6 May 1865 under the title "Public Responsibility and the Ballot." Nominally, this work represented an answer to opinions stated in the previous issue by John Stuart Mill, one of Wallace's idols.

... Mr. Mill truly says, that a voter is rarely influenced by "the fraction of a fraction of an interest, which he as an individual may have, in what is beneficial to the public," but that his motive, if uninfluenced by direct bribery or threats, is simply "to do right," to vote for the man whose opinions he thinks most true, and whose talents seem to him best adapted to benefit the country. The fair inference from this seems to be, that if you keep away from a man the influences of bribery and intimidation, there is no motive left but to do what he thinks will serve the public interest—in other words, "the desire to do right." Instead of drawing this inference, however, it is concluded that, as the "honest vote" is influenced by "social duty," the motive for voting honestly cannot be so strong "when done in secret, and when the voter can neither be admired for disinterested, nor blamed for selfish conduct." But Mr. Mill has not told us what motive there can possibly be to make the man, voting in secret, vote against his own conviction of what is right. Are the plaudits of a circle of admiring friends necessary to induce a man to vote for the candidate he honestly thinks the best; and is the fear of their blame the only influence that will keep him from "mean and selfish conduct," when no possible motive for such conduct exists, and when we know that, in thousands of cases, such blame does not keep him from what is much worse than "mean and selfish conduct," taking a direct bribe?

Perhaps, however, Mr. Mill means (though he nowhere says so) that "class interest" would be stronger than public interest—that the voter's share of interest in legislation that would benefit his class or profession, would overbalance his share of interest in the welfare of the whole community. But if this be so, we may assert, first, that the social influence of those around him will, in nine cases out of ten, go to increase and strengthen the ascendency of "class interests," and that it is much more likely that a man should be thus induced to vote for class interests as against public interests, than the reverse. In the second place, we maintain that any temporary influence whatever, which would induce a man to vote differently from what he would have done by his own unbiassed judgment, is bad—that a man has a perfect right to uphold the interests of his class, and that it is, on the whole, better for the community that he should do so. For, if the voter is sufficiently instructed, honest, and far-seeing, he will be convinced that nothing that is disadvantageous to the community as a whole can be really and permanently beneficial to his class or party; while, if he is less advanced in social and political knowledge, he will solve the problem the other way, and be fully satisfied that in advancing the interests of his class he is also benefiting the community at large. In neither case, is it at all likely, or indeed desirable, that the temporary and personal influence of others' opinions at the time of an election, should cause him to vote contrary to the convictions he has deliberately arrived at, under the continued action of those same influences, and which convictions are the full expression of his political knowledge and honesty at the time?

It seems to me, therefore, that if you can arrange matters so that every voter may be enabled to give his vote uninfluenced by immediate fear of injury or hope of gain (by intimidation or bribery), the only motives left to

> influence him are his convictions as to the effects of certain measures, or a certain policy, on himself as an individual, on his class, or on the whole community. The combined effect of these convictions on his mind will inevitably go to form his idea of "what is right" politically, that idea which, we quite agree with Mr. Mill, will in most cases influence his vote, rather than any one of the more or less remote personal interests which have been the foundation of that idea. From this point of view, I should be inclined to maintain that the right of voting is a "personal right" rather than a "public duty," and that a man is in no sense "responsible" for the proper exercise of it to the public, any more than he is responsible for the convictions that lead him to vote as he does. It seems almost absurd to say that each man is responsible to every or to any other man for the free exercise of his infinitesimal share in the government of the country, because, in that case, each man in turn would act upon others exactly as he is acted upon by them, and thus the final result must be the same as if each had voted entirely uninfluenced by others. What, therefore, is the use of such mutual influence and responsibility? You cannot by such means increase the average intelligence or morality of the country; and it must be remembered, that the character and opinions, which really determine each man's vote, have already been modified or even formed by the long-continued action of those very social influences which it is said are essential to the right performance of each separate act of voting. It appears to me that such influences, if they really produce any fresh effect, are a moral intimidation of the worst kind, and are an additional argument in favour of, rather than against, the ballot.
>
> ... it seems to me that in the days of standing armies, of an elaborate Poor Law, of State interference in education, of the overwhelming influence of wealth and the Priesthood, we have *not* arrived at that stage of general advancement and independence of thought and action in which we ought to give up so great and immediate a benefit to thousands as real freedom of voting, for the infinitesimal advantage to the national character which might be derived from the independent and open voting of the few who would feel it compatible with their duty to their families to struggle against unfair influence and unjust intimidation (S110 1865, 517).

The essence of this argument is that there is only one way to change materially the implications of a vote—at least in a positive way—and that is to evolve a voter who "is sufficiently instructed, honest, and far-seeing, [that] he will be convinced that nothing that is disadvantageous to the community as a whole can be really and permanently beneficial to his class or party." The underlying point at issue remains how to "raise the average intelligence or morality" of people. Slowly but surely, Wallace is coming to an answer on this matter, one in fact he had dimly recognized many years before in his first essays, and in the Sims letter referred to earlier: there being no merit to uninformed belief, people have to begin to take seriously that, as he later expressed it, "the thoughts we think and the deeds we do here will certainly affect our condition and the very form and organic expression of our personality hereafter" (S451 1892, 648). So, what kind of influence might cause them to "take seriously" that "hereafter"...?

Only ten days later, on 16 May 1865, Wallace attended a meeting of the Anthropological Society of London. There, the Revd J. W. Colenso delivered the paper "On the Efforts of Missionaries Among Savages." Wallace uttered a few brief comments on the spot (S111), but was unable to let the subject alone. The essay that emerged, "How to Civilize Savages" (S113), was printed in the 17 June 1865 issue of *The Reader*. Its tone and message may be gathered from the following lengthy excerpts:

> Do our missionaries really produce on savages an effect proportionate to the time, money, and energy expended? Are the dogmas of our Church adapted to people in every degree of barbarism, and in all stages of mental development? Does the fact of a particular form of religion taking root, and maintaining itself among a people, depend in any way upon race—upon those deep-seated mental and moral peculiarities which distinguish the European or Aryan races from the negro or the Australian savage? Can the savage be mentally, morally, and physically improved, without the inculcation of the tenets of a dogmatic theology? ...
>
> If the history of mankind teaches us one thing more clearly than another, it is this—that true civilization and a true religion are alike the slow growth of ages, and both are inextricably connected with the struggles and development of the human mind. They have ever in their infancy been watered with tears and blood—they have had to suffer the rude prunings of wars and persecutions—they have withstood the wintry blasts of anarchy, of despotism, and of neglect—they have been able to survive all the vicissitudes of human affairs; and have proved their suitability to their age and country by successfully resisting every attack, and by flourishing under the most unfavourable conditions.
>
> A form of religion which is to maintain itself and to be useful to a people, must be especially adapted to their mental constitution, and must respond in an intelligible manner to the better sentiments and the higher capacities of their nature. It would, therefore, almost appear self-evident that those special forms of faith and doctrine which have been slowly elaborated by eighteen centuries of struggle and of mental growth, and by the action and reaction of the varied nationalities of Europe on each other, cannot be exactly adapted to the wants and capacities of every savage race alike. Our form of Christianity, wherever it has maintained itself, has done so by being in harmony with the spirit of the age, and by its adaptability to the mental and moral wants of the people among whom it has taken root ...
>
> In the early Christian Church, the many uncanonical gospels that were written, and the countless heresies that arose, were but the necessary results of the process of adaptation of the Christian religion to the wants and capacities of many and various peoples. This was an essential feature in the growth of Christianity. This shows that it took root in the hearts and feelings of men, and became a part of their very nature. Thenceforth it grew with their growth, and became the expression of their deepest feelings and of

their highest aspirations; and required no external aid from a superior race to keep it from dying out ... In many places we have now had missions for more than the period of one generation. Have any self-supporting, free, and national Christian churches arisen among savages? If not—if the new religion can only be kept alive by fresh relays of priests sent from a far distant land—priests educated and paid by foreigners, and who are, and ever must be, widely separated from their flocks in mind and character—is it not the strongest proof of the failure of the missionary scheme? Are these new Christians to be for ever kept in tutelage, and to be for ever taught the peculiar doctrines which have, perhaps, just become fashionable among us? Are they never to become men, and to form their own opinions, and develop their own minds, under national and local influences? If, as we hold, Christianity is good for all races and for all nations alike, it is thus alone that its goodness can be tested; and they who fear the results of such a test can have but small confidence in the doctrines they preach.

But we are told to look at the results of missions. We are told that the converted savages are wiser, better, and happier than they were before—that they have improved in morality and advanced in civilization—and that such results can only be shown where missionaries have been at work. No doubt, a great deal of this is true; but certain laymen and philosophers believe that a considerable portion of this effect is due to the example and precept of civilized and educated men—the example of decency, cleanliness, and comfort set by them—their teaching of the arts and customs of civilization, and the natural influence of superiority of race. And it may fairly be doubted whether some of these advantages might not be given to savages without the accompanying inculcation of particular religious tenets. True, the experiment has not been fairly tried, and the missionaries have almost all the facts to appeal to on their own side; for it is undoubtedly the case that the wide sympathy and self-denying charity which gives up so much to benefit the savage, is almost always accompanied and often strengthened by strong religious convictions. Yet there are not wanting facts to show that something may be done without the influence of religion ... A missionary who is really earnest, and has the art (and the heart) to gain the affections of his flock, may do much in eradicating barbarous customs, and in raising the standard of morality and happiness. But he may do all this quite independently of any form of sectarian theological teaching, and it is a mistake too often made to impute all to the particular doctrines inculcated, and little or nothing to the other influences we have mentioned. We believe that the purest morality, the most perfect justice, the highest civilization, and the qualities that tend to render men good, and wise, and happy, may be inculcated quite independently of fixed forms or dogmas, and perhaps even better for the want of them. The savage may be certainly made amenable to the influence of the affections, and will probably submit the more readily to the teaching of one who does not, at the very outset, attack his rude superstitions. These will assuredly die out of themselves, when knowledge and morality and civilization have gained some influence over him; and

> he will then be in a condition to receive and assimilate whatever there is of goodness and truth in the religion of his teacher.
>
> Unfortunately, the practices of European settlers are too often so diametrically opposed to the precepts of Christianity, and so deficient in humanity, justice, and charity, that the poor savage must be sorely puzzled to understand why this new faith, which is to do him so much good, should have had so little effect on his teacher's own countrymen. The white men in our colonies are too frequently the true savages, and require to be taught and Christianized quite as much as the natives ... The savage may well wonder at our inconsistency in pressing upon him a religion which has so signally failed to improve our own moral character, as he too acutely feels in the treatment he receives from Christians. It seems desirable, therefore, that our Missionary Societies should endeavour to exhibit to their proposed converts some more favourable specimens of the effect of their teaching. It might be well to devote a portion of the funds of such societies to the establishment of model communities, adapted to show the benefits of the civilization we wish to introduce, and to serve as a visible illustration of the effects of Christianity on its professors. The general practice of Christian virtues by the Europeans around them would, we feel assured, be a most powerful instrument for the general improvement of savage races, and is, perhaps, the only mode of teaching that would produce a real and lasting effect.[10]

In Smith (2003–, Chapter 5), I write: "Wallace evidently has now reached the point of cogitating on exactly what it will take—what kinds of 'model institutions'—to deliver forms of instruction serving what might be termed 'believable example'; *i.e.*, that will provide a foundation for informed belief. Clearly, inculcation was not enough; further, and building on the thoughts presented in the 'Public Responsibility and the Ballot' letter earlier, neither were the opinions of the masses, which could not be depended on to 'increase the average intelligence or morality of the country.' " As Spencer had argued (and Wallace also believed), people should receive what they truly deserve, and this could not generate turns for the better until they bought into a belief system that helped them target decent goals. "How to Civilize Savages" is a powerful allegory for the ages, a look at the dilemma facing the whole of humankind, not savages alone.

At this point, an important new influence entered Wallace's life: spiritualism.

Wallace's Adoption of Spiritualism

The most extensive analyses of Wallace's adoption of spiritualism are by Kottler (1974) and Malinchak (1987).[11] Both investigators explore the matter under the assumption that Wallace did reverse himself on the applicability of natural selection to humankind and that this requires an explanation, possibly related to his spiritualism. Neither, however, dug far enough to come to any firm conclusions regarding the three outstanding questions surrounding his adoption of the belief: (1) When did Wallace first begin to investigate the subject? (2) At what point did he become a full believer? and most importantly, (3) What was his main reason for adopting it? Kottler does, however, thoroughly summarize the history of the spiritualism movement, survey Wallace's various "field" studies of spiritualistic phenomena, and describe how his efforts were received within the scientific community. Further, he reviews and discusses the various ways that spiritualism might have interacted with other factors to produce the alleged change of mind. He concludes: "I tend to believe Wallace was persuaded by his scientific as well as spiritual arguments against natural selection. Yet I remain convinced that Wallace's belief in the reality of psychical phenomena and their spiritualist interpretation created the initial doubts about natural selection and stimulated his rethinking, on grounds of utility, man's unique features" (Kottler 1974, 192).

Malinchak, meanwhile, resists coming to a full conclusion on the matter, merely offering the loaded observation that "It was only after Wallace engaged in his extensive studies in spiritualism and became convinced of the genuineness of spiritualistic phenomena that he began to inject quasi-religious notions of the guidance of higher intelligences in the development of the human mind into his scientific arguments" (Malinchak 1987, 109). At the same time, however, she sees no link between his conversion to spiritualism and his existing views on natural selection, instead referring the former to causes rooted in his earlier experiences with the supernatural, and a range of ambient social and intellectual trends.

But all of this begs the question of *whether* there was a reversal of position that needs explaining on the basis of his adoption of spiritualism—or any other factor—to begin with. I have just shown that in the year preceding the middle of 1865 Wallace had gone public with an escalating discussion on those elements of societal engagement that might lead to an elevation of social purpose—that is, to societal evolution. This sustained dialogue was clearly the central thing on his mind at that point. His published scientific work from this same period, April 1864 to June 1865, is rather uninteresting, consisting only of straightforward systematic treatments of Malay Archipelago birds (S102 1864, S112 1865) and land shells (S109 1865). It remains to be shown how his subsequent activities reveal a continuation of purpose, and to accomplish this we begin by returning to the three questions stated above, regarding his adoption of spiritualism.

In Smith (2003–) I provide a considerably more extensive review of available sources regarding Wallace's initial dealings with spiritualism than had previously been available. This establishes, seemingly once and for all, the following chronology. In *My Life* (S729 1905) he claims to have been aware of the spiritualism movement even while in the East (and I have confirmed that at least two publications he is known to have received during that period, *Athenaeum* and *Literary Gazette*, carried stories on the subject). There is some conflicting evidence as to how much, if any, attention he gave to the subject in the three-year period after his return to England in 1862, but the best available information leads me to think (as

Kottler also concluded) that he only began to take the matter seriously around June of 1865.

Malinchak (1987) suggests that this date might have reflected a simple accident of work schedule; specifically, that before that he had been too occupied with his collections to put his attention elsewhere. Slotten (2004) suggests the timing to be fallout from Wallace's jilting by his fiancee. Neither of these theories should be taken seriously. Malinchak is correct that for the time being his systematics work ceased, but after an interlude of about a year it recommenced, with years of as much publication activity as before (see Smith 2003–, Chapter 5). Chapter 10 of Slotten's (2004) otherwise excellent biography, concerning in part Wallace's spiritualism, is marred by omissions and errors in chronology and bibliography. His conclusion that "An emotional crisis, not an intellectual or a spiritual one, drove him into the embracing arms of mediums" is absurd—not only for the reasons being explained in this chapter (which Slotten does not even entertain), but for at least four additional reasons discussed in a note at the end of Chapter 5 of my "Alfred Russel Wallace: Evolution of an Evolutionist" (Smith 2003–).

However, Slotten may be correct in suggesting that the likely *immediate* catalyst for Wallace's investigations was his sister Fanny, who was a believer at that point, and with whom Wallace had been sharing a house around that time. Once the stage was set—and by this I mean the intellectual stage, as evidenced by his literary preoccupations of the previous year—it probably would not have required much convincing from someone so close to him to at least have a look.

Whatever the reason for Wallace's taking interest exactly when he did, the signal result was a nearly complete captivation of his attention for a full year. The unequivocal evidence of this is not only a nearly full and immediate cessation in his published output, but a nearly equally complete withdrawal from his professional involvements. The only substantial work he published between June 1865 and June 1866 was a systematic treatment of Malay Archipelago pigeons (S114), which appeared in print in *Ibis* in October 1865 (and could have been finished some months earlier). Between that date and the middle of May 1866, nothing whatsoever from his pen appeared in print—the longest unaccounted-for unproductive period in his entire career. Even more interestingly, apart from some short comments (S113a) offered at a 4 July 1865 meeting of the Ethnological Society, there is no record of his participating, either as presenter or commenter, in any other professional meeting over the same period. By contrast, in the preceding twelve-month period, twelve such commitments are known, and in the following twelve-month period, fourteen. Further, although he did attend the annual British Association for the Advancement of Science meetings in the late summer of 1865 (and took part in some committee work while there), he presented no paper that year—whereas in the two years preceding and following he presented a total of seven papers (and at least one each year). Clearly, for a full year he had taken a major "time out."[12]

While it is now apparent what was going on, the progression of Wallace's early engagement of spiritualism is more complicated than previous writers have recognized. Actually, three rather distinct and escalating levels in his interest can be identified, stages that might be termed his "Wallace the Seeker," "Wallace the Promoter," and "Wallace the Believer" periods. During the first of these, lasting from roughly June 1865 to December 1865, Wallace sought confirmation that his new object of attention was both demonstrably real, and underlain by a body of philosophy consistent with his "no merit to uninformed belief" theory of the basis of societal change. Many writers, including Kottler and Malinchak, have reviewed his seance attendance activities during this early period, but no attention has been given to the second matter, which ended up being more crucial: he might have been able to excuse (and apparently *did*) not encountering convincing physical manifestations at first, but this wouldn't have mattered had he quickly reached the conclusion that the philosophy of spiritualism was bereft of any logic relatable to the "refinement of informed belief" matter. Exhaustive literature reviews of newly dealt-with subjects were regular course for Wallace (see S741 1903, 176; Marchant 1975 [1916], 353, 363–64; S729 1905a, 2:100–01, 231, 233, 243, 350–51, 353), and in this instance his behavior was no different: he wanted to gain insight into spiritualism's history and objectives (S729 2:279–80; Malinchak 1987, 80–82). His own first writing on spiritualism, "The Scientific Aspect of the Supernatural" (S118 1866), straightforwardly attests to this, as it is nothing if not a detailed review of the subject's philosophy and literature.

By the middle of the fall of 1865 Wallace had attended a significant number of seances, but still hadn't witnessed any fully convincing phenomena. Further, he had thus far not been able to control the proceedings by having them staged in his own home. Then the perhaps single most important event in Wallace's post-Malay Archipelago life took place. The 1 December 1870 issue (Volume 1, Number 1) of the obscure newsletter *The Spiritual News* describes discussion that followed Wallace's first public address on spiritualism, "An Answer to the Arguments of Hume, Lecky, and Others, Against Miracles" (S174), presented during a soirée held late that same year. The host of the soirée, an entrepreneur and leading spiritualist named Benjamin Coleman, is quoted in the article as saying that "it was just five years ago" that he (Coleman) launched the series, and that at "the very first meeting held in that room in connection with Spiritualism, Mr. Wallace was present as a strong disbeliever." The meeting in question took place on 6 November 1865, so at that point it would appear Wallace was still in "Seeker" mode.

Wallace likely had reasons for attending the event that extended beyond its inaugural nature: it featured the first in a new series of lectures by the spiritualist sage Emma Hardinge (1823–99; Fig. 32). Hardinge, an Englishwoman who had spent many years in America (initially as an actress), was by 1865 one of the Movement's leading lights—a powerful communicator who spoke eloquently and extemporaneously, on subjects introduced from the audience. Portions of her 6 November 1865 lecture were later published in *The Spiritual Magazine*,[13] including such remarks as:

> In pointing to the analogy that exists between the great physical and spiritual laws of Earth, together with the modes in which they act, I have sought to shew you that all that man has called the supernatural, and classes as miracle, is but the out-working of an harmonious plan, which the mighty Spirit reveals through eternal laws; and the Spiritualism at which you marvel, and the Christianity before which you bow, are but parts of the same divine law and alternating life of order, which ever sees the day spring out of the darkest night ...
>
> ... By Chemistry, man learns through scientific processes, to dissolve and re-compose in changed form, every existing atom. Time, instruments, and material processes alone are asked for the chemistry of science to accomplish these results. To the Spirit (whose knowledge comprehends all laws revealed to man) such chemistry is possible, and truly is achieved, without the lapse of time, or the aid of human science yet known as such to Man ...
>
> ... Translated through the solemn utterance of dim antiquity all this is "Miracle"—in simple modern science, it is "Chemistry," requiring only knowledge to effect these changes; in modern spiritualistic phrase 'tis mediumship, or chemistry employing subtler forces to effect in yet more rapid time and simpler modes than man's, the self-same changes which man can make by science. To-day you listen to the tap, tap, of the electric telegraph of the soul; you translate into sentences that strange and grotesque form of telegraphy; you behold inscribed on the blank page the name of some beloved one written with no mortal hand; you feel the baptism of the falling water, you know not from whence; and the fragrance of flowers not gathered by mortal power appeals to your startled senses. You call this Spiritualism; and what is this but the chemistry of the spirit? ... (Anon. 1865, 531–2).

The apparent pivotal influence of Hardinge on Wallace (see discussion below, and in n. 14) has not previously been appreciated (none of the four most recent Wallace biographers—Raby 2001; Shermer 2002; Fichman 2004; or Slotten 2004—even mention her in this context). For Wallace, who was still trying to come to an understanding of the place of the higher sympathies in natural context, words such as those quoted above must have been revelatory. Hardinge gave several more lectures over the next six months, and their summary effect, when combined with the results of his literature review, was to turn him—not into a full believer—but into a sympathizer who now felt that the subject was worthy of investigation by the scientific research community. Still eschewing any professional commitments, he began to compose a monographic essay that pled for such attention. This was "The Scientific Aspect of the Supernatural" (S118 1866).

Slotten (2004, 245) places the publication of this work in late 1866, apparently unaware that it was actually ready for publication by no later than midsummer, and released in weekly installments shortly thereafter in the secularist periodical *The English Leader.* A note on page 9 of *The English Leader* issue of 21 July 1866 reports that they have received Wallace's manuscript and are ready to give it

Figure 32 Portrait of the spiritualist speaker Emma Hardinge (Britten).
The frontispiece to Emma Hardinge's book *Modern American Spiritualism* (1870).
Out of copyright.

"immediate attention." Probably not coincidentally the last lecture in Emma Hardinge's tour had been set for 24 June 1866 (per mention in the *National Reformer* issue of 1 July 1866)—that is, just in time for Wallace to hear or read before finishing up his manuscript. Concerning "The Scientific Aspect of the Supernatural" I have written:

> The orientation of this work is revealing. Conscious that he has not yet obtained satisfactorily definitive physical evidence, Wallace concentrates on literature review and producing a philosophical argument for investigating the phenomena. He begins by noting that our senses are limited, and that it is only through the accumulation of knowledge that we have elevated our understanding of physical processes above assumptions of the miraculous. He then argues that the so-called miracles of the past and present most likely represent non-miraculous aspects of natural process that we simply do not yet understand. Next he moves on to a consideration of cryptic forces in nature, and then to some of the recorded evidence of various spiritualistic phenomena. Finally, he treats the theory and moral teachings of spiritualism, drawing very heavily from the writings of Emma Hardinge to complete his review.[14] Entirely missing from the treatment are descriptions of any of his own investigations of the phenomena—which, of course, had so far only proved mildly corroborative. Nevertheless, he had done a passably good job of reducing a large and esoteric literature to a readable declaration of its legitimacy for study (Smith 2003–, Chapter 5).

Wallace's greater purpose at the point he submitted the essay for publication is revealed in the following excerpt from the work:

> Now here again we have a striking supplement to the doctrines of modern science. The organic world has been carried on to a high state of development, and has been ever kept in harmony with the forces of external nature, by the grand law of "survival of the fittest" acting upon ever varying organisations. In the spiritual world, the law of the "progression of the fittest" takes its place, and carries on in unbroken continuity that development of the human mind which has been commenced here (S118, 49–50).

From this passage several things seem evident. First and foremost, Wallace recognizes as "laws" (*i.e.*, not as theories or processes) both the "survival of the fittest" and the "progression of the fittest." He also recognizes them as applicable to different domains, yet connected in "unbroken continuity." Further, he speaks in terms of a "supplement," and not an "alteration," "revision," etc., to the "doctrines of modern science." Last and perhaps most interestingly, this turns out to be the first time in Wallace's published writings that he uses the term "survival of the fittest" (Smith 2008). (Some months later he first uses [at least in its modern sense] the term "evolution" in a letter on mimicry published in *Athenaeum* [S123 1866], just after the issuance of the pamphlet version of *The Scientific Aspect of the Supernatural* [see below]). Note in this context the famous letter in which Wallace suggested to Darwin that he adopt the term "survival of the fittest" as a way of conveying the essence of natural selection—it is dated 2 July 1866 (Marchant 1975 [1916], 140–43); that is to say, just as Wallace was readying his spiritualism essay for publication, and undoubtedly reflecting upon the implications of the "unbroken continuity" between the "progression of the fittest" and material nature.

Meanwhile, Wallace was beginning to resume his professional activities. From June 1866 on, and for the next few years, his rates of contribution to scientific meetings and publication of literary works closely approximated his pre-June 1865 efforts. Among the first stops was a short speech to the Anthropological Section of the annual British Association for the Advancement of Science meetings. Delivered on 23 August 1866, just as "The Scientific Aspect of the Supernatural" was being serialized in print, it features the following interesting admonition:

> The anthropologist must ever bear in mind that, as the object of his study is *man*, nothing pertaining to or characteristic of man can be unworthy of his attention. It will be only after we have brought together and arranged all the facts and principles which have been established by the various special studies to which I have alluded, that we shall be in a condition to determine the particular lines of investigation most needed to complete our knowledge of man, and may hope ultimately to arrive at some definite conclusions on the great problems which must interest us all—the questions of the origin, the nature, and the destiny of the human race. I would beg you to recollect also that *here* we must treat all these problems as purely questions of science, to be decided solely by facts and by legitimate deductions from facts. We can accept no conclusions as authoritative that have not been thus established.

> Our sole object is to find out for ourselves what is our true nature ... (S119, 93–94).

Can anyone doubt that Wallace was thinking of his spiritualism studies when he offered up these pointed remarks?

After seven weeks of continuations the final installment of "The Scientific Aspect of the Supernatural" was printed in *The English Leader* issue of 29 September 1866. Sometime during the fall, however, Wallace decided that the serialized format was not up to the goal of getting his message out, and arranged to have a pamphlet version released. Kottler (1974) and others (including Wallace himself, in his autobiography) make much of the fuss its distribution caused among his friends and acquaintances, but this story, amusing as it is, is less informative than the surrounding chronology of events. It was undoubtedly printed after 29 September; apart from this being likely *a priori*, the pamphlet version is based on the same typesetting as the magazine layout, but contains a few notes and edits (and an introduction) not present in the latter. The real question, however, is how *late* it might have been printed, and in turn what that might imply. In November of 1866, Wallace finally found a medium, Miss Nicholl (later Mrs Guppy), a friend of the family, who both produced convincing phenomena, and agreed to hold sessions, for free, in Wallace's own quarters. Slotten (2004) places Wallace's production and distribution of the pamphlet at a date significantly later than this event, but this is most unlikely. The very latest it could have been sent to the printers was early to mid-November, since a cover letter dated 22 November that presented a finished copy of the work to Thomas Huxley exists (Marchant 1975 [1916], 417–18). Consider Wallace's words in the work describing his purpose for producing it: "... Let us now return to the consideration of the probable nature and powers of those preter-human intelligences whose possible existence only it is my object to maintain ..." (S118, 7–8). This is no more than a plea—that study be given to a matter worthy of attention—and not the words of a full convert. As of its release Wallace was still in his "Promoter" stage; had the confirming manifestations occurred well before the date of release of the pamphlet (whatever date it was printed), he might have stopped its distribution, or at least added further commentary first—as he actually *did* later when he revised the essay for inclusion in *On Miracles and Modern Spiritualism* (S717 1875).[15] At the very least, as of 22 November he was still feeling that his *a priori* arguments were strong enough to merit a continued push—which, beginning at that time, included cover letter invitations to colleagues to take part in his family's now weekly sessions with Miss Nicholl.

Wallace's seances with Miss Nicholl in December 1866 and early 1867 produced some very remarkable manifestations (S126 1867, S132 1867, S137 1867), all at his home, and under Wallace's supervision. During this period he inevitably abandoned his "Promoter" role to turn "Believer." Also, inevitably, he begins to show evidence of this new influence in his writings. In a letter to the *Anthropological*

Review published in January 1867 he writes: "the principles of Mr. Darwin's *Origin of Species*, if applied to man with such modifications as are required by the great development and vast importance of his intellectual and moral rather than his mere animal nature, leads to the apparently paradoxical result that he is tending to become again as his progenitors once undoubtedly must have been, 'a single homogeneous race' " (S125, 105). What kind of "modifications," one wonders, is he referring to here? Similarly, in the unlikely setting of a monographic review of the butterfly family Pieridae presented to the Entomological Society on 18 February 1867, he says: "It is, therefore, no objection to a theory that it does not explain everything, but rather the contrary. A true theory will certainly enable us to understand many of the phenomena of life, but owing to our necessarily imperfect knowledge of past causes and events, there must always remain complicated knots that we cannot disentangle, and dark mysteries on which we can throw but a straggling ray of light" (S127, 309). This sounds very much like stage-setting to me.[16]

Evolution, à la Wallace

As 1867 proceeded and he witnessed further convincing seance phenomena, Wallace, being Wallace, surely began to think about possible venues for expressing himself fully on his new synthesis. The first opportunity to do so, at least in part, came when he published a review of the Duke of Argyll's anti-Darwinian book *The Reign of Law* in the fall of that year (S140). In it he writes:

> . . . why should we measure the creative mind by our own? Why should we suppose the machine too complicated to have been designed by the Creator so complete, that it would necessarily work out harmonious results? The theory of "continual interference" is a limitation of the Creator's power. It assumes that he could not work by pure law in the organic as he has done in the inorganic world; it assumes that he could not foresee the consequences of the laws of matter and mind combined—that results would continually arise which are contrary to what is best, and that he has to change what would otherwise be the course of nature in order to produce that beauty and variety and harmony, which even we, with our limited intellects, can conceive to be the result of self-adjustment in a universe governed by unvarying law. If we could not conceive the world of nature to be self-adjusting and capable of endless development, it would even then be an unworthy idea of a Creator to impute the incapacity of our minds to him; but when many human minds can conceive and can even trace out in detail some of the adaptations in nature as the necessary results of unvarying law, it seems strange that in the interests of religion any one should seek to prove that the System of Nature instead of being above, is far below our highest conceptions of it. I, for one, cannot believe that the world would come to chaos if left to Law alone. I cannot believe that there is in it no inherent power of developing beauty or variety, and that the direct action of the Deity is

> required to produce each spot or streak on every insect, each detail of structure in every one of the millions of organisms that live or have lived upon the earth. For it is impossible to draw a line. If any modifications of structure could be the result of law, why not all? If some self-adaptations could arise, why not others? If any varieties of colour, why not all the variety we see? No attempt is made to explain this except by reference to the fact that "purpose" and "contrivance" are everywhere visible, and by the illogical deduction that they could only have arisen from the direct action of some mind, because the direct action of our minds produces similar "contrivances;" but it is forgotten that adaptation, however produced, must have the appearance of design. The channel of a river looks as if made *for* the river although it is made *by* it; the fine layers and beds in a deposit of sand often look as if they had been sorted and sifted and levelled designedly; the sides and angles of a crystal exactly resemble similar forms designed by man; but we do not therefore conclude that these effects have, in each individual case, required the directing action of a creative mind, or see any difficulty in their being produced by natural Law (S140, 479–80).

It is not Wallace's object here to suggest that humankind is "above" natural law, but neither does he imply that river channels are the only kinds of things that display an "appearance of design." More to the point, a kind of "design" is envisioned that looks not to an anthropomorphic God, but instead to laws-based final causes. Actually, the review was an extraordinary accomplishment, managing simultaneously to again criticize the "continual interference" model of Creationists, resuscitate his "geographical Bauplan" model from the 1840s and 1850s (Smith 2003–), lay the groundwork for the final causes-flavored evolutionary cosmology evidenced later in *Man's Place in the Universe* (S728 1903) and *The World of Life* (S732 1910), and bring to the table some initial considerations on the workings of mind and spirit.

Any desire to expand further on this kind of thinking was thwarted for the time being when he realized that the time had finally come to put out a journal of his Eastern travels. Most of 1868 was consumed by the preparation of what would become *The Malay Archipelago* (S715 1869), but that year also produced some foreshadowing of his eventual split with Darwin over human evolution. Shortly before *Malay Archipelago* was finished, Wallace attended the annual meetings of the British Association for the Advancement of Science. There he sat in on a lecture by the Revd. F. O. Morris "On the Difficulties of Darwinism." After its conclusion he was cited as remarking: "With regard to the moral bearing of the question as to whether the moral and intellectual faculties could be developed by natural selection, that was a subject on which Mr. Darwin had not given an opinion. He (Mr. Wallace) did not believe that Mr. Darwin's theory would entirely explain those mental phenomena" (S142a 1868). It was the first public expression of his break with Darwin on the causes of evolution of the higher attributes of humankind—a break founded not on a reduction of his thoughts on natural selection, but instead on a wedding of the kind of thinking expressed in "Creation by Law" with his spiritualistic model of the "progression of the fittest."

Obviously, Wallace was now ready to square off against Darwin's views on the origin of the higher human faculties. The immediate problem became a venue within which he could fully express such thoughts, in writing.

For a while no such opportunity presented itself, but in the meantime a public discussion unfolded that must have left Wallace straining at the leash. An article entitled "On the Failure of 'Natural Selection' in the Case of Man" was published anonymously in *Fraser's Magazine* in September 1868, creating a considerable stir. Penned by William R. Greg (a writer on social issues who would become a key figure in the eugenics movement), it argued that in our society protection of the weak—the poor and the inferior in mind or body—had left natural selection an ineffectual agent for improvement. Greg uses Wallace's own reasoning as presented in his 1864 paper to the Anthropological Society (S93) as the basis for his argument, pointing to Wallace's observation that in humans selection had become refocused at the level of the mind.

Early on, reaction to the paper was generally favorable, but before long opinion turned vehemently critical. A few weeks later in *The Spectator* an anonymous writer opined that Greg's argument was flawed, because:

> ... The plan of God seems to be to ennoble the higher part of His universe at least, not so much by eliminating imperfection, as by multiplying graces and virtues. He balances the new evils peculiar to human life by infinitely greater weights in the scale of the good which is also peculiar to human life. "Natural selection" has its place and its function, doubtless, even amongst us. But over it, and high above it, is growing up a principle of supernatural selection, by our free participation in which we can alone become brethren of Christ and children of God (Anon. 1868, 1155).

Neither this position nor Greg's original manipulation of his thinking could have made Wallace do more than roll his eyes, if for contrasting reasons. In January 1869, however, another kind of evaluation of Greg's reasoning appeared, this time in a publication noted for its liberal views, the *Quarterly Journal of Science*. This writer concluded, again anonymously, that Greg and others had missed the point: selection was still going on, but its nature was changing as humankind evolved:

> ... So with the communities of civilized men—the struggle is between one society and another, whatever may be the bond uniting such society: and in the far distant future we can see no end to the possible combinations or societies which may arise amongst men, and by their emulation tend to his development. Moral qualities, amongst the others thus developed in the individual necessarily arise in societies of men, and are naturally selected, being a source of strength to the community which has them most developed: and there is no excuse for speaking of a failure of Darwin's law or of

> "supernatural" selection. We must remember what Alfred Wallace has insisted upon most rightly—that in man, development does not affect so much the bodily as the mental characteristics; the brain in him has become much more sensitive to the operation of selection than the body, and hence is almost its sole subject. At the same time it is clear that the struggle between man and man is going on to a much larger extent than the writer in "Fraser" allowed. The rich fool dissipates his fortune and becomes poor; the large-brained artizan does frequently rise to wealth and position; and it is a well-known law that the poor do not succeed in rearing so large a contribution to the new generation as do the richer. Hence we have a perpetual survival of the fittest. In the most barbarous conditions of mankind, the struggle is almost entirely between individuals: in proportion as civilization has increased among men, it is easy to trace the transference of a great part of the struggle little by little from individuals to tribes, nations, leagues, guilds, corporations, societies, and other such combinations, and accompanying this transference has been undeniably the development of the moral qualities and of social virtues (Anon. 1869).

This was the kind of thinking, relating selection to the societal role of the "higher attributes," that was bound to attract Wallace's favor, and it did: in a letter to Darwin dated 20 January 1869 he exclaims: "Have you seen in the last number of the *Quarterly Journal of Science* the excellent remarks on *Fraser's* article on Natural Selection failing as to Man? In one page it gets to the heart of the question, and I have written to the Editor to ask who the author is." A few lines later he adds: "Perhaps you have heard that I have undertaken to write an article for the *Quarterly* (!) on the same subject [*i.e.*, Lyell's *Principles of Geology*], to make up for that on 'Modern Geology' last year not mentioning Sir C. Lyell" (Marchant 1975 [1916], 190–91).

This was the opportunity Wallace had been waiting for. Since at least his writing of "The Scientific Aspect of the Supernatural" (S118) in the first half of 1866 the essential differences between "natural selection" and "evolution" had become more apparent to him, resulting in, as mentioned earlier, his relation of the concepts "survival of the fittest" to "progression of the fittest" in that work, and his concurrent letter recommending the use of the former term to Darwin. In the Lyell review he was given license to provide a thorough recap of Lyellian uniformitarianism, embed Darwinian principles within its context, and, once this foundation was established, introduce his new thoughts relating to the more embracing subject of evolution in general. Lyell's adoption of Darwinism, supposedly "the great distinguishing feature of this [new] edition" [of *Principles of Geology*], is only first dealt with two-thirds of the way through the essay—a bit strange, unless it is actually part of Wallace's agenda to stress that even the most thoroughly worked-out ideas may be subject to alteration or adjustment. After ten further pages recounting the history of evolutionary ideas and the various kinds of evidence for Darwinian natural selection in particular, he finally arrives at the culmination of his discussion: seven paragraphs explaining in brief why, after supporting all that has preceded, he yet feels that "more recondite" forces act to shape the moral and intellectual evolution of humankind: "Neither natural selection nor the more general theory of evolution can give any account whatever of the origin of sensational or conscious life ... the moral and higher intellectual nature of man is as unique a phenomenon as was conscious life on its first appearance in the world" (S146 1869, 391).

It is sometimes forgotten that the *Quarterly Review* article actually appeared several weeks after the issuance of *The Malay Archipelago* (S715), some of the last words of which are:

> ... We most of us believe that we, the higher races, have progressed and are progressing. If so, there must be some state of perfection, some ultimate goal, which we may never reach, but to which all true progress must bring us nearer. What is this ideally perfect social state towards which mankind ever has been, and still is tending? Our best thinkers maintain that it is a state of individual freedom and self-government, rendered possible by the equal development and just balance of the intellectual, moral, and physical parts of our nature,—a state in which we shall each be so perfectly fitted for a social existence, by knowing what is right, and at the same time feeling an irresistible impulse to do what we know to be right, that all laws and all punishments shall be unnecessary. In such a state every man would have a sufficiently well-balanced intellectual organization to understand the moral law in all its details, and would require no other motive but the free impulses of his own nature to obey that law ...
>
> ... although we have progressed vastly beyond the savage state in intellectual achievements, we have not advanced equally in morals. It is true that among those classes who have no wants that cannot be easily supplied, and among whom public opinion has great influence, the rights of others are fully respected. It is true, also, that we have vastly extended the sphere of those rights, and include within them all the brotherhood of man. But it is not too much to say, that the mass of our populations have not at all advanced beyond the savage code of morals, and have in many cases sunk below it. A deficient morality is the great blot of modern civilization, and the greatest hindrance to true progress ...
>
> During the last century, and especially in the last thirty years, our intellectual and material advancement has been too quickly achieved for us to reap the full benefit of it. Our mastery over the forces of nature has led to a rapid growth of population, and a vast accumulation of wealth; but these have brought with them such an amount of poverty and crime, and have fostered the growth of so much sordid feeling and so many fierce passions, that it may well be questioned, whether the mental and moral status of our population has not on the average been lowered, and whether the evil has not overbalanced the good ...
>
> This is not a result to boast of, or to be satisfied with; and, until there is a more general recognition of this failure of our civilization—resulting mainly

> from our neglect to train and develop more thoroughly the sympathetic feelings and moral faculties of our nature, and to allow them a larger share of influence in our legislation, our commerce, and our whole social organization—we shall never, as regards the whole community, attain to any real or important superiority over the better class of savages ... (S715 1891, 455–57).

This epilogue is exactly a continuation of the line of thought expressed in the 1864–65 works on "informed belief" discussed earlier. It perfectly complements the last seven paragraphs of the *Quarterly Review* article, which focus on the relation of "higher influences" to natural selection on individual human beings, as opposed to socially-mediated moral change. Here, in *The Malay Archipelago*, Wallace again reflects on what constitutes the "perfect social state" and how informed belief can contribute to its development; the main difference between it and his pre-1866 studies is that he now feels he recognizes a solution to the problem. In Smith (2003–, Chapter 5) I summarize:

> Wallace had by now come to the conclusion that the "Spirit Realm" described by spiritualist prophets such as Stainton Moses constituted a *natural* domain within which the trace of organic evolution was continued—in the same way the latter continued, was intimately linked with, and depended on, the inertia of continuing forms of inorganic evolution. The critical connection for Wallace would have been his recognition that, given the supposed nature of the spirit realm, the higher faculties of man did in fact have utility. But this was not a function contributing only to biological survival, and thus devolving from causes dictated by conditions of the immediate physical environment. Instead, the refinement of the higher faculties made possible a continuing elevation of function after the biological death of the individual within a purely psychic (or "will-expressed") domain of organization. Higher spiritual development meant a greater capacity for identifying (and setting into action) new causal forces contributing to the overall evolutionary progression (much as biological evolution had secondarily modified the evolution of physical systems such as the atmosphere).

Perhaps the most succinct statement of Wallace's social vision appears in his 1892 essay "Human Progress: Past and Future":

> ... I have endeavored to show, in the present article, that we are not limited to the depressing alternatives above set forth,—that education has the greatest value for the improvement of mankind,—and that selection of the fittest may be ensured by power and more effective agencies than the destruction of the weak and helpless. From a consideration of historical facts bearing upon the origin and development of human faculty I have shown reason for believing that it is only by a true and perfect system of education and the public opinion which such a system will create, that the special mode of selection on which the future of humanity depends can be brought into general action. Education and environment, which have so often stunted and debased human nature instead of improving it, are powerless to trans-

> mit by heredity either their good or their evil effects; and for this limitation of their power we ought to be thankful. It follows, that when we are wise enough to reform our social economy and give to our youth a truer, a broader, and a more philosophical training, we shall find their minds free from any hereditary taint derived from the evil customs and mistaken teaching of the past, and ready to respond at once to that higher ideal of life and of the responsibilities of marriage which will, indirectly, become the greatest factor in human progress (S445, 158–59).

In short, so long as we keep in mind that it is through the application of "intelligent conviction" that weaknesses can be eliminated, we and our derivative social systems can continue to evolve productively.

Change, Or No Change?

The interpretation of Wallace's motives and activities just given constitutes what I have termed the "no change of mind" theory of his personal evolution of thought. It should be noted that in applying this name I have been motivated, primarily, by trying to contrast the new understanding with the old approach based largely on the (I feel misguided) assumption that Wallace's 1858 model of natural selection was intended to cover all aspects of human evolution, just as it treated change in plants and nonhuman animals. Neither I nor Martin Fichman in his writings on the subject (Fichman 2001, 2004) have ever meant to imply that Wallace underwent *no* changes of position whatsoever during this period, just that there was never a *reversal* involved—at least, on his thoughts on man. It remains to take a quick look at some of the evidence that has been set forth to argue that there was.

Some supporters of the "change of mind" hypothesis (as reviewed by Kottler 1974 and Malinchak 1987) would have it that when in 1865 Wallace became acquainted with spiritualism, he found in this belief a way to reverse himself on his until-then materialist approach to natural selection. This scenario faces some serious difficulties, both in terms of its over-reliance on negative evidence, and its avoidance of certain clues to the contrary. In the first place, there is the central and questionable assumption that he held a position on which he could backtrack to begin with: as indicated earlier, there is nothing either in the Ternate essay itself or his later appraisals of it that suggests, even obliquely, that at that point he had embraced an understanding of natural selection meant to pertain to the higher human attributes.

Later, moreover, in 1875, Wallace actually *himself* directly dismissed the idea that the origin of his divergence from the views of Darwin on natural selection was due to his acceptance of spiritualism. In the Preface to the first edition of *On Miracles and Modern Spiritualism*, he seems pretty clear on this point:

> ... I am informed that, in an article entitled "Englische Kritiker und Anti-Kritiker des Darwinismus," published in 1861 [an error for 1871], he [Anton Dohrn] has put forth the opinion that Spiritualism and Natural Selection are incompatible, and that my divergence from the views of Mr. Darwin arises from my belief in Spiritualism. He also supposes that in accepting the spiritual doctrines I have been to some extent influenced by clerical and religious prejudices. As Mr. Dohrn's views may be those of other scientific friends, I may perhaps be excused for entering into some personal details in reply.
>
> From the age of fourteen ... Up to the time when I first became acquainted with the facts of Spiritualism, I was a confirmed philosophical sceptic, rejoicing in the works of Voltaire, Strauss, and Carl Vogt, and an ardent admirer (as I still am) of Herbert Spencer. I was so thorough and confirmed a materialist that I could not at that time find a place in my mind for the conception of spiritual existence, or for any other agencies in the universe than matter and force. Facts, however, are stubborn things. My curiosity was at first excited by some slight but inexplicable phenomena occurring in a friend's family, and my desire for knowledge and love of truth forced me to continue the inquiry. The facts became more and more assured, more and more varied, more and more removed from anything that modern science taught or modern philosophy speculated on. The facts beat me. They compelled me to accept them *as facts* long before I could accept the spiritual explanation of them; there was at that time "no place in my fabric of thought into which it could be fitted." By slow degrees a place was made; but it was made, not by any preconceived or theoretical opinions, but by the continuous action of fact after fact, which could not be got rid of in any other way. So much for Mr. Anton Dohrn's theory of the causes which led me to accept Spiritualism. Let us now consider the statement as to its incompatibility with Natural Selection.

He goes on to describe the "natural" basis of his study:

> Having, as above indicated, been led, by a strict induction from facts, to a belief—1stly, In the existence of a number of preterhuman intelligences of various grades and, 2ndly, That some of these intelligences, although usually invisible and intangible to us, can and do act on matter, and do influence our minds,—I am surely following a strictly logical and scientific course in seeing how far this doctrine will enable us to account for some of those residual phenomena which Natural Selection alone will not explain. In the 10th chapter [S165] of my *Contributions to the Theory of Natural Selection* I have pointed out what I consider to be some of those residual phenomena; and I have suggested that they may be due to the action of some of the various intelligences above referred to. This view was, however, put forward with hesitation, and I myself suggested difficulties in the way of its acceptance; but I maintained, and still maintain, that it is one which is logically tenable, and is in no way inconsistent with a thorough acceptance of the grand doctrine of Evolution, through Natural Selection, although implying (as indeed many of the chief supporters of that doctrine admit) that it is not

> the all-powerful, all-sufficient, and only cause of the development of organic forms (S717, 1901, vi–viii).[17]

Note in particular the words "... it is one which is logically tenable, and is in no way inconsistent with a thorough acceptance of the grand doctrine of Evolution ...", and that he simply feels natural selection "is not the all-powerful, all-sufficient, and only cause of the development of organic forms."

Perhaps more telling, however, are the opening notes to the revision of his "On the Origin of Human Races ..." that appeared in the collection *Contributions to the Theory of Natural Selection*:

> I had intended to have considerably extended this essay, but on attempting it I found that I should probably weaken the effect without adding much to the argument. I have therefore preferred to leave it as it was first written, with the exception of a few ill-considered passages which never fully expressed my meaning (S716 1870, viii).

Wallace deliberately uses the word "extended" rather than "reversed" or "changed" here; moreover, he is apparently concerned that any additions might actually "weaken" the gist of the argument. From this one can only conclude that he considered the original argument fundamentally sound. He eventually did decide to end the collection with an entirely new essay, but this only makes it more difficult to understand why, were he trying to express a "change of mind," he would have decided to leave this essay more or less as it was, or, for that matter, included it in the new work at all.

In Chapter 20 Benton quotes the "ill-considered passages" part of this passage, ignoring the most important part of it, the words "... which never fully expressed my meaning." He then attempts to interpret the changes in three particular passages as indicative of a reversal of opinion, but this is beside the point: whichever passages Wallace might be referring to, such text alterations as were made, were made not as a reversal of thought, but because they *apparently did not convey his full thoughts at the time.* This is precisely what Fichman and I are arguing: that in 1863–64 Wallace, temporarily beguiled by the writings of Spencer and distracted from his main course, made an exploratory attempt to describe human evolution in material, "Darwinistic" terms. Wallace himself owned up to such a diversion on two later occasions (S528 1896; S729 1905a, 1:104), discussing his lapse for a time into what he termed "individualist" thinking.

Then there is the famous 18 April 1869 letter from Wallace to Darwin, in which he states:

> I can quite comprehend your feelings with regard to my "unscientific" opinions as to Man, because a few years back I should myself have looked at them as equally wild and uncalled for ... My opinions on the subject have been modified solely by the consideration of a series of remarkable phenomena, physical and mental, which I have now had every opportunity

> of fully testing, and which demonstrate the existence of forces and influences not yet recognised by science (Marchant 1975 [1916], 200).

Taking Wallace at his word and no further, he is only saying here that the now "fully tested" phenomena have led him to a theory which "a few years back" he would have considered "wild and uncalled for" (*i.e.*, in the absence of presentable evidence). The key word here is "modified." Wallace was usually pretty good at wording things to express just what he meant, and this usage here—instead of "reversed," or even "changed," should not be willfully misconstrued. In his 1864 essay on man the main ground gained in 1858 regarding the role of necessary utility in evolutionary change is no more than *held*: no explanation is offered in either essay for how intellect or moral behavior *emerge*. That their *presence* influenced man in ways that would be subjected to the influence of natural selection he still did not doubt (nor did he in 1870, as expressed in the later version of the paper). Wallace had likely recognized for many years that man exhibited certain "above nature" qualities; "The Origin of Human Races ..." was his attempt to describe how these qualities, *once in existence*, could be expected to aid or retard natural selection. As of 1864, however, the manner of their own origin and the connection of this to a forward-moving evolutionary inertia was an issue he still had no handle on and deliberately avoided.

Kottler's (1974) conclusions regarding the changes Wallace made in 1870 to "The Origin of Human Races ..." are rather different:

> By 1870 Wallace was doubtful about natural selection's ability to produce such a future. The mediocre were, after all, the ones who reproduced most prolifically in civilized nations despite the fact that there was an indubitable advance, "on the whole a steady and a permanent one—both in the influence on public opinion of a high morality, and in general desire for intellectual evolution." Wallace was led to invoke an "... inherent progressive power of those glorious qualities which raise us so immeasurably above our fellow animals, and at the same time afford us the surest proof that there are other and higher existences than ourselves, from whom these qualities may have been derived, and towards whom we may be ever tending." The only other relevant change in the essay was Wallace's inclusion of the words "from some unknown cause" to explain the development of man's mind from its near-animal condition to the point at which it began to shield man's body from natural selection. Therefore this essay in its new form was contradictory. It still included passages describing natural selection's accumulation of slight variations in man's intellectual and moral nature leading to ever-higher human types. But in its final paragraph it referred to an inherent progressive power of development in man's intellectual and moral nature handed down from on high. With such an inherent power, man's intellectual and moral nature was independent of external conditions and the "chance" appearance of favorable variations. Therefore it was independent of and inexplicable by natural selection (Kottler 1974, 154).

There are two main problems with this assessment. First, there is again the weak presumption that before 1858—and 1864—Wallace had been treating the evolution of humankind's higher characters in the same basic fashion he had purely biological adaptations such as limb length or jaw strength. More importantly, however, Kottler attributes to Wallace the position that the influences received "from on high" both interrupt the physical operation of natural selection, and supersede it to the extent that their effects on humans are deterministic; *i.e.*, independent of individual free will. As I have argued (Smith 2003–, Chapter 6), "there is no reason to think that Wallace ever thought in these terms at any point in his career." In a late interview, he responded to a question about spiritual influence and continuity: "I do not mean that the control is absolute or that it is of the nature of interference. The control is evidently bound by laws as absolute and irrefragable as those which govern man and his universe. It is certainly dependent on us in a very large measure for its success. I believe we are influenced, not interfered with ..." (S746 1910). In my 2003 work I continue:

> As discussed earlier, it appears that in Wallace's version of natural selection the process operated by seizing upon—amplifying—the advantages accrued by any adaptational array that might pass into existence by the mere chance interaction of those forces underlying variation. Without initial aid "from on high" helping to expose humankind to subtle, unselfish activities "transcending time and space," selection for such activities would never come about. Still, the means by which such "aid" would manifest itself would yet lead to many dead-ends of application as human beings continued to act without a full appreciation of the longer term, larger scale, implications of those acts. Aid "from on high" might indeed be interpreted as a "progressive power" in operation, but it was no surer in its unfolding at any time or place than were the more rotely accumulated adaptations shaped through biological natural selection.

Returning now a final time to Benton's analysis in Chapter 20... Benton makes the best possible case for understanding Wallace's evolution through a "change of mind" approach, yet in the last analysis the only evidence he has to support his position is Wallace's later admission that he temporarily adopted "individualist" ways. But at the same time Wallace never later confessed to believing originally—that is before 1858 and for the several years thereafter—that he held the same views on humans as he did on animals and plants. Nothing he published between "On the Tendency" and "The Origin of Human Races" belies any such views, nor does any (to this point) known correspondence. Meanwhile, writings published by Wallace throughout the rest of his adult life give clear evidence of his belief in the existence and influence of "more recondite" forces, and his activities *c.*1864–69 inescapably demonstrate his interest in their special investigation. And spiritualism itself, it should be pointed out, is an evidence-based belief that embraces (à la Hardinge) Darwinistic understandings as they relate to materialistic biology.

Thus, the preponderance of evidence—and negative evidence—suggests that Wallace's adoption of Spencerian ways *c.*1863–64 was a temporary deviation from a more general, lifelong, track. Until such time as anyone can point to any kind of concrete evidence from the 1858 to 1864 period suggesting "On the Tendency" was meant to extend to all levels of human evolution, it is only reasonable to conclude that the "change of mind" theory represents the weaker of the two alternative interpretations of Wallace's personal evolution of thought.

* * *

Coda

In this 150th anniversary year of the public's introduction to natural selection, we might perhaps do more than just congratulate ourselves on a job well done, or acknowledge our respect for the concept's pioneers, Wallace and Darwin centrally among them, who made the whole trip possible. It would also be well, it seems, to pay some attention to what words such as "natural selection," "evolution," and "Darwinism" actually *mean*—and more importantly, to what they *don't* mean. "Darwinism"—a term championed by Wallace—has come to be understood, roughly, as the idea of "evolution by natural selection," yet despite this implied relationship it is commonly viewed as interchangeable with the idea of "evolution" itself. Now at this oversimplification I must protest (as many others have, including Wallace himself); not only does natural selection have no explicit connection to a *changing* outcome (the main reason why Wallace frequently described natural selection as a law, and as "*accumulating* variations," instead of *creating* them), but neither can we possibly defend the notion that natural selection is the only relevant factor in the unfolding of evolution, whether at the biological level alone, or any other level. Indeed, despite 150 years of study of the matter, the concept has proved largely ineffectual in coming to grips with some of the most central elements of biological change: for example, divergence, speciation, and the *origin* of variation.

To be sure, while these latter events are taking place, natural selection is always there—sometimes as the "ghost in the machine" lurking in the background, and sometimes standing right up front (as, for example, in the shaping of animal coloration patterns, as Caro *et al.* reviewed earlier)—but it would be most inaccurate to say that biological evolution is *exclusively* understandable as its product. I do not deny the basic premises behind the concept; in fact, I think they are absolutely unassailable. Still, validity does not equal universality—as Wallace incessantly argued. In Chapter 18 here I suggested, via Bateson, a scientific framework for contextualizing natural selection within the greater story (as the negative feedback part of the overall process, continually restoring stability and order to the disorder of variation), but even that framework leaves the rest of the process as a great black box whose contents are only in their earliest stages of being revealed.

Unlike Darwin, for Wallace (after 1866, at least—the date of his first usage of the term "survival of the fittest" in S118), the concept was never construed as a surrogate for the more general phenomenon. This is apparent from the vast expanse of his attention, such as we have documented here in this book. It was all evolution to him—the infusion of life energy into the inert, species changes, the emergence of human races, the political struggles within human societies, and last, but certainly not least, the exploratory voyages of spirit. And he was not the least bit shy about saying so, whether in that famous 2 July 1866 letter to Darwin, various other writings contrasting evolution with natural selection (*e.g.*; S165 1870, 333–34; S311 1879; S322 1880, 95–96; S649 1908, 1–12; S726 1898b, Chapter 13) or, perhaps most completely, in his 1900 article "Evolution":

> Evolution, as a general principle, implies that all things in the universe, as we see them, have arisen from other things which preceded them by a process of modification, under the action of those all-pervading but mysterious agencies known to us as "natural forces," or, more generally, "the laws of nature." More particularly the term evolution implies that the process is an "unrolling," or "unfolding"... The point to be especially noted here is, that evolution, even if it is essentially a true and complete theory of the universe, can only explain the existing conditions of nature by showing that it has been derived from some pre-existing condition through the action of known forces and laws. It may also show the high probability of a similar derivation from a still earlier condition; but the further back we go the more uncertain must be our conclusions, while we can never make any real approach to the absolute beginnings of things (S589 1901, 3–4).

What, then, of the success story everyone seems to read into the events surrounding that celebrated malarial fit Wallace rode out in early 1858? Well, it is mere history that his heroic early efforts in the field eventually shook loose the prize of natural selection from nature's cupboard, but this may not have been *the* full answer to Wallace's quest. In the mind of Alfred Russel Wallace, as distinct from that of Charles Robert Darwin, natural selection was more a *product* of evolution than it was its cause. True "Wallacism," it seems to me, must therefore be conceived of as, exactly, "*natural selection by evolution*." Surely, he would say, there must be greater forces at work, forces that conspire to achieve the fleeting interface between survival and non-survival that natural selection ultimately represents. It is the particular combinations of these, at different times and places, that lead to dinosauric gigantism, wingless birds, handsomely adorned but deadly frogs—and finally, so we may believe, those most elusive of worldly spirits: men and women of inspired hearts, and open minds.

Notes

1. Note especially the passage from "On the Habits of the Orang-utan of Borneo" (S26 1857) produced in Chapter 18.

2. This position is reflected in words later appearing in his autobiography *My Life*: "My paper written at Sarawak rendered it certain in my mind that the change had taken place by natural succession and descent—one species becoming changed either slowly or rapidly into another. But the exact process of the change and the causes which led to it were absolutely unknown and appeared almost unconceivable" (S729 1905a, 1:360).
3. Besides the well-known impressions made on Charles Lyell and Edward Blyth, the following words from the President of the Geological Society of London, William J. Hamilton, concluding his 15 February 1856 annual address, are noteworthy: "I must direct your attention to a paper published by Mr. Alfred Wallace on the law which has regulated the introduction of new species. Mr. Wallace is a naturalist of no ordinary calibre. His travels in South America and elsewhere are a sufficient guarantee of his high merits; he now writes from Sarawak, Borneo. From a careful examination of the actual distribution of existing forms of animal life, and the gradual but complete renewal of forms of life in successive geological epochs, he has deduced the following law: *Every species has come into existence coincident both in space and time with a pre-existing closely allied species*. The question is one of great importance, and deserving the careful investigation of every geologist ..." (Hamilton 1856, cxviii).
4. As also implied by McKinney (1972b, xii).
5. That Wallace actually was at least a little upset at being forestalled by Darwin, Hooker, and Lyell is suggested by his later drawing attention, in notes in published works, to the fact he wasn't shown the proofs to his 1858 essay before it was published—and on no fewer than five occasions, extending over a four decade period (Meyer 1870; S725 1891, S516 1895, S599 1903, S729 1905). This harping on a matter that otherwise might be viewed as peripheral to one of the most successful essays in the history of science seems a bit odd, if there is not something there to read between the lines.
6. Wallace's history as a "heretic personality" is well documented by Shermer (2002), and his keeping quiet on a matter of such interest to him for a full six years (1858–64) requires some explanation. It should be remembered that this was the man who had, in 1843, at the age of just twenty, audaciously sent a technical suggestion on lens preparation to one of the leading experts on the subject (Smith 2006)—not to mention that 1858 communication to Darwin ...
7. These included the Linnean, Anthropological, Geographical, Ethnological, Zoological, and Entomological Societies, and the British Association for the Advancement of Science. In less than eight months in 1862 he made at least five such appearances including comments and/or presentations; in 1863, ten; and in 1864, eighteen.
8. The original version of this paper was not published in full. This selection is taken from another reading of it he gave, at a meeting of the Ethnological Society of London on 24 January 1865 (S104 1866, 67).
9. In his first known publication he writes: "The correction of false ideas and incorrect opinions on well-known principles of science are not among the least benefits that would accrue from such a course as we have recommended. How many having imbibed a false opinion, and having embraced it for a time, as a certain and undoubted fact, are, on seeing it contradicted without a clear explanation, more apt to doubt the truth of the principle they have misunderstood, than willing to acknowledge that they have been so long in error. As the means of inciting to the acquirement of knowledge on all subjects, of creating a wish for information on what have been hitherto considered as abstruse

branches of knowledge, but which are frequently among the most interesting and generally useful,—and of inspiring a desire for diving deeper into its inexhaustible stores not yet exposed to the scrutinizing gaze of man, such an institution as this, conducted in the way we have described, will be invaluable" (S1a 1845, 69). The relation between informed belief and justice is a theme central to Wallace's writings throughout his life. See Smith (2003–, Chapter 1) for discussion.

10. There are striking similarities between opinions expressed in this work and those found in one of his earliest writings, from about 1843: [in speaking of the rural Welshman] "Their preachers, while they should teach their congregation moral duties, boldly decry their vices, and inculcate the commandments and the duty of doing to others as we would they should do unto us, here, as is too frequently the case throughout the kingdom, dwell almost entirely on the mystical doctrine of the atonement—a doctrine certainly not intelligible to persons in a state of complete ignorance, and which, by teaching them that they are not to rely on their own good deeds, has the effect of entirely breaking away the connection between their religion and the duties of their everyday life, and of causing them to imagine that the animal excitement which makes them groan and shriek and leap like madmen in the place of worship, is the true religion which will conduce to their happiness here, and lead them to heavenly joys in a world to come" (S623 1905, 221).
11. For some other relevant analyses, see: Barrow (1986), Blum (2006), Cremo (2003), DeCarvalho (1988/1989), Fichman (2004), Inglis (1992), Lamont (2004), Nelson (1988), Oppenheim (1985), and Pels (2003).
12. Wallace's mimicry studies were apparently one victim of this "time out." His paper "Mimicry, and Other Protective Resemblances Among Animals" (S134), published on 1 July 1867, was written in "1865–1866" according to his autobiography (S729 1905a, 1:407). Perhaps he began work on it in late 1864 (not so long after the Papilionidae paper, S96) or early 1865 but put it aside in mid-1865 to concentrate on his spiritualism investigations, only to resume in early or mid-1866: in August 1866 he presented related work at the annual British Association meetings (S121) and later in the year made some further comments (S123, S123a). In his autobiography he notes that as of 23 February 1867 he was still "preparing for publication" the "Mimicry" paper (S729 1905a, 2:3).
13. Anonymous 1865. The texts of Hardinge's programs were shortly thereafter compiled and published by F. Farrah in the spring of 1866 as the 122 page monograph *Extemporaneous Addresses.*
14. Wallace includes no fewer than 114 lines of Hardinge's writings in the pamphlet version of *The Scientific Aspect of the Supernatural*, on pages 50 through 54. This is more quoted material than he produces for anyone else in the essay, and more than twice as much as for any other one person, with the exception of Augustus De Morgan. Hardinge was a familiar figure to both the American and English spiritualist communities, and her publications were readily obtainable—and, in fact, Wallace's annotated copy of her pamphlet *On Ancient Magic and Modern Spiritualism*, published in 1865, still resides among the materials from his personal library held by the Library of the University of Edinburgh. Wallace's Hardinge quotations in *The Scientific Aspect of the Supernatural* are from the essay "Hades," one of her *Six Lectures on Theology and Nature*, published in Chicago in 1860, so he must have known that work as well.

15. Slotten (2004), morever, relays a "spirit influence" story concerning the printed copies of the pamphlets as they lay in wrapping paper at Wallace's house prior to their distribution—a story in which Miss Nicholl played a role. Given the earliest Friday in November on which she could have produced mind-changing manifestations—9 November (her first seance, held at least a week earlier than the second and also supposedly in November, was less impressive)—and the date of the Huxley communication, it is apparent that the contents of the pamphlet were not based on any knowledge of Nicholl. This is also apparent from remarks made by John Tyndall, one of the colleagues to whom Wallace sent a copy, who reported " 'deep disappointment' because it contained no record of my own experiments" (S729 1905a, 2:280).
16. One of Wallace's pet complaints was that a theory shouldn't have to explain everything. For other examples of this sentiment in his writings see: S89 (1864, 111), S165 (1870, 332–33), S173 (1870, 9), S382 (1885), and S649 (1908, 1). It is hardly surprising that he kept mentioning this: it is a predictable accompaniment to his frequently stated belief in the existence of "more recondite" natural influences.
17. In 1885 he further contextualizes his position in the essay "Are the Phenomena of Spiritualism in Harmony with Science?": "Science may be defined as knowledge of the universe in which we live—full and systematised knowledge leading to the discovery of laws and the comprehension of causes. The true student of science neglects nothing and despises nothing that may widen and deepen his knowledge of nature, and if he is wise as well as learned he will hesitate before he applies the term 'impossible' to any facts which are widely believed and have been repeatedly observed by men as intelligent and honest as himself. Now, modern Spiritualism rests solely on the observation and comparison of facts in a domain of nature which has been hitherto little explored, and it is a contradiction in terms to say that such an investigation is opposed to science. Equally absurd is the allegation that some of the phenomena of Spiritualism 'contradict the laws of nature,' since there is no law of nature yet known to us but may be apparently contravened by the action of more recondite laws or forces ..." (S379, 809).

References Cited

Works by Alfred Russel Wallace

Manuscript/Archival Materials

British Library: 1903 ms. "Man's Place in the Universe," manuscript of the 1903 book by the same title (S728). ADD MSS 46420, 263 folios.

Darwin Correspondence Project, Cambridge University: letter 2627 from Wallace to Darwin December? 1860; letter 5966 from Wallace to Darwin 1 March 1868; letter 6033 from Darwin to Wallace 21 March 1868.

Edinburgh University Library, Special Collections, Alfred Russel Wallace Library [ARWL]: Wallace's annotated copy of Edward Bellamy's *Looking backward.*

Linnean Society of London Archives: Wallace's Malay Archipelago journals, 4 notebooks, 13 June 1856–54 May 1861 (Ms. 178a–d, cited as MJa-d); Wallace's Natural history notebook, 1855–59 (Ms. 180); Wallace's Notebook on butterflies of the Malay archipelago, 1854–62 (Ms. 181); Wallace's Zoology notebook, chiefly entomological, including records of consignments to Stevens, 1855–58 (Ms. 179).

Natural History Museum (London): 1845–46 letters to Henry Walter Bates (**http://www.nhm.ac.uk/nature-online/collections-at-the-museum/wallace-collection/collecting.jsp**); Wallace-Henry Walter Bates correspondence, 1845–47 (NHM Archives WP1/3/11–14); Wallace family correspondence, 1855–59 (NHM Archives WP1/3/19–72).

Originally Published Items

For additional bibliographic information on the following items, and links to the full text of all of them, see: **http://www.wku.edu/~smithch/index1.htm**.

S1 1905 [c.1843]. The advantages of varied knowledge. In *My Life* (S729), 1:201–04. London: Chapman & Hall.

S1a 1845. An essay, on the best method of conducting the Kington Mechanic's Institution. In *The history of Kington*, ed. Richard Parry, 66–70. Kington: *s.n.*

S2 1847. Capture of *Trichius fasciatus* near Neath. *Zoologist* 5:1676.

S3 1849. Journey to explore the Province of Pará. *Annals and Magazine of Natural History,* 2nd ser., 3:74–75.

S4 1850. Journey to explore the natural history of South America. *Annals and Magazine of Natural History,* 2nd ser., 5:156–57.

S5 1850. On the umbrella bird (*Cephalopterus ornatus*), "Ueramimbé," L. G. *Proc Zool Soc Lond* 18:206–07.

S6 1850. Journey to explore the natural history of the Amazon River. *Annals and Magazine of Natural History*, 2nd ser., 6:494–96.
S7 1852. [untitled letter]. *Zoologist* 10:3641–43.
S8 1852. On the monkeys of the Amazon. *Proc Zool Soc Lond* 20:107–10.
S9 1853. Some remarks on the habits of the Hesperidae. *Zoologist* 11:3884–85.
S11 1853. On the Rio Negro. *J Roy Geogr Soc* 23:212–17.
S13 1853. On the habits of the butterflies of the Amazon Valley. *Trans Entomol Soc Lond*, n.s., 2, part VIII:253–64.
S14 1854. [untitled letter]. *Zoologist* 12:4395–97.
S20 1855. On the law which has regulated the introduction of new species. *Annals and Magazine of Natural History*, 2nd ser., 16:184–96. [reprinted with minor changes in S716 1870:1–25; and in S725 1891:3–19]
S21 1855. [untitled letter]. *Zoologist* 13:4803–07.
S22 1856. Borneo. *The Literary Gazette* (London), no. 2023:683–84.
S23 1856. Some account of an infant "orang-utan." *Annals and Magazine of Natural History*, 2nd ser., 17:386–90.
S24 1856. On the orang-utan or mias of Borneo. *Annals and Magazine of Natural History*, 2nd ser., 17:471–76.
S25 1856. Observations on the zoology of Borneo. *Zoologist* 14:5113–17.
S26 1856. On the habits of the orang-utan of Borneo. *Annals and Magazine of Natural History*, 2nd ser., 18:26–32.
S28 1856. Attempts at a natural arrangement of birds. *Annals and Magazine of Natural History*, 2nd ser., 18:193–216.
S31 1857. [untitled letter]. *Zoologist* 15:5414–16.
S33 1857. [untitled letter]. *Zoologist* 15:5652–57.
S35 1857. [untitled letter and postscript]. *Proc Entomol Soc Lond, 1856–1857*:91–93.
S37 1857. On the great bird of paradise, *Paradisea apoda*, Linn.; "*burong mati*" (*dead bird*) of the Malays; "*fanéhan*" of the natives of Aru. *Annals and Magazine of Natural History*, 2nd ser., 20:411–16.
S38 1857. On the natural history of the Aru Islands. *Annals and Magazine of Natural History*, 2nd ser., Supplement to vol. 20:473–85.
S40 1858. On the entomology of the Aru Islands. *Zoologist* 16:5889–94.
S41 1858. On the Arru Islands. *Proc Roy Geogr Soc Lond* 2:163–70.
S43 1858. On the tendency of varieties to depart indefinitely from the original type. *J Proc Linn Soc: Zool* 3:53–62. [reprinted with minor changes in S716 1870:26–44; and in S725 1891:20–33]
S44 1858. [untitled letter]. *Zoologist* 16:6120–24.
S45 1859. [untitled letters]. *Ibis* 1:111–13.
S48 1859. [untitled letter]. *Proc Zool Soc Lond* 27:129.
S50 1859. [untitled letter extracts]. *Proc Entomol Soc Lond, 1858–1859*:70.
S52 1859. Letter from Mr. Wallace concerning the geographical distribution of birds. *Ibis* 1:449–54.
S53 1860. On the zoological geography of the Malay Archipelago. *J Proc Linn Soc: Zool* 4:172-84.
S56 1860. Note on the sexual differences in the genus *Lomaptera*. *Proc Entomol Soc Lond, 1858–59*:107.

S62 1861. On the ornithology of Ceram and Waigiou. *Ibis* 3:283–91.

S67 1862. Narrative of search after birds of paradise. *Proc Zool Soc Lond, 1862*:153–61.

S78 1863. On the physical geography of the Malay Archipelago. *J Roy Geogr Soc* 33:217–34.

S82 1865. On the varieties of man in the Malay Archipelago. *Transactions of the Ethnological Society of London*, n.s., 3:196–215.

S83 1863. Remarks on the Rev. S. Haughton's paper on the bee's cell, and on the origin of species. *Annals and Magazine of Natural History*, 3rd ser., 12:303–09.

S89 1864. Remarks on the habits, distribution, and affinities of the genus *Pitta*. *Ibis* 6:100–14.

S93 1864. The origin of human races and the antiquity of man deduced from the theory of "natural selection." *Journal of the Anthropological Society of London* 2:clviii–clxx. [reprinted in modified form as "The development of human races under the law of natural selection" in S716 1870:303–31]

S96 1865. On the phenomena of variation and geographical distribution as illustrated by the Papilionidæ of the Malayan Region [read 17 March 1864]. *Trans Linn Soc Lond* 25, part I: 1–71. [reduced version reprinted as "The Malayan Papilionidae or swallow-tailed butterflies, as illustrative of the theory of natural selection" in S716 1870:130–200]

S102 1864. On the parrots of the Malayan Region, with remarks on their habits, distribution, and affinities, and the descriptions of two new species. *Proc Zool Soc Lond, 1864*:272–95.

S104 1866. On the progress of civilization in Northern Celebes [first read (but not printed in full) 19 September 1864; also read 24 January 1865, and printed in full the next year]. *Transactions of the Ethnological Society of London*, n.s., 4:61–70.

S109 1865. List of the land shells collected by Mr. Wallace in the Malay Archipelago, with descriptions of the new species by Mr. Henry Adams. *Proc Zool Soc Lond, 1865*:405–16.

S110 1865. Public responsibility and the ballot. *The Reader* 5:517.

S111 1865. [untitled discussion of "On the efforts of missionaries among savages," by Revd J. W. Colenso]. *Journal of the Anthropological Society of London* 3:cclxxxviii.

S112 1865. Descriptions of new birds from the Malay Archipelago. *Proc Zool Soc Lond, 1865*:474–81.

S113 1865. How to civilize savages. *The Reader* 5:671–72.

S113a 1865. [untitled discussion of "On craniology and phrenology in relation to ethnology" by Dr Cornelius Donovan]. *The Ethnological Journal*, no. 2:97.

S114 1865. On the pigeons of the Malay Archipelago. *Ibis*, n.s., 1:365–400.

S118 1866. The scientific aspect of the supernatural. *The English Leader* 2:59–60, 75–76, 91–93, 107–08, 123–25, 139–40, 156–57, 171–73. [in late 1866 reprinted with minor additions as pamphlet entitled *The scientific aspect of the supernatural: Indicating the desirableness of an experimental enquiry by men of science into the alleged powers of clairvoyants and mediums*]

S119 1867. Address [read 23 August 1866]. In *Report of the British Association for the Advancement of Science* 36, 93–94. London: John Murray, 1867.

S121 1866. On reversed sexual characters in a butterfly, and its interpretation on the theory of modification and adaptive mimicry. In *The British Association for the Advancement of Science. Nottingham meeting, August, 1866. Report of the papers, discussions, and general proceedings*, ed. William Tindal Robertson, 186–87. London: Thomas Forman, Nottingham & Robert Hardwicke, 1866.

S123 1866. Natural selection. *Athenaeum*, 1 December 1866:716–17.

S123a 1866. [untitled discussion regarding David Sharp's views on mimicry]. *J Proc Entomol Soc Lond 1866*: xlvi–xlviii.

S124 1867. Ice marks in North Wales (with a sketch of glacial theories and controversies). *Q J Sci* 4:33–51.

S125 1867. Mr. Wallace on natural selection applied to anthropology. *Anthropological Review* 5:103–05.

S126 1867. Postscript by Alfred R. Wallace [to an account of a séance written by Frances (Wallace) Sims, Wallace's sister]. *The Spiritual Magazine*, n.s., 2:51–52.

S127 1867. On the Pieridæ of the Indian and Australian regions. *Trans Entomol Soc Lond*, 3rd ser., 4, part III: 301–416.

S129 1867. [discussion concerning brilliant colors in caterpillar larvae]. *J Proc Entomol Soc Lond 1867*:lxxx–lxxxi.

S130 1867. Caterpillars and birds. *The Field, The Country Gentleman's Newspaper* 29:206.

S132 1867. Notes of a séance with Miss Nicholl at the house of Mr. A. S——, 15th May. *The Spiritual Magazine*, n.s., 2:254–55.

S134 1867. Mimicry, and other protective resemblances among animals. *Westminster Review*, n.s., 32, no. 1:1–43. [reprinted with additions in S716 1870:45–129; and in S725 1891:34–90]

S137 1867. [untitled letter concerning séance experiences]. *The Spiritual Magazine*, n.s., 2:349–50.

S139 1868. A theory of birds' nests: Shewing the relation of certain sexual differences of colour in birds to their mode of nidification. *Journal of Travel and Natural History* 1:73–89. [reprinted with additions in S716 1870:231–63; and in S725 1891:118–40]

S140 1867. Creation by law. *Q J Sci* 4:471–88. [reprinted with additions in S716 1870:264–302; and in S725 1891:141–66]

S142a 1868. [untitled discussion of "On the difficulties of Darwinism" by Revd. F. O. Morris]. *Athenæum*, 19 September 1868:373–74.

S146 1869. Sir Charles Lyell on geological climates and the origin of species. *Quarterly Review* 126:359–94.

S159 1870. The measurement of geological time. *Nature* 1:399–401, 452–55.

S161 1870. [untitled review of *Hereditary genius* by Francis Galton]. *Nature* 1:501–03.

S165 1870. The limits of natural selection as applied to man. In *Contributions to the theory of natural selection* (S716), 332–71.

S171 1871. On a diagram of the earth's eccentricity and the precession of the equinoxes, illustrating their relation to geological climate and the rate of organic change [abstract; read 20 September 1870]. In *Report of the British Association for the Advancement of Science* 40, 89. London: John Murray, 1871.

S173 1870. Man and natural selection. *Nature* 3:8–9.

S174 1870. An answer to the arguments of Hume, Lecky, and others, against miracles. *The Spiritualist* (London) 1:113–16.

S176 1870. The difficulties of natural selection. *Nature* 3:85–86.

S184 1871. The theory of glacial motion. *Nature* 3:309–10.

S186 1871. [untitled review of *The descent of man and selection in relation to sex* by Charles Darwin]. *The Academy* 2:177–83.

S231 1873. Free-trade principles and the coal question. *The Daily News* (London), 16 September 1873:6.

S257 1876. Address [read 6 September 1876 as President of Section D, Biology]. In *Report of the British Association for the Advancement of Science* 46, 100–119. London: John Murray, 1877.

S272 1877. The colours of animals and plants. *Macmillan's Magazine* 36:384–408, 464–71.

S302 1879. Animals and their native countries. *Nineteenth Century* 5:247–59.

S304 1879. Colour in nature. *Nature* 19:501–05.

S306 1879. Reciprocity the true free trade. *Nineteenth Century* 5:638–49.

S311 1879. [untitled review of *Evolution, old and new* by Samuel Butler]. *Nature* 20:141–44.

S313 1879. Glacial epochs and warm polar climates. *Quarterly Review* 148:119–35.

S318 1879. The protective colours of animals. In *Science for all*, ed. Robert Brown, 2:128–37. London: Cassell, Petter, Galpin.

S319 1879. Protective mimicry in animals. In *Science for all*, ed. Robert Brown, 2:284–96. London: Cassell, Petter, Galpin.

S322 1880. The origin of species and genera. *Nineteenth Century* 7:93–106.

S329 1880. How to nationalize the land: A radical solution of the Irish land problem. *Contemporary Review* 38:716–36.

S338 1881. [untitled review of *Studies in the theory of descent pt. II* by August Weismann]. *Nature* 24:457–58.

S348 1881. [untitled review of *Vignettes from nature* by Grant Allen]. *Nature* 25:381–82.

S353 1882. Dr. Fritz Müller on some difficult cases of mimicry. *Nature* 26:86–87.

S359 1883. Difficult cases of mimicry. *Nature* 27:481–82.

S364 *n.d. How land nationalisation will benefit householders, labourers, and mechanics.* London: Land Nationalisation Society Tract, no. 3.

S365 *n.d. The "why" and the "how" of land nationalisation.* London: Land Nationalisation Society Tract, no. 6.

S368 1883. [untitled letter]. *The Vaccination Inquirer and Health Review* 5:160.

S371 *n.d. How to experiment in land nationalisation.* London: Land Nationalisation Society Tract, no. 8.

S374 1885. *(To members of Parliament and others.) Forty-five years of registration statistics, proving vaccination to be both useless and dangerous.* London: E. W. Allen.

S378 1885. The colours of arctic animals. *Nature* 31:552.

S379 1885. Are the phenomena of spiritualism in harmony with science? *The Medium and Daybreak* (London) 16:809–10.

S382 1885. The "Journal of Science" on spiritualism. *Light* (London) 5:327–28.

S385 *n.d. State-tenants* versus *freeholders.* London: Land Nationalisation Society Tract, no. 15.

S389 1886. Romanes *versus* Darwin. An episode in the history of the evolution theory. *Fortnightly Review*, n.s., 40:300–16.

S403 *n.d. Land lessons from America.* London: Land Nationalisation Tract, no. 18.

S412 *n.d.* Introductory note to *A colonist's plea for land nationalisation*, by Arthur J. Ogilvy. London: Land Nationalisation Society Tract, no. 23.

S424 1890. [untitled review of *The colours of animals: Their meaning and use especially considered in the case of insects* by Edward B. Poulton]. *Nature* 42:289–91.

S425 1890. Birds and flowers. *Nature* 42:295.

S427 1890. Human selection. *Fortnightly Review*, n.s., 48:325–37.

S431 1891. A British convert/A distinguished convert. *The New Nation* 1:50, 135.

S432 1891. Modern biology and psychology. *Nature* 43:337–41.
S441 1891. English and American flowers. *Fortnightly Review*, n.s., 50:525–34, 796–810.
S445 1892. Human progress: Past and future. *The Arena* 5:145–59.
S446 1892. H. W. Bates, the naturalist of the Amazons. *Nature* 45:398–99.
S450 1892. Presidential Address. In *Report of the Land Nationalisation Society 1891–92*, 15–26. London: Land Nationalisation Society Tract, no. 48.
S451 1892. Spiritualism. In *Chambers's encyclopædia*, 9:645–49. London and Edinburgh: William and Robert Chambers.
S459 1892. Note on sexual selection. *Natural Science* 1:749–50.
S462 1893. The glacier theory of alpine lakes. *Nature* 47:437–38.
S481 1893. The ice age and its work. *Fortnightly Review*, n.s., 54:616–33, 750–74.
S483 1893. The programme of land nationalisers. *Land and Labor*, no. 50:1–2.
S494 1894. What are zoological regions? *Nature* 49:610–13.
S495 1894. President's Address. In *The Local Government (England and Wales) Act, 1894, relating to parish councils, etc.*, 1–11. London: Land Nationalisation Society Tract, no. 57.
S497 1894. Woman and natural selection. *Humanitarian*, n.s., 4:315.
S498 1894. Economic and social justice. In *Vox clamantium; the gospel of the people*, "by writers, preachers & workers brought together by Andrew Reid," 166–97. London: A. D. Innes.
S499 1894. Panmixia and natural selection. *Nature* 50:196–97.
S501 1894. A new book on socialism. *Land and Labor*, no. 57:52–54.
S507 1894. The social economy of the future. In *The New Party described by some of its members*, new ed., ed. Andrew Reid, 177–211. London: Hodder Brothers.
S512. *Suggestions for solving the problem of the unemployed, etc., etc.* London: Land Nationalisation Society Tract, no. 64 (1894). [reprinted with changes and additions in 1897 as "Re-occupation of the land: The only solution to the problem of the unemployed" in *Forecasts of the coming century by a decade of writers*, ed. Edward Carpenter, 9–26. Manchester: The Labour Press]
S516 1895. [untitled discussion presented in Meyer (1895)]. *Nature* 52:415.
S527 1896. The problem of utility: Are specific characters always or generally useful? *J Linn Soc: Zool* 25:481–96.
S528 1896. [untitled letter to Keir Hardie]. *Labour Leader* 8:251.
S535 1897. On the colour and colour-patterns of moths and butterflies. *Nature* 55:618–19.
S545 1900. Justice, not charity, as the fundamental principle of social reform [originally read, and printed, in 1898]. In *Studies scientific and social* (S727), 2:521–28.
S551 1898. The vaccination question. *The Times* (London), 1 September 1898:10.
S589. Evolution. *The Sun* (New York), 23 December 1900:4–5. [reprinted nearly verbatim in 1901 in *The progress of the century*, 3–29. New York and London: Harper & Brothers]
S599 1903. The dawn of a great discovery (my relations with Darwin in reference to the theory of natural selection). *Black and White* 25:78–79.
S602 1903. Man's place in the universe. *The Independent* (New York) 55:473–83.
S622 1905. If there were a socialist government—how should it begin? *The Clarion* (London), 18 August 1905:5.
S623 1905 [*c.*1843]. The South-Wales farmer. In *My Life* (S729), 1:206–22.
S649. Evolution and character. *Fortnightly Review*, n.s., 83:1–24 (1908). [reprinted in 1912 in *Character and life; a symposium*, ed. Percy L. Parker, 3–50. London: Williams & Norgate]

S655 1909. *The remedy for unemployment.* London: The Clarion Press, Pass On Pamphlets, no. 8.

S656 1909. [untitled remarks on receiving the Darwin-Wallace Medal in 1908]. In *The Darwin-Wallace celebration held on Thursday, 1st July 1908, by the Linnean Society of London*, 5–11. London: printed for the Linnean Society by Burlington House, Longmans, Green & Co.

S713 1853. *Palm trees of the Amazon and their uses.* London: John Van Voorst.

S714. Editions cited:

1853. *A narrative of travels on the Amazon and Rio Negro.* London: Reeve & Co.

1889. [same title]. 2nd ed. London: Ward, Lock & Co.

1911. *Travels on the Amazon.* London: Ward, Lock & Co.

1969. *A narrative of travels on the Amazon and Rio Negro.* New York: Haskell House. [reprint of 1889 2nd ed.]

S715. Editions cited:

1869a. *The Malay Archipelago; the land of the orang-utan and the bird of paradise.* 2 vols. London: Macmillan.

1869b. [same title]. New York: Harper & Brothers.

1883. [same title]. 8th ed. London and New York: Macmillan.

1891. [same title]. 10th ed., revised. London and New York: Macmillan.

1962. [same title]. Reprint of 10th ed. New York: Dover.

1989. [same title]. John Bastin, ed. Reprint of 1869 Harper ed. Singapore: Oxford University Press.

http://www.papuaweb.org/dlib/bk/wallace/cover.html [same title]. based on 1891 10th ed.

S716. Editions cited:

1870. *Contributions to the theory of natural selection. A series of essays.* London and New York: Macmillan.

1871. [same title]. 2nd ed., with corrections and additions. London and New York: Macmillan.

S717. Editions cited:

1875. *On miracles and modern spiritualism.* London: James Burns.

1881. [same title]. 2nd ed. London: Trübner & Co.

1901. *Miracles and modern spiritualism.* Reprint of 1896 rev. 3rd ed. London: Nichols & Co.

S718 1876. *The geographical distribution of animals; with a study of the relations of living and extinct faunas as elucidating the past changes of the earth's surface.* 2 vols. London: Macmillan.

S719 1878. *Tropical nature, and other essays.* London and New York: Macmillan.

S721. Editions cited:

1880. *Island life: Or, the phenomena and causes of insular faunas and floras, including a revision and attempted solution of the problem of geological climates.* London: Macmillan.

1892. [same title]. 2nd and rev. ed. London and New York: Macmillan.

S722. Editions cited:

1882. *Land nationalisation; its necessity and its aims.* London: Trübner & Co.

1906. [same title]. 4th ed. (reprint of 1892 new ed.). London: Swan Sonnenschein.

S723 1885. *Bad times: An essay on the present depression of trade.* London: Macmillan.

S724 1889. *Darwinism; an exposition of the theory of natural selection with some of its applications.* London and New York: Macmillan.

S725. Editions cited:

1891. *Natural selection and tropical nature; essays on descriptive and theoretical biology.* London and New York: Macmillan.

1998. [same title]. Reprint of 1891 ed. New York: Kessinger.

S726. Editions cited:

1898a. *The wonderful century; its successes and its failures.* London: Swan Sonnenschein.

1898b. [same title]. New York: Dodd, Mead.

1901. [same title]. 4th ed. (reprint of 1898 ed.). London: Swan Sonnenschein.

1970. [same title]. Reprint of 1898 Swan Sonnenschein ed. Farnborough, UK: Gregg.

S727 1900. *Studies scientific and social.* 2 vols. London and New York: Macmillan.

S728. Editions cited:

1903a. *Man's place in the universe; a study of the results of scientific research in relation to the unity or plurality of worlds.* London: Chapman & Hall.

1903b. [same title]. New York: McClure, Phillips.

1904. [same title]. 4th ed., revised. London: Chapman & Hall.

S729. Editions cited:

1905a. *My Life; a record of events and opinions.* 2 vols. London: Chapman & Hall.

1905b. [same title]. 2 vols. New York: Dodd, Mead.

1906. [same title]. 2 vols. Reprint of 1905 Dodd, Mead ed. New York: Dodd, Mead.

1908. [same title]. New ed., condensed and revised. London: Chapman & Hall.

1969. [same title]. Reprint of 1905 Chapman & Hall ed. Westmead, Farnborough, Hants: Gregg.

S730 1907. *Is Mars habitable? A critical examination of Professor Percival Lowell's book "Mars and its canals," with an alternative explanation.* London: Macmillan.

S731 1908. *Notes of a botanist on the Amazon and Andes* [original material by Richard Spruce, "edited and condensed by Alfred Russel Wallace"]. 2 vols. London: Macmillan.

S732 1910. *The world of life; a manifestation of creative power, directive mind and ultimate purpose.* London: Chapman & Hall.

S733. Editions cited:

1913. *Social environment and moral progress.* London and New York: Cassell & Co.

2007. [same title]. Reprint of 1913 New York Cassell & Co. ed. Whitefish, MT: Kessinger.

S734 1913. *The revolt of democracy.* London and New York: Cassell & Co.

S734a 2002. *Peixes do Rio Negro/Fishes of the Rio Negro* [organization, introductory text, and translation by Mônica de Toledo-Piza Ragazzo]. São Paulo: Editora da Universidade de São Paulo, Imprensa Oficial do Estado.

S736 1893. Anon. Woman and natural selection [interview]. *The Daily Chronicle* (London), 4 December 1893:3.

S737 1894. Tooley, Sarah A. Heredity and pre-natal influences. An interview with Dr. Alfred Russel Wallace [interview]. *Humanitarian,* n.s., 4:80–88.

S738 1898. "A. D." A visit to Dr. Alfred Russel Wallace, F.R.S. [interview]. *The Bookman* (London) 13:121–24.

S741 1903. Dawson, Albert. A visit to Dr. Alfred Russel Wallace [interview]. *The Christian Commonwealth* 23:176–77.

S745 1909. Rann, Ernest H. Dr. Alfred Russel Wallace at home [interview]. *The Pall Mall Magazine* 43:274–84.

S746 1910. Begbie, Harold. New thoughts on evolution [interview]. *The Daily Chronicle* (London), 3 November 1910:4, and 4 November 1910:4.
S750 1912. Rockell, Frederick. The last of the great Victorians. Special interview with Dr. Alfred Russel Wallace [interview]. The *Millgate Monthly* 7, part 2: 657–63.

Works by Other Authors

Agassiz, L. 1840. On the polished and striated surfaces of the rocks which form the beds of glaciers in the Alps. *Proc Geol Soc Lond* 3:321–22.
Alatalo, R. V., and J. Mappes. 1996. Tracking the evolution of warning signals. *Nature* 382:708–10.
Albright, H. M. 1971. *Origins of National Park Service administration of historic sites.* Philadelphia: Eastern National Parks & Monument Association. **http://www.nps.gov/history/history/online_books/albright/index.htm**
Allen, D. E. 1976. *The naturalist in Britain.* London: Allen Lane.
——. 1996. Tastes and crazes. In *Cultures of natural history,* eds. N. Jardine, J. A. Secord, and E. C. Spary, 394–407. Cambridge, UK: Cambridge University Press.
——. ed. 2001. *Naturalists and society.* Aldershot: Ashgate.
Allen, Grant. 1879. *The colour-sense; its origin and development. An essay in comparative psychology.* London: Trübner.
Allen, G. E. 1975. *Life science in the twentieth century.* New York: Wiley.
Amundsen, T., and H. Pärn. 2006. Female coloration: Review of functional and nonfunctional hypotheses. In Hill and McGraw 2006b, 280–345.
Anderson, N. M. 1991. Cladistic biogeography of marine water striders (Insecta: Hemiptera) in the Indo-Pacific. *Australian Systematic Botany* 4:151–63.
Andersson, M. B. 1994. *Sexual selection.* Princeton, NJ: Princeton University Press.
Anon. 1865. Miss Emma Hardinge. *The Spiritual Magazine* 6:529–43.
Anon. 1868. Natural and supernatural selection. *The Spectator* 41:1154–55.
Anon. 1869. The alleged failure of natural selection in the case of man. *Q J Sci* 6:152–53.
Anon. 1870. The land question in England. *Westminster Review* 38:233–62.
Anon. 1871. *Report on spiritualism of the committee of the London Dialectical Society, together with the evidence, oral and written, and a selection from the correspondence.* London: Longmans.
Argyll, Duke of. 1893. Glacier action. *Nature* 47:389.
Arnold, David. 1993. *Colonizing the body: State medicine and epidemic disease in nineteenth-century India.* Berkeley, CA: University of California Press.
Bacot, A. 1912. Protective resemblance. *Proc Entomol Soc Lond 1912*:cxiv.
Badyaev, A. V., and G. E. Hill. 2003. Avian sexual dichromatism in relation to phylogeny and ecology. *Annu Rev Ecol Syst* 34:27–49.
Baker, D. B. 2001. Alfred Russel Wallace's record of his consignments to Samuel Stevens, 1854–1861. *Zoologische Mededeelingen* 75:254–341.
Barkow, J. H., L. Cosmides, and J. Tooby. 1992. *The adapted mind.* New York: Oxford University Press.
Barnes, H. E., and H. Becker. 1961. *Social thought from lore to science.* 3 vols. New York: Dover.
Barnett, C. A., M. Bateson, and C. Rowe. 2007. State-dependent decision making: Educated predators strategically trade off the costs and benefits of consuming aposematic prey. *Behavioral Ecology* 18:645–51.

Baron-Cohen, Simon. 2003. *The essential difference: The truth about the male and female brain.* Reading, MA: Perseus Books.

Barrow, J. D., and F. J. Tipler. 1986. *The anthropic cosmological principle.* New York: Oxford University Press.

Barrow, L. 1986. *Independent spirits: Spiritualism and English plebeians, 1850–1910.* London: Routledge and Kegan Paul.

Barton, G. 2002. *Empire forestry and the origins of environmentalism.* Cambridge, UK: Cambridge University Press.

Barton, Ruth. 1987. John Tyndall, pantheist; a rereading of the Belfast Address. *Osiris* 3:111–34.

——. 2003. "Men of science": Language, identity and professionalization in the mid-Victorian scientific community. *Hist Sci* 41:73–119.

Bastin, J. 1989. Introd. to *The Malay Archipelago: The land of the orang-utan, and the bird of paradise,* by A. R. Wallace. Singapore and Oxford, UK: Oxford University Press.

Bates, H. W. 1843. Notes on coleopterous insects frequenting damp places. *The Zoologist* 1:114–15.

——. 1862. Contributions to an insect fauna of the Amazon valley: Lepidoptera: Heliconidae. *Trans Linn Soc Lond* 23:495–566.

——. 1863. *The naturalist on the River Amazons.* 2 vols. London: John Murray.

——. 1864. *The naturalist on the River Amazons.* 2nd ed. London: John Murray.

——. 1892. *The naturalist on the River Amazons.* London: John Murray. [reprint of the unabridged ed.]

——. 1910. *The naturalist on the River Amazons.* Popular ed. London: John Murray.

Bateson, G. 1972. *Steps to an ecology of mind.* San Francisco: Chandler Publishing Co.

——. 1979. *Mind and nature: A necessary unity.* New York: Dutton.

Baudrillard, J. 1994. The system of collecting. In *The cultures of collecting,* eds. J. Elsner and R. Cardinal, 7–24. London: Reaktion Books.

Beatty, C. D., K. Beirinckx, and T. N. Sherratt. 2004. The evolution of Müllerian mimicry in multispecies communities. *Nature* 431:63–67.

Bebel, August. 1885. *Woman in the past, present and future.* London: Modern Press.

Beccaloni, J. 2002. Library acquires important collection of Wallaceana. *Waterhouse Times* 37:10–11. **http://www.nhm.ac.uk/resources/nature-online/collections-at-the-museum/wallace/pdf/Waterhouse-Ties-37–2002.pdf**

Becker, L. C. 1977. *Property rights: Philosophic foundations.* London: Routledge and K. Paul.

Beddall, B. G. 1968. Wallace, Darwin, and the theory of natural selection: A study in the development of ideas and attitudes. *J Hist Biol* 1:261–323.

——. 1988a. Darwin and divergence: The Wallace connection. *J Hist Biol* 21:1–68.

——. 1988b. Wallace's annotated copy of Darwin's *Origin of Species. J Hist Biol* 21:265–89.

Beddard, F. E. 1895. *Animal coloration: An account of the principle facts and theories relating to the colours and markings of animals.* 2nd ed. London: Swan Sonnenschein.

Behe, M. J. 1996. *Darwin's black box: The biochemical challenge to evolution.* New York: Free Press.

Behrens, R. R. 1988. The theories of Abbott H. Thayer: Father of camouflage. *Leonardo* 21:291–96.

——. 1999. The role of artists in ship camouflage during World War I. *Leonardo* 32:53–59.

——. 2002. *False colors: Art, design, and modern camouflage.* Dysart, IA: Bobolink Books.

Bellamy, Edward. 1924. *Equality.* New York: D. Appleton-Century Co.

——. 1927. *Looking backward, 2000–1887.* New York: Vanguard Press.

——. 1960 [1888]. *Looking backward, 2000–1887.* New York: New American Library.

Belt, T. 1874. *The naturalist in Nicaragua.* London: John Murray.

Benjamin, A. 9 August 2007. Pygmy elephants face uncertain future. *Guardian Unlimited.* **http://www.guardian.co.uk/environment/2007/aug/09/endangeredspecies.conservation/print**

Benson, S. B. 1936. Concealing coloration among some desert rodents of the southwestern United States. *Univ Calif Publs Zool* 40:1–70.

Benton, T. 1991. Biology and social science: Why the return of the repressed should be given a (cautious) welcome. *Sociology* 25:1–29.

——. 1993. *Natural relations: Ecology, animal rights, and social justice.* London: Verso.

——. 1997. Where to draw the line?: Alfred Russel Wallace in Borneo. *Studies in Travel Writing,* no. 1:96–116.

——. 1999a. Sustainable development and accumulation of capital: Reconciling the irreconcilable? In *Fairness and futurity: Essays on environmental sustainability and social justice,* ed. A. Dobson, 199–229. Oxford, UK: Oxford University Press.

——. 1999b. Evolutionary psychology and social science: A new paradigm or just the same old reductionism? *Adv Human Ecol* 8:65–98.

Berenbaum, M. 2001. Plant-herbivore interactions. In *Evolutionary ecology: Concepts and case studies,* eds. C. W. Fox, D. A. Roff, and D. J. Fairbairn, 303–14. New York: Oxford University Press.

Berendzen, R., R. Hart, and D. Seeley. 1976. *Man discovers the galaxies.* New York: Science History Publications.

Berlocher, S. H., and J. L. Feder. 2002 Sympatric speciation in phytophagous insects: Moving beyond controversy? *Ann Rev Entomol* 47:773–815.

Berry, Andrew, ed. 2002. *Infinite tropics: An Alfred Russel Wallace anthology.* London and New York: Verso.

——. 2003. Reasons for being nice and having sex. *London Review of Books* 25, 6 February 2003:36.

Bhattacharya, S., M. Harrison, and M. Worboys. 2005. *Fractured states.* New Delhi: Orient Longman.

Birkhead, T. R. 2000. *Promiscuity: An evolutionary history of sperm competition.* Cambridge, MA: Harvard University Press.

Blair, W. F. 1955. Mating call and stage of speciation in the *Microhyla olivacea-M. carolinensis* complex. *Evolution* 9:469–80.

Blaisdell, M. L. 1992. *Darwinism and its data; the adaptive coloration of animals.* New York: Garland.

Blom, Philipp. 2003. *To have and to hold: An intimate history of collectors and collecting.* Woodstock, NY: The Overlook Press.

Blum, D. 2006. *Ghost hunters.* New York: Penguin.

Blum, M. S. 1981. *Chemical defenses of arthropods.* New York: Academic Press.

Bonney, T. G. 1871. On the formation of "cirques," and their bearing upon theories attributing the excavation of mountain valleys mainly to the action of glaciers. *Q J Geol Soc Lond* 27:312–24.

Bonney, T. G. 1873. Lakes of the north-eastern Alps, and their bearing on the glacier-erosion theory. *Q J Geol Soc Lond* 29:382–95.

——. 1874. Notes on the Upper Engadine and the Italian valleys of Monte Rosa, and their relation to the glacier-erosion theory of lake-basins. *Q J Geol Soc Lond* 30:479–89.

——. 1892. The Lake of Geneva (review). *Nature* 47:5–6.

——. 1893a. Some lake basins in France. *Nature* 47:341.

——. 1893b. Do glaciers excavate? *The Geographical Journal* 1:481–99 (and following discussion to page 504).

——. 1896. *Ice-work, present and past.* London: Kegan Paul, Trench, Trübner.

——. 1902. Alpine valleys in relation to glaciers. *Q J Geol Soc Lond* 58:699.

Booth, C. L. 1990. Evolutionary significance of ontogenetic colour change in animals. *Biol J Linn Soc* 40:125–63.

Boulenger, G. A. 1895. Descriptions of four new batrachians discovered by Mr. Charles Hose in Borneo. *Annals and Magazine of Natural History,* 6th ser., 16:169–71.

Bowers, M. D. 1990. Recycling plant natural products for insect defense. In *Insect defenses: Adaptive mechanisms and strategies of prey and predators,* eds. D. L. Evans and J. O. Schmidt, 353–86. Albany, NY: State University of New York Press.

Bowler, P. J. 1976. Alfred Russel Wallace's concepts of variation. *Journal of the History of Medicine and Allied Sciences* 31:17–29.

Bowmaker, J. K. 1980. Colour vision in birds and the role of oil droplets. *Trends in Neurosciences* 3:196–99.

Bowman, S. E. 1986. *Edward Bellamy.* Boston: Twayne Publishers.

Bowman, S. E., *et al.* 1962. *Edward Bellamy abroad; an American Prophet's influence.* New York: Twayne Publishers.

Brackman, A. C. 1980. *A delicate arrangement; the strange case of Charles Darwin and Alfred Russel Wallace.* New York: Times Books.

Breuker, C. J., and P. M. Brakefield. 2002. Female choice depends on size but not symmetry of dorsal eyespots in the butterfly *Bicyclus anynana. Proc Roy Soc Lond* B 269:1233–39.

Brewster, David. 1854. *More worlds than one.* London: John Murray.

Brigham, A. 1997. "Rose Hill": Adapted from nature. The making of a new townscape—the first residential estate in Dorking. *Surrey History* 5:194–211.

Brindley, John. 1840. *The marriage system of socialism, freed from the misrepresentations of its enemies, etc.* Chester: T. Thomas.

——. c.1840. *The immoralities of socialism: Being an exposure of Mr. Owen's attack upon marriage.* Birmingham: Bull & Turner.

Brodie, E. D. 1989. Genetic correlations between morphology and antipredator behaviour in natural populations of the garter snake *Thamnophis ordinoides. Nature* 342:542–43.

Bronstein, J. L. 1999. *Land reform and working-class experience in Britain and the United States, 1800–1862.* Stanford, CA: Stanford University Press.

Brooks, D. R., and E. O. Wiley. 1986. *Evolution as entropy: Toward a unified theory of biology.* Chicago: University of Chicago Press.

Brooks, J. L. 1984. *Just before the Origin: Alfred Russel Wallace's theory of evolution.* New York: Columbia University Press.

Brower, A. V. Z. 1995. Locomotor mimicry in butterflies? A critical review of the evidence. *Phil Trans Roy Soc Lond* B 347:413–25.

Brower, L. P., W. N. Ryerson, L. L. Coppinger, and S. C. Glazier. 1968. Ecological chemistry and the palatability spectrum. *Science* 161:1349–50.

Brown, D. M., and C. A. Toft. 1999. Molecular systematics and biogeography of the cockatoos (Psittaciformes: Cacatuidae). *The Auk* 116:141–57.

Brown, K. S. Jr., C. F. Klitzke, C. Berlingeri, and P. E. R. dos Santos. 1995. Neotropical swallowtails: Chemistry of food plant relationships, population ecology, and biosystematics. In *Swallowtail butterflies: Their ecology and evolutionary biology*, eds. J. M. Scriber, Y. Tsubaki, and R. C. Lederhouse, 405–46. Gainesville, FL: Scientific Publishers.

Browne, E. Janet. 1980. Darwin's botanical arithmetic and the "Principle of Divergence," 1854–1858. *J Hist Biol* 13:53–89.

——. 1992. A science of empire: British biogeography before Darwin. *Revue d'Histoire des Sciences* 45:453–75.

——. 1995. *Voyaging*. Vol. 1 of *Charles Darwin*. New York: Knopf.

——. 2002. *The power of place*. Vol. 2 of *Charles Darwin*. Princeton, NJ: Princeton University Press.

Bulmer, M. 2005. The theory of natural selection of Alfred Russel Wallace FRS. *Notes & Records of the Royal Society* 59:125–36.

Burchfield, J. D. 1974. Darwin and the dilemma of geological time. *Isis* 65:300–21.

——. 1975. *Lord Kelvin and the age of the earth*. New York: Science History Publications.

Burkhardt, F., and S. Smith, eds. 1991. *The correspondence of Charles Darwin. Volume 7, 1858–1859. Supplement to the correspondence 1821–1857*. Cambridge, UK: Cambridge University Press.

——, eds. 1993. *The correspondence of Charles Darwin. Volume 8, 1860*. Cambridge, UK: Cambridge University Press.

Burnett, John, *et al.* 1886. *The claims of labour. A course of lectures*. Edinburgh: Edinburgh Co-Operative Print Co.

Burtt, E. D. 1951. The ability of adult grasshoppers to change colour on burnt ground. *Proc Roy Entomol Soc Lond* A 26:45–48.

Burtt, E. H. Jr., ed. 1979. *The behavioral significance of color*. New York: Garland STPM Press.

Butchart, S. H. M., A. J. Stattersfield, J. Baillie, L. A. Bennun, S. N. Stuart, H. R. Akçakaya, C. Hilton-Taylor, and G. M. Mace. 2005. Using Red List indices to measure progress to the 2010 target and beyond. *Phil Trans Roy Soc Lond* B 360:255–68.

Butlin, R. 1989. Reinforcement of premating isolation. In *Speciation and its consequences*, eds. D. Otte and J. A. Endler, 158–79. Sunderland, MA: Sinauer Associates.

Buxton, E. N. 1902. *Two African trips, with notes and suggestions on big game preservation in Africa*. London: E. Stanford.

Byatt, A. S. 1992. Morpho Eugenia. In *Angels and insects*, 3–160. London: Chatto & Windus.

Camerini, J. R. 1993. Evolution, biogeography and maps: An early history of Wallace's Line. *Isis* 84:700–27.

——. 1996. Wallace in the field. *Osiris*, 2nd ser., 11:44–65.

——. 1997. Remains of the day: Early Victorians in the field. In Lightman 1997, 354–77.

——, ed. 2002. *The Alfred Russel Wallace reader: A selection of writings from the field*. Baltimore: Johns Hopkins University Press.

Caro, T. 1995. Pursuit-deterrence revisited. *Trends in Ecology and Evolution* 10:500–03.

——. 2005a. The adaptive significance of coloration in mammals. *BioScience* 55:125–36.

——. 2005b. *Antipredator defenses in birds and mammals*. Chicago: University of Chicago Press.

Caro, T., C. M. Graham, C. J. Stoner, and J. K. Vargas. 2003. Correlates of horn and antler shape in bovids and cervids. *Behavioral Ecology and Sociobiology* 55:32–41.

Carpenter, Edward, *et al.* 1897. *Forecasts of the coming century.* Manchester: Labour Press.

Carson, R. 1962. *Silent spring.* Boston: Houghton Mifflin.

Carter, Brandon. 1974. Large number coincidences and the anthropic principle in cosmology. In *Confrontation of cosmological theories with observational data*, ed. M. S. Longair, 291–98. Dordrecht: D. Reidel.

Catton, W. R. Jr., and R. E. Dunlap. 1978. Environmental sociology: A new paradigm. *American Sociologist* 13:41–49.

CBD (Secretariat of the Convention on Biological Diversity). 1992. *Convention on biological diversity.* New York and Geneva: United Nations. **http://www.cbd.int/convention/convention.shtml**

Chaisson, E. 2001. *Cosmic evolution: The rise of complexity in nature.* Cambridge, MA: Harvard University Press.

[Chambers, Robert]. 1844 (and many later editions). *Vestiges of the natural history of creation.* London: John Churchill.

——. 1994 [1844]. *Vestiges of the natural history of creation and other evolutionary writings.* Ed. J. A. Secord. Chicago: University of Chicago Press.

Chase, M. 1979. A Victorian scientist at Grays: 1872–1876. *Panorama: The Journal of the Thurrock Local History Society*, no. 22:8–17.

——. 1985. *The people's farm: English radical agrarianism, 1775–1840.* Oxford, UK: Clarendon Press.

——. 1991. Out of radicalism: The mid-Victorian freehold land movement. *English Historical Review* 106:319–45.

——. 1996. We wish only to work for ourselves: The Chartist land plan. In *Living and learning: Essays in honour of J. F. C. Harrison*, eds. M. Chase and I. Dyck, 133–48. Aldershot, Hants: Scolar Press.

——. 2003. "Wholesome object lessons": The Chartist land plan in retrospect. *English Historical Review* 118:59–85.

Cheng R.-C., and I.-M. Tso. 2007. Signaling by decorating webs: Luring prey or deterring predators? *Behavioral Ecology* 18:1085–91.

Chiao, C. C., E. J. Kelman, and R. T. Hanlon. 2005. Disruptive body patterning of cuttlefish (*Sepia officinalis*) requires visual information regarding edges and contrast of objects in natural substrate backgrounds. *Biological Bulletin* 208:7–11.

Claeys, G. 1987a. Justice, independence, and industrial democracy: The development of John Stuart Mill's views on socialism. *Journal of Politics* 49:122–47.

——. 1987b. *Machinery, money, and the millennium: From moral economy to socialism, 1815–1860.* Princeton, NJ: Princeton University Press.

——. 1988. From "politeness" to "rational character": The critique of culture in Owenite socialism, 1800–1850. In *Working class and popular culture*, eds. L. H. van Voss and F. van Holthoon, 19–32. Amsterdam: Stichting Beheer IISG.

——, ed. 1993. *Selected works of Robert Owen.* 4 vols. London: W. Pickering.

——. 2000. The "survival of the fittest" and the origins of Social Darwinism. *J Hist Ideas* 61:223–40.

——, ed. 2001. *The Chartist movement in Britain, 1838–1850.* 6 vols. London: Pickering & Chatto.

——, ed. 2006. *Owenite socialism: Pamphlets and correspondence.* 10 vols. London: Routledge.

Clemens, Florence. 1937. Conrad's Malaysian fiction: A new study in sources with an analysis of factual material involved. PhD diss., Ohio State University.

Clements, Harry. 1983. *Alfred Russel Wallace: Biologist and social reformer.* London: Hutchinson.

Clerke, A. M. 1885. *A popular history of astronomy during the nineteenth century.* Edinburgh: A. & C. Black.

——. 1890. *The system of the stars.* London: Longmans, Green.

——. 1904. Life in the universe. *Edinburgh Review* 200:59–74.

Clodd, Edward 1892. Memoir of the author. In *The naturalist on the River Amazons*, by Henry Walter Bates. London: John Murray.

Cluysenaar, Anne. 2001. Stilled. *Scintilla* (Usk) 5:55–56.

Cock, A. G. 1977. Bernard's symposium—the species concept in 1900. *Biol J Linn Soc* 9:1–30.

Colgrove, J. 2005. "Science in a democracy": The contested status of vaccination in the progressive era and the 1920s. *Isis* 96:167–91.

Colp, R. Jr. 1992. "I will gladly do my best." How Charles Darwin obtained a Civil List pension for Alfred Russel Wallace. *Isis* 83:3–26.

Colwell, R. K. 2000. A barrier runs through it ... or maybe just a river. *Proc Natl Acad Sci* 97:13470–72.

Conrad, Joseph. 1920. *Lord Jim; a romance.* Garden City, NY: Doubleday, Doran.

——. 2002. *An outcast of the Islands*, rev. ed. Ed. J. H. Stapes, with notes by Hans van Marle. Oxford, UK: Oxford University Press.

Coope, G. R. 2004. Several million years of stability among insect species because of, or in spite of, Ice Age climatic instability? *Phil Trans Roy Soc Lond* B 359:209–14.

Cott, H. B. 1940. *Adaptive coloration in animals.* London: Methuen.

Coulter, H. L., and B. L. Fisher. 1991. *A shot in the dark: Why the P in the DPT vaccination may be hazardous to your child's health.* Garden City, NY: Avery.

Coyne, J. A. 1974. The evolutionary origin of hybrid inviability. *Evolution* 28:505–06.

Coyne, J. A., and H. A. Orr. 1989. Patterns of speciation in *Drosophila. Evolution* 43:362–81.

——. 2004. *Speciation.* Sunderland, MA: Sinauer Associates.

Cracraft, J. 1985. Historical biogeography and patterns of differentiation within the South American avifauna: Areas of endemism. *Ornithological Monographs* 36:49–84.

Cranbrook, Earl of, D. M. Hill, C. J. McCarthy, and R. P. Prys-Jones. 2005. A. R. Wallace, collector: Tracing his vertebrate specimens. Part I. In *Wallace in Sarawak—150 years later. An international conference on biogeography and biodiversity*, eds. A. A. Tuen and I. Das, 8–34. Kota Samarahan: Institute of Biodiversity and Environmental Conservation, Universiti Malaysia Sarawak.

Cranbrook, Earl of, D. M. Hill, C. J. McCarthy, R. P. Prys-Jones, and L. Tomsett (in prep.). A. R. Wallace, collector: Tracing his vertebrate specimens from Sarawak. Part II.

Creighton, C. 1888 Vaccination. *Encyclopaedia Britannica.* 9th ed. Edinburgh: Adam and Charles Black.

——. 1889. *Jenner and vaccination, a strange chapter of medical history.* London: Sonnenschein.

Cremo, M. A. 2003. *Human devolution: A Vedic alternative to Darwin's theory.* Los Angeles: Bhaktivedanta Book Publishing.

Croizat, L. 1962. *Space, time, form: The biological synthesis.* Caracas: published by the Author.

Croll, J. 1864. Of the physical cause of the change of climate during geological epochs. *Philosophical Magazine,* 4th ser., 28:121–37.

——. 1868. On geological time, and the probable date of the glacial and Upper Miocene Period. *Philosophical Magazine,* 4th ser., 35:363–84; 36:141–54, 362–86.

——. 1870. On the cause of the motion of glaciers. *Philosophical Magazine,* 4th ser., 40:210 ff.

——. 1875. *Climate and time in their geological relations.* London: Daldy, Isbister.

——. 1885. *Discussions on climate and cosmology.* Edinburgh: Adam & Charles Black.

Cronin, H. 1991. *The ant and the peacock.* Cambridge, UK: Cambridge University Press.

Crookshank, E. M. 1889. *History and pathology of vaccination, Vol. 1.* London: H. K. Lewis.

Cross, W. R. 1950. *The burned-over district; the social and intellectual history of enthusiastic religion in western New York, 1800–1850.* Ithaca, NY: Cornell University Press.

Crowe, M. J. 1986. *The extraterrestrial life debate, 1750–1900: The idea of a plurality of worlds from Kant to Lowell.* Cambridge, UK: Cambridge University Press.

Curle, Richard. 1934. Joseph Conrad: Ten years after. *Virginia Quarterly Review* 10:420–35.

Curwen, E. C., ed. 1940. *The journal of Gideon Mantell, surgeon and geologist: Covering the years 1818–1852.* London: Oxford University Press.

Cuthill, I. C., M. Stevens, J. Sheppard, T. Maddocks, C. A. Párraga, and T. S. Troscianko. 2005. Disruptive coloration and background pattern matching. *Nature* 434:72–74.

Cuthill, I. C., M. Stevens, A. M. M. Windsor, and H. J. Walker. 2006. The effects of pattern symmetry on detection of disruptive and background-matching coloration. *Behavioral Ecology* 17:828–32.

Darwin, C. 1859. *On the origin of species by means of natural selection: Or, the preservation of favoured races in the struggle for life.* London: John Murray.

——. 1871. *The descent of man, and selection in relation to sex.* 2 vols. London: John Murray.

——. 1985. *The origin of species.* Harmondsworth: Penguin. [reprint of 1968 ed.]

Darwin, C., and A. R. Wallace. 1858. On the tendency of species to form varieties; and on the perpetuation of varieties and species by natural means of selection. *J Proc Linn Soc: Zool* 3:45–62.

Darwin, F., ed. 1887a. *The life and letters of Charles Darwin.* 3 vols. London: John Murray.

——, ed. 1887b. *The life and letters of Charles Darwin.* 2nd ed. 3 vols. London: John Murray.

——, ed. 1902. *Charles Darwin: His life told in an autobiographical chapter, and in a selected series of his published letters.* London: John Murray.

——, ed. 1969 [1888]. *The life and letters of Charles Darwin.* 3 vols. New York: Johnson Reprint Corporation. [facsimile reprint of the 1888 "7th thousand revised" John Murray ed.]

Darwin, F., L. Darwin, and H. Darwin. 1859. [records of Coleoptera at Down]. *Entomologist's Weekly Intelligencer* 6:99.

Darwin, F., and A. C. Seward, eds. 1903. *More letters of Charles Darwin.* 2 vols. London: John Murray.

Darwin, G. H. 1903. Radio-activity and the age of the sun. *Nature* 68:496.

Davies, P. C. W. 1982. *The accidental universe.* Cambridge, UK: Cambridge University Press.

——. 2007. *Cosmic jackpot: Why our universe is just right for life.* Boston: Houghton-Mifflin.

Davison, G. W. H., reviser. 1999. *The birds of Borneo.* 4th ed. Kota Kinabalu: Natural History Publications (Borneo).

De Duve, C. 1995. *Vital dust: Life as a cosmic imperative.* New York: Basic Books.
Dean, R. 1995. Owenism and the Malthusian population question, 1815–1835. *History of Political Economy* 27:579–97.
DeCarvalho, R. J. 1988/1989. Methods and manifestations: The Wallace-Carpenter debate over spiritualism. *Journal of Religion and Psychical Research* 11:183–94; 12:20–25.
Delabecque, A. 1892. *Atlas des lacs Français.* Paris: Ministère des Travaux Publics.
Desmond, A. 1989. *The politics of evolution: Morphology, medicine, and reform in radical London.* Chicago: University of Chicago Press.
——. 2001. Redefining the x axis: "Professionals," "amateurs" and the making of mid-Victorian biology—a progress report. *J Hist Biol* 34:3–50.
Desmond, A., and James Moore. 1991. *Darwin.* New York: Warner Books.
Diamond, J. M. 2007. An incomparable life. *Nature* 449:659–60.
Diamond, J. M., and M. E. Gilpin. 1983. Biogeographic umbilici and the origin of the Philippine avifauna. *Oikos* 41:307–21.
Dick, S. J. 1982. *Plurality of worlds: The origins of the extraterrestrial life debate from Democritus to Kant.* Cambridge, UK: Cambridge University Press.
——. 1996. *The biological universe: The twentieth-century extraterrestrial life debate and the limits of science.* Cambridge, UK: Cambridge University Press.
——. 2000. Cosmotheology: Theological implications of the new universe. In *Many worlds: The new universe, extraterrestrial Life, and the theological implications,* ed. S. J. Dick, 191–210. Philadelphia: Templeton Foundation Press.
Dick, S. J., and J. E. Strick. 2004. *The living universe: NASA and the development of astrobiology.* New Brunswick, NJ: Rutgers University Press.
Dickens, P. 2004. *Society & nature: Changing our environment, changing ourselves.* Cambridge, UK: Polity.
Dickerson, R. E., E. D. Merrill, R. C. McGregor, W. Schultze, E. H. Taylor, and A. W. Herre. 1923. *Distribution of life in the Philippines.* Manila: Bureau of Printing.
Dixey, F. A. 1908a. Recent developments in the theory of mimicry. In *Report of the seventy-seventh meeting of the British Association for the Advancement of Science,* 736–37. London: John Murray.
——. 1908b. On Müllerian mimicry and diaposematism. *Trans Entomol Soc Lond 1908:* 559–83.
Dobler, S. 2001. Evolutionary aspects of defense by recycled plant compounds in herbivorous insects. *Basic and Applied Ecology* 2:15–26.
Dobzhansky, T. G. 1937. *Genetics and the origin of species.* New York: Columbia University Press.
——. 1940. Speciation as a stage in evolutionary divergence. *American Naturalist* 74: 312–21.
Douglas, R. 1979. *Land, people, & politics: A history of the land question in the United Kingdom, 1878–1952.* London: Allison and Busby.
Doyle, Arthur Conan. 1998. *The lost world.* Ed. Ian Duncan. Oxford, UK: Oxford University Press.
Duffels, J. P., and H. Turner. 2002. Cladistic analysis and biogeography of the cicadas of the Indo-Pacific subtribe Cosmopsaltriina (Hemiptera: Cicadoidea: Cicadidae). *Systematic Entomology* 27:235–61.

Dugdale, R. L. 1977. *"The Jukes"; a study in crime, pauperism, disease, and heredity.* 3rd ed., revised. New York: G. P. Putnam's Sons.

Durant, J. R. 1979. Scientific naturalism and social reform in the thought of Alfred Russel Wallace. *Brit J Hist Sci* 12:31–58.

Durbach, N. 2000. "They might as well brand us": Working-class resistance to compulsory vaccination in Victorian England. *Social History of Medicine* 13:45–62.

——. 2005. *Bodily matters: The anti-vaccination movement in England, 1853–1907.* Durham, NC: Duke University Press.

Durkheim, E. 1964. *The rules of sociological method.* 8th ed. Toronto: Collier Macmillan.

Dyson, F. 1988. *Infinite in all directions.* New York: Harper and Row.

Eagle, J. V., and G. P. Jones. 2004. Mimicry in coral reef fishes: Ecological and behavioural responses of a mimic to its model. *J Zool* 264:33–43.

Edmunds, M. 1974a. *Defence in animals: A survey of anti-predator defences.* Burnt Mill: Longman.

——. 1974b. Significance of beak marks on butterfly wings. *Oikos* 25:117–18.

Edmunds, M., and R. A. Dewhirst. 1994. The survival value of countershading with wild birds as predators. *Biol J Linn Soc* 51:447–52.

Endersby, J. 2003. Escaping Darwin's shadow. *J Hist Biol* 36:385–403.

——. 2008. *Imperial nature: Joseph Hooker and the practices of Victorian science.* Chicago: University of Chicago Press.

Endler, J. A. 1978. A predator's view of animal color patterns. *Evolutionary Biology* 11:319–64.

——. 1981. An overview of the relationships between mimicry and crypsis. *Biol J Linn Soc* 16:25–31.

——. 1984. Progressive background matching in moths, and a quantitative measure of crypsis. *Biol J Linn Soc* 22:187–231.

——. 1991. Interactions between predators and prey. In *Behavioural ecology an evolutionary approach,* 3rd ed., eds. J. R. Krebs and N. B. Davies, 169–96. Oxford, UK: Blackwell Scientific.

Endler, J. A., and P. W. Mielke, Jr. 2005. Comparing entire colour patterns as birds see them. *Biol J Linn Soc* 86:405–31.

Erwin, T. L. 2004. The biodiversity question: How many species of terrestrial arthropods are there? In *Forest canopies,* 2nd ed., eds. M. D. Lowman and H. B. Rinker, 259–69. Amsterdam: Elsevier.

Exnerová, A., P. Stys, E. Fučiková, S. Veselá, K. Svádová, M. Prokopová, V. Jarošik, R. Fuchs, and E. Landová. 2007. Avoidance of aposematic prey in European tits (Paridae): Learned or innate? *Behavioral Ecology* 18:148–56.

Fagan, M. B. 2007. Wallace, Darwin, and the practice of natural history. *J Hist Biol* 40:601–35.

Farber, P. L. 2000. *Finding order in nature: The naturalist tradition from Linnaeus to E. O. Wilson.* Baltimore: Johns Hopkins University Press.

Farber, P. L., and E. Mayr. 1986. The historical impact of ornithology on the biological sciences. *Acta: XIX Congressus Internationalis Ornithologici,* 2:2716–23. Ottawa: University of Ottawa Press.

Farley, M., P. Keating, and O. Keel. 1987. La vaccination à Montréal dans la second moitié du XIXe siècle: Pratiques, obstacles et resistances. In *Sciences & médecine au Québec: Perspectives sociohistoriques,* eds. M. Fournier *et al.,* 87–127. Québec: Institut Québécois de Recherche sur la Culture.

Farrell, B. D. 1998. "Inordinate fondness" explained: Why are there so many beetles? *Science* 281:553–57.

Farrell, B. D., and C. Mitter. 1998. The timing of insect/plant diversification: Might *Tetraopes* (Coleoptera: Cerambycidae) and *Asclepias* (Asclepiadaceae) have co-evolved? *Biol J Linn Soc* 64:553–77.

Felsenstein, J. 1981. Skepticism towards Santa Rosalia, or why are there so few kinds of animals? *Evolution* 35:24–38.

Ferns, P. N., and S. A. Hinsley. 2004. Immaculate tits: Head plumage pattern as an indicator of quality in birds. *Animal Behaviour* 67:261–72.

Ferns, P. N., and A. Lang. 2003. The value of immaculate mates: Relationships between plumage quality and breeding success in shelducks. *Ethology* 109:521–32.

Fichman, M. 2001. Science in theistic contexts: A case study of Alfred Russel Wallace on human evolution. *Osiris* 16:227–50.

——. 2004. *An elusive Victorian: The evolution of Alfred Russel Wallace.* Chicago: University of Chicago Press.

Fichman, M., and J. E. Keelan. 2007. Resister's logic: The anti-vaccination arguments of Alfred Russel Wallace and their role in the debates over compulsory vaccination in England, 1870–1907. *Studies in History and Philosophy of Science Part C: Studies in History and Philosophy of Biological and Biomedical Sciences* 38:585–607.

Fisher, R. A. 1927. On some objections to mimicry theory; statistical and genetic. *Trans Entomol Soc Lond* 75:269–78.

——. 1930. *The genetical theory of natural selection.* Oxford, UK: Clarendon Press.

——. 1958. *The genetical theory of natural selection.* New York: Dover.

Flammarion, C. 1903. The earth and man in the universe. *The Independent* (New York) 55:958–68.

Forel, F. A. 1892. *Le Léman. Monographie limnologique. Tome premier.* Lausanne: F. Rouge.

Forsman, A., and S. Appelqvist. 1998. Visual predators impose correlational selection on prey color pattern and behavior. *Behavioral Ecology* 9:409–13.

——. 1999. Experimental manipulation reveals differential effects of colour pattern on survival in male and female pygmy grasshoppers. *Journal of Evolutionary Biology* 12:391–401.

Forsman, A., K. Ringblom, E. Civantos, and J. Ahnesjö. 2002. Coevolution of color pattern and thermoregulatory behavior in polymorphic pygmy grasshoppers *Tetrix undulata. Evolution* 56:349–60.

Fox, D. L. 1976. *Animal biochromes and structural colours.* 2nd ed. Berkeley, CA: University of California Press.

Fox, H. M., and G. Vevers. 1960. *The nature of animal colors.* New York: Macmillan.

France, P., and W. St. Clair, eds. 2002. *Mapping lives: The uses of biography.* Oxford, UK: Oxford University Press.

Fraser, S., A. Callahan, D. Klassen, and T. N. Sherratt. 2007. Empirical tests of the role of disruptive coloration in reducing detectability. *Proc Roy Soc Lond* B 274:1325–31.

Freeman, R. B. 1977. *The works of Charles Darwin: An annotated bibliographical handlist.* 2nd ed. Folkstone: Dawson.

Fry, I. 1996. On the biological significance of the properties of matter: L. J. Henderson's theory of the fitness of the environment. *J Hist Biol* 29:155–96.

Fryer, J. C. F. 1914. An investigation by pedigree breeding into the polymorphism of *Papilio polytes*, Linn. *Phil Trans Roy Soc Lond* B 204:227–54.

Futuyma, D. J. 1998. *Evolutionary biology.* 3rd ed. Sunderland, MA: Sinauer Associates.

Gaffney, M. 1997. Alfred Russel Wallace's campaign to nationalize land: How Darwin's peer learned from John Stuart Mill and became Henry George's ally. *American Journal of Economics and Sociology* 56:609–15.

Gaffney, M., and F. Harrison. 1994. *The corruption of economics.* London: Shepheard-Walwyn.

Galton, F. 1865. Hereditary talent and character. *Macmillan's Magazine* 12:157–66, 318–27.

——. 1869. *Hereditary genius: An inquiry into its laws and consequences.* London: Macmillan.

——. 1908. *Memories of my life.* London: Methuen.

——. 2001. "Kantsaywhere" and "The Donoghues of Dunno Weir." *Utopian Studies* 12:188–233.

Gamberale, G., and B. S. Tullberg. 1998. Aposematism and gregariousness: The combined effect of group size and coloration on signal repellence. *Proc Roy Soc Lond* B 265:889–94.

Gamberale-Stille, G. 2000. Decision time and prey gregariousness influence attack probability in naïve and experienced predators. *Animal Behaviour* 60:95–99.

Gander, R., ed. 1998. *Alfred Russel Wallace at school, 1830–37.* Rustington, W. Sussex.

Gardner, Howard. 1999. *Intelligence reframed: Multiple intelligences for the 21st century.* New York: Basic Books.

Gardner, J. N. 2003. *Biocosm: The new scientific theory of evolution: Intelligent life is the architect of the universe.* Makawao, Maui, HI: Inner Ocean.

——. 2007. *Intelligent universe: AI, ET, and the emerging mind of the cosmos.* Franklin Lakes, NJ: New Page Books.

Garwood, E. J. 1910. Features of alpine scenery due to glacial protection. *Geographical Journal* 36:310–36.

Gascon, C., J. R. Malcolm, J. L. Patton, and six others. 2000. Riverine barriers and the geographic distribution of Amazonian species. *Proc Natl Acad Sci* 97:13672–77.

Gauld, A. 1968. *The founders of psychical research.* London: Routledge and Kegan Paul.

Gay, H. 1996. Invisible resource: William Crookes and his circle of support, 1871–81. *Brit J Hist Sci* 2:311–36.

Gayon, J. 1998. *Darwinism's struggle for survival: Heredity and the hypothesis of natural selection.* Cambridge, UK: Cambridge University Press.

Geikie, A. 1905a. *Landscape in history, and other essays.* New York: MacMillan.

——. 1905b. The centenary of Hutton's "Theory of the Earth" (Presidential Address to the British Association for the Advancement of Science, Edinburgh meeting 1892). In Geikie 1905a, 158–197.

——. 1905c. Geological time (Presidential Address to the Geological Section of the British Association for the Advancement of Science, Dover meeting 1899). In Geikie 1905a, 198 ff.

Geikie, J. 1874. *The great ice age and its relation to the antiquity of man.* London: W. Isbister.

——. 1877. *The great ice age and its relation to the antiquity of man.* 2nd ed. London: Daldy, Isbister & Co.

——. 1881. *Prehistoric Europe; a geological sketch.* London: Edward Stanford.

——. 1894. *The great ice age and its relation to the antiquity of man.* 3rd ed. London: Edward Stanford.

George, H. 1893. *A perplexed philosopher: Being an examination of Mr. Herbert Spencer's various utterances on the land question, with some incidental reference to his synthetic philosophy.* London: Kegan Paul, Trench, Trübner.

George, Wilma. 1964. *Biologist philosopher; a study of the life and writings of Alfred Russel Wallace.* London: Abelard-Schuman.

——. 1979. Alfred Russel Wallace, the gentle trader: Collecting in Amazonia and the Malay Archipelago 1848–1862. *Journal of the Society for the Bibliography of Natural History* 9:503–14.

George, W. H. 2001. The Barking connections of Sir Alfred Russel Wallace (1823–1913). *Barking & District Historical Society Newsletter* May 2001:2–4.

Gerould, J. H. 1916. Mimicry in butterflies. *American Naturalist* 50:184–92.

Ghiselin, M. T. 1969. *The triumph of the Darwinian method.* Berkeley: University of California Press.

Gibbs, John. 1856. *Compulsory vaccination briefly considered, in its scientific, religious and political aspects.* London: Sotheran and Willis.

Gillham, N. W. 2001. *A life of Sir Francis Galton: From African exploration to the birth of eugenics.* Oxford, UK: Oxford University Press.

Gilliard, E. T. 1969. *Birds of paradise and bower birds.* London: Weidenfeld and Nicolson.

Gittleman, J. L., and P. H. Harvey. 1980. Why are distasteful prey not cryptic? *Nature* 286:149–50.

Godfrey, D., J. N. Lythgoe, and D. A. Rumball. 1987. Zebra stripes and tiger stripes: The spatial frequency distribution of the pattern compared to that of the background is significant in display and crypsis. *Biol J Linn Soc* 32:427–33.

Goldsmith, T. H. 1990. Optimization, constraint, and history in the evolution of eyes. *Q Rev Biol* 65:281–322.

Golinski, J. 1998. *Making natural knowledge: Constructivism and the history of science.* Cambridge, UK: University of Cambridge Press.

Goodwin, T. W. 1984. *The biochemistry of carotenoids.* London: Chapman & Hall.

Gould, S. J. 1980. Wallace's fatal flaw. *Natural History* 89, January:26–40.

——. 1993. A special fondness for beetles. *Natural History* 102, January:4 + .

——. 1994. A foot soldier for evolution. In *Eight little piggies: Reflections in natural history,* 439–56. New York: W. W. Norton.

——. 2002. *The structure of evolutionary theory.* Cambridge, MA: Belknap Press of Harvard University Press.

Grafen, A. 1990. Sexual selection unhandicapped by the Fisher process. *J Theor Biol* 144:473–516.

Grant, R. E. 1827. Notice regarding the ova of the *Pontobdella muricata,* Lam. *Edinburgh Journal of Science* 7:160–61.

Grant, V., 1966. The selective origin of incompatibility barriers in the plant genus *Gilia. American Naturalist* 100:99–118.

Gray, J. 1979. John Stuart Mill on the theory of property. In *Theories of Property: Aristotle to the present: Essays,* eds. A. Parel and T. Flanagan, 251–70. Waterloo, ON: Wilfrid Laurier University Press.

Green, L. 1995. *Alfred Russel Wallace: His life and work.* Bengeo, Hertford: Occasional Paper no. 4, Hertford and Ware Local History Society.

Greene, J. C. 1981. Darwin as a social evolutionist. In *Science, ideology, and world view: Essays in the history of evolutionary ideas*, 95–127. Berkeley: University of California Press.

Greer-Wooten, B. 1972. *The role of general systems theory in geographical research.* Toronto: York University Department of Geography, Discussion Paper 3.

[Greg, W. R.]. 1868. On the failure of natural selection in the case of man. *Fraser's Magazine* 78:353–62.

Gregory, P. G., and D. J. Howard. 1993. Laboratory hybridization studies of *Allonemobius fasciatus* and *A. socius* (Orthoptera: Gryllidae). *Annals of the Entomological Society of America* 86:694–701.

Griffith, N. S. 1986. *Edward Bellamy: A bibliography.* Metuchen, NJ: Scarecrow Press.

Griffith, S. C., and S. R. Pryke. 2006. Benefits to females of assessing color displays. In Hill and McGraw 2006b, 233–79.

Grove, R. H. 1995. *Green imperialism: Colonial expansion, tropical island Edens, and the origins of environmentalism, 1600–1860.* Cambridge, UK: Cambridge University Press.

Guilford, T. 1986. How do warning colors work? Conspicuousness may reduce recognition errors in experienced predators. *Animal Behaviour* 34:286–88.

——. 1990. The Evolution of Aposematism. In *Insect defenses: Adaptive mechanisms and strategies of prey and predators*, eds. D. L. Evans and J. O. Schmidt, 23–61. Albany, NY: State University of New York Press.

Hadwen, W. R. 1896. *The case against vaccination: An address. Gloucester: Gloucester Anti-Vaccination Society.*

Hailman, J. P. 1977. *Optical signals: Animal communication and light.* Bloomington, IN: Indiana University Press.

Hall, J. P. W., and D. J. Harvey. 2002. The phylogeography of Amazonia revisited: New evidence from riodinid butterflies. *Evolution* 56:1489–97.

Hall, R., and J. D. Holloway, eds. 1998. *Biogeography and geological evolution of SE Asia.* Leiden: Backhuys.

Hamilton, W. J. 1856. The anniversary address of the President. *Q J Geol Soc Lond* 12:xxvi–cxix.

Hanlon, R. T., and J. B. Messenger. 1988. Adaptive coloration in young cuttlefish (*Sepia officinalis* L.): The morphology and development of body patterns and their relation to behavior. *Phil Trans Roy Soc Lond* B 320:437–87.

Hardin, G. 1960. *Nature and man's fate.* London: Cape.

Hardinge, Emma. 1860. *Six lectures on theology and nature.* Chicago: *s.n.*

——. 1866. *Extemporaneous addresses.* London: F. Farrah.

Hardy, Anne. 1983. Smallpox in London: Factors in the decline of the disease in the nineteenth century, *Medical History* 27:111–38.

——. 2000. "Straight back to barbarism": Antityphoid inoculation and the Great War, 1914. *Bull Hist Med* 74:265–90.

Hardy, A. C. 1984. *Darwin and the spirit of man.* London: Collins.

Harold, A. S. 2005. Peixes do Rio Negro. Fishes of the Rio Negro. *Copeia*, no. 1:212–14.

Harris, E. E. 1991. *Cosmos and anthropos: A philosophical interpretation of the anthropic cosmological principle.* Atlantic Highlands, NJ: Humanities Press International.

——. 1992. *Cosmos and theos: Ethical and theological implications of the anthropic cosmological principle.* Atlantic Highlands, NJ: Humanities Press International.

Harrison, J. F. C. 1969. *Quest for the new moral world; Robert Owen and the Owenites in Britain and America.* New York: Scribner.
Hawkins, B. A., R. Field, H. V. Cornell, and nine others. 2003b. Energy, water, and broad-scale geographic patterns of species richness. *Ecology* 84:3105–17.
Hawkins, B. A., and E. E. Porter. 2003a. Relative influences of current and historical factors on mammal and bird diversity patterns in deglaciated North America. *Global Ecology and Biogeography* 12:475–81.
——. 2003b. Water-energy balance and the geographic pattern of species richness of western Palearctic butterflies. *Ecological Entomology* 28:678–86.
Hawkins, B. A., E. E. Porter, and J. A. F. Diniz. 2003a. Productivity and history as predictors of the latitudinal diversity gradient of terrestrial birds. *Ecology* 84:1608–23.
Hawlena, D., R. Boochnik, Z. Abramsky, and A. Bouskila. 2006. Blue tail and striped body: Why do lizards change their infant costume when growing up? *Behavioral Ecology* 17:889–96.
Haynes, R. 1982. The *Society for Psychical Research, 1882–1982: A history.* London: Macdonald.
Heffernan, W. C. 1978. The singularity of our inhabited world: William Whewell and A. R. Wallace in dissent. *Journal of the History of Ideas* 39:81–100.
Heiling, A. M., K. Cheng, L. Chittka, A. Goeth, and M. E. Herberstein. 2005. The role of UV in crab spider signals: effects on perception by prey and predators. *Journal of Experimental Biology* 208:3925–31.
Heiling, A. M., M. E. Herberstein, and L. Chittka. 2003. Crab spiders manipulate flower signals. *Nature* 421:334.
Henderson, L. J. 1970 [1913]. *The fitness of the environment; an inquiry into the biological significance of the properties of matter.* Gloucester, MA: Peter Smith.
Hennig, W. 1966. *Phylogenetic systematics.* Urbana, IL: University of Illinois Press.
Herschel, J. F. W. 1869. *Outlines of astronomy.* 10th ed. London: Longmans, Green.
Hespenheide, H. A. 1975. Reversed sex-limited mimicry in a beetle. *Evolution* 29:780–83.
Hill, G. E. 2002. *A red bird in a brown bag: The function and evolution of colorful plumage in the house finch.* New York: Oxford University Press.
——. 2006. Female mate choice for ornamental coloration. In Hill and McGraw 2006b, 137–200.
Hill, G. E., and K. J. McGraw, eds. 2006a. *Bird coloration. Vol. 1: Mechanisms and measurements.* Cambridge, MA: Harvard University Press.
——, eds. 2006b. *Bird coloration. Vol. 2: Function and evolution.* Cambridge, MA: Harvard University Press.
Hill, J. A., ed. 1932. *Letters from Sir Oliver Lodge, psychical, religious, scientific and personal.* London: Cassell.
Hill, M. 1997. *The man who said NO!: The life of Henry George.* London: Othila Press.
Hill, R. I., and J. F. Vaca. 2004. Differential wing strength in *Pierella* butterflies (Nymphalidae, Satyrinae) supports the deflection hypothesis. *Biotropica* 36:362–70.
Himes, N. E. 1928. The place of John Stuart Mill and of Robert Owen in the history of English neo-Malthusianism. *Quarterly Journal of Economics* 42:627–40.
Hobbs, W. H. 1911. *Characteristics of existing glaciers.* New York: MacMillan.
Hobson, J. A. 1897. The influence of Henry George in England. *Fortnightly Review* 68:835–44.

Hodge, M. J. S., and G. Radick. 2003. *The Cambridge companion to Darwin.* Cambridge, UK: Cambridge University Press.

Houde, A. E. 1997. *Sex, color, and mate choice in guppies.* Princeton, NJ: Princeton University Press.

Houston, A. I., M. Stevens, and I. C. Cuthill. 2007. Animal camouflage: Compromise or specialize in a 2 patch-type environment. *Behavioral Ecology* 18:769–75.

Houston, Amy. 1997. Conrad and Alfred Russel Wallace. In *Conrad: Intertexts and appropriations: Essays in memory of Yves Hervouet,* eds. G. M. Moore, O. Knowles, and J. H. Stape, 29–48. Amsterdam: Rodopi.

Howard, D. J. 1993. Reinforcement: Origins, dynamics and fate of an evolutionary hypothesis. In *Hybrid zones and the evolutionary process,* ed. R. G. Harrison, 46–69. New York: Oxford University Press.

——. 1999. Conspecific sperm and pollen precedence and speciation. *Annu Rev Ecol Syst* 30:109–32.

Howland, Marie. 1874. *Papa's own girl: A novel.* New York: J. P. Jewett.

Hoyle, F. 1983. *The intelligent universe.* New York: Holt, Rinehart and Winston.

Hoyt, W. G. 1976. *Lowell and Mars.* Tucson, AZ: University of Arizona Press.

Huheey, J. E. 1980. The question of synchrony or "temporal sympatry" in mimicry. *Evolution* 34:614–16.

——. 1988. Mathematical models of mimicry. *American Naturalist* 131:S22-S41.

Huhta, E., S. Rytkonen, and T. Solonen. 2003. Plumage brightness of prey increases predation risk: An among-species comparison. *Ecology* 84:1793–99.

Hunt, Toby, J. Bergsten, Z. Levkanicova, and thirteen others. 2007. A comprehensive phylogeny of beetles reveals the evolutionary origins of a superradiation. *Science* 318:1913–16.

Hunt, Tristram. 2004. A revolutionary who won over Victorian liberals. *New Statesman* 17, 20 September:16–17.

Huxley, Leonard, ed. 1901. *Life and letters of Thomas Henry Huxley.* 2 vols. New York: Appleton.

Huxley, T. H. 1859. Darwin on the origin of species. *The Times* (London), 26 December:8–9.

——. 1879. [Address to the Department of Anthropology]. In *Report of the Forty-Eighth Meeting of the British Association for the Advancement of Science; held at Dublin in August 1878,* 573–78. London: John Murray.

——. 1880. The coming of age of *The origin of species. Science* 1:15–20.

——. 1894. *Man's place in nature, and other anthropological essays.* London: Macmillan.

——. 1909. Geological reform. In *Discourses biological and geological: Essays,* 308–42. New York: Appleton.

——. 1971 [1863]. *Man's place in nature.* Ann Arbor, MI: University of Michigan Press.

Inglis, B. 1992. *Natural and supernatural; A history of the paranormal from earliest times to 1914,* rev. ed. Bridport, Dorset, UK: Prism Press.

Irons, J. C., ed. 1896. *Autobiographical sketch of James Croll, with memoir of his life and work.* London: Edward Stanford.

Irwin, R. E. 1994. The evolution of plumage dichromatism in the New World blackbirds: Social selection on female brightness? *American Naturalist* 144:890–907.

Isaac, N. J. B., J. Mallet, and G. M. Mace. 2004. Taxonomic inflation: Its influence on macroecology and conservation. *Trends in Ecology and Evolution* 19:464–69.

Jardine, N., J. A. Secord, and E. C. Spary, eds. 1996. *Cultures of natural history.* Cambridge, UK: Cambridge University Press.

Johnson, N. A. 2000. Gene interactions and the origin of species. In *Epistasis and the evolutionary process*, eds. J. B. Wolf, E. D. Brodie III, and M. J. Wade, 197–212. New York: Oxford University Press.

——. 2006. The evolution of reproductive isolating barriers. In *Evolutionary genetics: Concepts and case studies*, eds. C. W. Fox and J. B. Wolf, 374–86. New York: Oxford University Press.

Johnson, N. A., and M. J. Wade. 1995. Conditions for soft selection favoring the evolution of hybrid inviability. *J Theor Biol* 176:493–99.

Johnson, S. G. 1991. Effects of predation, parasites, and phylogeny on the evolution of bright coloration in North American male passerines. *Evolutionary Ecology* 5:52–62.

Joly, J. 1903. Radium and the geological age of the earth. *Nature* 68:526.

Jones, Greta. 2002. Alfred Russel Wallace, Robert Owen and the theory of natural selection. *Brit J Hist Sci* 35:73–96.

Jones, L., and J. Bowes. 1840. *Report of the discussion on marriage as advocated by Robert Owen, between L. Jones and J. Bowes.* Liverpool: J. Stewart.

Kain, R. J. P., and H. C. Prince. 1985. *The tithe surveys of England and Wales.* Cambridge, UK: Cambridge University Press.

Karl, F. R., and L. Davies, eds. 1986. *The collected letters of Joseph Conrad. Vol. 2, 1898–1902.* Cambridge, UK: Cambridge University Press.

Keelan, J. E. 2004. The Canadian anti-vaccination leagues, 1872–1892. PhD diss., University of Toronto.

——. 2005. Empirical quagmires: Canadian anti-vaccinationists and their arguments. *Wellcome History*, Issue 30:2–4.

Kelly, J. K., and M. A. F. Noor. 1996. Speciation by reinforcement: A model derived from studies of *Drosophila. Genetics* 143:1485–97.

Kelman, E. J., R. J. Baddeley, A. J. Shohet, and D. Osorio. 2007. Perception of visual texture and the expression of disruptive camouflage by the cuttlefish, *Sepia officinalis. Proc Roy Soc Lond* B 274:1369–75.

Kelvin, W. T. Lord 1895. On the age of the earth. *Nature* 51:224–27.

Kemp, D. J. 2007. Female butterflies prefer males bearing bright iridescent ornamentation. *Proc Roy Soc Lond* B 274:1043–47.

Kerr, J. T. 2001. Butterfly species richness patterns in Canada: Energy, heterogeneity, and the potential consequences of climate change. *Conservation Ecology* 5: Article no. 10.

Kerr, J. T., and D. J. Currie. 1999. The relative importance of evolutionary and environmental controls on broad-scale patterns of species richness in North America. *Ecoscience* 6:329–37.

Kerr, J. T., and L. Packer. 1997. Habitat heterogeneity as a determinant of mammal species richness in high-energy regions. *Nature* 385:252–54.

Kettlewell, H. B. D. 1955. Selection experiments on industrial melanism in the Lepidoptera. *Heredity* 9:323–42.

Keynes, R. D., ed. 1988. *Charles Darwin's Beagle diary.* Cambridge, UK: Cambridge University Press.

Kilner, R. M. 2006. The evolution of egg colour and patterning in birds. *Biological Reviews* 81:383–406.

Kiltie, R. A. 1988. Countershading: Universally deceptive or deceptively universal? *Trends in Ecology and Evolution* 3:21–23.

Kimball, R. T. 2006. Hormonal control of coloration. In Hill and McGraw 2006a, 431–68.

Kingdon, J. 1982. *East African mammals: An atlas of evolution in Africa. Vol. 3, Part C, Bovids.* London: Academic Press.

——. 2006. The zebra's stripes. In *The encyclopedia of mammals*, 2nd ed., ed. D. W. Macdonald, 690. New York: Facts on File Inc.

Kingsland, S. 1978. Abbott Thayer and the protective coloration debate. *J Hist Biol* 11:223–44.

Kirby, K. R., W. F. Laurance, A. K. Albernaz, G. Schroth, P. M. Fearnside, S. Bergen, E. M. Venticinque, and C. da Costa. 2006. The future of deforestation in the Brazilian Amazon. *Futures* 38:432–53.

Kitson Clark, G. S. R. 1967. *The critical historian.* London: Heinemann.

Kleiner, S. A. 1985. Darwin's and Wallace's revolutionary research programme. *Brit J Philos Sci* 36:367–92.

Knapp, S. 1999. *Footsteps in the forest: Alfred Russel Wallace in the Amazon.* London: Natural History Museum.

——. 2003. Conservation: Dynamic diversity. *Nature* 422:475.

Kostal, R. W. 1994. *Law and English railway capitalism, 1825–1875.* Oxford, UK: Clarendon Press.

Kottler, M. J. 1974. Alfred Russel Wallace, the origin of man, and spiritualism. *Isis* 65:44–92.

——. 1985. Charles Darwin and Alfred Russel Wallace: Two decades of debate over natural selection. In *The Darwinian heritage*, ed. D. Kohn, 367–432. Princeton, NJ: Princeton University Press.

Krebs, R. A., and D. A. West. 1988. Female mate preference and the evolution of female-limited Batesian mimicry. *Evolution* 42:1101–04.

Ladiges, P. Y., C. J. Humphries, and L. W. Martinelli, eds. 1991. *Austral biogeography.* Melbourne: CSIRO.

Lamont, P. 2004. Spiritualism and a mid-Victorian crisis of evidence. *The Historical Journal* 47:897–920.

Larsen, Anne. 1996. Equipment for the field. In *Cultures of natural history*, eds. N. Jardine, J. A. Secord, and E. C. Spary, 358–77. Cambridge, UK: Cambridge University Press.

Laurent, J. 2005. Henry George: Evolutionary economist? In *Henry George's legacy in economic thought*, ed. J. Laurent, 73–98. Cheltenham, UK: Edward Elgar.

Leibowitz, A. H. 1994. Selection for hybrid inviability through kin selection. *J Theor Biol* 170:163–74.

Lenton, T. M. 1998. Gaia and natural selection. *Nature* 394:439–47.

Levy, J. H., ed. 1890. *A symposium on the land question.* London: T. F. Unwin.

Lewontin, R. C. 1984. Adaptation. In *Conceptual issues in evolutionary biology*, ed. E. Sober, 235–51. Cambridge, MA: MIT Press.

Lightman, B., ed. 1997. *Victorian science in context.* Chicago: University of Chicago Press.

——. 2004. Interpreting agnosticism as a nonconformist sect: T. H. Huxley's "New Reformation." In *Science and dissent in England, 1688–1945*, ed. Paul Wood, 197–214. Aldershot, Hampshire, UK: Ashgate.

Lindström, L., R. V. Alatalo, A. Lyytinen, and J. Mappes. 2001. Strong antiapostatic selection against novel rare aposematic prey. *Proc Natl Acad Sci* 98:9181–84.

Lindström, L., R. V. Alatalo, J. Mappes, M. Riipi, and L. Vertainen. 1999. Can aposematic signals evolve by gradual change? *Nature* 397:249–51.

Link, K. 2005. *The vaccine controversy: The history, use, and safety of vaccination.* Westport, CT: Praeger.

Lipow, Arthur. 1982. *Authoritarian socialism in America. Edward Bellamy & the nationalist movement.* Berkeley: University of California Press.

Loewenberg, B. J. 1959. *Darwin, Wallace, and the theory of natural selection.* Cambridge, MA: Arlington Books.

Lorch, P. D., and M. R. Servedio. 2007. The evolution of conspecific gamete precedence and its effect on reinforcement. *Journal of Evolutionary Biology* 20:937–49.

Lövei, G. L., and K. D. Sunderland. 1996. Ecology and behavior of ground beetles (Coleoptera: Carabidae). *Annu Rev Entomol* 41:231–56.

Lovelock, J. E. 1988. *The ages of Gaia: A biography of our living earth.* New York: Norton.

Low, H., 1848. *Sarawak: Its inhabitants and productions.* London: Richard Bentley.

Lowell, P. 1895. *Mars.* Boston: Houghton Mifflin.

——. 1906. *Mars and its canals.* New York: Macmillan.

——. 1908. *Mars as the abode of life.* New York: Macmillan.

Lubbock, J. 1882. *Ants, bees, and wasps, a record of observations on the habits of the social Hymenoptera.* 2nd ed. London: Kegan Paul, Trench.

Lundrigan, B. 1996. Morphology of horns and fighting behavior in the family Bovidae. *Journal of Mammalogy* 77:462–75.

Lyell, C. 1830–33. *Principles of geology.* 3 vols. London: John Murray.

——. 1845. *Travels in North America.* London: John Murray.

Lyell, C., and J. D. Hooker. 1858. Introd. to On the tendency of species to form varieties; and on the perpetuation of varieties and species by natural means of selection, by C. Darwin and A. R. Wallace. *J Proc Linn Soc: Zool* 3:45–46 [45–62].

Lyytinen, A., P. M. Brakefield, L. Lindström, and J. Mappes. 2004. Does predation maintain eyespot plasticity in *Bicyclus anynana? Proc Roy Soc Lond* B 271:279–83.

MacDougall, A., and M. S. Dawkins. 1998. Predator discrimination error and the benefits of Müllerian mimicry. *Animal Behaviour* 55:1281–88.

MacNair, E. W. 1957. *Edward Bellamy and the Nationalist Movement, 1889 to 1894.* Milwaukee: Fitzgerald Co.

MacNair, M. R. 1987. A model for the spread of genes for post-mating reproductive isolation by natural selection. *J Theor Biol* 125:105–15.

Malinchak, M. 1987. Spiritualism and the philosophy of Alfred Russel Wallace. PhD diss., Drew University.

Mallet, J. 1995. A species definition for the Modern Synthesis. *Trends in Ecology and Evolution* 10:294–99.

——. 1999. Causes and consequences of a lack of coevolution in Müllerian mimicry. *Evolutionary Ecology* 13:777–806.

——. 2004. Poulton, Wallace and Jordan: How discoveries in *Papilio* butterflies led to a new species concept 100 years ago. *Systematics and Biodiversity* 1:441–52.

——. 2008a [in review]. Hybridization, ecological races, and the nature of species: Empirical evidence for the ease of speciation. *Phil Trans Roy Soc Lond B.*

——. 2008b [in review]. Introd. to *On the phenomena of variation and geographical distribution as illustrated by the Papilionidae of the Malayan Region,* by A. R. Wallace.

Reprint ed., eds. C.-L. Chan and G. W. Beccaloni. Kota-Kinabalu, Malaysia: Natural History Publications (Borneo).

——. 2008c [in review]. Mayr's view of Darwin: Was Darwin wrong about speciation? *Biol J Linn Soc.*

Mallet, J., and M. Joron. 1999. Evolution of diversity in warning color and mimicry: Polymorphisms, shifting balance, and speciation. *Annu Rev Ecol Syst* 30:201–33.

Marchant, J. 1913. The life story of the Author. In *The revolt of democracy*, by A. R. Wallace, vii–xlv. London: Cassell.

——. 1916a. *Alfred Russel Wallace; letters and reminiscences.* 2 vols. London: Cassell.

——. 1916b. *Alfred Russel Wallace; letters and reminiscences.* New York: Harper.

——. 1975 [1916]. *Alfred Russel Wallace; letters and reminiscences.* New York: Arno Press. [reprint of 1916 Harper ed.]

Marshall, G. A. K. 1908. On diaposematism, with reference to some limitations of the Müllerian hypothesis of mimicry. *Trans Entomol Soc Lond 1908*:93–142.

Marshall, J. L., M. L. Arnold, and D. J. Howard. 2002. Reinforcement: The road not taken. *Trends in Ecology and Evolution* 17:558–63.

Martin, D. E. 1981. *John Stuart Mill and the land question.* Hull, UK: University of Hull Publications.

Martin, Paul. 1999. *Popular collecting and the everyday self: The reinvention of museums?* London: Leicester University Press.

Martin, T. E., and A. V. Badyaev. 1996. Sexual dichromatism in birds: Importance of nest predation and nest location for females versus males. *Evolution* 50:2454–60.

Maruyama, M. 1963. The second cybernetics: Deviation-amplifying mutual causal processes. *American Scientist* 51:164–79.

Maugham, W. S. 1933. Neil MacAdam. In *Ah King: Six stories.* London: William Heinemann.

Maunder, E. W. 1903. The Earth's place in the universe. *Journal of the British Astronomical Association* 13:227–34.

Maxwell, D. C. E. 1891. *Stepping-stones to socialism.* Hull: W. Andrew.

Maynard Smith, J. 1975. *The theory of evolution.* 3rd ed. Harmondsworth: Penguin.

——. 1976. Sexual selection and the handicap principle. *J Theor Biol* 57:239–42.

——. 1978. The handicap principle—a comment. *J Theor Biol* 70:251–52.

——. 1991. Theories of sexual selection. *Trends in Ecology and Evolution* 6:146–51.

Maynard Smith, J., and D. Harper. 2003. *Animal signals.* New York: Oxford University Press.

Mayr, E. 1942. *Systematics and the origin of species.* New York: Columbia University Press.

——. 1944. Wallace's line in the light of recent zoogeographic studies. *Q Rev Biol* 19:1–14.

——. 1963. *Animal species and evolution.* Cambridge, MA: Belknap Press of Harvard University Press.

——. 1982. *The growth of biological thought: Diversity, evolution, and inheritance.* Cambridge, MA: Belknap Press of Harvard University Press.

McCarthy, D. 2005. Biogeography and scientific revolutions. *The Systematist* 25:3–12.

McDougall, H. 1992 [1882]. *Sketches of our life at Sarawak.* Singapore: Oxford University Press.

McGraw, K. J. 2006a. Mechanics of carotenoid-based coloration. In Hill and McGraw 2006a, 177–242.

——. 2006b. Mechanics of melanin-based coloration. In Hill and McGraw 2006a, 243–94.

——. 2006c. Mechanics of uncommon colors in birds: Pterins, porphyrins, and psittacofulvins. In Hill and McGraw 2006a, 354–98.

McGraw, K. J., G. E. Hill, and R. S. Parker. 2005. The physiological costs of being colorful: Nutritional control of carotenoid utilization in the American goldfinch, *Carduelis tristis*. *Animal Behaviour* 69:653–60.

McKinney, H. L. 1969. Wallace's earliest observations on evolution: 28 December 1845. *Isis* 60:370–73.

——. 1972a. *Wallace and natural selection*. New Haven: Yale University Press.

——. 1972b. Introd. to Wallace's *A narrative of travels on the Amazon and Rio Negro* (S714), 2nd ed., v–xiii. New York: Dover.

McOuat, G. R. 1996. Species, rules and meaning: The politics of language and the ends of definitions in 19th century natural history. *Studies in History and Philosophy of Science, Part A* 27:473–519.

——. 2001. Cataloguing power: Delineating "competent naturalists" and the meaning of species in the British Museum. *Brit J Hist Sci* 34:1–28.

McWilliams Tullberg, R. 1998. *Women at Cambridge*, rev. ed. Cambridge, UK: Cambridge University Press.

MEA (Millennium Ecosystem Assessment). 2005. *Ecosystems and human well-being: Biodiversity synthesis*. Washington, DC: World Resources Institute.

Medearis, J. 2005. Labor, democracy, utility, and Mill's critique of private property. *American Journal of Political Science* 49:135–49.

Meiri, S., and G. M. Mace. 2007. New taxonomy and the origin of species. *PLoS Biology* 5:e194.

Melosi, M. V., ed. 1980. *Pollution and reform in American cities, 1870–1930*. Austin, TX: University of Texas Press.

Merilaita, S. 1998. Crypsis through disruptive coloration in an isopod. *Proc Roy Soc Lond* B 265:1059–64.

——. 2003. Visual background complexity facilitates the evolution of camouflage. *Evolution* 57:1248–54.

Merilaita, S., and J. Lind. 2005. Background-matching and disruptive coloration, and the evolution of cryptic coloration. *Proc Roy Soc Lond* B 272:665–70.

——. 2006. Great tits (*Parus major*) searching for artificial prey: Implications for cryptic coloration and symmetry. *Behavioral Ecology* 17:84–87.

Merilaita, S., A. Lyytinen, and J. Mappes. 2001. Selection for cryptic coloration in a visually heterogeneous habitat. *Proc Roy Soc Lond* B 268:1925–29.

Merilaita, S., J. Tuomi, and V. Jormalainen. 1999. Optimization of cryptic coloration in heterogeneous habitats. *Biol J Linn Soc* 67:151–61.

Metcalfe, I., J. M. B. Smith, M. Moorwood, and I. Davidson, eds. 2001. *Faunal and floral migrations and evolution in SE Asia-Australasia*. Lisse, Netherlands: A. A. Balkema.

Meyer, A. B. 1870. *Charles Darwin und Alfred Russel Wallace: Ihre ersten Publicationen über die "Entstehung der Arten" nebst einer Skizze ihres Lebens und einem Verzeichniss ihrer Schriften*. Erlangen: E. Besold.

——. 1895. How was Wallace led to the discovery of natural selection? *Nature* 52:415.

Michaux, B. 1991. Distributional patterns and tectonic development in Indonesia: Wallace reinterpreted. *Australian Systematic Botany* 4:25–36.

Michaux, B. 1994. Land movements and animal distributions in East Wallacea (eastern Indonesia, Papua New Guinea and Melanesia). *Palaeogeography, Palaeoclimatology, Palaeoecology* 112:323–43.

——. 1995. Distributional patterns in West Wallacea and their relationship to regional tectonic structures. *Sarawak Museum Journal* 69:163–79.

——. 1996. The origin of southwest Sulawesi and other Indonesian terranes: A biological view. *Palaeogeography, Palaeoclimatology, Palaeoecology* 122:167–83.

Middleton, G. V. 2004. J. W. Spencer: His life in Canada, and his work on preglacial river valleys. *Geoscience Canada* 31:49–56.

Midgley, M. 1992. *Science as salvation: A modern myth and its meaning.* London: Routledge.

Mill, J. S. 1965 [1848]. *Principles of political economy, with some of their applications to social philosophy.* In *Collected works of John Stuart Mill, Vols. 2–3.* Ed. M. Robson. Toronto: University of Toronto Press.

——. 1989. *On liberty; with the subjection of women; and chapters on socialism.* Ed. S. Collini. Cambridge, UK: Cambridge University Press.

Miller, Russell. 1983. *Continents in collision.* Alexandria, VA: Time-Life Books.

Milner, R. 1990. Darwin for the prosecution, Wallace for the defense: Part I: How two great naturalists put the supernatural on trial. *North Country Naturalist* 2:19–35.

Milnes, A. 1897. Statistics of small-pox and vaccination. *J Roy Stat Soc* 60:552–612.

Mittelbach, G. G., D. W. Schemske, H. V. Cornell, and nineteen others. 2007. Evolution and the latitudinal diversity gradient: Speciation, extinction and biogeography. *Ecology Letters* 10:315–31.

Monod, Jacques. 1971. *Chance and necessity: An essay on the natural philosophy of modern biology.* New York: Knopf.

Moore, James. 1982. Charles Darwin lies in Westminster Abbey. *Biol J Linn Soc* 17:97–113.

——. 1997. Wallace's Malthusian moment: The common context revisited. In Lightman 1997, 290–311.

Morgan, A. E. 1944. *Edward Bellamy.* New York: Columbia University Press.

Morrison, Frances. 1838. *The influence of the present marriage system upon the character and interest of females contrasted with that prepared by Robert Owen.* Manchester: A. Heywood.

Muensterberger, W. 1994. *Collecting: An unruly passion: Psychological perspectives.* Princeton, NJ: Princeton University Press.

Muir, J. 1894. *The mountains of California.* New York: Century Co.

Müller, F. 1878. Über die Vortheile der Mimicry bei Schmetterlingen. *Zoologischer Anzeiger* 1:54–55.

——. 1879. Ituna and Thyridia; a remarkable case of mimicry in butterflies. *Trans Entomol Soc Lond* 1879:xx–xxix.

Müller, S. 1839–40. Over de zoogdieren van den Indischen Archipel. In *Verhandelingen over de natuurlijke geschiedenis der Nederlandsche overzeesche bezittingen, Vol. 2, Zoologie,* ed. C. J. Temminck, 1–57. Leiden: S. en J. Luchtmans & C. C. van der Hoek.

Mundy, N. I. 2006. Genetic basis of color variation in wild birds. In Hill and McGraw 2006a, 469–506.

Muona, J. 1991. The Eucnemidae of South-east Asia and the Western Pacific—a biogeographical study. *Australian Systematic Botany* 4:165–82.

Murchison, R. I. 1839. *The Silurian system.* London: John Murray.

Murray, J. J. 1972. *Genetic diversity and natural selection.* New York: Hafner.

National Biodiversity Network's Species Dictionary. 2008. http://nbn.nhm.ac.uk/nhm/

Needham, A. E. 1974. *The significance of zoochromes.* Berlin: Springer.

Nelson, G. J., and N. I. Platnick. 1981. *Systematics and biogeography: Cladistics and vicariance.* New York: Columbia University Press.

Nelson, G. J., and D. E. Rosen, eds. 1981. *Vicariance biogeography: A critique.* New York: Columbia University Press.

Nelson, G. K. 1988. Modern spiritualist conception of ultimate reality. *Ultimate Reality and Meaning* 11:102–14.

Newcomb, Simon. 1902. *The stars: A study of the universe.* London: John Murray.

Newman, E. 1854. President's address to the Entomological Society. *Zoologist* 12:4223–24.

Newschaffer, C. J., L. A. Croen, J. Daniels, and eleven others. 2007. The epidemiology of autism spectrum disorders. *Annu Rev Public Health* 28:235–58.

Newton, I. 1718. *Opticks, or, a treatise of the reflections, refractions, inflections and colours of light.* 2nd ed., with additions. London: printed for W. and J. Innys.

NHM (Natural History Museum), London. Wallace Collection. **http://www.nhm.ac.uk/nature-online/collections-at-the-museum/wallace-collection/introduction.jsp.**

Noakes, R. J. 1999. Telegraphy is an occult art: Cromwell Fleetwood Varley and the diffusion of electricity to the other world. *Brit J Hist Sci* 32:421–59.

——. 2002. "Instruments to lay hold of spirits": Technologizing the bodies of Victorian spiritualism. In: *Bodies/Machines*, ed. I. R. Morus, 125–64. Oxford, UK: Berg.

——. 2004. Spiritualism, science, and the supernatural in mid-Victorian Britain. In *The Victorian supernatural*, eds. N. Brown, C. Burdett, and P. Thurschwell, 23–43. Cambridge, UK: Cambridge University Press.

Noor, M. A. F. 1999. Reinforcement and other consequences of sympatry. *Heredity* 83: 503–08.

Norris, K. S., and C. H. Lowe. 1964. An analysis of background color-matching in amphibians and reptiles. *Ecology* 45:565–80.

Numbers, R. L. 2006. *The creationists: From scientific creationism to intelligent design.* Expanded ed. Cambridge, MA: Harvard University Press.

O'Brien, E. M. 1998. Water-energy dynamics, climate, and prediction of woody plant species richness: An interim general model. *Journal of Biogeography* 25:379–98.

Offer, A. 1981. *Property and politics, 1879–1914: Landownership, law, ideology, and urban development in England.* Cambridge, UK: Cambridge University Press.

O'Hanlon, R. 1984. *Joseph Conrad and Darwin: The influence of scientific thought on Conrad's fiction.* Edinburgh: Salamander Press.

O'Hara, R. J. 1986. Diagrammatic classifications of birds, 1819–1901: Views of the natural system in 19th-century British ornithology. In *Acta: XIX Congressus Internationalis Ornithologici*, 2:2746–59. Ottawa: University of Ottawa Press.

——. 1988. Homage to Clio, or, toward an historical philosophy for evolutionary biology. *Syst Zool* 37:142–55.

——. 1991. Representations of the natural system in the nineteenth century. *Biology and Philosophy* 6:255–74.

——. 1992. Telling the tree: Narrative representation and the study of evolutionary history. *Biology and Philosophy* 7:135–60.

Ohsaki, N. 1995. Preferential predation of female butterflies and the evolution of Batesian mimicry. *Nature* 378:173–75.

——. 2005. A common mechanism explaining the evolution of female-limited and both-sex Batesian mimicry in butterflies. *Journal of Animal Ecology* 74:728–34.

Omland, K. E., and C. M. Hofmann. 2006. Adding color to the past: Ancestral-state reconstruction of coloration. In Hill and McGraw 2006b, 417–54.

Oppenheim, J. 1985. *The other world; spiritualism and psychical research in England, 1850–1914.* Cambridge, UK: Cambridge University Press.

Ortíz-Barrientos, D., and M. A. F. Noor. 2005. Evidence for a one-allele assortative mating locus. *Science* 310:1467.

Ortolani, A., and T. M. Caro. 1996. The adaptive significance of color patterns in carnivores: Phylogenetic tests of classic hypotheses. In *Carnivore behavior, ecology and evolution, Vol. 2*, ed. J. L. Gittleman, 132–88. Ithaca, NY: Comstock.

Osborn, H. F. 1894. *From the Greeks to Darwin; an outline of the development of the evolution idea.* New York and London: Macmillan.

Osorio, D., and A. D. Ham. 2002. Spectral reflectance and directional properties of structural coloration in bird plumage. *Journal of Experimental Biology* 205:2017–27.

Ospovat, Dov. 1981. *The development of Darwin's theory.* Cambridge, UK: Cambridge University Press.

Otte, D., and J. A. Endler, eds. 1989. *Speciation and its consequences.* Sunderland, MA: Sinauer Associates.

Ottow, R. 1993. Why John Stuart Mill called himself a socialist. *History of European Ideas* 17:479–83.

Owen, A. 1990. *The darkened room: Women, power, and spiritualism in late Victorian England.* Philadelphia: University of Pennsylvania Press.

Owen, Robert. 1834. *Marriage. A lecture delivered at the Institution, 14, Charlotte Street, on Sunday evening, Nov. 30, 1934.* London: *s.n.*

——. 1836. *The book of the New Moral World, Part One.* London: Effingham Wilson.

——. 1841. *The book of the New Moral World, Part Six.* London: Effingham Wilson.

——. 1993a [1835]. Lecture on the marriages of the priesthood of the old immoral world. In *Selected works of Robert Owen*, ed. G. Claeys, Vol. 2, 280–81. London: Pickering.

——. 1993b [1826]. The social system. In *Selected works of Robert Owen*, ed. G. Claeys, Vol. 2, 67–68. London: Pickering.

——. 1993c [1841]. A development of the principles and plans on which to establish self-supporting home colonies. In *Selected works of Robert Owen*, ed. G. Claeys, Vol. 2, 359–61. London: Pickering.

Owen, Robert Dale. 1840. *Address on the hopes and destinies of the human species.* London: J. Watson.

Párraga, C. A., T. Troscianko, and D. J. Tolhurst. 2002. Spatiochromatic properties of natural images and human vision. *Current Biology* 12:483–87.

Paul, D. B. 1995. *Controlling human heredity, 1865 to the present.* Amherst, NY: Humanity Books.

Pearson, K. 1924. *The life, letters and labours of Francis Galton. Vol. 2, Researches of middle life.* Cambridge, UK: University Press.

Peirce, C. S. 1901. [untitled book review of Wallace's *Studies scientific and social* (S727 1900)]. *Nation* 72:36–37.

Pels, Peter. 2003. Spirits of modernity: Alfred Wallace, Edward Tylor, and the visual politics of fact. In *Magic and modernity: Interfaces of revelation and concealment*, eds. B. Meyer and P. Pels, 241–71. Stanford, CA: Stanford University Press.

Pickering, W. H. 1903. Man's place in the universe. *The Independent* (New York) 55:597–600.

Pinheiro, C. G. 2007. Asynchrony in daily activity patterns of butterfly models and mimics. *Journal of Tropical Ecology* 23:119–23.

Pinker, S. 1997. *How the mind works.* New York: Norton.

Pitman, N. C. A., J. W. Terborgh, M. R. Silman, P. Nuñez V., D. A. Neill, C. E. Cerón, W. A. Palacios, and M. Aulestia. 2001. Dominance and distribution of tree species in upper Amazonian terra firme forests. *Ecology* 82:2101–17.

Platnick, N. I. 1991. On areas of endemism. In *Austral biogeography*, eds. P. Y. Ladiges, C. J. Humphries, and L. W. Martinelli, vii–viii. Melbourne: CSIRO.

Plummer, A. 1971. *Bronterre: A political biography of Bronterre O'Brien, 1804–1864.* London: Allen and Unwin.

Pocock, J. G. A. 1975. *The Machiavellian moment: Florentine political thought and the Atlantic Republican tradition.* Princeton, NJ: Princeton University Press.

Polaszek, A., and the Earl of Cranbrook, 2006. Insect species described from Alfred Russel Wallace's Sarawak collections. *Malayan Nature Journal* 57:433–62.

Popham, E. J. 1941. The variation in the colour of certain species of *Arctocorisa* (Hemiptera, Corixidae) and its significance. *Proc Zool Soc Lond* A 111:135–59.

Poulton, E. B. 1887. The experimental proof of the protective value of colour and markings in insects in reference to their vertebrate enemies. *Proc Zool Soc Lond 1887*:191–274.

——. 1888. Notes in 1887 upon Lepidopterous larvae, &c. *Trans Entomol Soc Lond 1888*:595–96.

——. 1890. *The colours of animals: Their meaning and use, especially considered in the case of insects.* 2nd ed. London: Kegan Paul, Trench, Trübner.

——. 1896. A naturalist's contribution to the discussion upon the age of the Earth. In *Annual Report of the British Association for the Advancement of Science*, 66:808–28. London: John Murray.

——. 1926. Protective resemblance borne by certain African insects to the blackened areas caused by grass fires. *Proceedings of the 3rd International Congress of Entomology*, Vol. 2:433–51.

Powell, R. A. 1982. Evolution of black-tipped tails in weasels: Predator confusion. *American Naturalist* 119:126–31.

Prendergast, D. K., and W. M. Adams. 2003. Colonial wildlife conservation and the origins of the Society for the Preservation of the Wild Fauna of the Empire (1903–1914). *Oryx* 37:251–60.

Proctor, R. A. 1870. *Other worlds than ours.* London: Longmans, Green.

Prum, R. O. 1999. The anatomy and physics of avian structural colours. In *Proceedings of the 22nd International Ornithological Congress*, 1633–53. Johannesburg: BirdLife South Africa.

——. 2006. Anatomy, physics, and evolution of structural colors. In Hill and McGraw 2006a, 295–353.

Prum, R. O., and S. Williamson. 2002. Reaction-diffusion models of within-feather pigmentation patterning. *Proc Roy Soc Lond* B 269:781–92.

Punnett, R. C. 1915. *Mimicry in butterflies.* Cambridge, UK: University Press.

Pycraft, W. P. 1925. *Camouflage in nature.* London: Hutchinson & Co.

Qureshi, S. 2004. Displaying Sara Baartman, the "Hottentot Venus." *History of Science* 42:233–57.

Raby, P. 1996. *Bright paradise: Victorian scientific travellers.* London: Chatto & Windus.

——. 2001. *Alfred Russel Wallace. A life.* London: Chatto & Windus; Princeton, NJ: Princeton University Press.

Racheli, L., and T. Racheli. 2004. Patterns of Amazonian area relationships based on raw distributions of papilionid butterflies (Lepidoptera: Papilioninae). *Biol J Linn Soc* 82:345–57.

Ramsay, A. C. 1862. On the glacial origin of certain lakes in Switzerland, the Black Forest, Great Britain, Sweden, North America, and elsewhere. *Q J Geol Soc Lond* 18:185–205.

Randall, J. E. 2005. A review of mimicry in marine fishes. *Zoological Studies* 44:299–328.

RCOV (Royal Commission on Vaccination). 1890. Testimony of A. R. Wallace presented before the Royal Commission on vaccination on 26 February, 5 and 12 March, 21 May 1890. In *Third report of the Royal Commission appointed to inquire into the subject of vaccination,* 6–35, 121–31. London: Her Majesty's Stationery Office.

——. 1891. Testimony of Edgar March Crookshank presented before the Royal Commission on vaccination on 9 July, 30 August, 12 and 19 November 1890. In *Fourth report of the Royal Commission appointed to inquire into the subject of vaccination,* 1–110. London: Her Majesty's Stationery Office.

——. 1897. *Sixth report of the Royal Commission appointed to inquire into the subject of vaccination.* London: Her Majesty's Stationery Office.

Rees, M. J. 1997. *Before the beginning: Our universe and others.* Reading, MA: Addison-Wesley.

——. 2000. *Just six numbers: The deep forces that shape the universe.* New York: Basic Books.

——. 2001. *Our cosmic habitat.* Princeton, NJ: Princeton University Press.

Regan, B. C., C. Julliot, B. Simmen, F. Viénot, P. Charles-Dominique, and J. D. Mollon. 2001. Fruits, foliage and the evolution of primate colour vision. *Phil Trans Roy Soc Lond* B 356:229–83.

Richards, E. 1989. Huxley and woman's place in science: The "woman question" and the control of Victorian anthropology. In *History, humanity and evolution: Essays for John C. Greene,* ed. James Moore, 253–84. Cambridge, UK: University Press.

Riipi, M., R. V. Alatalo, L. Lindström, and J. Mappes. 2001. Multiple benefits of gregariousness cover detectability costs in aposematic aggregations. *Nature* 413:512–14.

Robertson, K. A., and A. Monteiro. 2005. Female *Bicyclus anynana* butterflies choose males on the basis of their dorsal UV-reflective eyespot pupils. *Proc Roy Soc Lond* B 272:1541–46.

Rohwer, S., S. D. Fretwell, and D. M. Niles. 1980. Delayed maturation in passerine plumages and the deceptive acquisition of resources. *American Naturalist* 115:400–37.

Romanes, G. J. 1886. Physiological selection: An additional suggestion on the origin of species. *J Linn Soc: Zool* 19:337–411.

Rooney, C. J. 1985. *Dreams and visions: A study of American utopias, 1865–1917.* Westport, CT: Greenwood Press.

Roper, T. J. 1994. Conspicuousness of prey retards reversal of learned avoidance. *Oikos* 69:115–18.

Rose, S. P. R. 1997. *Lifelines: Biology, freedom, determinism.* London: Allen Lane.

Ross, A. M. 1893. *Memoirs of a reformer, 1832–1892.* Toronto: Hunter, Rose.

Ross, Hugh. 1993. *The creator and the cosmos: How the greatest scientific discoveries of the century reveal God.* Colorado Springs: NavPress.
Rothschild, M. 1983. *Dear Lord Rothschild. Birds, butterflies and history.* London: Hutchinson.
——. 1993. Phytochemical selection of aposematic insects. *Phytochemistry* 33:1037.
Rowe, C., L. Lindström, and A. Lyytinen. 2004. The importance of pattern similarity between Müllerian mimics in predator avoidance learning. *Proc Roy Soc Lond* B 271:407–13.
Rowland, H. M., E. Ihalainen, L. Lindström, J. Mappes, and M. P. Speed. 2007a. Co-mimics have a mutualistic relationship despite unequal defences. *Nature* 448:64–67.
Rowland, H. M., M. P. Speed, G. D. Ruxton, M. Edmunds, M. Stevens, and I. F. Harvey. 2007b. Countershading enhances cryptic protection: An experiment with wild birds and artificial prey. *Animal Behaviour* 74:1249–58.
Rudwick, M. J. S. 1985. *The great Devonian controversy.* Chicago: University of Chicago Press.
——. 2005. *Bursting the limits of time: The reconstruction of geohistory in the age of revolution.* Chicago: University of Chicago Press.
Ruse, Michael. 1979. *The Darwinian revolution: Science red in tooth and claw.* Chicago: University of Chicago Press.
——. 1980. Charles Darwin and group selection. *Annals of Science* 37:615–30.
——. 1999. *The Darwinian revolution: Science red in tooth and claw.* 2nd ed. Chicago: University of Chicago Press.
Rutherford, Ernest. 1904. *Radio-activity.* Cambridge, UK: University Press.
Ruxton, G. D. 2002. The possible fitness benefits of striped coat coloration for zebra. *Mammal Review* 32:237–44.
Ruxton, G. D., T. N. Sherratt, and M. P. Speed. 2004a. *Avoiding attack: The evolutionary ecology of crypsis, warning signals and mimicry.* Oxford, UK: Oxford University Press.
Ruxton, G. D., M. P. Speed, and D. J. Kelly. 2004b. What, if anything, is the adaptive function of countershading? *Animal Behaviour* 68:445–51.
Ruxton, G. D., M. P. Speed, and T. N. Sherratt. 2004c. Evasive mimicry: When (if ever) could mimicry based on difficulty of capture evolve? *Proc Roy Soc Lond* B 271:2135–42.
Sargant, W. L. 1860. *Robert Owen and his social philosophy.* London: Smith, Elder and Co.
Scarpelli, G. 1992. "Nothing in nature that is not useful"; The anti-vaccination crusade and the idea of *harmonia naturae* in Alfred Russel Wallace. *Nuncius* 7:109–30.
Schaefer, M. H., and N. Stobbe. 2006. Disruptive coloration provides camouflage independent of background matching. *Proc Roy Soc Lond* B 273:2427–32.
Schor, Naomi. 1994. Collecting Paris. In *The cultures of collecting,* eds. J. Elsner and R. Cardinal, 252–74. London: Reaktion Books.
Schuh, R. T., and G. M. Stonedahl. 1986. Historical biogeography in the Indo-Pacific: A cladistic approach. *Cladistics* 2:237–55.
Schuler, W., and E. Hesse. 1985. On the function of warning coloration: A black and yellow pattern inhibits prey-attack by naive domestic chicks. *Behavioral Ecology and Sociobiology* 16:249–55.
Schulte, J. A. II, J. Melville, and A. Larson. 2003. Molecular phylogenetic evidence for ancient divergence of lizard taxa on either side of Wallace's line. *Proc Roy Soc Lond* B 270: 597–603.

Sclater, P. L. 1858. On the general geographical distribution of the members of the class Aves. *J Proc Linn Soc: Zool* 2:130–45.

Secord, J. A. 1986. *Controversy in Victorian geology: The Cambrian-Silurian dispute.* Princeton, NJ: Princeton University Press.

——. 2003. *Victorian sensation: The extraordinary publication, reception, and secret authorship of Vestiges of the natural history of creation.* Chicago: University of Chicago Press.

Sedgwick, Adam. 1843. Letter 1. In *Geology of the Lake District, in three letters addressed to W. Wordsworth. n.p.*: Kendal.

Senar, J. C. 2006. Color displays as intrasexual signals of aggression and dominance. In Hill and McGraw 2006b, 87–136.

Servedio, M. R., and M. A. F. Noor. 2003. The role of reinforcement in speciation: Theory and data meet. *Annu Rev Ecol Syst* 34:339–364.

Severin, T. 1997. *The Spice Islands voyage: In search of Wallace.* London: Little, Brown.

Sharpe, R. B., 1906. 3. Birds. In *The history of the collections contained in the Natural History departments of the British Museum*, ed. E. R. Lankester, 79–515. London: British Museum.

Shermer, M. 2002. *In Darwin's shadow: The life and science of Alfred Russel Wallace: A biographical study on the psychology of history.* Oxford, UK: Oxford University Press.

Sherratt, T. N., and C. D. Beatty. 2003. The evolution of warning signals as reliable indicators of prey defense. *American Naturalist* 162:377–89.

Sherratt, T. N., D. Pollitt, and D. M. Wilkinson. 2007. The evolution of crypsis in replicating populations of web-based prey. *Oikos* 116:449–60.

Sherratt, T. N., A. Rashed, and C. D. Beatty. 2004. The evolution of locomotory behavior in profitable and unprofitable simulated prey. *Oecologia* 138:143–50.

——. 2005. Hiding in plain sight. *Trends in Ecology and Evolution* 20:414–16.

Sherry, Norman. 1966. *Conrad's Eastern world.* Cambridge, UK: University Press.

Shipley, Mrs J. B. 1890. Bebel's bricks or Bellamy's? *Liberty* (Boston and New York) 7, 21 June 1890:3.

Silberglied, R. E., A. Aiello, and D. M. Windsor. 1980. Disruptive coloration in butterflies: Lack of support in *Anartia fatima. Science* 209:617–19.

Sillén-Tullberg, B. 1985. Higher survival of an aposematic than of a cryptic form of a distasteful bug. *Oecologia* 67:411–15.

——. 1988. Evolution of gregariousness in aposematic butterfly larvae: A phylogenetic analysis. *Evolution* 42:293–305.

Silva, J. M. C. da, and D. C. Oren. 1996. Application of parsimony analysis of endemicity in Amazonian biogeography: An example with primates. *Biol J Linn Soc* 59:427–37.

Simpson, G. G. 1977. Too many lines: The limits of the Oriental and Australian zoogeographic regions. *Proceedings of the American Philosophical Society* 121:107–20.

Slotten, R. A. 2004. *The heretic in Darwin's court: The life of Alfred Russel Wallace.* New York: Columbia University Press.

Smith, C. H. 1984. The dynamics of animal distribution: An evolutionary/ecological model. PhD diss., University of Illinois **http://www.wku.edu/~smithch/DISS/dissertation.htm**

——. 1986. A contribution to the geographical interpretation of biological change. *Acta Biotheoretica* 35:229–78. **http://www.wku.edu/~smithch/essays/SMITH86.htm**

——. 1989. Historical biogeography: Geography as evolution, evolution as geography. *New Zealand Journal of Zoology* 16:773–85. **http://www.wku.edu/~smithch/essays/SMITH89.htm**

——, ed. 1991. *Alfred Russel Wallace: An anthology of his shorter writings.* Oxford, UK: Oxford University Press.

——. 1992/1999. *Alfred Russel Wallace on spiritualism, man, and evolution: An analytical essay.* Torrington, CT. **http://www.wku.edu/~smithch/essays/ARWPAMPH.htm**

——. 2000– . The Alfred Russel Wallace Page. **http://www.wku.edu/~smithch/index1.htm**

——. 2003. Wallace's "second moment": Intelligent conviction and the course of human evolution. **http://www.wku.edu/~smithch/essays/WALLMO.htm**

——. 2003– . Alfred Russel Wallace: Evolution of an evolutionist. **http://www.wku.edu/~smithch/wallace/chsarwp.htm**

——, ed. 2004a. *Alfred Russel Wallace: Writings on evolution, 1843–1912.* 3 vols. Bristol, UK: Thoemmes Continuum.

——. 2004b. Wallace's unfinished business. *Complexity* 10:25–32.

——. 2004c. Alfred Russel Wallace on man: A famous "change of mind"—or not? *History and Philosophy of the Life Sciences* 26:257–70.

——. 2006. Reflections on Wallace. *Nature* 443:33–34.

——. 2008, forthcoming. What's in a word? On reading—and misreading—Alfred Russel Wallace. *The Linnean.*

Smith, K. G. V. 1987. Darwin's insects: Charles Darwin's entomological notes, with an introduction and comments by Kenneth G. V. Smith. *Bulletin of the British Museum (Natural History) Historical Series* 14:1–143.

Smith, N. G. 1966. Evolution of some arctic gulls (*Larus*): An experimental study of isolating mechanisms. *Ornithological Monographs* 4:1–99.

Smith, R. A. 1982. *The expanding universe: Astronomy's "great debate," 1900–1931.* Cambridge, UK: Cambridge University Press.

Söderqvist, T. 1996. Existential projects and existential choice in science: Science biography as an edifying genre. In *Telling lives in science*, eds. M. Shortland and R. R. Yeo, 45–84. New York: Cambridge University Press.

Soper, K. 1995. *What is nature?: Culture, politics and the non-human.* Oxford, UK: Blackwell.

Sowan, P. W., and J. I. Byatt. 1974. Alfred Russel Wallace [1823–1913]; his residence in Croydon [1878–81] and his membership of the Croydon Microscopical and Natural History Club. *Proceedings of the Croydon Natural History and Scientific Society* 15:83–97.

Speed, M. P. 2000. Warning signals: receiver psychology and predator memory. *Animal Behaviour* 60:269–78.

Speed, M. P., D. J. Kelly, A. M. Davidson, and G. D. Ruxton. 2005. Countershading enhances crypsis with some bird species but not others. *Behavioral Ecology* 16:327–34.

Spencer, H. 1851. *Social statics: Or, the conditions essential to human happiness specified, and the first of them developed.* London: John Chapman.

——. 1891. *Justice: Being part IV of The principles of ethics.* London: Williams and Norgate.

Spencer, J. W. W. 1881. Discovery of the preglacial outlet of the basin of Lake Erie into that of Lake Ontario. *Proceedings of the American Philosophical Society* 19:300–37.

——. 1890a. The deformation of Iroquois beach and birth of Lake Ontario. *American Journal of Science*, 3rd ser., 40:443–51.

——. 1890b. Origin of the basins of the Great Lakes of America. *Q J Geol Soc Lond* 46:521–31, 533.

Srygley, R. B. 1994. Locomotor mimicry in butterflies? The associations of positions of centres of mass among groups of mimetic, unprofitable prey. *Phil Trans Roy Soc Lond* B 343:145–55.

Stack, D. A. 2003. *The first Darwinian Left: Socialism and Darwinism 1859–1914.* Gretton, Cheltenham, UK: New Clarion Press.

——. 2008. *Queen Victoria's skull: George Combe and the mid-Victorian mind.* London: London Books.

Stepan, N. 1982. *The idea of race in science: Great Britain 1800–1960.* London: Macmillan.

Stephens, S. G. 1950. The internal mechanism of speciation in *Gossypium. Botanical Review* 16:115–49.

Stevens, M. 2005. The role of eyespots as anti-predator mechanisms, principally demonstrated in the Lepidoptera. *Biological Reviews* 80:573–88.

——. 2007. Predator perception and the interrelation between different forms of protective coloration. *Proc Roy Soc Lond* B 274:1457–64.

Stevens, M., and I. C. Cuthill. 2006. Disruptive coloration, crypsis and edge detection in early visual processing. *Proc Roy Soc Lond* B 273:2141–47.

Stevens, M., I. C. Cuthill, C. A. Párraga, and T. Troscianko. 2006a. The effectiveness of disruptive coloration as a concealment strategy. *Progress in Brain Research* 155:49–65.

Stevens, M., I. C. Cuthill, A. M. M. Windsor, and H. J. Walker. 2006b. Disruptive contrast in animal camouflage. *Proc Roy Soc Lond* B 273:2433–38.

Stocking, G. W. Jr. 1987. *Victorian anthropology.* New York: Free Press.

Stoner, C. J., O. R. P. Bininda-Emonds, and T. M. Caro. 2003b. The adaptive significance of coloration in lagomorphs. *Biol J Linn Soc* 79:309–28.

Stoner, C. J., T. M. Caro, and C. M. Graham. 2003a. Ecological and behavioral correlates of coloration in artiodactyls: Systematic analyses of conventional hypotheses. *Behavioral Ecology* 14:823–40.

Strauss, D. 2001. *Percival Lowell: The culture and science of a Boston Brahmin.* Cambridge, MA: Harvard University Press.

Stresemann, E. 1975. *Ornithology: From Aristotle to the present.* Cambridge, MA: Harvard University Press.

Strickland, Hugh. 1841. On the true method of discovering the natural system in zoology and botany. *Annals and Magazine of Natural History* 6:184–94.

Stuart-Fox, D. M., A. Moussalli, G. R. Johnston, and I. P. F. Owens. 2004. Evolution of color variation in dragon lizards: Quantitative tests of the role of crypsis and local adaptation. *Evolution* 58:1549–59.

Suess, E. 1904–24. *The face of the earth.* 5 vols. Trans. H. B. C. Sollas, ed. W. J. Sollas. Oxford, UK: Clarendon Press.

Summers, K., and M. E. Clough. 2001. The evolution of coloration and toxicity in the poison frog family (Dendrobatidae). *Proc Natl Acad Sci* 98:6227–32.

Sumner, F. B., and H. S. Swarth. 1924. The supposed effects of the color tone of the background upon the coat color of mammals. *Journal of Mammalogy* 5:81–113.

Swaddle, J. P., and I. C. Cuthill. 1994. Female zebra finches prefer males with symmetric chest plumage. *Proc Roy Soc Lond* B 258:267–71.

Swedenborg, E. 1843–44. *The animal kingdom, considered anatomically, physically and philosophically.* 2 vols. Trans. J. G. Wilkinson. London: W. Newberry.

Sword, G. A. 1999. Density-dependent warning coloration. *Nature* 397:217.

[Tait, P. G.] 1869. Geological time. *North British Review* 50:215–33.

Taylor, M. W. 1992. *Man versus the State: Herbert Spencer and late Victorian individualism.* Oxford, UK: Clarendon Press.

Thayer, A. H. 1896. The law which underlies protective coloration. *The Auk* 13:124–29.

——. 1903. Protective coloration in its relation to mimicry, common warning colours, and sexual selection. *Trans Entomol Soc Lond 1903*:553–69.

Thayer, G. H. 1909. *Concealing-coloration in the animal kingdom: An exposition of the laws of disguise through color and pattern: Being a summary of Abbott H. Thayer's discoveries.* New York: Macmillan.

Thomas, J. L. 1983. *Alternative America: Henry George, Edward Bellamy, Henry Demarest Lloyd, and the adversary tradition.* Cambridge, MA: Belknap Press of Harvard University Press.

Thomas, O., 1906. 1. Mammals. In *The history of the collections contained in the Natural History departments of the British Museum*, ed. E. R. Lankester, 2, Chapter 1:3–78. London: British Museum.

Thomson, W. (Lord Kelvin). 1862. On the age of the sun's heat. *MacMillan's Magazine* 5:288–93.

——. 1863. On the secular cooling of the Earth. *Philosophical Magazine*, 4th ser., 25:1–14.

Treves, A. 1997. Primate natal coats: A preliminary analysis of distribution and function. *American Journal of Physical Anthropology* 104:47–70.

Tso, I., C. Liao, R. Huang, and E. Yang. 2006. Function of being colorful in web spiders: Attracting prey or camouflaging oneself? *Behavioral Ecology* 17:606–13.

Turner, H. H. 1903. Man's place in the universe; a reply to Dr. Wallace. *Fortnightly Review* 73:598–605.

Turner, H., P. Hovencamp, and P. C. van Welzen. 2001. Biogeography of Southeast Asia and the West Pacific. *Journal of Biogeography* 28:217–30.

Turner, J. R. G. 1977. Butterfly mimicry: The genetical evolution of an adaptation. *Evolutionary Biology* 10:163–206.

——. 1978. Why male butterflies are non-mimetic: Natural-selection, sexual selection, group selection, modification and sieving. *Biol J Linn Soc* 10:385–432.

Van Buskirk, J., J. Aschwanden, I. Buckelmüller, S. Reolon, and S. Rüttiman. 2004. Bold tail coloration protects tadpoles from dragonfly strikes. *Copeia*, no. 3:599–602.

Van Oosterzee, P. 1997. *Where worlds collide: The Wallace line.* Kew, Victoria, Australia: Reed.

Vetter, J. 1999. Contemplating man under all his varied aspects: The anthropological work of Alfred Russel Wallace, 1843–70. MPhil thesis, University of Oxford, Oxford, UK.

Vorobyev, M. 2003. Coloured oil droplets enhance colour discrimination. *Proc Roy Soc Lond* B 270:1255–61.

Wade, M. J. 1985. Soft selection, hard selection, kin selection, and group selection. *American Naturalist* 125:61–73.

Wade, M. J., H. Patterson, N. W. Chang, and N. A. Johnson. 1994. Postcopulatory, prezygotic isolation in flour beetles. *Heredity* 72:163–67.

Waldron, J. 1988. *The right to private property.* Oxford, UK: Clarendon Press.

Wallace, B. 1988. Selection for the inviability of sterile hybrids. *Journal of Heredity* 79:204–10.

Walsh, B. D. 1863. Notes by Benj. D. Walsh. *Proc Entomol Soc Philadelphia* 2:182–272.

Walsh, B. D. 1864. On phytophagic varieties and phytophagic species. *Proc Entomol Soc Philadelphia* 3:403–30.

Wandell, B. A. 1995. *Foundations of vision.* Sunderland, MA: Sinauer Associates.

Ward, P. D., and D. Brownlee. 2000. *Rare earth: Why complex life is uncommon in the universe.* New York: Copernicus.

Ward, S. B. 1976. Land reform in England and Wales 1880–1918. PhD thesis, University of Reading.

Weber, M. 1949. *The methodology of the social sciences.* Trans. and ed. E. A. Shils. Glencoe, IL: Free Press.

Wedgwood, B., and H. C. Wedgwood. 1980. *The Wedgwood circle, 1730–1897: Four generations of a family and their friends.* Westfield, NJ: Eastview Editions.

Wegener, A. 1915. *Die Entstehung der Kontinente und Ozeane.* Braunschweig: F. Vieweg.

Whewell, W. 2001 [1853]. *Of the plurality of worlds.* Ed. M. Ruse. Chicago: University of Chicago Press.

White, C. M., and M. D. Bruce. 1986. *The birds of Wallacea (Sulawesi, the Moluccas & Lesser Sunda Islands, Indonesia): An annotated check-list.* London: British Ornithologists' Union.

White, Michael, and J. R. Gribbin. 1995. *Darwin: A life in science.* London: Simon & Schuster.

Widdicombe, T., and H. S. Preiser, eds., 2002. *Revisiting the legacy of Edward Bellamy (1850–1898).* Lewiston, NY: Edward Mellen Press.

Wiley, E. O., and D. R. Brooks. 1982. Victims of history—a nonequilibrium approach to evolution. *Systematic Zoology* 31:1–24.

Williams-Ellis, A. 1966. *Darwin's moon: A biography of Alfred Russel Wallace.* London: Blackie.

Willis, K. J., L. Gillson, and T. M. Brncic. 2004. How "virgin" is virgin forest? *Science* 305:943–44.

Willis, K. J., L. Gillson, and S. Knapp, eds. 2007. Biodiversity hotspots through time: Using the past to manage the future: special issue. *Phil Trans Roy Soc Lond* B 362:169–333.

Wilson, E. O. 1994. *Naturalist.* Washington, DC: Island Press.

Wilson, J. G. 2000. *The forgotten naturalist: In search of Alfred Russel Wallace.* Kew, Victoria: Arcadia.

Wilson, K. A., M. F. McBride, M. Bode, and H. P. Possingham. 2006. Prioritizing global conservation efforts. *Nature* 440:337–40.

Wilson, P. 1993. Anthropic principle. In *Encyclopedia of cosmology: Historical, philosophical, and scientific foundations of modern cosmology,* ed. N. S. Hetherington, 11–17. New York: Garland.

Wiltshire, D. 1978. *The social and political thought of Herbert Spencer.* Oxford, UK: Oxford University Press.

Wolpert, L. 1992. *The unnatural nature of science.* London: Faber and Faber.

Worster, D. 1994. *Nature's economy: A history of ecological ideas.* 2nd ed. New York: Cambridge University Press.

Wright, W. B. 1937. *The Quaternary ice age.* London: MacMillan.

Wyszecki, G., and W. S. Stiles. 1982. *Color science: Concepts and methods, quantitative data and formulae.* 2nd ed. New York: John Wiley.

Yeo, R. 1993. *Defining science: William Whewell, natural knowledge and public debate in early Victorian Britain.* Cambridge, UK: Cambridge University Press.
Young, A. M. 1979. The evolution of eyespots in tropical butterflies in response to feedings on rotten fruit: An hypothesis. *Journal of the New York Entomological Society* 87:66–77.
Young, R. M. 1969. Malthus and the evolutionists: The common context of biological and social theory. *Past and Present,* no. 43:109–45.
Zahavi, A. 1975. Mate selection—a selection for a handicap. *J Theor Biol* 53:205–14.
——. 1977. The cost of honesty (further remarks on the handicap principle). *J Theor Biol* 67:603–05.
Zeki, S. 1993. *A vision of the brain.* Oxford, UK: Blackwell Scientific.

INDEX